NanoScience and Technology

Springer
*Berlin
Heidelberg
New York
Barcelona
Hong Kong
London
Milan
Paris
Singapore
Tokyo*

Physics and Astronomy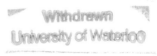

http://www.springer.de/phys/

NANOSCIENCE AND TECHNOLOGY

Series Editors: K. von Klitzing   R. Wiesendanger

The series NanoScience and Technology is focused on the fascinating nano-world, mesoscopic physics, analysis with atomic resolution, nano and quantum-effect devices, nanomechanics and atomic-scale processes. All the basic aspects and technology-oriented developments in this emerging discipline are covered by comprehensive and timely books. The series constitutes a survey of the relevant special topics, which are presented by leading experts in the field. These books will appeal to researchers, engineers, and advanced students.

Sliding Friction
Physical Principles and Applications   2nd Edition
By B. N. J. Persson

Scanning Probe Microscopy
Analytical Methods
Editor: R. Wiesendanger

Mesoscopic Physics and Electronics
Editors: T. Ando, Y. Arakawa, K. Furuya, S. Komiyama, and H. Nakashima

Series homepage – http://www.springer.de/phys/books/nst/

Bo N. J. Persson

Sliding Friction

Physical Principles and Applications

Second Edition
With 313 Figures
Including 3 Color Figures

 Springer

Dr. Bo N. J. Persson
Institut für Festkörperforschung
Forschungszentrum Jülich
52425 Jülich, Germany

Series Editors:
Professor Dr., Dres. h. c. Klaus von Klitzing
Max-Planck-Institut für Festkörperforschung, Heisenbergstrasse 1
70569 Stuttgart, Germany

Professor Dr. Roland Wiesendanger
Institut für Angewandte Physik, Universität Hamburg, Jungiusstrasse 11
20355 Hamburg, Germany

ISSN 1434-4904

ISBN 3-540-67192-7 2nd Edition Springer-Verlag Berlin Heidelberg New York

ISBN 3-540-63296-4 Springer-Verlag Berlin Heidelberg New York

Library of Congress Cataloging-in-Publication Data. Persson, B.N.J. Sliding Friction : physical principles and applications/ Bo N.J. Persson. – 2nd ed. p. cm. – (Nanoscience and technology) Includes bibliographical references and index. ISBN 3540671927 (alk. paper) 1. Friction. 2. Surfaces (Physics) 3. Friction–Industrial applications. 4. Lubrication and lubricants. I. Title. II. Series QC197.P47 2000   531'.1134–dc21   00-027993

This work is subject to copyright. All rights are reserved, whether the whole or part of the material is concerned, specifically the rights of translation, reprinting, reuse of illustrations, recitation, broadcasting, reproduction on microfilm or in any other way, and storage in data banks. Duplication of this publication or parts thereof is permitted only under the provisions of the German Copyright Law of September 9, 1965, in its current version, and permission for use must always be obtained from Springer-Verlag. Violations are liable for prosecution under the German Copyright Law.

Springer-Verlag is a company in the BertelsmannSpringer publishing group.
© Springer-Verlag Berlin Heidelberg 1998, 2000
Printed in Germany

The use of general descriptive names, registered names, trademarks, etc. in this publication does not imply, even in the absence of a specific statement, that such names are exempt from the relevant protective laws and regulations and therefore free for general use.

Typesetting: Data conversion by PTP-Protago, Berlin
Cover concept: eStudio Calamar Steinen
Cover design: *design & production*, Heidelberg

Printed on acid-free paper
SPIN: 10760490   57/3144/ba   5 4 3 2 1 0

Preface to the Second Edition

Our knowledge of sliding friction continues to develop rapidly, and many important new results have been discovered since the first edition of this book was published. This edition includes several new topics, e.g., experimental and theoretical results related to the friction on superconductors, experimental studies and computer simulations of the layering transition, nanoindentation, wear in combustion engines, rubber wear, the effect of humidity on sliding friction, rolling and sliding of carbon nanotubes and the friction dynamics of granular materials.

I would like to thank P. Ballone, L. Bocquet, G. Heinrich, F. Mugele, V.L. Popov, M. Salmeron and R. Superfine for useful discussions and for supplying many figures for this book. I also thank Angela Lahee, Claus-Dieter Bachem and Claus Ascheron from Springer-Verlag for their comments and improvements to the text.

Jülich, January 2000 *Bo N.J. Persson*

Preface

Sliding friction is one of the oldest problems in physics and certainly one of the most important from a practical point of view. It has been estimated that the losses in the USA resulting from ignorance of tribology amount to 6% of the gross national product, a whopping $420 billion annually. The ability to produce durable low-friction surfaces and lubricant fluids has become an important factor in the miniaturization of moving components in many technological devices. These include magnetic storage and recording systems (computer disk heads), miniature motors, and many aerospace components. In short, the modern world depends upon the smooth and satisfactory operation of countless tribological systems – systems which are often taken for granted. In spite of this, many aspects of sliding friction are still not well understood, and an understanding of friction on an atomic level is just starting to emerge.

Sliding friction is not just a nuisance. Without friction there would be no violin music and it would be impossible to walk or drive a car. In some cases one wants to maximize the friction rather than to minimize it. This is the case, e.g., for the friction between the tires of a car and the road during braking, and much work has been devoted to this topic.

Sliding friction has been studied intensively for several hundred years and by many of the brightest scientists, e.g., Leonardo da Vinci, Coulomb, and Reynolds. The topic is of great theoretical interest and involves fundamental physics, e.g., questions related to the origin of irreversibility and adiabaticity, the role of self-organized criticality and, in the case of boundary lubrication, dynamical phase transitions in molecularly thin lubrication layers. This book presents a general introduction to the fundamental aspects of sliding friction with the emphasis on new experimental and theoretical results obtained during the last few years. We focus on general physical effects and a number of derivations of particular important results are included. In order to illustrate the "universal" nature of sliding friction, I have included a number of applications not usually contained in books on tribology, e.g., sliding of flux line lattices and charge density waves and a section on the dynamics of earthquakes. I believe that this book can be read by anyone having a basic knowledge of classical and fluid mechanics and a minimum knowledge of statistical and solid-state physics. The sections marked with an asterisk may be skipped on a first reading.

I would like to emphasize that this book is a rather personal view on sliding friction. Some of the material presented is original, e.g., the theory of time dependent plastic deformations (Sect. 11.4). Some of the topics considered are not very well understood at present, and the final picture may differ from the presentation in this book. It is my sincere hope, however, that the book will stimulate the reader to think about sliding friction and perhaps develop some of the ideas or unsolved problems mentioned here. Finally, I hope that the book will transmit some of the beauty which I find is inherent in the subject of friction, and which I have experienced during the process of writing this book.

This book is organized as follows. The first third (Chaps. 1–7) presents a broad overview of sliding friction. The second third (Chap. 8) contains a detailed discussion of the sliding of adsorbate layers on surfaces, which is of fundamental importance for many important applications, e.g., boundary lubrication and the question of what is the correct hydrodynamic boundary condition at a solid-fluid interface. Chapters 9–13 focus on boundary lubrication and dry friction dynamics. Finally, Chap. 14 considers in detail several "novel" sliding systems which, in addition to being of great interest in their own right, illustrate the concepts and results developed in the earlier sections.

I would like to thank G. Baumann, T. Baumberger, A.D. Berman, C. Caroli, S.C. Colbeck, U. Dürig, D. Eigler, J. Gimzewski, S. Granick, H.D. Grohmann, G. Gudehus, J.N. Israelachvili, B. Kasemo, K. Knothe, J. Krim, U. Landman, E. Meyer, T. Nattermann, A. Nitzan, G. Reiter, M.O. Robbins, M. Rodahl, M. Rosso, U.D. Schwarz, D. Schumacher, J.B. Sokoloff, E. Tosatti, A.I. Volokitin, R. Wiesendanger, E.D. Williams and Ch. Wöll for useful discussions and for supplying many figures for this book. I also thank S. Granick, G. Gudehus, B. Jones, and D. Lamoen for useful comments on the manuscript. Special thanks go to Angela Lahee and Claus-Dieter Bachem from Springer-Verlag for critical comments and improvements of the text, and to Helmut Lotsch and Roland Wiesendanger for offering me the opportunity to publish the book in the new Springer series "NanoScience and Technology".

Jülich, May 1997

Bo N.J. Persson

Contents

1. Introduction 1
2. Historical Note 9
3. Modern Experimental Methods and Results 17
   3.1 Surface Forces Apparatus 19
   3.2 Friction Force Microscopy 26
   3.3 Quartz-Crystal Microbalance 31
4. Surface Topography and Surface Contaminants 37
5. Area of Real Contact: Elastic and Plastic Deformations . 45
   5.1 Microscale Junctions 45
   5.2 Nanoscale Junctions 54
   5.3 Area of Real Contact for Polymers 77
6. Sliding on Clean (Dry) Surfaces 93
7. Sliding on Lubricated Surfaces 101
   7.1 Hydrodynamic Lubrication 102
   7.2 Fluid Rheology 110
   7.3 Elastohydrodynamics 119
   7.4 On the Thickness of Lubrication Films 120
   7.5 Wetting and Capillarity 134
   7.6 Introduction to Boundary Lubrication 148
   Appendix: Elastohydrodynamics 162
8. Sliding of Adsorbate Layers 171
   8.1 Brownian Motion Model 172
   8.2 Electronic and Phononic Friction 175
   8.3 Sliding Friction in the Low-Density Limit 198
       8.3.1 Analytical Results for a 1D-Model 198
       8.3.2 A Case Study: Chaotic Motion for a 3D-Model ... 205
   8.4 Linear Sliding Friction 208

	8.5	A Case Study: Xe on Silver	215
	8.6	Applications	228
		8.6.1 Boundary Conditions in Hydrodynamics	228
		8.6.2 The Layering Transition	236
		8.6.3 Spreading of Wetting Liquid Drops	239
	8.7	Two-Dimensional Fluids*	243
		8.7.1 2D Hydrodynamics	243
		8.7.2 Relation Between Sliding Friction and Diffusion	245
	8.8	Solid–Fluid Heat Transfer	248
	8.9	Non-linear Sliding Friction	254
		8.9.1 Dynamical Phase Transitions in Adsorbate Layers: Lennard–Jones Model	255
		8.9.2 Dynamical Phase Transitions in Adsorbate Layers: Frenkel–Kontorova Model	275
	8.10	Role of Defects at High Sliding Velocity and for Small Amplitude Vibrations*	278
		8.10.1 Model	278
		8.10.2 High Sliding Velocities	280
		8.10.3 Small Amplitude Vibrations	284
	8.11	Friction and Superconductivity: a Puzzle*	286
	8.12	Layering Transition: Nucleation and Growth	297
9.	**Boundary Lubrication**	313
	9.1	Relation Between Stress σ and Sliding Velocity v	314
	9.2	Inertia and Elasticity: "Starting" and "Stopping"	319
	9.3	Computer Simulations of Boundary Lubrication	322
	9.4	Origin of Stick-Slip Motion and of the Critical Velocity v_c	324
	9.5	Comparison with Experiments and Discussion	329
10.	**Elastic Interactions and Instability Transitions**	335
	10.1	Elastic Instability Transition	335
	10.2	Elastic Coherence Length	341
	10.3	Sliding of Islands, Big Molecules, and Atoms	345
	10.4	Sliding Friction: Contribution from Defects*	354
11.	**Stress Domains, Relaxation, and Creep**	363
	11.1	The Model	364
	11.2	Critical Sliding State at Zero Temperature	367
	11.3	Relaxation and Creep at Finite Temperature	370
	11.4	Time-Dependent Plastic Deformation in Solids	382
12.	**Lubricated Friction Dynamics**	395
	12.1	Small Corrugation: Shear-Melting and Freezing	396
	12.2	Large Corrugation: Interdiffusion and Slip at Interface	405

13. Dry Friction Dynamics 415
13.1 A Case Study: Creep and Inertia Motion 416
13.2 Memory Effects: Time Dependence of Contact Area 418
13.3 Theory 421
13.4 Non-linear Analysis and Comparison with Experiments 430

14. Novel Sliding Systems 435
14.1 Dynamics of Earthquakes 435
14.2 Sliding on Ice and Snow 439
14.3 Lubrication of Human and Animal Joints 445
14.4 Sliding of Flux-Line Systems and Charge Density Waves ... 447
    14.4.1 Flux-Line Systems 447
    14.4.2 Charge Density Waves 462
14.5 Frictional Coulomb Drag
Between Two Closely Spaced Solids 465
14.6 Muscle Contraction* 470
14.7 Internal Friction
and Plastic Stick-Slip Instabilities in Solids 480
14.8 Rolling Resistance 486
14.9 Friction Dynamics for Granular Materials 492

15. Outlook 497

References 499

Subject Index 513

1. Introduction

Friction is usually introduced and studied at an early stage during courses on classical mechanics and general physics. Thus one may think that friction is a simple and well understood subject. However, nothing could be more wrong. The two fundamental forces of Nature which manifest themselves on a macroscopic length scale are the electromagnetic and the gravitational forces. On a subatomic level other force fields come into play and contribute to the interaction between the fundamental particles of Nature, namely the quarks and leptons. However, only the electromagnetic forces are of direct relevance to the mechanical and friction properties of solids which is our concern in this book.

In this section we first discuss some basic questions related to the notion of friction. Next we describe some simple properties of friction which have been found to hold remarkably well for a large class of systems under quite general conditions. Finally, we give two examples which illustrate the practical use of friction.

On the Notion of Friction

The fundamental forces of Nature share certain properties which are not possessed by friction. For example, the equation of motion for particles interacting via electromagnetic forces can be formulated using the Lagrangian or Hamiltonian formalism and, in fact, all the forces of Nature seem to obey certain gauge-symmetries [1.1]. Furthermore, all the fundamental interactions in Nature can be described by Hamiltonian and Lagrangian functions which are analytical functions of the dynamical variables.

The friction forces observed for macroscopic bodies are ultimately due to the electromagnetic forces between the electrons and nuclear particles. Thus an exact treatment of the interaction between two solids would consider the coupling between all the electrons and nuclei using microscopic equations of motion for these particles (quantum electrodynamics). This, of course, is a hopelessly complicated problem. The concept of friction is a substitute for such a microscopic approach. Thus, the friction force which acts on a block that is sliding on a substrate is usually considered to be a function only of the center of mass velocity $\dot{x}(t)$ of the block, or, more precisely, a functional of $\dot{x}(t)$, since the friction force in general depends not only on the instantaneous velocity but on the velocity of the block at all earlier times (memory effects). Neglecting wear processes, friction arises from the transfer of collective translational kinetic energy into nearly random heat motion. Friction can formally be considered as resulting from the process of eliminating, or "integrating out", microscopic degrees of freedom in the following manner:

Consider a solid block on a substrate. Practically all surfaces of macroscopic bodies are rough, at least on a microscopic scale, and the interaction between the two solids occurs in the regions where surface asperities "touch" (we will show in Sect. 14.5 that the region where no "direct" contact between the solids occur, i.e., where the surfaces are separated by more than, say, 10Å, gives an entirely negligible contribution to the friction force). In each such contact area a layer of "foreign" molecules (e.g., grease) may occur. The basic idea now is to eliminate short-distance processes in order to obtain an effective equation of motion for the center of mass of the block. The first step is to study the motion of the "lubrication" layer under the influence of the forces from the surface asperities. The general solution to this problem depends on the motion of the surface asperities (which may be of the stick-slip nature even if the center of mass of the block moves steadily). Next, consider the motion of the surface asperities which are coupled by elastic forces to the center of mass motion of the block, as well as to the "lubrication" layer in the contact areas. The latter coupling can now be eliminated by substituting the solution for the motion of the lubrication molecules (which depends on the motion of the surface asperities) into the equation of motion for the surface asperities. The next step is to consider the equation of motion for the center of mass of the block, which is coupled to the surface asperities via elastic forces. These coupling terms can be eliminated by substituting the solution for the motion of the asperities into the equation of motion for the center of mass $x(t)$. This will, in general, give an equation of motion for $x(t)$ which contains memory effects. This approach is completely general, but may involve more steps than indicated above. For example, large surface asperities may have smaller surface asperities where the important interaction occurs; this would introduce another step in the hierarchy of length scales. It is also important to note that even if the motion of the center of mass is steady, on some smaller length scale rapid motion involving stick-slip will *always* occur, since, as will be discussed in Sect. 10.1, if this were not the case, the kinetic friction force would vanish in proportion to the sliding velocity $\dot{x}$ as $\dot{x} \to 0$, in contrast to experimental observations for the sliding of a block on a substrate (neglecting thermally activated creep motion, e.g., assuming zero temperature). One of the fundamental problems in friction is to discover the origin of the microscopic stick-slip motion.

This approach to sliding friction is similar to the renormalization group approach to second-order (or continuous) phase transitions [1.2]. In the latter case it is well-known that integrating out the "short" wavelength fluctuations will result in an effective Hamiltonian where various correlation functions are non-analytical functions of some of the parameters of the theory. Similarly, neglecting thermally activated creep, the friction force is in general a non-analytical function of the sliding velocity $\dot{x}(t)$ even though the fundamental (electromagnetic) interactions are analytical. For example, the kinetic friction force is usually finite as $\dot{x} \to 0$, but will change sign as $\dot{x}$ changes sign, which implies non-analyticity. Such non-analyticity can only arise after the elimination of infinite number of degrees of freedom, and will, strictly speaking, only occur for an infinite system.

The elimination of degrees of freedom to obtain effective equations of motion for the "large-scale" and "long-time" behavior of particle systems is completely general and very widely used. Thus even the most fundamental equations of physics governing the interactions between the quarks and leptons (the theory of general relativity, quantum electrodynamics, the theory of weak interaction and quantum chromodynamics) are now believed to be "effective" theories, obtained from a more fundamental theory, e.g., string theory, by integrating out degrees of freedom [1.1].

The problem of a block sliding on a solid substrate is very complicated and a systematic reduction of the degrees of freedom as described above can only be performed approximately. However, for simpler systems the approach outlined above can be performed essentially exactly. Thus consider the frictional drag force which acts on a small spherical particle moving with a constant velocity v in a fluid. In this case the first step would be to integrate out the short-wavelength (thermal) fluctuations in the liquid. This will, in the long-wavelength and long-time limit, result in the Navier–Stokes equations of fluid dynamics. The next step is to calculate the friction force acting on the particle by solving the Navier–Stokes equations with the correct boundary condition on the (moving) sphere. This gives the famous Stokes formula in which the friction force is proportional to the velocity v (see, e.g., [1.3]). As another example, consider an atom or molecule moving on the surface of a solid. This motion is coupled to the excitations of the substrate, and will result in a friction force acting on the atom (or molecule) which again (for small v) is proportional to the velocity v [1.4]. This problem is equivalent to a particle coupled to an infinite set of harmonic oscillators and can be solved by first integrating the equation of motion for the oscillators (which depends on the motion of the particle) and then substituting the result into the equation of motion of the particle (which is coupled to the oscillators). This approach has also been applied in studies of the tunneling of particles that are coupled by friction to the surrounding [1.5].

Let us now address the following problem: How large must a solid be in order for friction to be a well-defined concept [1.6]. It is clear that for a finite solid the phonon modes and the low-energy electronic excitations do not form a continuum but rather discrete levels, and if the spacing between the levels becomes larger than the broadening which results from the (ever-present) coupling to the surrounding, then one expects a strong reduction in the friction when two such small solids are slid relative to each other. This effect has, in fact, been observed for the damping of vibrations in molecules adsorbed on small metallic particles. Thus the lifetime of the C-O stretch vibration for CO adsorbed on small Pt particles is much longer than for CO adsorbed on the surface of a macroscopic Pt crystal [1.7]. Since the damping of the C-O stretch vibration most likely involves low-energy electronic excitations (the C-O frequency is much higher than the highest phonon frequency of Pt and decay by multiphonon emission is therefore very slow [1.8]), the effect probably results from the relatively large spacing between the electronic levels in the small Pt particles compared with the near continuum of electronic levels in a macroscopic Pt crystal.

A decrease in the sliding friction for small solids is also expected from classical physics [1.9]. The basic idea is that if the solids in question have sufficiently weak coupling to the environment, the primary mechanism for the friction is due to ergodic behavior within the solids themselves caused by the phenomenon of chaos. At sufficiently low temperatures and for sufficiently weak interface interactions, a transition can take place from chaotic (and hence ergodic) behavior to non-chaotic behavior, in which there will be no dissipation. This effect has been observed in many computer simulations, the most famous being the Fermi–Pasta–Ulam result [1.10], which showed that there is a transition from a regime in which the vibrational modes are damped, because of the interaction of the modes caused by anharmonicity, to one in which there is no damping.

Finally, let us comment on the concept of reversibility. One of the basic assumptions in thermodynamics is that if a change occurs slowly enough, no friction will occur. Thus a reversible engine is one for which every process is reversible in the sense that, by infinitesimal changes, we can make the engine go in the opposite direction [1.11]. This is possible only if the friction vanishes as the sliding velocity $v \to 0$. Now, this assumption may seem contrary to everyday experience, since a finite force is usually necessary in order to slide a block on a substrate. In fact, at zero temperature, if the block–substrate interaction is strong enough (which is always the case in practice), *elastic instabilities* will occur at the interface between the block and the substrate (Sect. 10.1), and independent of how slowly the block is pulled on the substrate, rapid processes will occur at the interface (e.g., molecular groups may undergo local stick-slip motion) and the friction force will remain finite as $v \to 0$. However, for any temperature $T > 0$K, thermally activated processes will remove the elastic instabilities if the sliding velocity is low enough and the friction will vanish linearly with v as $v \to 0$. Of course, if the temperature is very low, thermal excitation over the barriers will occur extremely slowly and extremely low sliding velocities would be necessary in order for the sliding friction to vary linearly with v. However, this does not change the conclusion that, if the changes occur slowly enough, friction can, *in principle*, be neglected and the engine would operate reversibly.

Elementary Aspects of Sliding Friction

The coefficient of friction μ between two solids is defined as F/L where F denotes the friction force and L the load or the force normal to the surface (Fig. 1.1). There is a very simple law concerning μ which is amazingly well obeyed. This law states that μ is independent of the apparent area of contact. This means that for the same load L the friction forces will be the same for a small block as for a large one. A corollary is that μ is independent of the load or, equivalently, the friction force is proportional to the load as illustrated in Fig. 1.2 for paper on paper. The physical explanation is that the area of *real* atomic contact between two solids is usually proportional to the load (Chap. 5). A second interesting observation is that the coefficient of friction μ is often nearly velocity independent unless the sliding velocity is very low, where thermally activation becomes important, or very high. Finally, contrary to common opinion, for typical engineering surfaces μ is

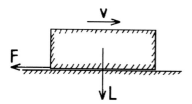

Fig. 1.1. The friction force F and the normal force (the load) L for sliding contact.

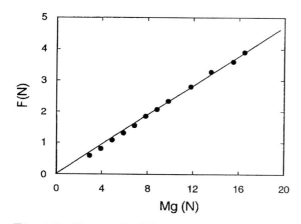

Fig. 1.2. The steady sliding friction force F at $v = 10\,\mu\mathrm{m/s}$ as a function of the loading force $L = Mg$ for a paper-paper interface. The dynamical friction coefficient is $\mu = 0.24$. From [1.19].

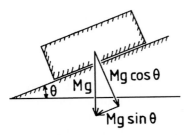

Fig. 1.3. A solid block on an inclined substrate.

usually nearly independent of the surface roughness, unless the surfaces are very rough or very smooth; in both these cases the friction forces are larger than in the intermediate roughness region.

That μ is approximately independent of the load L can be demonstrated by a simple experiment. We set up a plane, inclined at a small angle θ, and place a block of weight Mg on the plane; see Fig. 1.3. We then tilt the plane to a steeper angle, until the block just begins to slide under its own weight. The component of the weight parallel to the plane is $Mg\sin\theta$, and this must equal the friction force F. The component of the weight normal to the surface is $L = Mg\cos\theta$. With these values $\mu = F/L = \tan\theta$. If this law were exact, an object would start to slide at some definite inclination, independent of the weight Mg, as is indeed usually observed.

We emphasize that the "laws" presented above are empirical and result from an extremely complex set of events. When we proceed to the detailed study of sliding friction we will discover and discuss some of the limitations of these laws.

Two Applications

This book is concerned almost entirely with fundamental scientific aspects of friction. Nevertheless, sliding friction has very many practical applications and forms the basis for understanding many different phenomena. We close this section by giving two simple illustrations of the practical use of sliding friction.

An interesting application of the Coulomb friction law $F = \mu L$ is in the calculation of the minimum speed of a vehicle from the length of its skid marks. The skid distance d can be obtained by the condition that the initial kinetic energy $Mv^2/2$ is entirely "dissipated" by the friction between the road and the wheels during the skid. The friction force is $F = \mu L$ where the load $L = Mg$. Thus if μ can be taken as a constant during skid $Mv^2/2 = Fd$ or

$$Mv^2/2 = Mg\mu d$$

so that

$$v = (2d\mu g)^{1/2} \ .$$

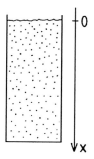

Fig. 1.4. A container filled with sand.

Hence, if the Coulomb law is obeyed, the initial velocity is entirely determined by the coefficient of friction and the length of the skid marks. This simple result does not hold perfectly since the coefficient of friction μ tends to increase with increasing velocity. Nevertheless, properly applied measurements of the skid length provide a conservative estimate of the speed of the vehicle at the moment when the brakes were first applied, and it has become a standard procedure for data of this kind to be obtained at the scene of a traffic accident.

As a second application we show that friction explains why the pressure at the bottom of a container filled with sand is essentially independent of the filling height [1.12]. This phenomenon is due to the fact that the forces are transmitted to the side walls such that the lower parts of the contents do not have to carry the weight of the part above. This is why the flow velocity in an hour-glass does not depend on the filling height. The constant pressure is in marked contrast to an incompressible liquid for which the pressure increases linearly with degree of filling.

To derive this result for a cylindrical container of radius R one starts by calculating the pressure increase $p(x)$ from the top ($x = 0$) to the bottom; see Fig. 1.4. Consider a horizontal slice of thickness Δx of weight $\pi R^2 \Delta x \rho g$ where ρ is the mass density which is assumed to be independent of x. The weight is partly compensated by the friction at the container walls. This friction is proportional to the normal force $L = 2\pi R \Delta x p(x)$. The total force on the slice Δx must equal zero so that

$$\pi R^2 [p(x + \Delta x) - p(x)] = \pi R^2 \Delta x \rho g - 2\pi R \Delta x \mu p(x) ,$$

or

$$\frac{dp}{dx} = \rho g - \frac{2\mu p}{R} ,$$

which can be integrated to give

$$p = \frac{\rho g R}{2\mu} \left(1 - e^{-2\mu x/R}\right) .$$

Near the surface $x = 0$ the pressure increases linearly with x as in an incompressible fluid, but at a depth comparable to the radius of the container the pressure approaches a constant value. This model explains only the height dependence of the average pressure. In granular media pressure fluctuations are significant and experiments have shown that the forces are concentrated on an irregular network of contacts [1.12]. Also, one expects thermally activated creep to occur, which results in a very slow increase of the pressure at the bottom of the container.

A large number of books on tribology and friction have been published during the last fifty years. Here I would in particular like to mention the books by Bowden and Tabor [1.13], More [1.14], and Rabinowicz [1.15], and three recent conference proceedings [1.16–18].

2. Historical Note

Tribology is the science and technology of interacting solid surfaces in relative motion. The word *tribology* is based upon the Greek word *tribos*, meaning rubbing. The topics covered by this word are in the main ancient and well-known and include the study of *lubricants, lubrication, friction, wear,* and *bearings*. Since surface interactions dictate or control the functioning of practically every device developed by man to enhance the quality of life, tribology has been of central importance for thousands of years, even if this has not always been generally recognized.

This section presents a brief survey of the evolution of our knowledge about friction. For a complete and very fascinating presentation, see D. Dowson's *History of Tribology* [2.1].

It is the engineering aspect of friction which has the longest pedigree. The first practical aspects of friction have their roots in prehistory. More than 400,000 years ago, our hominid ancestors in Algeria, China, and Java were making use of friction when they chipped stone tools. By 200,000 B.C., Neanderthals had achieved a clear mastery of friction, generating fire by the rubbing of wood on wood and by the striking of flint stones. The second application showing understanding of friction phenomena, namely the use of liquid lubricants to minimize the work required to transport heavy objects, dates back more than 4000 years.

The value of lubricants was appreciated at an early stage in the Sumerian and Egyptian civilizations. A chariot from about 1400 B.C. was found in the tomb of Yuaa and Thuiu, along with traces of the original lubricant on the axle. At an even earlier stage lubricants had been used to reduce friction between sledges used for the transportation of building blocks and the wooden logs or planks on which they moved. Thus, the use of lubricant, possibly grease, oil, or mud, for this purpose, was recorded as early as 2400 B.C.

The Egyptian method of moving large stone statues is depicted in the well-known painting from a grotto at El-Bershed dated about 1880 B.C. shown in Fig. 2.1. The colossus is secured on a sledge but there are no rollers or levers. A most interesting feature of this painting is that it shows an officer standing at the front of the pedestal pouring lubricant from a jar onto the ground immediately in front of the sledge.

Another interesting story related to the early use of lubricants comes from the building of the Pyramids in the third millennium B.C. The horizontal

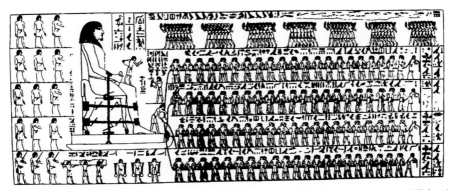

Fig. 2.1. Transporting an Egyptian colossus. Painting from the tomb of Tehuti-Hetep, El-Bershed (c. 1880 B.C.). From [2.1].

joints in Egyptian buildings often had a mean thickness of only 0.5 mm, i.e., the Egyptian masons could cut stones to within 0.25 mm over lengths of about 1.8 m. The blocks of stone were huge, often weighing tens of tons. Dry joints would have been perfectly adequate under these circumstances, but the traces of mortar between the layers of stones suggest that the builders used the "squeeze-film" mode of lubrication to slide the block accurately into place. Hydrated calcium sulphate (gypsum) was used to form the thin bed of viscid mortar, the efficiency of squeeze-film action being demonstrated by the precision of the final structure.

The first recorded systematic study in the field of tribology is due to Leonardo da Vinci, who not only performed experimental studies of friction, but also developed bearing materials, studied wear, and presented ingenious schemes for rolling-element bearings. His genius is best illustrated by his consideration of the application of mechanical principles to practical problems.

The sketches shown in Fig. 2.2 demonstrate that Leonardo da Vinci measured the force of friction between objects on both horizontal and inclined surfaces. Cords attached to the object to be moved were allowed to pass over fixed rollers or pulleys to weights which gave a measure of the resisting force. The torque on a roller placed in a semicircular section trough or half-bearing was measured in the same way. The illustration in Fig. 2.2b is most interesting since it clearly relates to Leonardo da Vinci's studies of the influence of the apparent contact area upon friction resistance.

Concerning the effect of apparent area of contact and the normal force upon friction, Leonardo da Vinci wrote: "The friction made by the same weight will be of equal resistance at the beginning of its movement although the contact may be of different breadth and length." And, "Friction produces double the amount of effort if the weight be doubled." These observations are essentially the first two laws of friction usually attributed to Amonton, but it seems entirely equitable to attribute them also to Leonardo.

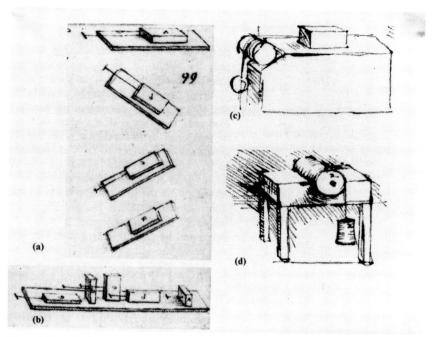

Fig. 2.2a–d. Leonardo da Vinci's studies of friction. Sketches from the *Codex Atlanticus* and the *Codex Arundel* showing experiments to determine: (**a**) the force of friction between horizontal and inclined planes; (**b**) the influence of the apparent contact area upon the force of friction; (**c**) the force of friction on a horizontal plane by means of a pulley; (**d**) the friction torque on a roller and half bearing. From [2.1].

Note that nowhere does Leonardo refer to the "force" of friction. The phenomenon of force, though evident in everyday life, was not adequately appreciated and defined by Renaissance mathematicians and engineers, and it was a further 200 years before Isaac Newton resolved the situation.

Leonardo also introduced the coefficient of friction as the ratio between the friction force and the normal force (load). For polished and smooth surfaces he concluded that "every friction body has a resistance of friction equal to one-quarter of its weight." This conclusion is incorrect, but the value $\mu = 1/4$ is quite realistic for the materials which were commonly used in bearings at the end of the Middle Ages and in the early part of the Renaissance.

Leonardo da Vinci's contributions to sliding friction were made about 200 years before the publication (1687) of Newton's *Principia*, which formed the basis for all the subsequent studies of sliding friction, and in particular for the comprehensive study by Charles Augustin Coulomb (1736–1806). He is perhaps best known for his contributions in the fields of electricity and magnetism, but to tribologists his name is practically synonymous with friction.

Table 2.1. Static friction force as a function of the time of stationary contact, as obtained by C.A. Coulomb. From [2.1].

t (min.)	F_s (arb. units)
0	5.02
2	7.90
4	8.66
9	9.25
26	10.36
60	11.86
960	15.35

Coulomb investigated the influence of five main factors upon friction, namely,

a) The nature of the materials in contact and their surface coatings.
b) The extent of the surface area.
c) The normal pressure (or force).
d) The length of time that the surfaces remained in stationary contact.
e) Ambient conditions such as temperature, humidity, and even vacuum.

He used his experimental data to construct empirical equations relating the force of friction to the variables listed above. The procedure is well illustrated by the report of his study of the influence of time t of stationary contact on the static friction force F_s on two pieces of well-worn oak lubricated with tallow. His results were presented in tabular form as in Table 2.1. Coulomb found that the observed values of friction force could be represented accurately by an equation of the form $F_s = A + Bt^\alpha$ where $\alpha \approx 0.2$. However, it is interesting to note that a similarly good fit to the experimental data is achieved by $F = A + B\ln(t)$, a relation which has now been shown to be of very general validity (Chaps. 5 and 13). Figure 2.3 shows F_s as a function of $\ln(t)$ (the $t = 0$ data point in Table 2.1 would fall on the straight line if the actual measurement occurred at $t = 12$s, as indicated by the arrow in Fig. 2.3.).

The variation of the static friction force with the time of stationary contact was explained by Coulomb by considering the surfaces of fibrous materials to have the form shown in Fig. 2.4. The surfaces of wood were considered to be covered by flexible, elastic fibers like the hairs on a brush. When such surfaces were brought together the bristles penetrated each other and since it took a finite time for meshing to become complete it was quite likely that the static friction would also increase with the time of stationary contact. Coulomb considered that the application of a tangential force would cause the elastic fibers to fold over as shown in Fig. 2.4. After a certain amount of deformation, which depended only upon the size of the asperities, or fibers, sliding would take place as the opposing asperities slipped out of mesh. This

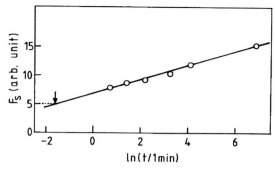

Fig. 2.3. The static friction force as a function of the natural logarithm of the time of stationary contact. *Circles*: experimental data points from Table 2.1.

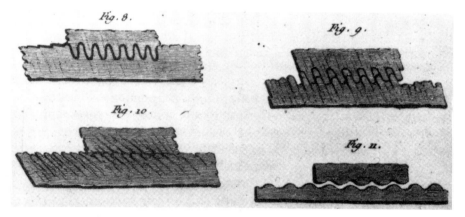

Fig. 2.4. C.A. Coulomb's representation of rough surfaces (Coulomb, 1785). From [2.1].

picture is remarkably similar to the modern picture of sliding of surfaces with grafted chain monolayers; see Sects. 3.1, 7.6, and 12.2.

Coulomb summarized many of his results in the *friction law*

$$F = \mu L ,\tag{2.1}$$

where F is the friction force and L the normal force. He found that the friction coefficient μ is usually nearly *independent* not only of L but also of the sliding velocity (as long as it is not too high or too low), the contact area, and the surface roughness.

The theory of *hydrodynamic lubrication* is based on the equations of motion of viscous fluids, developed by Euler, Bernoulli, Poiseuille, Navier, and

Stokes during the period 1730–1845. Some important lubricants used during the Industrial Revolution were:

a) *Sperm oil,* which was obtained from the large cavity in the head of the sperm-whale, and was highly prized for its ability to reduce friction and for its stability.
b) *Lard oil,* obtained from the fat of pigs, was held to be an excellent lubricant. It was much cheaper than sperm oil and consequently much more widely used.
c) *Olive oil.* The best olive oil used for lubrication was said to be just as good as sperm oil. It has a much higher viscosity than sperm oil, thus enabling it to be used for heavier duties.
d) *Mineral oil.* The role of mineral oil in industrial processes was initially quite small, but from these insignificant beginnings was to grow a great industry capable of providing vast quantities of hydrocarbon lubricants at very low cost.

Gustav Adolph Hirn performed the first systematic studies of hydrodynamic lubrication in 1847. He found that at high sliding velocity, v, the coefficient of friction is directly proportional to v when the temperature, and hence the viscosity, is held constant. He also found that when the speed was reduced to a certain value the lubricant is expelled resulting in direct contact between the surfaces, and the friction force once more becomes enormous (boundary lubrication). He discovered the effect of running-in upon bearing friction and pointed out that lubricated bearings must be run continuously for a certain time before a steady value of friction, lower than the initial value, is attained.

Nikolai Pavlovich Petrov studied the effect of the lubricant upon the friction of journal bearings (1883). He realized that in hydrodynamic lubrication, the friction between the sliding surfaces was to be found in simple hydrodynamics rather than in the contemporary concepts of interlocking asperities. He derived a simple expression, now called *Petrov's law,* for the friction force on a journal bearing.

Beauchamp Tower conducted experiments in England to determine the friction of the journal of a railway-car wheel, and observed the generation of hydrodynamic pressure in the first partial journal bearing. The pressures generated even in these early experiments of 1885 were sufficiently large to displace plugs or stoppers placed in oil holes in the bearing casing. Osborne Reynolds was intrigued by Tower's results and showed that they could be well explained by applying the principles of fluid mechanics. In 1886 Reynolds published his classical theory of hydrodynamic lubrication, which is widely used in the design of modern machinery.

Hydrodynamic or *fluid film* lubrication is one of at least two distinct modes of lubrication. It was the English biologist W.B. Hardy who clarified the situation and in 1922 introduced the term *boundary lubrication.* He found that very thin adsorbed layers, perhaps only about 10Å thick, were sufficient

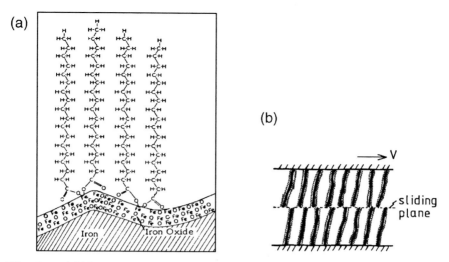

Fig. 2.5. (a) Fatty acids bind to metal oxides via their polar "head". (b) Schematic representation of W.B. Hardy's concept of boundary lubricating films.

to cause two glass surfaces to slide over each other with very low friction. He concluded that under such circumstances the lubrication depended wholly upon the *chemical* constitution of the fluid, but not on its viscosity, and that a good lubricant was a fluid which was adsorbed by the solid surfaces. His experiment showed that, of paraffins, alcohols and fatty acids, the last give the lowest sliding friction in boundary lubrication experiments. Fatty acid molecules have a polar head and a long hydrocarbon tail. The polar head tends to bind relatively strongly to polar surfaces such as metal oxides (Fig. 2.5a), and Hardy pictured a sliding interface as consisting of solid surfaces covered by monolayers of fatty acids with the hydrocarbon tail pointing away from the surfaces; the sliding occurs at the plane where the hydrocarbon tails meet as illustrated in Fig. 2.5b.

The correct explanation of why, in dry friction and boundary lubrication, the friction coefficient μ in (2.1) is independent of the (apparent) contact area and of the normal load L was not presented until around 1940 when Bowden and Tabor pointed out that, because of surface roughness, there was a crucial difference between the apparent and the real area of contact, and that it was the real area alone which determined the magnitude of the friction force. Since the real area of contact could be shown to be proportional to the load, the friction "law" (2.1) followed naturally. The reason why the friction coefficient μ is independent of the sliding velocity v (for small v) was also first realized during this century. At first, this result is not at all obvious since it is well-known that when a "particle" moves slowly (velocity v) in some medium it usually experiences a friction or drag force F *proportional* to the velocity

v. For example, the friction force on a small spherical particle moving slowly in a fluid is given by the famous Stokes formula and is proportional to the velocity v. But Tomlinson (1929) showed that a velocity-independent friction force results if, during sliding, *rapid processes* occur somewhere in the system even if the center of mass of the block moves *arbitrarily slowly* relative to the substrate. (This statement is actually not quite correct. At any nonzero temperature, at low enough velocity v, the molecular groups at the sliding interface will surmount the energy barriers which inhibit the sliding motion by thermal excitation, rather than by being pulled over the barriers by the external driving force. This will result in creep motion and a friction force proportional to v as the velocity v tends to zero. However, in most practical sliding problems, v is so large that creep motion can be neglected and the friction force is nearly velocity independent.) One fundamental problem in sliding friction, which is addressed in this book, is to understand the microscopic origin of these rapid processes, and to relate them to the macroscopic motion of the block.

In the period from 1960 to 1987, very little progress was made in the area of fundamental studies of sliding friction. During this period most scientists with an interest in surface science studied simpler phenomena such as the adsorption of atoms and molecules on single crystal surfaces in an ultra-high vacuum. But during the last few years sliding friction has become a "hot" topic. New experimental techniques, such as the *friction force microscope,* have been developed which allow well-defined experiments to be performed on well-characterized model systems. Simultaneously, the development of powerful computers now allows simulations (e.g., numerical integration of Newton's equation) to be performed on realistic sliding systems. Furthermore, it is now realized that sliding friction may exhibit universal properties and that it forms the basis for the understanding of other important topics of current interest, e.g., models for earthquakes.

3. Modern Experimental Methods and Results

Practically all sliding friction devices have an interface where the friction force is generated, a finite sliding mass M, and some elastic properties usually represented by a spring k_s as in Fig. 3.1. The spring does not need to be an external spring but could represent the overall elastic properties of the sliding device. In most sliding friction experiments the free end of the spring moves with a constant velocity v_s, but sometimes it varies with time. The force in the spring as a function of time is the basic quantity registered in most of these experiments. It is important to note that, due to inertia, during acceleration the spring force is not equal to the friction force acting on the block.

Figure 3.2 shows the *spring force* as a function of time for three different cases. Assuming that sliding starts with the spring at its natural length, the spring force initially increases linearly with time while the block is stationary. When the spring force reaches a critical value F_a, the so-called static friction force, the block starts to move. The motion is either of the steady type as in Fig. 3.2a, in which case the spring force equals the kinetic friction force F_b, or else stick-slip motion occurs, where the block alternates between stick and slip. In the latter case the kinetic friction force cannot be directly deduced from the stick-slip oscillations, unless the nature of the sliding dynamics is understood (Sect. 7.6.). One fundamental problem in sliding friction is to map out (and understand) the dynamical phase diagram, i.e., to determine the regions in the (k_s, v_s) plane where steady and stick-slip motion occur, and also the nature of the transitions between stick-slip and steady sliding, e.g., whether the transitions are continuous or discontinuous. It is an experimental observation that stick-slip always disappears if the spring k_s is stiff enough, or if the sliding velocity v_s is high enough; see Fig. 3.3. Hence, the kinetic friction force F_b can always be measured by making the spring k_s so stiff that the

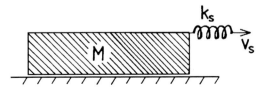

Fig. 3.1. A block with mass M sliding on a flat substrate. A spring (spring constant k_s) is connected to the block and the "free" end of the spring moves with the velocity v_s.

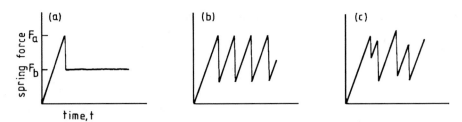

Fig. 3.2. Sliding dynamics of the block in Fig. 3.1 for three different cases: (**a**) steady sliding; (**b**) periodic stick-slip motion; (**c**) chaotic motion.

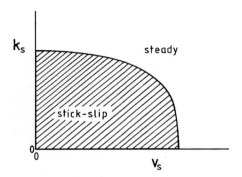

Fig. 3.3. General form of a kinetic phase diagram. The dashed region in the (k_s, v_s) plane denotes the values of k_s and v_s for which the motion is of the stick-slip nature.

motion is steady. Note also that during stick-slip motion the oscillations in the spring force are not always periodic as in Fig. 3.2b, but sometimes behave chaotically as in Fig. 3.2c. Non-periodic stick-slip motion may result from various sources, e.g., from the intrinsic nonlinear properties of the equations of motion, or simply from surface inhomogeneities, e.g., randomly distributed areas of "dirt", with an average spacing on the order of, or longer than, the linear size L of the block. [If the average separation between the surface inhomogeneities were much smaller than L, the fluctuations in the friction force would be negligible. From random-walk arguments, the fluctuation δF in the friction force F due to surface inhomogeneities with the concentration n, is $\delta F \propto (nA)^{-1/2}$ where A is the surface area.]

It is important to note that the stick-slip motion considered above involves the whole block, i.e., it refers to the center of mass motion of the block. But stick-slip can occur on many different length scales. Thus, even if the motion of the center of mass is steady (or smooth), *local* stick-slip motion usually occurs at the interface between the block and the substrate, involving, e.g., individual molecular groups or surface asperities. In fact, if this were not the case, the kinetic friction force F_k would depend linearly on the sliding velocity v as $v \to 0$ (Sect. 10.1), in contrast to the experimental observations

$F_k \approx$ const. (neglecting thermally activated creep motion). During steady sliding these local (rapid) slip events occur incoherently, and since a large number of "units" are involved, with respect to the center of mass motion, the fluctuations will nearly average out. However, the local stick-slip events may be probed indirectly, e.g., by studying the elastic waves (sound waves) emitted from the sliding interface.

At least three different experimental techniques are used today to gain fundamental insight into sliding friction, namely the surface forces apparatus, friction force microscopy and the quartz crystal microbalance.

3.1 Surface Forces Apparatus

This is a modern version of the sliding configuration shown in Fig. 3.1. It consists of two curved, molecularly smooth surfaces immersed in a liquid or in an atmosphere of controllable vapor pressure or humidity [3.1]. When the surfaces are brought into contact the solids deform elastically and a small circular contact area is formed, with a typical diameter D of order 30 μm, see Fig. 3.4. The surfaces are separated by a lubricant film which could be as thin as a single monolayer. The separation between the surfaces can be measured to within ~ 1 Å by studying multiple beam interference fringes from the junction. Typically, the sliding surfaces are made from thin mica sheets which are glued onto curved glass surfaces. Mica is a layer silicate, which can be cleaved to produce atomically smooth surfaces, i.e., without a single step, over macroscopic areas. Hence it is an ideal surface for well-defined model studies. The thickness of the lubricant film can be varied by changing the load, and, by applying a tangential force, sliding friction can be studied for a lubrication film of a given thickness.

Sliding friction studies have been performed on two different classes of adsorption systems as illustrated in Fig. 3.5. The first class consists of "inert" molecules, e.g., hydrocarbon fluids or silicon oils. In these cases the adsorbate–substrate interaction potential is relatively weak and, more importantly, the *lateral corrugation* of the adsorbate–substrate interaction potential is very small. It has been suggested for these systems that, during the transition from stick to slip, the lubricant film goes from a solid pinned state to a

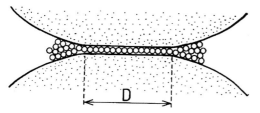

Fig. 3.4. Contact area between two curved surfaces in the surface force apparatus.

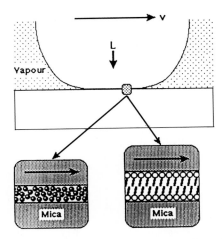

Fig. 3.5. Schematic representation of the contact region of two lubricated mica surfaces in a surface force apparatus experiment. From [3.3].

fluid state. As an example [3.2,3], Fig. 3.6a shows the spring force as a function of time, at several spring velocities v_s, for a 12Å thick hexadecane ($C_{16}H_{34}$) film. Note that for $v_s < v_c^+$, where $v_c^+ \approx 0.3\,\mu m/s$, stick-slip oscillations occur while an *abrupt* transition to steady sliding occurs when v_s is increased above v_c^+ (note: v_c and v_c^+ denote the spring velocities at the transition between stick-slip and steady sliding, when the spring velocity is reduced from the steady sliding region, and increased from the stick-slip region, respectively). The second class of systems includes grafted chain molecules, e.g., fatty acids on mica or on metal oxides. In these cases the corrugation of the adsorbate–substrate interaction potential is much stronger, and usually no fluidization of the adsorbate layers occurs; the sliding now occurs at the plane where the grafted chains from the two surfaces meet. As an example [3.3,4], Fig. 3.6b shows the spring force for DMPE monolayers on mica. DMPE is a grafted chain molecule (as in Fig. 2.5) which forms a close packed crystalline monolayer on mica. Again, for $v_s < v_c^+$ stick-slip motion is observed while only steady sliding is observed for $v_s > v_c^+$, where $v_c^+ \approx 0.1\,\mu m/s$ in the present case. However, note that the amplitude of the stick-slip spikes decreases *continuously* as v_s increases towards v_c^+ reflecting a qualitatively different type of sliding scenario than that of the hexadecane film in Fig. 3.6a. Although it is not clear from Fig. 3.6b, the transition from stick-slip to steady sliding at v_c^+ is discontinuous and hysteresis occurs as a function of v_s, i.e., $v_c < v_c^+$. We now discuss these two classes of sliding systems in more detail.

Large Corrugation

The decrease in the amplitude of the stick-slip spikes in Fig. 3.6b with increasing v_s can be understood as follows. Let us consider the molecules at

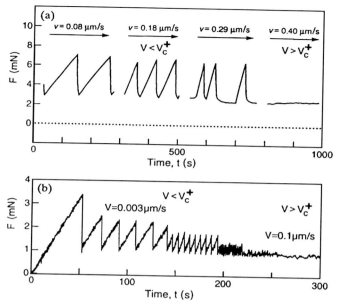

Fig. 3.6. The spring force F as a function of time t for (**a**) a hydrocarbon liquid (hexadecane) and (**b**) for mica surfaces coated with end-grafted chain molecules (DMPE). When the spring velocity increases beyond v_c^+ the sliding motion becomes steady, where $v_c^+ \approx 0.3\,\mu\text{m/s}$ in case (**a**) and $v_c^+ \approx 0.1\,\mu\text{m/s}$ in case (**b**). From [3.3].

the interface during a stick time period. The minimum free energy configuration may involve some interdiffusion of the hydrocarbon chains from the two monolayer films, or some other molecular rearrangement process. However, a large (on the scale of the thermal energy $k_B T$) free energy barrier may separate the interdiffused state from the state where no interdiffusion has occurred and a long time τ may be necessary in order to reach the thermal equilibrium state. For low spring velocity v_s the system spends a long time in the pinned state and nearly complete interdiffusion may occur, leading to a "large" force necessary to break the pinning. Thus "high" stick-slip spikes occur at "low" spring velocity. As the spring velocity increases, the system will spend less time in the pinned state before the critical stress (necessary for initiating sliding) has been reached, resulting in less interdiffusion and a smaller static friction force. Thus the amplitude of the stick-slip spikes are expected to decrease with increasing spring velocity.

The discussion above can be made more transparent by a diagrammatic representation. Let us assume that after the return to the pinned state the static friction force depends only on the time of stationary contact t, $F_0 = F_0(t)$. We assume that $F_0(0) = F_k$, where F_k is the kinetic friction force at low sliding velocity just before stick, and that $F_0(t)$ increases monotonically

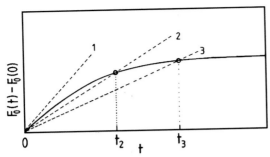

Fig. 3.7. *Solid line*: variation of the static friction force with the stopping time t. *Dashed lines*: the variation of the spring force for three different spring velocities (1)–(3), $v_1 > v_c$ but $v_c > v_2 > v_3$.

with the time t of stationary contact, as indicated in Fig. 3.7. The dashed lines in Fig. 3.7 show the increase in the *spring force*, $k_s v_s t$, as a function of the contact time for three different cases. In case (1) the spring force increases faster with t than the initial linear increase of the static friction force; hence the motion of the block will not stop and no stick-slip motion will occur. If the spring velocity v_s is lower than the critical velocity v_c [cases (2) and (3)] determined by the initial slope of the $F_0(t)$ curve [$k_s v_c = dF_0/dt(t=0)$], the spring force will be smaller than the static friction force $F_0(t)$ until t reaches the value t_2 [case (2)] or t_3 [case (3)], at which time slip starts. Thus in these cases stick-slip motion occurs. It is plausible from Fig. 3.7 that, if $F_0(t)$ has the form indicated in the figure, the amplitude of the stick-slip oscillations decreases *continuously* as the spring velocity approaches v_c from below, as indeed is observed in the experiment (Fig. 3.6b). However, an accurate analysis (Chap. 12) shows that the transition from stick-slip to steady sliding may be discontinuous.

If τ is short enough stick-slip disappears over the whole (k_s, v_s)-plane. In this case we may say that the boundary lubrication film is in a liquid-like state and an arbitrary small external force will lead to a finite sliding velocity. This is illustrated in Fig. 3.8a which shows the spring force as a function of the spring velocity for monolayer films of CaABS which has a larger separation between the chains than DMPE, leading to higher chain mobility and lower barriers to interdiffusion (Fig. 3.8b).

Finally, let us note that abrupt transitions have been observed during *steady* sliding, where the friction force drops from a "high" value for $v_s < v_c^*$ to a "low" value for $v_s > v_c^*$. Such dynamical (first-order) phase transitions imply that the configuration of the grafted chain molecules changes abruptly at $v_s = v_c^*$, see Fig. 3.9. Dynamical phase transitions between two different steady sliding states have also been observed in computer simulations; the physical nature and origin of one such transition will be discussed in Sect. 8.9.

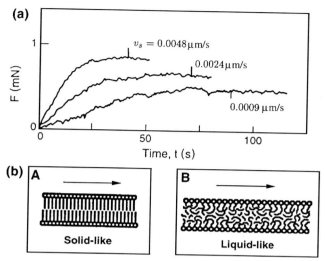

Fig. 3.8. (a) The spring force as a function of time for liquid-like grafted monolayer films (CaABS molecules). (b) Schematic representation of the lubrication film for (*A*) solid-like (such as DMPE in Fig. 7.35a) and (*B*) liquid-like (such as CaABS) boundary lubrication films. From [3.4].

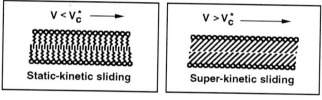

Fig. 3.9. Transition to super-kinetic sliding occurring above some critical velocity v_c^*. In the super-kinetic regime the grafted chain molecules shear align into a new configuration resulting in very low friction. From [3.3].

Small Corrugation

Let us consider the second class of systems illustrated by the hexadecane film in Fig. 3.6a. Here the corrugation of the adsorbate–substrate interaction potential is very weak and we believe that the transition from stick to slip involves (neglecting creep motion) a dynamical phase transition where the thin (e.g., one or two monolayer) lubricant film makes a transition from a *solid* state, which pins the surfaces together, to a *fluid* state where the surfaces slide relative to each other, see Fig. 3.10. [The solid structure in Fig. 3.10 has a perfectly commensurate (with respect to the substrate) crystalline structure. This is unlikely to be the case in most practical cases, where instead small crystalline regions or even a glassy state are likely to occur at stick.

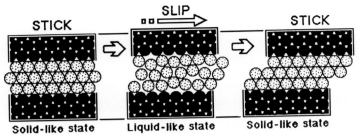

Fig. 3.10. During stick-slip motion the lubrication film alternates between a frozen solid state and a liquid or fluidized state. Such dynamical phase transitions have been observed in computer simulations (Sects. 8.9 and 9.3). From [3.2].

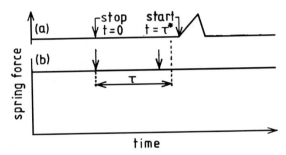

Fig. 3.11. Stop-start experiments.

Furthermore, it is very likely that, even during stick, a nonuniform stress distribution exists in the lubrication film and the elastic solids in the contact region; see Chap. 11.] The transition from slip to stick involves a nucleation process, where solid islands are nucleated in the fluid phase. Support for this picture comes from computer simulations (Sect. 8.9, Chap. 9) and from stop-start experiments [3.2] of the following type: Assume that the spring velocity $v_s > v_c^+$ so that steady sliding occurs. Suppose now that the spring velocity is abruptly reduced to zero at time $t = 0$, kept at zero for a time period τ^*, and then abruptly increased back to its original value v_s. If τ^* is below some characteristic time τ (which depends on the external load, the temperature, and v_s) steady sliding continues for $t > \tau^*$ while if $\tau^* > \tau$ a stick-slip spike occurs, see Fig. 3.11. This can be loosely interpreted as implying that the system needs a certain time τ during "stop" to nucleate the pinned state.

The transition from steady to stick-slip motion for the hexadecane film in Fig. 3.6a can be understood as follows [3.5,6]: Let us assume that, after the return to the pinned state, the static friction force depends only on the time of stationary contact t, $F_0 = F_0(t)$. We assume that $F_0(0) = F_k$, where F_k is the kinetic friction force at low sliding velocity just before stick, and that $F_0(t)$ increases monotonically with the time t of stationary contact, as indicated in

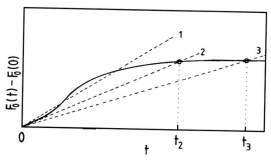

Fig. 3.12. *Solid line*: variation of the static friction force with the stopping time t. *Dashed lines*: the variation of the spring force for three different (1)–(3) spring velocities, $v_1 > v_c$, $v_2 = v_c$ and $v_c > v_3$.

Fig. 3.12. The dashed lines in Fig. 3.12 show the increase in the *spring force*, $k_s v_s t$, as a function of the contact time for three different cases. In case (1) the spring force increases faster with t than the initial linear increase of the static friction force; hence the motion of the block will not stop and no stick-slip motion will occur. If the spring velocity v_s is equal to [case (2)] or lower than [case (3)] the critical velocity v_c, determined by the initial slope of the $F_0(t)$ curve [$k_s v_c = dF_0/dt(t=0)$], the spring force will be smaller than the static friction force $F_0(t)$ until t reaches the value t_2 [case (2)] or t_3 [case (3)], at which time slip starts. In these cases stick-slip motion will occur. It is plausible from Fig. 3.12 that, if $F_0(t)$ has the form indicated in the figure, the amplitude of the stick-slip oscillations disappears *abruptly* as the spring velocity approaches v_c from below, i.e., the transition is *discontinuous*, and also that the amplitude of the stick-slip spikes is nearly independent of v_s, in agreement with experiment, see Fig. 3.6a. A more accurate analysis (presented in Chap. 12) shows, however, that the transition from stick-slip to steady sliding can be more complicated than suggested by the discussion above.

The solid pinned state which occurs at "stick" for the system studied in Fig. 3.6a is only stable if the temperature is below some critical value T_c. If $T > T_c$ the lubrication layer is always in a fluid state, implying that no stick-slip occurs and that an arbitrarily low external shear force will give rise to a nonzero sliding velocity. Such liquid-like behavior is observed [3.2] for the system of Fig. 3.6a if the temperature is increased from $T = 17°C$ to $T = 25°C$; thus the melting temperature T_c of the hexadecane film is somewhere between 17 and 25°C.

3.2 Friction Force Microscopy

Recently, sliding friction has been probed on an atomistic scale using the friction force microscope [3.7]. Here a very sharp tip, e.g., made from tungsten or diamond, is slid over a substrate (Fig. 3.13). Typical normal forces are of order $L \sim 10^{-7} - 10^{-9}$N and the contact area between the tip and the substrate has a diameter of order $D \sim 10-30$Å. Sometimes the force–distance curves exhibit stick-slip behavior on an atomic scale. As an example, Fig. 3.14 shows results for the sliding of a tungsten tip on a mica surface. The tip is first brought into repulsive interaction with the substrate. During sliding, initially the sliding force increases linearly with the parallel displacement of the tip because of elastic deformations of the tip. At some critical force the top of the tip jumps a distance corresponding to the size of the substrate unit cell. This process repeats itself periodically as the tip is moved laterally over the surface. In the measurements shown in Fig. 3.14 the perpendicular force was rather high, which resulted in a large contact area. The reason why atomic scale resolution is nevertheless obtained may be the trapping of a mica flake on the tip, which, during sliding, could "jump" between commensurate binding positions on the substrate.

Usually atomic scale fluctuations in the friction force are not resolved, but rather the changes in the friction properties over larger length scales are probed. To illustrate this, consider Fig. 3.15a which shows the friction force when a silicon oxide tip is slid over a Si(110) surface which has 2 μm-wide stripes of SiO_2 adjacent to 4 μm wide stripes of hydrogen-passivated silicon. Higher friction (bright area) is observed on Si-H whereas the friction on SiO_2 is lower by a factor of two. Figure 3.15b shows the dependence of the friction force on the load. Note that the friction force increases linearly with the load on both surface areas with the friction coefficient equal to 0.3 on SiO_2 and 0.6 on Si-H.

A linear dependency of the friction force on the load, which is nearly universally observed for macroscopic bodies, is in general not found or expected

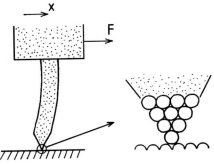

Fig. 3.13. A sharp tip made from, e.g., tungsten or diamond, sliding over a substrate (schematic).

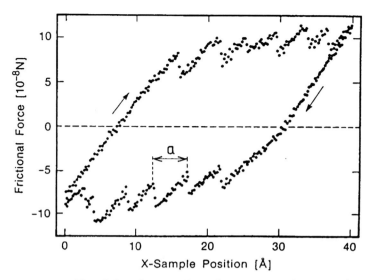

Fig. 3.14. The sliding force in the x-direction as a function of sample position as a tungsten tip is scanned back and forth on a mica surface. The substrate lattice constant $a \approx 5\text{Å}$. From [3.7].

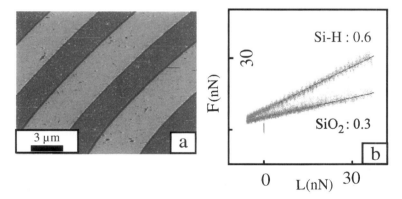

Fig. 3.15. (a) Friction force map of Si/SiO_2 grid etched in 1:100 HF/H_2O. The bright area corresponds to high friction and is observed on the Si-H whereas the friction on SiO_2 is lower by a factor of two. (b) The relation between the friction force F and the load L on the SiO_2 and Si-H areas in (a). Note the linear dependency $F = \mu L$ where $\mu = 0.6$ on Si-H and $\mu = 0.3$ on SiO_2. From [3.17].

in friction force microscopy measurements for the following reason: In most macroscopic experiments the area of real contact can be shown to be proportional to the load (Chap. 5) and the local pressure in the contact area is a material constant (e.g., the plastic yield stress). This implies a linear relation between the friction force and the load. However, in the friction force microscopy measurements mainly elastic deformation occurs in the contact area between the tip and the substrate which results in a non-linear relation between the contact area δA and the load. For example, if the tip–substrate adhesion forces can be neglected, then (for a smooth tip) from elastic continuum theory $\delta A \sim L^{2/3}$ (Chap. 5). As a consequence, the (average) pressure P in the contact area will depend on the load so that the friction force, which is the integral of the friction shear stress σ over the contact area δA, will in general be some non-linear function of the load L. For example, if $\sigma = \sigma_0 + bP$ and $\delta A \sim L^{2/3}$ one gets a friction force of the form $\sim aL^{2/3} + bL$ where a and b are constants. The fact that the shear stress depends on the pressure P is well known from classical friction measurements. For example, if one compares the friction properties of tin surfaces (hardness $\sim 70\text{N/mm}^2$) with hard steel surfaces (hardness $\sim 7000\text{N/mm}^2$) lubricated with typical fatty acids, the area supporting the load for the tin surfaces will be ~ 100 times larger than the area for the steel surfaces. Thus, the area of lubricant film to be sheared will be ~ 100 times greater. The observed coefficients of friction, however, does not differ by more than a factor of two. This implies that the shear strength of lubricant films is roughly proportional to the pressure P to which they are subjected [3.8]. Such a pressure dependence of σ can be understood on the basis of the following microscopic picture: during sliding, molecular groups at the interface flip from one configuration to another. The flips are accompanied by local expansions of the system. The barrier to sliding (or the activation barrier involved in the flips) will have a contribution from the work done against the external pressure P during the local expansions (Sect. 7.2).

Let us present a measurement which illustrates this non-linear dependence of the friction force on the load. Figure 3.16 shows two friction force maps obtained (with a silicon oxide tip) from an island film of C_{60} on a GeS(001) surface. In (a) the normal load $L = 6.7$nN while in (b) $L = 30$nN. Note that the contrast has changed: in (a) the bright (high friction) areas are the dendritically shaped C_{60} islands, while the GeS substrate is dark (low friction). In (b) the contrast is opposite, i.e., the C_{60} islands now have lower friction than the GeS substrate. This can only come about if the friction force varies non-linearly with the load on one or both of the two surface areas. Indeed, a detailed study shows that while the friction force varies nearly linearly with the load on the GeS area, it varies approximately as $L^{2/3}$ on the C_{60} islands.

In macroscopic friction experiments it is usually found that the friction force vanishes continuously when the load $L \to 0$ as implied by Coulomb's friction law (2.1). This is usually not the case in experiments involving a single asperity contact area, such as friction force microscopy with smooth

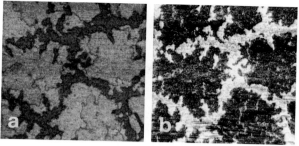

Fig. 3.16. Friction force maps of crystalline C_{60} islands on a GeS(001) substrate. Dark areas represent low friction and bright areas high friction. The normal force L was (a) $L = 6.7$nN and (b) 30nN. From [3.18].

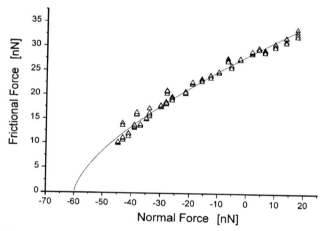

Fig. 3.17. The friction force as a function of the normal load for an amorphous carbon tip on a mica substrate under atmospheric conditions. From [3.19].

probing tips. To illustrate this, Fig. 3.17 shows the friction force as a function of the normal force (the load L) for a silicon tip covered by amorphous carbon sliding on a mica substrate under normal atmospheric conditions. Note that the friction force is finite for $L = 0$ and for small negative L. This arises from the adhesion forces between the tip and the high-energy mica substrate, which give a finite contact area also for small negative load. In the present case, when L is reduced to ~ -45nN the tip jumps away from contact. According to contact theories (Sect. 5.2), *the contact area remains finite the whole way down to jump-off*, and this explains why the friction force does not go to zero continuously at this point. The reason why adhesion usually does not show up in macroscopic friction measurements is due to surface roughness which results in many contact points (junctions): Since the asperities in contact with

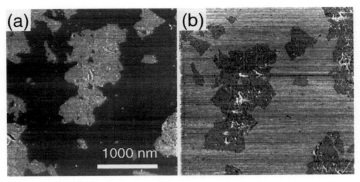

Fig. 3.18. Friction force maps on MoS$_2$ platelets deposited on mica in different humidities: (**a**) 0%, and (**b**) 40%. From [3.17].

the substrate have different sizes and different amounts of elastic deformation, as the load is reduced the junctions will break (or pop) one after another, and when the load vanishes the area of real contact will be essentially zero. Thus the friction force vanishes continuously as $L \to 0$.

Sliding friction is known to be extremely sensitive to surface properties, often depending mainly on the top layer of atoms or molecules. This has also been observed with the friction force microscope. As an example, Fig. 3.18 shows friction force maps (with a silicon oxide tip) from MoS$_2$-platelets deposited on mica in different humidities. In (a) the measurements have been performed in dry nitrogen and the friction force is highest on the MoS$_2$ platelets (bright area). At 40% humidity (Fig. 3.18b) the contrast has switched and the friction is highest on the mica surface. This turns out to be due mainly to a strong increase in the friction of mica (which is hydrophilic) with increasing humidity, while the friction of MoS$_2$ and other hydrophobic surfaces (e.g., Al$_2$O$_3$) is nearly independent of the humidity.

Finally, let us consider sliding friction for a silicon oxide tip on NaCl, where the friction force map exhibits step-site-specific contrast. Figure 3.19 shows a friction force map on a NaCl(001) surface with a finite density of steps. Note that both down and upward steps lead to increased friction. This indicates that there are rather large barriers for the motion of the tip atoms at a step, not only during the displacement of the tip from a lower terrace towards an upper terrace, but also when the tip is displaced in the reverse direction. The latter barrier, known as the Schwoebel barrier, plays an important role in thin-film growth.

The friction forces probed in the studies above are due to stick-slip motion on an atomic level. That is, the tip first deforms elastically while it is pinned at the tip–substrate interface, and when the applied force is high enough the tip displaces rapidly a distance which could be as short as a single substrate lattice spacing. As discussed in Sect. 10.1, this gives rise to a (time-averaged) friction force which, to a first approximation, is velocity independent (neglect-

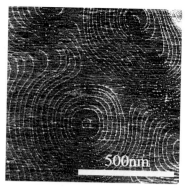

Fig. 3.19. Friction force map on NaCl(001). Increased friction is observed at the step sites. From [3.17].

ing thermally activated creep motion), as is indeed observed experimentally. It is important to note that if the velocity of the tip at the tip–substrate interface were on the order of the drive velocity at all stages of sliding, then the friction force would be proportional to the drive velocity, contrary to what is observed experimentally.

3.3 Quartz-Crystal Microbalance

Recently it has been found that molecules and atoms weakly adsorbed on the surface of a solid can *slip* relative to the surface if the solid performs mechanical vibrations. [3.9–11] The slip occurs as a result of the force of inertia, F, acting on the adsorbates during the vibrational motion of the crystal. The force of inertia is extremely weak (typically $\sim 10^{-9}$eV/Å per atom) and cannot, by itself, move adsorbates over the lateral barriers of the binding potential (which, for physisorption systems, would requires forces of order $\sim 10^{-3}$eV/Å). Thus the only role of the external force F is to slightly tilt the potential surface so that the diffusion barrier is slightly lower in the direction of the external force than in the opposite direction, see Fig. 3.20. This will lead to a slow thermally activated drift of the adsorbate layer in the direction of F, where the drift velocity v is *proportional* to the driving force F. It follows that at low temperature the drift velocity should exhibit an activated temperature dependence, $v \propto \exp(-\Delta E/k_B T)$, and, in particular, that v should vanish at zero temperature. This may be the case even if the adsorbate layer forms an incommensurate (IC) structure, since such a structure is always pinned by surface defects, with a pinning barrier which may be too large to be overcome by the external force (Sect. 8.5).

Figure 3.21 shows the experimental set-up. Thin, smooth (or rough) silver and gold films are evaporated onto two surfaces of a quartz-crystal microbal-

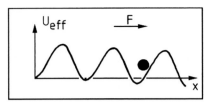

Fig. 3.20. Under the influence of a weak external force F, an adsorbate on a substrate experiences a corrugated potential slightly "tilted" in the direction of F. The adsorbate can move on the surface by *thermal excitation* over the barriers. Since the barriers are slightly lower in the direction of F, the particle will jump slightly more frequently in the direction of F which results in a slow drift motion in the direction of F.

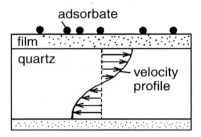

Fig. 3.21. A quartz-crystal microbalance (QCM) setup for measuring sliding friction of adsorbate layers.

ance (QCM). By applying a voltage to the quartz crystal (which is piezoelectric) it will vibrate with a characteristic frequency $\omega_0 \sim 10^7 \text{s}^{-1}$ and with an amplitude of order 100Å. If the driving force (i.e., the applied voltage) is suddenly removed the amplitude of the oscillations will decay with increasing time

$$u \propto \cos(\omega_0 t) e^{-\gamma t/2},$$

and both the resonance frequency ω_0 and the damping γ can be determined by analyzing the decay of the amplitude of vibration with increasing time. If molecules are adsorbed on the surface of the metal film this gives rise to a small shift in the resonance frequency $\Delta \omega$ of the crystal ("mass load"), and an increase in the damping $\Delta \gamma$, both being proportional to the adsorbed mass, and these can be measured even for a small fraction of an adsorbate monolayer.

The QCM measurements probe the *linear sliding friction* $\bar{\eta}$ defined as follows: Consider a solid adsorbate slab and assume that, in addition to the periodically corrugated adsorbate–substrate potential, a weak external force $\boldsymbol{F}$ acts on each adsorbate molecule parallel to the surface. This will lead to a drift motion so that

$$mN_1\bar{\eta}\langle\dot{\boldsymbol{r}}\rangle = N\boldsymbol{F},\qquad(3.1)$$

where N and N_1 are the total number of adsorbates and the number of adsorbates in the first layer, in direct contact with the substrate, respectively. In (3.1), $\boldsymbol{v} = \langle\dot{\boldsymbol{r}}\rangle$ is the drift velocity of an adsorbate and $\langle\ldots\rangle$ stands for thermal average. This definition of $\bar{\eta}$ takes into account the fact that the friction force is generated at the interface between the substrate and the adsorbate layer. If $P_1 = N_1/N$ denotes the fraction of the adsorbates in the first layer, in direct contact with the substrate, then for coverages up to a monolayer $P_1 = 1$, and in this case the definition above is valid even when the adsorbate layer is in a 2D-fluid state. It is also convenient to introduce the slip time $\tau = 1/P_1\bar{\eta}$ with the following physical meaning: Suppose the adsorbate layer is sliding with the steady-state velocity $\boldsymbol{v}_0$ as the driving force $\boldsymbol{F}$ is suddenly (at $t = 0$) set equal to zero. The drift motion of the adsorbate layer for $t > 0$ satisfies the Newton equation

$$Nm\dot{\boldsymbol{v}} + N_1 m\bar{\eta}\boldsymbol{v} = 0.\qquad(3.2)$$

The solution to (3.2) is $\boldsymbol{v} = \boldsymbol{v}_0\exp(-t/\tau)$, i.e., τ is the time it takes for the drift velocity to decay to $1/e$ of its initial value.

In the QCM measurements, $\boldsymbol{F}$ is the force of inertia acting on the adsorbates. By treating the QCM as a (damped) harmonic oscillator, we show in Sect. 8.6.1 that the adsorbate induced frequency shift $\Delta\omega$ and damping $\Delta\gamma$ are given by

$$\Delta\omega = -N_1\frac{m}{M}\frac{\omega\bar{\eta}^2 P_1}{\omega^2 + (\bar{\eta}P_1)^2},\qquad(3.3)$$

$$\Delta\gamma = 2N_1\frac{m}{M}\frac{\omega^2\bar{\eta}}{\omega^2 + (\bar{\eta}P_1)^2}.\qquad(3.4)$$

In these equations, M is the mass of the vibrating crystal and ω is the frequency of vibration. Note that (3.3) and (3.4) give

$$\Delta\gamma/\omega = -2\Delta\omega\tau.\qquad(3.5)$$

Using these equations, the quantities Nm and τ (or $\bar{\eta}$) can be immediately obtained from $\Delta\omega$ and $\Delta\gamma$. As an example, Fig. 3.22 shows the result for the sliding friction $\bar{\eta} = 1/\tau P_1$ as a function of Xe-coverage on a Ag(111) surface at the temperature $T = 77.4$K (where a multilayer Xe-film would be in a solid state) [3.12]. For the IC solid monolayer case $\bar{\eta} \sim 10^9 \text{s}^{-1}$, so that the frictional shear stress which acts on a Xe-film sliding at $v = 1$cm/s is only $mn_1\bar{\eta}v \approx 10 \text{N/m}^2$ (where $n_1 = N_1/A$ is the number of adsorbates per unit area). This should be compared with typical shear stresses involved in sliding of a steel block on a steel surface under boundary lubrication conditions, where the shear stress in the contact areas is of order $\sim 10^8 \text{N/m}^2$, or the shear stress necessary to slide an C_{60} island on a NaCl surface (Sect. 10.3), which is $\sim 10^5 \text{N/m}^2$. [Note, however, that in the QCM measurements the

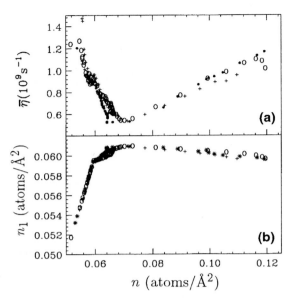

Fig. 3.22. (a) QCM measurements of the sliding friction $\bar{\eta} = 1/\tau P_1$ for Xe on a silver film at $T = 77.4$ K (1 monolayer = 0.05970 atoms/Å^2). (b) The number of adsorbates per unit area in the first layer in direct contact with the substrate, n_1, as a function of the total number of adsorbates per unit area, n. From [3.12].

friction force is *proportional* to the drift velocity, while in the latter cases the friction force (at low sliding velocities) is velocity independent. The reason why the friction force is proportional to the driving force $\boldsymbol{F}$ in QCM measurements is the small magnitude of F which implies linear response. The fact that the drift velocity is nevertheless rather high (typically $v \sim 1$cm/s) is due to the fact that $k_{\rm B}T$ is large compared with the lateral barriers of the adsorbate–substrate interaction potential.]

As a second example, consider benzene on Cu(111) (Fig. 3.23). In this case, beyond bilayer coverage, the sliding friction increases linearly with the coverage, up to at least ~ 20Å thick films. This implies that there is a contribution to the sliding friction from internal excitations in the film, e.g., phonons. The origin of the friction in QCM experiments will be addressed in Sect. 8.5, but some possibilities are summarized in Fig. 3.24. Table 3.1 gives the sliding friction $\bar{\eta}$ for several physisorption systems deduced [using (3.3-5)] from QCM data for incommensurate monolayer films.

The importance of the electronic friction channel (a) in Fig. 3.24 has recently been demonstrated directly by QCM measurements using lead films which become superconducting at $T_{\rm c} \approx 7$K. The sliding friction for incommensurate N$_2$-monolayers on the Pb film was observed to decrease abruptly (see Fig. 3.25) when the lead film is cooled below $T_{\rm c}$. In the present system, at least half of the sliding friction in the normal state must be of electronic

3.3 Quartz-Crystal Microbalance   35

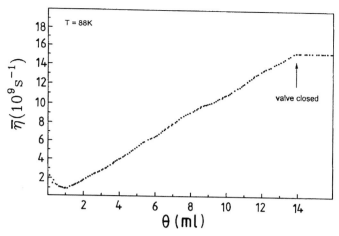

Fig. 3.23. QCM measurements of the sliding friction $\bar{\eta} = 1/\tau P_1$ for benzene on a copper film at $T = 88\,\mathrm{K}$. From [3.20].

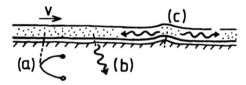

Fig. 3.24. Some mechanisms which contribute to sliding friction $\bar{\eta}$. (*a*) and (*b*) illustrate direct transfer of translational kinetic energy into electronic (electron–hole pair) and phononic excitations in the substrate. In (*c*) a surface defect excites phonons in the sliding layer which may be thermalized in the film before the energy is finally transferred to the substrate via process (*a*) or (*b*).

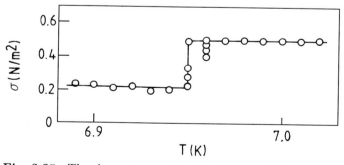

Fig. 3.25. The shear stress σ (for the sliding velocity $v = 1\,\mathrm{cm/s}$) versus temperature for an incommensurate monolayer of N_2 physisorbed on a Pb(111) surface. From [3.21].

Table 3.1. Sliding friction at incommensurate monolayer coverages deduced from QCM data from [3.9,13–16].

System	$\bar{\eta}(s^{-1})$
H_2O/Ag	3×10^8
C_6H_{12}/Ag	5×10^8
N_2/Au	3×10^8
Xe/Au	1×10^9
Xe/Ag	1×10^9
Kr/Au	1×10^8

origin. This result is all the more remarkable since the evaporated lead film is certainly not perfect and crystalline, and moreover was exposed to air for 10–15 minutes. The resulting oxide layer should have reduced the coupling between the sliding molecular film and the metal conduction electrons, while enhancing the coupling to phononic excitations in the sliding film, by virtue of a larger atomic corrugation than that of a clean metal surface.

4. Surface Topography and Surface Contaminants

Most surfaces of solids are rough, at least on a microscopic scale. Most engineering surfaces are covered by asperities having slopes in the range $5° - 10°$. Unlubricated metal surfaces encountered in an industrial environment will, in general, be covered by a whole series of surface films, as shown in Fig. 4.1. Working outwards from the metal interior, we first encounter an oxide layer, produced by reaction of oxygen from the air with the metal; this is present on all metals except gold. Next will come an adsorbed layer derived from the atmosphere, the main constitutes of this layer being water and oxygen molecules. Outermost, there will usually be grease or oil films. Note also that the metal just below the oxide layer is in general harder than that in the bulk (it is "work hardened").

Metal surfaces of this type generally have initial coefficients of friction in the range 0.1–0.3 when slid together. Higher values (typically of order unity) are reached if the surfaces continue to be slid over each other, since under these conditions the grease film, which is the one with the most drastic influence on the friction, will eventually be worn off.

The topography of surfaces can be studied by several different techniques. The *profilometer* is perhaps the most widely used method of measuring roughness. It employs a stylus, usually with a rounded end, of some very hard material, e.g., diamond. Such a stylus closely resembles a phonograph needle. In a roughness measurement the stylus is moved across the surface, and its vertical movements are amplified electrically. Figure 4.2 shows the surface topography of three different steel surfaces. Note the magnification of the vertical scale relative to the horizontal scale.

Scanning tunneling microscopy (STM) and *atomic force microscopy* (AFM) are more recent developments which allow studies of surfaces on an

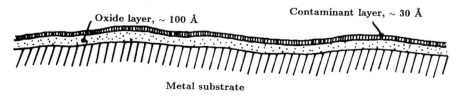

Fig. 4.1. A metal surface with an oxide film and adsorbed molecules (schematic).

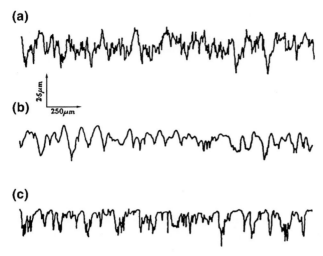

Fig. 4.2. Profiles of mild steel specimen after three surface treatments. (a) Surface ground only. (b) Surface ground and then lightly polished. (c) Surface ground and then lightly abraded. From [4.6,7].

atomic level. In STM a sharp metal tip is brought to a distance of a few Ångstroms from the substrate. A voltage is applied between the tip and the substrate and the tip is scanned along the surface. At the same time it is displaced perpendicular to the surface in such a way that the tunneling current is kept constant. This results in a contour map with atomic scale resolution. Figure 4.3 shows STM pictures of three different silver surfaces. Note that atomic steps can be detected in Fig. 4.3a. One limitation of STM is that it requires a substrate with relatively high conductivity. In AFM a sharp tip is again brought into close proximity of the surface to be studied. The tip is scanned along the surface and displaced perpendicular to the surface in such a way that the force between the tip and the substrate stays constant. The resolution of AFM is usually slightly less than that of STM, but sometimes atomic scale resolution is obtained. The AFM can be used for both insulating and conducting surfaces.

The techniques described above are not very convenient for assessing a substantial area for departures from smoothness. For this latter purpose it is more convenient to resort to an optical technique. There are many methods of examining surfaces optically to gain an insight into their roughness. Generally, a beam of light is shone on the surface and the reflected beam is analyzed. Irregularities on the surface show up as features in the reflected beam.

Recently a great effort has been devoted to the study and classification of surface roughness. It has been found that a variety of surfaces and interfaces occurring in nature are well represented by a kind of roughness associated with *self-affine* fractal scaling [4.1]. Here roughness is characterized by pa-

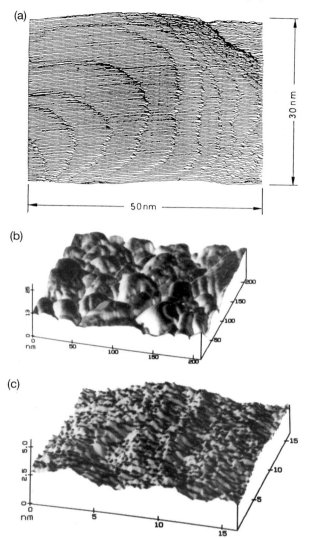

Fig. 4.3. STM topography of three different silver surfaces, prepared by evaporation on a glass substrate. From (**a**) [4.8] and (**b**), (**c**) [4.9].

rameters that refer not to the roughness itself, but to the fashion in which the roughness *changes* when the observation scale itself changes. A self-affine surface has the property that if we make a scale change that is *different for each direction*, then interfaces do not change their morphology.

Sliding friction for macroscopic bodies is usually, to a good approximation, independent of the nature of the surface roughness, unless the surfaces are very rough or very smooth. For very rough surfaces, "interlocking" of surface asperities will occur leading to an increased friction. For very smooth clean surfaces adhesion will lead to a very large contact area and thus again to a high friction. For example, if two single crystals with perfectly clean and flat surfaces are brought in (commensurate) contact, a new single crystal will be formed. However, for most engineering surfaces the friction force is independent of the surface roughness. The physical origin of this will be discussed later but is related to the fact that the area of *real* contact between rough surfaces usually is independent of the nature of the roughness. The situation is different, however, when the contact area is very small as in friction force microscopy (Sect. 3.2). In this case it has been found that the shape of the tip and its atomic-scale roughness can have a strong effect on the relation between friction force and the load.

A Case Study: Ceramic Ball Bearings

As an example of the state of the art in preparing smooth surfaces, let us consider ceramic ball bearings [4.2]. Recent development work has concentrated on hybrid bearings, consisting of steel rings with ceramic balls. Hybrid bearings are already in service in highly demanding applications such as machine tool spindles and their usage is becoming more widespread. There are many types of ceramic materials having different compositions, microstructure, and properties but silicon nitride has proved to have the best combination of physical and mechanical properties for use in bearings. Current annual production of silicon nitride balls world-wide is estimated at more than a million.

Silicon nitride balls are made from powder. Methods have been developed in which the powder is formed into relatively porous balls, which are then compacted. Figure 4.4a illustrates the different stages. The input material is usually high purity silicon nitride powder with a mean particle size of less than 1 µm. Silicon nitride is a covalent compound and cannot be densified by heat and pressure alone since the material dissociates before becoming fully bonded. It is necessary to blend additives such as yttrium oxide into the powder to promote densification by liquid phase sintering. Due to the hardness of silicon nitride (indentation hardness $\sim 16 \times 10^9 \text{N/m}^2$) polishing of balls is limited to lapping using diamond abrasives. Examples of rough-lapped and finished balls are shown in Fig. 4.4b.

The surface roughness of silicon nitride balls produced in a normal way is in the range 50–80Å, but balls with a roughness ~ 20Å can also be produced. This level is equivalent to that of the best quality steel balls. The deviation from spherical form for a ball with a diameter on the order of a few centimeters is below 1000Å. Scanning laser confocal profilometry has been used to examine balls. Figure 4.5 shows the surface topography of (a) an unpolished ball, (b) a rough-lapped ball, and (c) a finished ball.

A very low level of surface roughness is not always necessary. A rougher surface can have advantages in retaining lubricants. The high hardness of

Fig. 4.4. (a) Stages in forming of ceramic balls. (b) Rough lapped and finished silicon nitride balls. From [4.2].

the material means that silicon nitride balls are less susceptible to dents, scratches, and indentations than steel balls. The inert surfaces of silicon nitride balls lead, when compared to steel balls, to a strongly reduced tendency to form cold welded junctions. This is one of the reasons why ball bearings with ceramic balls have a much longer lifetime than those with steel balls.

The local stresses in the contact areas in bearings can exceed $4 \times 10^9 \mathrm{N/m^2}$. A fine, uniform microstructure is therefore essential for good rolling contact fatigue resistance. The fatigue resistance of silicon nitride balls is considerably higher than that of steel balls.

A Case Study: Formation of Hard Layers on Rail Tracks

The physical properties of solids in the vicinity of their surfaces often differ from their bulk properties. Thus in addition to oxide and contaminant layers, metals usually have work-hardened layers which sometimes extend many micrometers into the bulk. These layers may result from production processes, which often lead to smaller grain sizes and a higher density of dislocations at the surface than in the bulk, or they may result from the external forces

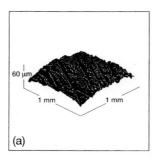 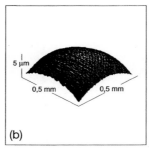 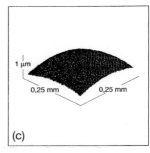

Fig. 4.5. Surface topography of silicon nitride balls. (a,b) rough lapped balls and (c) finished ball. From [4.2].

(e.g., pulsating load) which the solid experiences during practical use. As an illustration of the latter let us consider the surface modification of rail tracks arising from the load cycling associated with the wheel–rail interaction [4.3].

When a rail first goes into use, plastic deformation occurs because of the load cycling. After a run-in time (typically a few months) a thin ($\sim 10\,\mu m$) hard coating is formed on the steel surface, with a hardness that is roughly four times higher than that of the underlying pearlite. The hardening is caused by several effects, e.g., an increase in the density of dislocations, a decrease of the grain sizes, and by mechanically induced transformations of pearlite into a modification comparable with martensite [4.4,5]. Figure 4.6 shows a section of a steel track after run-in, where the surface finish has reached a steady state. (The steady state is reached when the rate of growth of the hard layer into the underlying pearlite, equals the rate at which the upper surface wears.)

Figure 4.6 shows that below the hard layer, the pearlite base metal is deformed plastically by the external stresses associated with the pulsating wheel-rail interaction, while the hard layer has nearly no visible structure. However, note that the hard layer is very brittle, and in Fig. 4.6 one can observe several vertical cracks, which may result in the removal of wear fragments, as can be seen in one place in the figure. Note that the pearlite base metal, which is plastically soft, has been squeezed into the wear "hole".

New rails have a randomly distributed surface roughness, and the height within a length of $\sim 10\,cm$ fluctuates by about $\sim 10\,\mu m$. Wheel and rail form a dynamical system with a variety of mechanical resonances. When a train is running on the track the natural surface roughness of the rail will excite different vibrational modes, which give rise in turn to periodic force fluctuations and periodic wear processes on the rail and finally result in the formation of quasi-periodic corrugations. For certain frequency ranges, typically between 1100 and 1300 Hz and between 1500 and 1800 Hz the dynamic system is more prone to corrugation. For vehicle speeds between 30 and 60 m/s corrugation

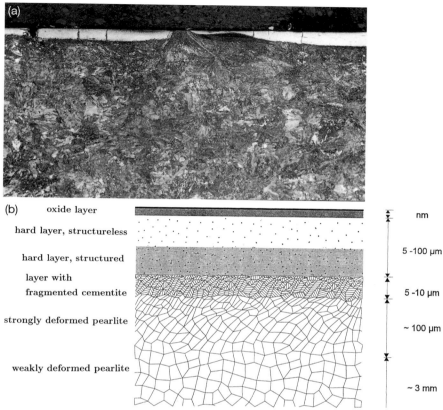

Fig. 4.6. (a) A section of a well run-in rail (vertical cut). (b) Schematic representation of (a) showing the different layers. From [4.3].

wavelengths between 2 and 6 cm will be amplified continuously. A steady state with a maximum depth of $\sim 120\,\mu\text{m}$ for a wavelength of 3 cm is typically reached after a few years, resulting in a frightful noise when the wheel runs over the corrugated rail. Therefore the rails usually have to be ground as soon as the corrugation depth is $\sim 80\,\mu\text{m}$ in order to reduce the surface roughness, at a worldwide annual cost of several \$100 million. Much research has been devoted to understanding the origin of the wear process (which involves the hard layer), and trying to reduce the wear [4.3].

A Case Study: Surface Layers and Wear in Combustion Engines

Wear during the running-in of a combustion engine is usually much smaller than the amplitude of the surface roughness of the cylinder and piston ring. The latter is typically of order a few μm (rms roughness amplitude) and

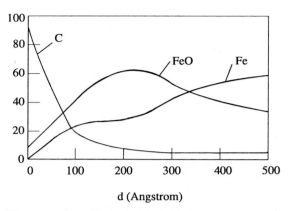

Fig. 4.7. Distribution of elements in the surface region of a combustion engine cylinder. The distribution (in atomic %) is shown as a function of the distance d from the surface of the cylinder. From [4.10].

the wear resulting from the running-in is orders of magnitude smaller. The running-in of an engine has two effects [4.10]:

(a) Formation of a thin surface layer. During the running-in, thin graphite-like nano-layers (100–1000 Å thick) are formed on the surfaces of both the piston ring and cylinder of combustion engines. These layers consist mainly of carbon originating from the fuel, the lubricant and the steel. The change in chemical composition of surface layers during running-in has been called *tribomutation*. In the course of the following use (many thousands of hours), the thickness and the composition of the nano-layer does not change. This means that the surface layer is constantly worn out and created again.

(b) Macro-form adaption, resulting from wear and plastic flattening of the asperities. In the initial state, the typical height of the surface asperities is about 4 μm but after ∼ 15 h of use, it decreases to ∼ 1 μm. However, the mesoscopic roughness (micron size asperities) does not seem to play an important role for the wear rate, which is determined mainly by the properties of the thin surface layer.

Figure 4.7 shows the distribution of elements in the surface region of a run-in engine cylinder. Within the first 100 Å carbon dominates, followed by a layer of oxidized iron (FeO) extending several 100 Å into the wall.

In a good engine, the wear rate is of order 10 Å/h or less. The sizes of wear particles vary from several nanometers to a micrometer. The wear probably results from micro-fracture (crack propagation) and detachment of surface particles. However, the separated particles may suffer many "impacts" with the surfaces and may be incorporated many times into the surfaces, before being permanently removed from them.

5. Area of Real Contact: Elastic and Plastic Deformations

The friction force equals the shear stress integrated over the area ΔA of *real* contact. Because of surface roughness, the area of real contact is usually much smaller than the apparent area of contact. In this section we discuss the physical processes which determine the area of real contact and present some experimental methods which have been used to estimate ΔA. In most practical applications, the diameter of the contact areas (junctions) are on the order of $\sim 10\,\mu\mathrm{m}$. However, the present drive towards microsystems, e.g., micromotors, has generated a great interest in the nature of nanoscale junctions. The physical processes which determine the formation and behavior of nanoscale junctions are quite different from those of microscale junctions. We consider first microscale junctions and then nanoscale junctions.

5.1 Microscale Junctions

Elastic Deformations

In the surface force apparatus two crossed glass cylinders covered with atomically smooth mica sheets are pressed together to form a small circular contact area with a radius r_0 which depends on the load F. In these measurements the force F is so small that only elastic deformation occurs and the radius r_0 as a function of the load F can be calculated using the theory of elasticity. This problem was first solved by H. Hertz. A simple derivation of the size of the contact area formed when two homogeneous and isotropic elastic bodies of arbitrary shape are pressed together can be found in the book by Landau and Lifshitz [5.1]. Here we only quote the results for two spheres with radius R and R'. The contact area $\Delta A = \pi r_0^2$ is a circular region with radius

$$r_0 = \left(\frac{RR'}{R+R'}\right)^{1/3} (\kappa F)^{1/3}, \qquad (5.1)$$

where

$$\kappa = \frac{3}{4}\left(\frac{1-\nu^2}{E} + \frac{1-\nu'^2}{E'}\right),$$

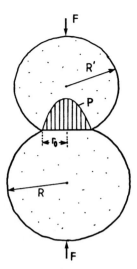

Fig. 5.1. Elastic deformation and pressure distribution in the contact area between two elastic solids.

with E and ν the elastic modulus and Poisson number, respectively. The distance h by which the two spheres approach each other (i.e., h is the difference between $R + R'$ and the distance between the centers of mass of the two bodies after contact) is given by

$$h = \left(\frac{R+R'}{RR'}\right)^{1/3} (\kappa F)^{2/3} . \tag{5.2}$$

The pressure distribution in the contact area has the form (Fig. 5.1)

$$P(x,y) = P_0 \frac{3}{2}\left[1 - \left(\frac{r}{r_0}\right)^2\right]^{1/2}, \tag{5.3}$$

where $P_0 = F/\pi r_0^2$ is the average pressure and $r = (x^2+y^2)^{1/2}$ is the distance from the center of the circular contact area. Note that the pressure goes continuously to zero at the periphery of the contact area, i.e., for $r = r_0$, and that the pressure at the center $r = 0$ is a factor $3/2$ higher than the average pressure P_0.

Plastic Deformations

As discussed in Chap. 4, most real surfaces have roughness on many different length scales. Thus when a block is put on a substrate, the actual contact area will not be the whole bottom surface (area A) of the block, but the area of *real* contact, ΔA, is usually much smaller than A. Initially, as the block is lowered towards the substrate, a single contact area (junction) occurs, where

Table 5.1. The penetration hardness (Brinell or Vickers), σ_c, for a number of materials. From [5.2].

Material	σ_c (10^9 N/m^2)
lead	0.04
aluminium	0.15
polystyrene	0.2
silver	0.25
iron	0.70
steel	1–7
glass	4–6
Fe$_2$O$_3$	11
quartz	11
SiC	24
diamond	80

the perpendicular pressure becomes so high as to give rise to local plastic deformation. As the block is further lowered more junctions are formed until the area of real contact ΔA is so large that the load $L = Mg$ is balanced by the contact pressure integrated over ΔA. If one assumes that each junction is in a state of incipient plastic flow, this gives $Mg = \sigma_c \Delta A$, where the penetration hardness σ_c is the largest compressive stress that the material can bear without plastic yielding. Note that σ_c is a material parameter which has been measured for a large number of materials using indentation experiments.

In an indentation hardness test, a pyramid, cone, or ball, made from a very hard material, e.g., diamond or tungsten carbide, is pressed into a flat surface. The yield stress σ_c of the substrate is defined as the ratio $\sigma_c = L/\Delta A$ where ΔA is the contact area, or the projected area of the impression at the surface. Table 5.1 gives the penetration hardness σ_c for a number of different materials. Note that "hardness" is completely unrelated to the elastic properties of a material. Thus rubber, for example, is elastically very soft but plastically hard. Metals, on the other hand, are elastically very hard but can be plastically relatively soft. The physical origin of elasticity is the stretching (or compression) of bonds or it may have an entropic origin as for rubber, while plasticity has its origin in the motion (and creation) of imperfections (e.g., dislocations). Thus, even a relatively high concentration of imperfections usually has negligible influence on the elastic modulus E, whereas it will affect the plastic yield stress σ_c.

As an illustration of how the plastic yield stress can be used to estimate the area of real contact, assume that a steel cube of side 10 cm is put on a steel table. The penetration hardness of steel is (Table 5.1) $\sigma_c \sim 10^9$ N/m^2 and since the load $L = Mg \sim 100$ N this gives the area of real contact $\delta A \sim 0.1$ mm^2, i.e., only a fraction 10^{-5} of the apparent area. Various experiments have shown that the diameter of a junction is on the order of $\sim 10\,\mu m$ so that about 1000 junctions are expected at the block–substrate interface.

Brittle Fracture

Some solids, such as glass, when exposed to large stresses undergo brittle fracture accompanied by very little plastic deformation. Brittle fracture is due to minute crack-like defects. Although the ideal strength is very high for a glassy solid whose atoms are held together with strong bonds, the severe stress concentration at sharp cracks reduces this strength to ordinary observed low values. The local tensile stress at an atomically sharp crack tip is proportional to $(l/a)^{1/2}$ where l is the width of the crack and a an atomic dimension. For an ideal brittle solid (no plasticity) during brittle fracture the bonds at the crack edges are "popped" like the teeth of a zip (Fig. 5.2). However, brittle fracture of surface asperities is rather unimportant in determining the area of real contact between two solids for the following two reasons: First, small surface asperities are unlikely to have any cracks, at least not cracks which are large enough to give rise to the high local stresses necessary for brittle fracture. In fact, it is known that macroscopic samples of glass, e.g., glass fibers with a length between 1 cm and 1 m, have a strength which (on the average) decreases monotonically with increasing size of the system. The reason is that "large" cracks, e.g., 1 mm wide, only occur with a large probability in large samples. It is clear that the probability of finding a "large" crack at a small surface asperity is very small. Secondly, very few materials, if any, are expected to exhibit brittle fracture under the very high compressive stress at a junction. Experiments confirm this conclusion: Visual inspection of a wear track on quartz made by a tungsten-carbide rider reveals an appearance similar to that of scratches on metals, suggesting that the mode of deformation has been similar [5.2].

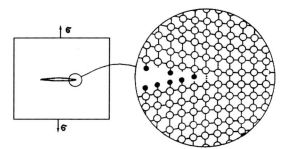

Fig. 5.2. During ideal brittle fracture the bonds at the tip of the growing crack are broken. The high-energy surfaces of the freshly created crack regions are either reconstructed or, more likely, the surface dangling bonds are "passivated" by foreign atoms, e.g., oxygen or hydrogen from the atmosphere if the crack starts on the surface of the solid. Thus the crack formation is usually irreversible: if the external load is removed, the crack will not heal out. Note also that the high stress at the crack edges may enhance thermally activated flow (creep) and stress-induced corrosion.

Area of Real Contact: Elastic and Plastic Deformations

For most "natural" surfaces the area of real contact can be accurately estimated by assuming that each junction is in a state of incipient plastic flow so that $\Delta A = L/\sigma_c$. However, for very smooth surfaces this is no longer true: the pressure in a large fraction of the junctions is below the penetration hardness, and the area of real contact is larger than predicted by $\Delta A = L/\sigma_c$. Furthermore, even if the formula $\Delta A = L/\sigma_c$ is accurate for newly prepared surfaces, after repeated sliding over the same area this may no longer be true. (Thus engineers like to use their machines not only next day but preferably next year as well – with the same surfaces. It is unlikely that the area of real contact is determined by *irreversible* plastic deformations the millionth time we make contact.) Rather, after repeated sliding, surface asperities will be smoothened out (by plastic deformation), and, finally, mainly elastic deformation will occur in the contact areas. But even in these cases it is usually found that the area of real contact is proportional to the load L. At first, this may seem a paradox, since we know that the area of real contact between two smooth surface asperities varies nonlinearly ($\propto F^{2/3}$) with the force F with which they are squeezed together; see (5.1). The explanation of the paradox is related to the finite asperity height distribution, as can be seen from the following simple model calculation due to Greenwood [5.3].

Let us approximate a surface as consisting of N elastic spherical caps with the same radius R but with different heights, which is pressed against a rigid flat plane, see Fig. 5.3. Let $N\Phi(z)dz$ be the number of spheres with heights between z and $z+dz$, i.e., $\Phi(z)$ is the height probability distribution function. All asperities of height greater than d will make contact, and will be compressed a distance $h = z - d$. Assuming elastic deformations, using (5.1) and (5.2), the asperity contact area equals $\pi r_0^2 = \pi Rh = \pi R(z-d)$ and the asperity load equals $R^{1/2}h^{3/2}/\kappa$. Thus the total number of contacts is

$$\Delta N = \int_d^\infty dz N\Phi(z) ,$$

the area of real contact

$$\Delta A = \int_d^\infty dz N\Phi(z)\pi R(z-d) ,$$

Fig. 5.3. Contact between a rough surface and a rigid plane. All asperities of height z greater than the separation d make contact.

and the total load

$$L = \int_d^\infty dz N\Phi(z) R^{1/2}(z-d)^{3/2}/\kappa \ .$$

Now, most engineering surfaces, such as those in Fig. 4.2, have a nearly Gaussian height distribution, i.e.,

$$\Phi(z) = \left(\frac{1}{2\pi l^2}\right)^{1/2} e^{-(z-z_0)^2/2l^2} \ ,$$

where l is the rms width of the probability distribution $\Phi(z)$. However, for what follows it does not matter whether the distribution is exactly Gaussian: the only important feature of the height distribution is that the number of asperities of a given height falls off rapidly with increasing height. In such a case we only need $\Phi(z)$ for z very close to $z = d$ and we can expand

$$(z-z_0)^2 = [(z-d) + (d-z_0)]^2 \approx (d-z_0)^2 + 2(d-z_0)(z-d) \ ,$$

so that

$$\Phi(z) \approx A e^{-\lambda(z-d)} \ ,$$

where

$$A = \left(\frac{1}{2\pi l^2}\right)^{1/2} e^{-(d-z_0)^2/2l^2} \ , \qquad \lambda = \frac{d-z_0}{l^2} \ .$$

Using this expression for $\Phi(z)$ in the formulas above gives

$$\Delta N = NA/\lambda$$

$$\Delta A = \pi N R A/\lambda^2$$

$$L = (2\pi)^{1/2} N R^{1/2} A/(\kappa \lambda^{5/2}) \ .$$

Note that the last equation yields

$$\left(\frac{d-z_0}{l}\right)^2 = 2\ln\left(\frac{Nl^{3/2}R^{1/2}}{\kappa L}\right) - 5\ln\left(\frac{d-z_0}{l}\right) \ .$$

From this equation it is clear that $d-z_0$ is very insensitive to variation of the load L and the number of surface asperities N (which is proportional to the total surface area), and in most practical cases the separation $d-z_0$ will be $\sim 2l - 3l$. The same is true for any rapidly decaying probability distribution, i.e., not only for a Gaussian. Hence, λ can be treated as a constant and, to a good approximation, the area of real contact is *proportional* to the load and the average size of micro contacts $\Delta A/\Delta N$ is *independent* of the load L. The origin of these results is that with increasing load L, new contacts are continuously forming, initially with zero size and zero pressure, and the growth and formation exactly balance. This analysis of surface contact goes back to Archard who summarized all that matters: "If the primary result of increasing the load is to cause existing contact areas to grow, then the area of real contact will not be proportional to the load. But if the primary result is to form new areas of contact, then the area and load will be proportional."

It is clear that even if elastic deformation were to occur in most contact areas, at some places the local pressure will be high enough to give rise to plastic deformation. Within the model studied above, it is easy to estimate the fraction of the area of real contact where plastic deformation has taken place. Consider an asperity of height z. The asperity contact area is $\delta A = \pi R(z-d)$ and the asperity load $\delta L = R^{1/2}(z-d)^{3/2}/\kappa$. Thus the average pressure in the contact region $P(z) = [(z-d)/R]^{1/2}/(\pi\kappa)$. If $P > \sigma_c$, where σ_c is the indentation hardness, the junction deforms plastically. Thus all junctions with $z > z_1$ will deform plastically, where z_1 is defined by $P(z_1) = \sigma_c$ or $z_1 = d + (\pi\kappa)^2 R\sigma_c^2$. Thus the fraction of the junctions which are plastically deformed is

$$\Delta = \frac{\int_{z_1}^{\infty} dz \Phi(z)}{\int_d^{\infty} dz \Phi(z)} = e^{-\lambda(z_1-d)} \equiv e^{-\psi} ,$$

where

$$\psi = \pi^2 \kappa^2 (d-z_0) R\sigma_c^2/l^2 .$$

Since typically $d - z_0 \sim 2.5l$ we get

$$\psi = 2.5\pi^2 \kappa^2 (R/l)\sigma_c^2 \approx 20(R/l)(\sigma_c/E)^2 .$$

Note that the fraction of junctions which have undergone plastic deformation does not depend on the load. Since for steel $E \sim 1 \times 10^{11} \text{N/m}^2$ and $\sigma_c \sim 1 \times 10^9 \text{N/m}^2$ we get $\psi \sim 0.002(R/l)$. Thus, if $R < 100l$ essentially all junctions will be in a state of incipient plastic flow so that $\Delta A \approx L/\sigma_c$. Experiments have shown that in typical engineering applications, $R/l \sim 10$ for rough surfaces (grit-blasted), ~ 100 for ground surfaces, and $\sim 10^4$ or more for smooth surfaces (ground and polished) [5.4]. It is clear that, except for polished surfaces, one expects essentially all junctions to be in a state of incipient plastic flow so that $\Delta A \approx L/\sigma_c$.

The concepts discussed above are fundamental for tribology. Elastic contacts may, ultimately, fail by fatigue, but can be expected to give useful service before they do. Elastic strains are unlikely to crack the surface oxide, the fundamental protective layer of a metal.

Experimental Implications

The discussion above shows that in most practical cases the area of real contact is proportional to the load. This has a number of important consequences.

a) Electrical Resistivity

Measurements of the electrical resistance of an interface are frequently made, either for their intrinsic interest (e.g., to study the functioning of electric contacts) or because they give information on the geometry of the interface. The electric conductance G of a circular contact (radius r_0) is given by $G = 2r_0/\rho$, where ρ is the electrical resistivity of the material. Note that $G \propto r_0$ rather than $\propto r_0^2$. This is a consequence of the fact that the resistance is produced *near* the constriction. If N junctions occur at the interface, all with the same radius r_0, then the area of real contact $\Delta A = N\pi r_0^2$ and the conductance $G = 2Nr_0/\rho$. Thus, in order to determine ΔA from a contact resistivity measurement, it is necessary to know the radius r_0 of the junctions. Conversely, if one assumes that ΔA is known, e.g., via $\Delta A = L/\sigma_c$, then it is possible to estimate the radius of a junction from the measured contact resistivity. In practice, metal surfaces are usually covered by thin insulating oxide films which may lead to important modifications of the contact conductance from the ideal value $G = 2Nr_0/\rho$. The formula above for the conductance is only valid if the electron mean free path is short compared with the radius r_0 of the junction. This is in general not the case for nanoscale junctions, where another formula for the conductance, based on ballistic quantum transport, must be used; see (5.8).

b) Heat Transfer

The thermal resistance of the interface between two solids depends on the area of real contact. If the block initially has a higher temperature than the substrate, the temperature of the block will decrease until the two solids finally have the same temperature. If the experiment is performed at low temperature (so that black-body radiation between the two solids can be neglected) and under ultrahigh vacuum conditions (so that heat flow through the air gap between the solids can be neglected) then the total heat flow between the solids must occur via the area of real contact. Since the area of real contact is often very small, the thermal resistance is usually very high. This fact is well-known in the low-temperature community. Thus, one leading experimentalist on low-temperature physics writes [5.5]: "The surfaces have to be well prepared and strong enough bolts, made, for example, from hardened BeCu, have to be used. These are tightened in a controlled way until they almost yield, which supplies sufficient force for what is almost a cold weld to be produced between the two parts. This can rip up oxide layers and can then make an intimate metallic contact."

c) Sliding Friction

The fact that the area of real contact is proportional to the load explains immediately why the friction force in most cases is proportional to the load, and forms the basis for the theory of friction of Bowden and Tabor which will be discussed in Chap. 6.

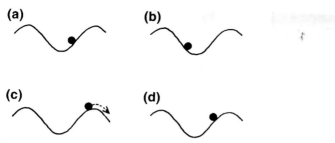

Fig. 5.4a–d. Schematic picture of the pinning barriers experienced by four dislocations at a junction in a state of incipient plastic flow. The dislocation in (**c**) is close to the top of the pinning barrier and a slight increase in the shear stress at the junction would pull the dislocation over the barrier inducing local plastic flow. Alternatively, the dislocation can overcome the pinning barrier by thermal excitation. The latter process leads to a slow increase of the junction area even if the external load is constant.

Critical State and Perpendicular Creep

In a crystalline material such as iron or copper, plastic deformations are due to the motion of dislocations. The dislocations are always pinned, e.g., by foreign impurities, and a minimum force is necessary in order to pull a dislocation over its pinning barrier. This minimum force determines the macroscopic yield stress σ_c. The pinning barriers are usually so high that thermal excitation over the barriers is negligible in a first approximation, and the yield stress is only weakly dependent on temperature.

However, thermal excitations are very important for sliding friction in that they give rise to a *slow* increase of the contact area with increasing time of stationary contact. This effect comes about as follows: Since the perpendicular stress in a contact area (junction) immediately after a block has been put on a substrate equals the critical value σ_c, it follows that at the junction interface some dislocations are located near the top of the pinning barrier, see Fig. 5.4, so that an arbitrarily small increase in the perpendicular stress will lead to motion of some dislocations and to further plastic deformation. But the dislocations which are close to the top of a barrier (Fig. 5.4c), can "jump" over the barrier by thermal excitation, leading to local plastic flow. This will produce a slow, thermally induced, creep motion which will increase the contact area with increasing time. This effect has been observed for a wide variety of solids, e.g., stone, paper, metal, and ice.

Creep also occurs in non-crystalline solids (no dislocations), where the creep motion results from other types of thermally activated processes. As an example, Fig. 5.5 shows the slow increase in the contact area between rough surfaces of acrylic plastic with increasing time of stationary contact at 10 MPa normal stress [5.6]. The red area is the area of real contact after 1 s while the yellow and blue areas is the increase in the area of real contact after 100 s and 10,000 s, respectively. Note that the area of real contact increases by growth of

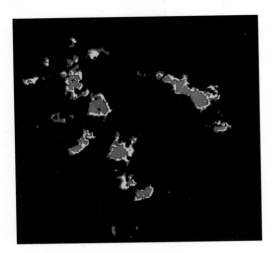

Fig. 5.5. Contact area between rough acrylic plastic with increasing time of stationary contact at 10 MPa normal stress. The red area is the area of real contact after 1 s while the yellow and blue areas is the increase in the area of real contact after 100 s and 10,000 s, respectively. From [5.6]. The figure is prepared by overlaying digitized video images taken after the indicated time of contact. The area increases by growth of existing contacts, appearance of new contacts and contact coalescence.

existing contacts, appearance of new contacts, and contact coalescence. A detailed study (see Sect. 13.2) shows that the area of real contact increases with the time t of stationary contact as $\Delta A \propto \ln(1 + t/\tau_0)$, where τ_0 is a constant.

The increase in the contact area with the time of stationary contact will give rise to a steady increase in the static friction force with contact time. This effect is extremely important for sliding dynamics at low sliding velocities (Chap. 13), e.g., it may be the reason why earthquakes occur (Sect. 14.1). Note, however, that this effect is absent in the measurements using the surface forces apparatus, where only elastic deformation occurs.

5.2 Nanoscale Junctions

It is well-known that the theoretical shear strength of a perfect crystal is much higher, typically by a factor of about 100, than the observed yield stress. For a cubic crystal the ideal yield stress is $\sigma_c \sim G/2\pi$, where $G = E/2(1+\nu)$ is the shear modulus. This corresponds to the shear stress necessary to slide two planes of atoms past each other, see Fig. 5.6. However, plastic deformation in macroscopic crystals usually occurs by motion of dislocations, which requires shear stresses about one hundred times smaller than the ideal yield stress. The concentration of dislocations in a crystal varies strongly from case to

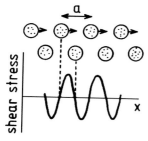

Fig. 5.6. Relative shear of two planes of atoms in a uniformly strained perfect crystal. If the relation between the shear stress and the displacement is approximated by a sine function, then $\sigma \approx (G/2\pi)\sin(2\pi x/a)$ where $G = E/2(1+\nu)$ is the shear modulus. For a small displacement x this formula gives $\sigma = G\gamma$ where the strain $\gamma = x/a$. The critical stress σ_c at which the lattice becomes unstable is thus $\sigma_c = G/2\pi$.

case. But even in an annealed single crystal the average separation between two nearby dislocations is relatively small, typically of order $1\,\mu$m. Hence, since in most traditional engineering applications metallic junctions have a diameter of about $10\,\mu$m, the use of the macroscopic yield stress should be a good approximation when estimating the area of real contact.

This situation may change drastically in future engineering applications involving microsystems. As an example, Fig. 5.7a shows a rotary micromotor driven electrostatically, that is, with attractive forces generated by applying a voltage across a narrow gap separating the rotor from a fixed stator. The motor is less than $100\,\mu$m in diameter and just a few micrometers thick. Micromotors have been operated at speeds exceeding 20,000 rpm; at these speeds friction and wear remain challenges for devices that must operate continuously. In microsystems the area of real contact is so small that plastic flow caused by motion of (pre-existing) dislocations is unlikely to occur and the yield stress may now be similar to the ideal value $\sim G/2\pi$. Let us analyze the situation in more detail.

Consider a block resting on a substrate. Mechanical stability requires at least three contact areas (junctions). Thus, the *smallest possible* contact area consists of three junctions of atomic dimension, giving the total contact area $\Delta A \sim 3a^2$, where a is of the order of a few Ångstroms. If M is the mass of the block, the local pressures in the contact areas is of order $Mg/3a^2$. The maximum mass the block can have without plastic yielding in the contact areas is then determined by $Mg/3a^2 = \sigma_c$ where $\sigma_c \sim G/2\pi$. This gives $M \sim Ga^2/2g$. For steel $G \sim 10^{11}\,\text{N/m}^2$ giving $M \sim 10^{-9}\,\text{kg}$ corresponding to a cube with the side $\sim 50\,\mu$m. Based on this calculation one may be tempted to conclude that microsystems will have just a few atomic sized contact areas.

However, the estimate above has neglected the influence of adhesive forces between the surfaces of the block and the substrate. These forces cannot

Fig. 5.7. (a) A six-stator polysilicon micromotor. [Photo courtesy of R.S. Müller, University of California, Berkeley.] (b) A magnetically driven micromotor. The rotor diameter is 150 μm, and the gears have diameters of 77, 100, and 150 μm. [Photo courtesy of H. Guckel, University of Wisconsin.]

be neglected and have two important effects. First, the overall attraction between the block and the substrate will enhance the normal force that the block exerts on the substrate. For a micrometer sized (or smaller) object this latter effect may be much more important than the contribution to the normal force from the weight of the block. To see this, let us consider two flat surfaces separated by the distance d. The van der Waals force (per unit area) between the surfaces is (we neglect retardation effects) [5.7]

$$F_{\rm vdW} = 3\hbar\bar{\omega}/16\pi^2 d^3$$

with

$$\bar{\omega} = \int_0^\infty du \frac{[\epsilon_1(iu) - 1][\epsilon_2(iu) - 1]}{[\epsilon_1(iu) + 1][\epsilon_2(iu) + 1]},$$

where $\epsilon_1(\omega)$ and $\epsilon_2(\omega)$ are the dielectric functions of the two materials. In a typical case $\hbar\bar{\omega} \sim 10\,{\rm eV}$. Let us determine the separation d at which this force equals the gravitational force on the block. Assume that the block is a rectangular disk with thickness h (Fig. 5.8) so that the gravitational force per unit area equals $\rho g h$. This will equal the van der Waals force when $d = (3\hbar\bar{\omega}/16\pi^2 \rho g h)^{1/3}$. If the block is 1 μm thick this gives $d \sim 1\,\mu{\rm m}$. Since $F_{\rm vdW} \propto 1/d^3$, if the separation between the surfaces were instead 1000 Å, then the van der Waals force would be a factor of 10^3 more important than the gravitational force. We conclude that in many cases the gravitational force can be neglected when determining the area of real contact in sliding systems of micrometer size. In addition to the van der Waals interaction,

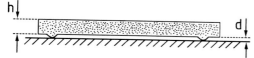

Fig. 5.8. A cylindrical disk with thickness h a distance d from a substrate.

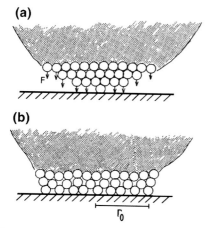

Fig. 5.9. (a) A rigid tip in atomic contact with a (rigid) substrate. The vertical arrows indicate the force acting on the tip atoms from the substrate. (b) For a real tip with finite elastic (and plastic) properties the tip atoms in the vicinity of the substrate will move into direct contact with the substrate over a region with radius r_0.

electrostatic interaction between net charges, e.g., generated during the sliding process (similar to the electric charging of the hair during combing), may make an important contribution to the adhesion force. Similar, if capillary bridges occur they will enhance the adhesion and friction forces (Sect. 7.5).

A second effect of the adhesion between the two surfaces is that atomic sized contacts may be impossible, even if the external load were to vanish. That is, the adhesion forces tend to increase the contact areas so that instead of atomic-sized junctions, the smallest junctions may have a radius of order 10–100 Å. The basic physics is illustrated in Fig. 5.9. Suppose that a tip (say a surface asperity of a micrometer sized block) is moved towards a (rigid) substrate and that the tip atoms are not allowed to move in response to the substrate forces. In this case, when the block is resting on the substrate, the geometry may be as in Fig. 5.9a. However, the atoms in the vicinity of the outermost tip atom in contact with the substrate will feel a strong attraction from the substrate; this interaction may be of the van der Waals type or, if the separation to the surface is very small (at most a few Ångstoms), of a chemical nature. If the tip atoms are now allowed to respond to the attraction they will move into direct contact with the substrate over a region with a typical radius of $r_0 \sim 10-100$ Å, see Fig. 5.9b. Simultaneously, energy will be stored in the elastic displacement field in the tip, but this increase in the total energy is more than compensated for by the decrease in the energy resulting from the tip–substrate bonding.

Elastic Deformations

If the contact area resulting from the tip–substrate adhesion (see Fig. 5.9b) is large enough, it is possible to describe the deformation field using the elastic continuum approximation. In the so-called JKR theory an elastic sphere (radius R) is brought into contact with a flat substrate. It is assumed that a bonding with free energy per unit area, $\Delta\gamma$, occurs between the surfaces of the sphere and the substrate when direct contact occurs. As a result of the surface forces (as manifested by $\Delta\gamma$), an adhesion neck is formed (Fig. 5.9b) and the contact radius remains finite even if no external load (such as the weight of the block) is applied. Furthermore, the contact cannot be broken in a continuous fashion, but breaks abruptly due to a mechanical instability at a critical tensile pull-off force given by [5.8]

$$F_c = (3\pi/2) R \Delta\gamma . \tag{5.4}$$

The minimum contact radius at which the instability occurs is given by

$$r_c = (3\pi\kappa/2)^{1/3} \left(R^2 \Delta\gamma\right)^{1/3} , \tag{5.5}$$

and the (average) tensile stress in the contact area at rupture is

$$\frac{F_c}{\pi r_c^2} = \left(\frac{3\Delta\gamma}{2\pi^2 \kappa^2 R}\right)^{1/3} . \tag{5.6}$$

The general relation between the contact radius r_0 and the loading force F is given by

$$r_0 = (\kappa R)^{1/3} \left\{ F + 2F_c \left[1 + (1 + F/F_c)^{1/2} \right] \right\}^{1/3} . \tag{5.7}$$

Note that $r_0 \to r_c$ as $F \to -F_c$. The JKR theory reduces to the Hertz theory when $\Delta\gamma \to 0$. A plot of the contact radius as a function of the loading force is shown in Fig. 5.10a. Note that the adhesion force (5.4) is independent of the elastic properties of the solids; this remarkable prediction is in agreement with experimental observations.

The JKR theory is valid when $r_c^2 \gg R r_a$, where r_c is given by (5.5) and where r_a is an atomic-scale radius which characterizes the extent of the short-ranged attractive interaction responsible for the adhesion energy $\Delta\gamma$. Thus the JKR theory is accurate for elastically soft materials or when the adhesion energy is large. In the opposite limit, $r_c^2 \ll R r_a$, the so called DMT theory must be used [5.9]. Intermediate cases have been studied by numerical calculations [5.10] and an analytical interpolation formula has been derived in [5.11]. For an overview of elastic contact theories, see [5.8].

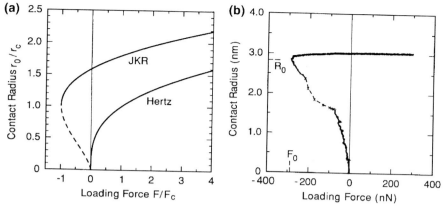

Fig. 5.10. (a) Contact radius as a function of the loading force for an elastic contact between a sphere and a flat substrate. In contrast to the Hertzian theory, adhesive interactions are accounted for in the JKR theory. (b) Contact radius as a function of the loading force measured for an Au contact made by indenting a sharp tip into a flat substrate. The dashed section of the curve indicates a mechanical instability. From [5.14]

Elastic contact theories, such as the JKR theory, break down when the maximum stress reaches the yield stress of the material (note: for a nanoscale junction this is not the macroscopic yield stress but, as discussed above, a much higher stress). According to (5.6), the maximum tensile stress during pull-off scales as $R^{-1/3}$ with the radius of the tip. Thus, for a sufficiently large radius of the sphere the tensile stress will not exceed the yield stress and the contact breaks by elastic fracture. For gold the critical radius for elastic fracture is of order 750 Å. Figure 5.10b shows the relation between the contact radius and the loading force for a case where plastic deformation occurs during the pull-off. One can identify a maximum force of adhesion, $F_c = 290$ nM, and a corresponding contact radius, $r_c = 28$ Å. These values for F_c and r_c are compatible with the JKR prediction with $E = 2 \times 10^{11}$ Pa. The retraction occurs by plastic deformation but is interrupted by a mechanical instability (dashed line in Fig. 5.10b) similar to the fracture of the elastic contact in the JKR theory.

Plastic Deformations: A Case Study

Plastic deformations of metals on length scales beyond the average separation between two nearby dislocations usually occurs by motion of dislocations. The plastic yield processes in nanoscale junctions are quite different [5.12,13]. To understand this, let us calculate the pressure needed to *generate* a dislocation loop with radius R by applying a pressure P within a circular area of radius R (Fig. 5.11). The energy of the dislocation loop is of order $(GRb^2/2)\ln(R/r_a)$ where b is the lattice constant (or Burger vector) and r_a a cut off distance of the order of a few Ångstroms. The work necessary to create the dislocation

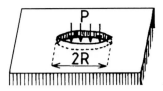

Fig. 5.11. A dislocation loop formed by applying a pressure P within a circular area (radius R) on the surface of a solid.

is of order $\pi R^2 b P$. Thus,

$$P \sim (Gb/2\pi R)\ln(R/r_\mathrm{a}).$$

For a nanoscale junction, $R \sim 10\,\text{Å}$ so that for gold $P \sim 5 \times 10^9\,\text{N/m}^2$. This is much higher than the macroscopic yield stress of gold which is about $10^8\,\text{N/m}^2$. Thus metals (and other materials) are *much harder* on a nanometer scale than measured in macroscopic measurements. Furthermore, it is not clear that plastic deformations on the nanometer scale always (or usually) occur by the generation and motion of dislocations, although scanning tunneling microscopy studies of nanoscale indentation footprints have shown that dislocations are generated in some cases; see Fig. 5.12.

In this section we take a look at one of the present frontiers in adhesion physics. We consider the plastic deformation of nanoscale gold junctions. One must expect, however, that nanoscale yielding properties of hard and brittle materials might be different from the ones observed with Au, but very little is known about this topic. The results presented below have been obtained by *Dürig* and coworkers [5.14].

The experiments were performed at room temperature and under high vacuum conditions in order to avoid capillary effects due to water. The force applied during indentation and retraction of the tip was measured simultaneously with the electrical conductance of the junction. The contact area was estimated from the measured conductance, using the Landauer formula, which is valid for narrow junctions:

$$G = (2e^2/h)N_\mathrm{c} \tag{5.8}$$

where N_c is the number of conduction channels supported by the junction. Note that N_c depends not only on the size of the junction but also on its shape. However, in the present case N_c is roughly equal to the number of gold atoms which fit into the narrowest cross section of the junction. Equation (5.8) is only valid if the transmission probability for each of the channels is close to unity, but theoretical calculations suggest that this is a good approximation.

Each indentation experiment consisted of an approach–retraction cycle, which starts with the tip $\sim 20\,\text{Å}$ above the surface; it is then ramped 100 Å towards the sample and subsequently retracted by twice the approach distance at a rate of $20\,\text{Å/s}$.

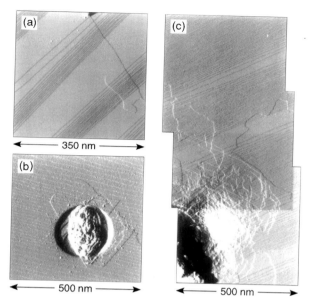

Fig. 5.12. Contact footprint recorded after indentation of an Ir tip into a Pt(111) substrate. Indentation depth is approximately 1 nm, 2 nm and 5 nm for (**a,b**) and (**c**), respectively. The pattern of quasi-parallel rows of monatomic steps is intrinsic to the substrate and reflects the local miscut of the surface. A hillock several nanometers high is always produced at the point where contact was made. In addition, monatomic steps with more or less random orientation are observed. These steps originate from vertical displacements of the sample surface that are produced by dislocations within the bulk with Burgers vectors perpendicular to the surface. From [5.22].

We now describe, with reference to Figs. 5.13 and 5.14, the nature of the tip deformation as a function of the tip–substrate separation, first on approaching the surface and then during the retraction. Figure 5.13 shows data for three representative examples of the conductance (a) and (b) and applied force (c) as a function of the tip–substrate displacement. Approach and retraction are indicated by arrows. Figure 5.14 shows a schematic view of the junction at points (A–D) in Fig. 5.13.

a) Approach

1) A connective neck between the tip and the substrate is formed spontaneously when the tip–substrate spacing is of the order of one atomic diameter (Fig. 5.14A). The neck formation is driven by adhesive surface forces which give rise to the overall attractive tip–substrate force observed.

2) The adhesion neck is under strong tensile stress, typically of the order of $(5-10) \times 10^9$ Pa, leading to plastic flow and a rapid increase in the neck diameter with indentation depth.

62   5. Area of Real Contact: Elastic and Plastic Deformations

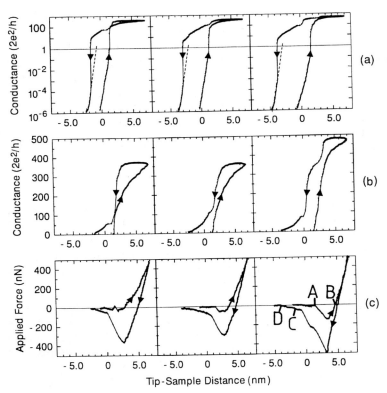

Fig. 5.13. Conductance (**a,b**) and applied force (**c**) as a function of the tip–sample displacement. Data for three representative examples are shown. Approach and retraction are indicated by arrows. From [5.14].

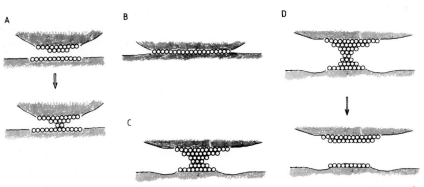

Fig. 5.14. Artist's view of the tip deformation at points (A–D) in Fig. 5.13.

3) When the neck diameter has reached 40–60 Å (Fig. 5.14B), the compressive loading phase starts. In the compressive loading regime the growth of the contact area is slower and proportional to the indentation depth, as for a Hertzian contact. Clearly, the mechanism of neck deformation must be different from that in the tensile stress regime. This is to be expected based on the qualitative difference in the distribution of the shear stress. Under compressive loading the maximum of the shear stress is within the substrate at a depth comparable to the contact radius. For tensile loading, in contrast, the maximum of the shear stress is located close to the periphery of the contact, which makes it easy for the contact to grow.

b) Retraction

4) Unlike indentation, the neck diameter remains practically constant as the load is removed during retraction. The mechanical work performed during indentation is recovered to a large degree in the retraction phase.

5) As the tip is retracted further, tensile stress is built up, yet no significant change of the contact diameter can be observed until the tensile stress exceeds a critical value of about 10^{10} Pa.

6) Yielding sets in smoothly as can be seen from the gradual decrease of the conductance. The shrinking of the neck increases rapidly as the neck is strained. Simultaneously, the tensile force acting on the tip decreases at a somewhat lower rate than that at which the neck contracts. As a consequence tensile stress escalates to the point at which this stress can no longer be sustained, and the neck contracts abruptly in a catastrophic event. In some cases a cascade of such events can be observed. The process can be considered as partial fracture, whereby a large amount of elastic strain energy is released in a single event. The vigor of the fracture depends on the size of the neck. Typically, large diameter necks (diameter 100 Å or more) at the yield point completely break in a single event and the conductance drops to zero. On the other hand, no fracture is observed for small necks (diameter 40 Å or less). Hence, one obtains a lower critical diameter of the order of 40–50 Å where fracture occurs.

7) Adhesion necks of subcritical diameter all deform in the same characteristic manner: (a) An almost perfect exponential decrease of the conductance is observed as the neck is elongating, and (b) irrespective of the initial neck size the connective neck breaks at a conductance value of either 3 or 4 (Fig. 5.14C shows a junction close to the point of rupture). Thus, one concludes that the cross section of the smallest sustainable contacts comprise either three or four atoms. Furthermore, as the initial conditions have no effect on the final deformation process, the plastic flow is a local phenomenon involving essentially only atoms within a limited range of the constriction.

Molecular dynamics simulations suggest that plastic yielding of nanometer sized contacts under tensile stress involves a series of structural transformations between ordered and disordered states of a small number of atomic

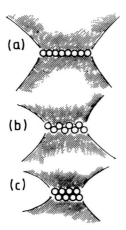

Fig. 5.15a. During retraction the tip elongates by a sequence of atomic rearrangement processes in one or a few layers at the most narrow region of the constriction.

layers adjacent to the constriction [5.15]. In each transformation the neck shrinks as one atomic layer is added, see Fig. 5.15a. This picture explains in a natural way why the neck cross section decreases in an exponential manner with the tip–substrate distance. Recent experiments have indeed shown that the contact area decreases stepwise giving rise to plateaus in the conductance as a function of the tip–substrate separation.

8) When the adhesion neck finally breaks, a gap of width about 10 Å opens between tip and substrate (this result is deduced from the measured tunneling conductance). The gap is too wide to result from elastic contraction alone, which can be calculated to give about 1 Å displacement. On the other hand, the width of the gap corresponds almost exactly to the thickness of the transformation zone, suggesting that the atoms in this zone have a high mobility and are migrating out of the gap region as soon as the neck breaks, see Fig. 5.14D.

The yield stress for nanojunctions is found to be of order 6×10^9 Pa, which is independent of the contact diameter. For comparison, the macroscopic yield stress for Au is on the order of 2×10^8 Pa. Thus, the rigidity of nanoscale Au necks is about a factor of 30 larger than that for macroscopic objects, but still about a factor of three lower than the value expected as a result of coherent slip (Fig. 5.6).

Nanoindentation

The experiments described above involve the indentation of a metal tip on a metal surface. Since the bond between two different metals is usually stronger than the internal bonding of the softer of the two metals, one expects the formation of a neck during retraction, as observed above for the iridium tip on the gold substrate. This neck-formation may be prevented if one or

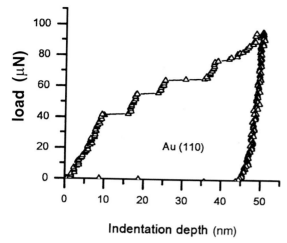

Fig. 5.15b. Load–displacement curve for Au single crystal. From [5.28].

both of the metal surfaces are protected by, e.g., oxide layers, or organic contamination films (see below). Alternatively, if the tip is made from an inert material, e.g., diamond passivated by hydrogen or oxygen, no neck formation is expected. This is illustrated in Fig. 5.15b, which shows the result of an experiment where a diamond indenter with a blunt end (radius of curvature approximately 2050 Å) is pushed against a gold (110) surface [5.28]. Note the series of yield events separated by elastic loading. If the unloading occurs before the first yield event, no hysteresis is observed, as expected when only elastic deformation occurs. Thus the Au surface deforms elastically between each yield event. We also note that the observed yield stress ($\sim$ 5 GPa) in these nanoscale experiments is much higher than the macroscopic yield stress of gold. The reason for this was discussed above.

The local yield processes described above involve small volume elements with a diameter of about 10–100 Å. Such small yield events cannot easily be detected directly in plastic deformation under most practical conditions, e.g., during the elongation of a macroscopic solid bar. (It might, however, be possible to probe the yield events indirectly by registering the elastic waves in a solid block during slow plastic deformation. Associated with each local yield event should be an emitted elastic wave pulse which may be registered by a sound detector with high time-resolution.) Nevertheless, plastic deformation, which on a macroscopic length scale may seem to be a continuous flow process, will, on the nanometer scale, involve very discontinuous and rapid slip or flip events, separated by time periods where only elastic loading occurs. This microscopic picture of plastic yield forms the basis for a fundamental treatment of time-dependent plastic deformation (e.g., creep and relaxation), see Sect. 11.4.

Adsorbate Covered Surfaces:
Passivation of Tip–Substrate Bonding

The scenario presented above is for clean metal surfaces. The adhesion properties change dramatically when the surface is covered by, e.g., an organic overlayer. For example, if a monolayer of n-hexadecanethiol ($C_{16}H_{33}SH$) is deposited on a gold surface, it forms an inert spacer for the tip (Fig. 5.16), preventing it from coming in contact with the metallic Au substrate, even when the local pressure under the tip is in excess of 5×10^9 Pa. Hence, such alkane layers act as excellent lubricants for nanometer scale structures by suppressing strong adhesive interactions that would lead to the spontaneous formation of adhesion necks and, thus, to wear and associated high friction.

It is known that $C_{16}H_{33}SH$ binds strongly to gold surfaces via the sulfur head group. The inert hydrocarbon tail of the molecule points away from the surface with a $\sim 30°$ tilt from the surface normal. Figure 5.17 shows the distance–force relation when a tungsten tip with radius of curvature $R \sim 5000$ Å is pushed into a gold surface covered by a $C_{16}H_{33}S$ monolayer film [5.16]. Note the following:

1) Prior to tip contact, there is no evidence of appreciable attractive (negative) forces, consistent with the low surface energy of the methyl-terminated self-assembled monolayer (the contribution from the van der Waals interaction to the tip–sample adhesion is below the detection limit of the experiment).
2) The force–distance curve exhibits hysteresis during the compression and decompression of the monolayer film between the tip and the substrate.
3) There is no evidence for an adhesive interaction between the tip and the Au substrate at any stage in the indentation cycle. Thus, the monolayer film completely passivates the gold surface and even the highest tip force (corresponding to a local pressure in the contact area of order 5×10^9 Pa) cannot penetrate or squeeze out the monolayer film.

The hysteresis in Fig. 5.17 indicates the existence of a relatively long relaxation or rearrangement time: the motion of the tip is so fast that the compression of the film during the approach stage and the decompression during the retraction period cannot occur adiabatically. If the process occurred adiabatically the elastic energy stored up in the film during compression would be transferred back to the tip during decompression, and no hysteresis would occur in Fig. 5.17. Quite generally, adiabaticity requires that the longest molecular rearrangement time τ, is much shorter than the interaction time τ^* between the tip and the monolayer film so that the monolayer is always very close to thermal equilibrium during all stages of the tip motion.

Note that the elastic properties of the monolayer film are determined not only by the elastic properties of the bonds between the atoms in the film, but may be mainly of entropic origin: Confining the molecules to a smaller volume will reduce the configurational entropy (as compared with a free film) and

Fig. 5.16. An indentation experiment on a gold surface passivated by a monolayer of $C_{16}H_{33}SH$.

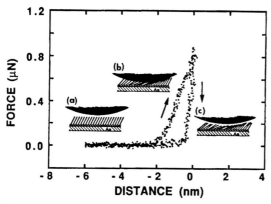

Fig. 5.17. The tip force as a function of the tip–sample separation. The rate of perpendicular displacement is 50 Å/s. The zero separation is arbitrarily assigned to the point at which the motion changes direction. From [5.16].

hence will yield an effective repulsion between the tip and the substrate. The relaxation time τ also has an entropic contribution: During compression, the alkane backbones are forced closer to the surface and may become disordered and entangled with one another. The subsequent reorganization of the film requires concerted chain motions and a significant time τ.

The hysteresis in Fig. 5.17 is of fundamental importance for sliding friction. During sliding, junctions are formed and broken at a rate which is proportional to the sliding velocity. It is clear that if the time τ^* involved in the formation and breaking of a junction is much longer than the relaxation time τ the sliding will be almost adiabatic and the friction force small. On the other hand, if the times are of comparable magnitude a large contact hysteresis occurs, leading to large energy dissipation and to a large friction force.

The molecular rearrangement processes associated with the transition from the compressed monolayer film to the thermal equilibrium film may involve large (compared with the thermal energy k_BT) free energy barriers, which must be overcome by thermal excitation. Thus the relaxation time τ often depends sensitively on the temperature [typically $\tau \propto \exp(\epsilon/k_BT)$], and

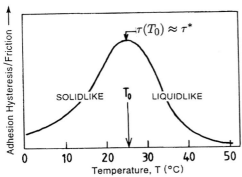

Fig. 5.18. Schematic diagram representing the trends observed in the boundary friction and adhesion energy hysteresis of a variety of different surfactant monolayers. From [5.23].

it is often possible to vary τ by several orders of magnitude by simply varying the temperature. At high enough temperature τ is very short, $\tau/\tau^* \ll 1$ and the film is "liquid-like" *with respect to the experiment in question*. At low enough temperature τ is very long, $\tau/\tau^* \gg 1$ and the film is "solid-like" with respect to the experiment in question. In both cases very small hysteresis and friction are expected. On the other hand, when $\tau \approx \tau^*$, large hysteresis and friction prevail. Thus, the adhesion hysteresis and the sliding friction vary non-linearly with the temperature, often peaking at some particular temperature, T_0, where $\tau(T_0) \approx \tau^*$, see Fig. 5.18. Experiments have shown that the whole bell-shaped curve is shifted along the temperature axis when the time τ^* is varied, e.g., by changing the sliding velocity in a friction experiment. For example, an increase in the temperature (which implies a decrease in τ) produces the same effect as a decrease in the sliding speed (which results in an increase in τ^*), since in both cases τ/τ^* will decrease.

Finally, let us note that only when the monolayer film binds strongly to the substrate will it withstand the huge local pressures which may occur at the tip-substrate interface. Thus, the $C_{16}H_{33}SH$ film protects the gold substrate even when the pressure is so high as to plastically deform the gold substrate, see Fig. 5.19. However, in another experiment, where the gold surface was contaminated by a thin (unknown) organic film, the tungsten tip already penetrated the contamination layer at a relatively low tip force. The resulting force-distance curve is shown in Fig. 5.20, where the initial increase of the force on approach corresponds to the repulsive barrier caused by the contaminants. When the tip has penetrated the contaminant layer, the tip-substrate interaction is initially strongly attractive, as in the earlier studies with a gold tip on a gold substrate.

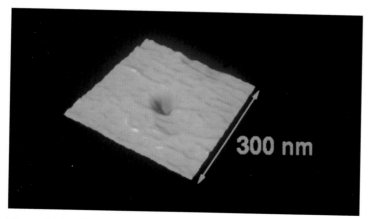

Fig. 5.19. Constant repulsive force image of a gold surface covered by a $CH_3(CH_2)_{21}SH$ monolayer film. The indentation has resulted from a loading cycle with a peak load of 57μN. The deformation has a width and depth of approximately 510 Å and 28 Å, respectively. The passivation by the monolayer film is complete: The tip–substrate interaction during the indentation is purely repulsive and no transfer of gold from the substrate to the tungsten tip occurred. From [5.16].

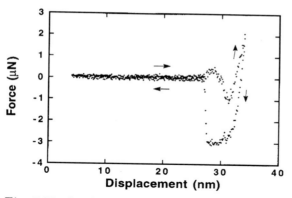

Fig. 5.20. Loading cycle for a tungsten tip on a gold surface without the monolayer film but with a layer of weakly bound organic contaminants. From [5.16].

Contact Hysteresis: Microscopic Model

In this section we study a simple model which illustrates in detail the basic physics behind the diagram in Fig. 5.18. Assume that a layer of large molecules is adsorbed on a flat substrate. The molecules can take two different configurations, A and B, say a spherical and an oblate spheroidal shape as indicated in Fig. 5.21a. Assume for simplicity that both states have the same energy and vibrational entropy. Thus, in thermal equilibrium, on the average half of the molecules will be in state A and half in state B. Assume

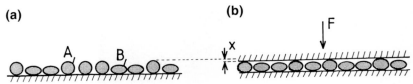

Fig. 5.21. (a) A surface with a layer of molecules which can take two different shapes, a spherical state A and an oblate spheroidal state B. In thermal equilibrium the molecules fluctuate between the two different states. (b) An external surface squeezing the monolayer against the substrate will have two effects: First, the fraction of the molecules in state A is reduced and, secondly, the molecules in state A are elastically deformed.

that a molecule must overcome an energy barrier of height ϵ when going from state A to state B. Thus, due to thermal "kicks" from the surroundings, a molecule will fluctuate randomly in time between the two configurations with the average rate $w = \nu \exp(-\beta\epsilon)$ where ν is the prefactor and $1/\beta = k_B T$. Consider now an indentation experiment where a flat rigid surface is pushed towards the system in Fig. 5.21a. During the whole approach–retraction cycle we assume that the upper surface only interacts with the molecules in state A, i.e., the displacement of the upper solid is so small that no direct contact with the molecules in state B occurs. After contact, the molecules in state A are elastically deformed, and the fraction of molecules in state A is decreased while the number of molecules in state B is increased, see Fig. 5.21b. The latter follows from the fact that because of the presence of the external force from the upper surface, the barrier for jumps from state A to state B is reduced while the barrier for the transition from state B to state A is unaffected.

Let P_A and P_B denote the probabilities that a molecule is in state A or in state B, respectively, so that $P_A + P_B = 1$. If w_B and w_A denote the jumping rates from $B \to A$ and from $A \to B$, respectively, then we have

$$\frac{dP_A}{dt} = w_B P_B - w_A P_A \; . \tag{5.9}$$

Substituting $P_B = 1 - P_A$ in (5.9) gives

$$\frac{dP_A}{dt} = w_B - [w_A + w_B] P_A \; .$$

This equation is easy to integrate:

$$P_A(t) = P_A(0) e^{-\int_0^t dt' [w_A(t') + w_B(t')]} \\ + \int_0^t dt' w_B(t') e^{-\int_{t'}^t dt'' [w_A(t'') + w_B(t'')]} \; . \tag{5.10}$$

Now, assume that the potential barrier separating state A from state B has the form (Fig. 5.22a)

5.2 Nanoscale Junctions

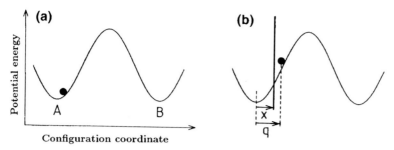

Fig. 5.22. (a) The plot of the potential energy of a molecule along the "reaction" coordinate q between state A and state B. The "particle" (*black dot*) represents the state of the system. (b) The upper surface is represented by an infinite barrier (*vertical line*) at $q = x$, which confines the motion of the "particle" to the region $q > x$ on the right of the infinite barrier.

$$U = U_0 \left(1 - \cos kq\right) , \tag{5.11}$$

where $k = 2\pi/a$, where a is the separation between the two minima in Fig. 5.22a. The "particle" (black dot) in Fig. 5.22 represents the state of the system. With (5.11) the barrier height for the transition $B \to A$ equals $\epsilon = 2U_0$ and

$$w_B = \nu e^{-\beta 2 U_0} \equiv w . \tag{5.12}$$

Note that w is independent of time. When a molecule in state A is squeezed by the upper surface, the barrier to state B is reduced. The vertical line in Fig. 5.22b represents an infinite barrier (corresponding to the surface of the upper solid) at $q = x$, which confines the motion of the particle to the region $q > x$ to the right of the barrier. Thus, the effective barrier equals $2U_0 - U(x) = U_0(1+\cos kx)$ where x is the displacement of the upper surface towards the lower surface ($x = 0$ corresponds to the point of initial contact with the molecules in state A). Thus

$$w_A = \nu e^{-\beta U_0(1+\cos kx)} = w e^{\beta U_0(1-\cos kx)} , \tag{5.13}$$

where we have neglected the dependence of the prefactor ν on x (the x-dependence of ν is unimportant compared with the x-dependence of the activation barrier). We now assume that the upper surface moves towards the lower surface with the speed v until $x = vt$ reaches $x_0 = vt_0$ at which point the retraction starts (with same speed v). Inserting (5.13) into (5.10) gives, for the approach period,

$$P_A(t) = P_A(0) \exp\left\{-w \int_0^t dt' \left(1 + e^{\beta U_0[1-\cos(kvt')]}\right)\right\}$$
$$+ w \int_0^t dt' \exp\left\{-w \int_{t'}^t dt'' \left(1 + e^{\beta U_0[1-\cos(kvt'')]}\right)\right\} \tag{5.14}$$

where $P_A(0) = 0.5$ is the initial probability that a molecule is in state A. Similarly, during the retraction time $t_0 < t < 2t_0$ (where t_0 is the time at which the retraction starts):

$$P_A(t) = P_A(t_0) \exp\left\{-w \int_{t_0}^{t} dt' \left(1 + e^{\beta U_0(1-\cos[kv(2t_0-t')])}\right)\right\}$$
$$+ w \int_{t_0}^{t} dt' \exp\left\{-w \int_{t'}^{t} dt'' \left(1 + e^{\beta U_0(1-\cos[kv(2t_0-t'')])}\right)\right\}. \quad (5.15)$$

Now, note that $\tau = 1/w$ is a relaxation time corresponding to the average time the molecule spends in state B before flipping to state A due to a thermal fluctuation. Similarly, we may introduce the time $\tau^* = 1/kv = a/(2\pi v)$ which characterizes the motion of the upper surface. Writing $t' = t\xi$, $t'' = t\xi'$ and $\bar{x} = kvt$, (5.14) takes the form

$$P_A(t) = P_A(0) \exp\left\{-\bar{x}\frac{\tau^*}{\tau} \int_0^1 d\xi \left(1 + e^{\beta U_0[1-\cos(\bar{x}\xi)]}\right)\right\}$$
$$+ \bar{x}\frac{\tau^*}{\tau} \int_0^1 d\xi \exp\left\{-\bar{x}\frac{\tau^*}{\tau} \int_\xi^1 d\xi' \left(1 + e^{\beta U_0[1-\cos(\bar{x}\xi')]}\right)\right\}. \quad (5.16)$$

In a similar way one can express $P_A(t)$ during the retraction in the dimensionless variables τ^*/τ and βU_0. If there are N molecules at the interface, on the average of $NP_A(t)$ molecules will be in the state A and the force F that the molecules exert on the upper surface equals

$$F = NP_A U'(x) = NP_A kU_0 \sin kx. \quad (5.17)$$

In Fig. 5.23 we show the force F (in units of NkU_0) when the upper surface first moves towards the lower surface with the velocity v until $\bar{x} = 0.4$, i.e., $x = 0.2a/\pi \approx 0.064a$, at which point the retraction starts. Note the hysteresis in the force–distance plot. In the calculation we have used $\beta U_0 = 15$ and $\tau^*/\tau = 2$. Figure 5.24 shows the external work

$$\Delta E = \oint dx F \quad (5.18)$$

necessary for the indentation-retraction cycle. Note that $\Delta E \to 0$ as $\tau^*/\tau \to 0$ and ∞. The physical origin of this behavior is as follows: When $\tau^* \ll \tau$ the motion of the upper surface is so rapid that the molecules in state A do not have enough time to jump over the barrier to state B, i.e., the molecules A will deform purely elastically and the elastic energy stored up in the molecules during approach is returned in full to the upper solid during retraction. Hence no adhesion hysteresis occurs. On the other hand, when $\tau^* \gg \tau$ the motion is so slow that the system is always very close to thermal equilibrium. Thus the motion is adiabatic and again the energy stored up during the compression phase is returned completely to the upper solid during

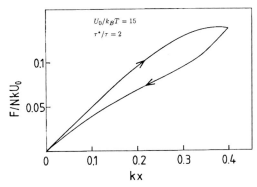

Fig. 5.23. The force F as a function of the displacement $\bar{x} = kx$ during an approach and retraction cycle (constant speed v). In the calculation $\beta U_0 = 15$ and $\tau^*/\tau = 2$ and the turning point $\bar{x}_0 = 0.4$.

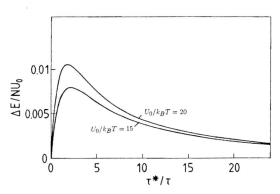

Fig. 5.24. The adhesion hysteresis as a function of τ^*/τ for $\beta U_0 = 15$ and 20.

the decompression phase. The largest adhesion hysteresis occurs when $\tau \approx \tau^*$. Since $\tau \propto \exp(2U_0/k_{\rm B}T)$ depends strongly on the temperature, Fig. 5.24 corresponds to a temperature–adhesion hysteresis diagram of the form shown in Fig. 5.18.

The model above illustrates the physical origin of another well-known effect. Many materials, when exposed to rapidly (in time) varying forces behave as nearly perfect elastic materials, while they behave as (high viscosity) fluids when exposed to slowly varying forces. One famous example which illustrates this effect is the remarkable toy polymer (silly putty) which behaves as a perfect elastic solid when dropped on the floor but which slowly flows out to a pancake-shaped object when left on a table for a few minutes. In order to exhibit this behavior, the relaxation time τ associated with changes in the shape must be much longer than the time of contact $\tau^* \sim 0.01\,{\rm s}$ during an elastic collision between the solid and the floor, while it must be much shorter than the time $\tau^* \sim 1000\,\mathrm{s}$ it takes for the object to flow out to a pancake-shaped object.

Contact Hysteresis: Phenomenological Model

If the displacement $x(t)$ in the study above is small enough the system responds linearly. By linear we mean, in general, that if the strain $\varepsilon(t)$ is found as a function of time for one stress history $\sigma(t)$, then for another stress history $\lambda\sigma(t)$, where λ is a constant, the strain will be $\lambda\varepsilon(t)$. It has been found that the small-strain viscoelasticity of many solids can be relatively accurately represented by simple rheological models [5.17]. These models do not explain the microscopic origin of the viscoelastic behaviour, but are very useful for practical calculations and as a guide for qualitative thinking. We show in this section how the response of the monolayer film discussed in the last section can be described by a simple rheological model.

Figure 5.25 shows a rheological model of the monolayer film in Fig. 5.21. The springs and dashpot in Fig. 5.25 correspond to the elastic and viscous components of the film. If x_1 and x_2 denote the elongation of the two springs, then if inertia effects can be neglected (usually the case), the response to an applied force $F(t)$ will be

$$k_1 x_1 = F, \tag{5.19}$$

$$k_2 x_2 + \gamma \dot{x}_2 = F, \tag{5.20}$$

where γ is a damping parameter characterizing the dashpot. The relation between the total displacement $x = x_1 + x_2$ and the driving force F is easily obtained from (5.19) and (5.20). If we introduce the Fourier transforms

$$x(\omega) = \int dt\, x(t) \mathrm{e}^{\mathrm{i}\omega t},$$

$$x(t) = \frac{1}{2\pi} \int d\omega\, x(\omega) \mathrm{e}^{-\mathrm{i}\omega t};$$

and similarly for F, we get

$$x(\omega) = x_1(\omega) + x_2(\omega) = \left(\frac{1}{k_1} + \frac{1}{k_2 - \mathrm{i}\omega\gamma}\right) F(\omega)$$

or

$$K(\omega) x(\omega) = F(\omega), \tag{5.21}$$

where

$$K(\omega) = \frac{k_1(k_2 - \mathrm{i}\omega\gamma)}{k_1 + k_2 - \mathrm{i}\omega\gamma}. \tag{5.22}$$

The parameters k_1 and γ/k_2 can be related to the parameters that enter in the microscopic model as follows. Let us first consider a rapid indentation,

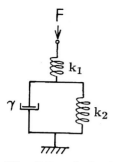

Fig. 5.25. A rheological model of the monolayer film in Fig. 5.21.

where thermal excitation over the barrier separating state A from state B (see Fig. 5.22) can be neglected. In this case the system behaves purely elastically, and if N_1 denotes the number of molecules in state A (which will not change during the rapid indentation), then for a small indentation the elastic energy $U(x)$ [see (5.11)] can be expanded to quadratic order in x:

$$\Delta E \approx N_1 U_0 k^2 x^2 / 2$$

and the force $F = N_1 U_0 k^2 x$. Similarly, during a rapid indentation the dashpot will not displace so that $x_2 = 0$ and $F = k_1 x$. Thus,

$$k_1 = N_1 U_0 k^2 \ .$$

In the present case $N_1 = N/2$ and we get

$$k_1 = N U_0 k^2 / 2 \ . \tag{5.23}$$

Next, note that when $F = 0$ the solution to (5.20) is of the form $x_2(t) \sim \exp(-k_2 t/\gamma)$. This time dependence must be the same as that of the probability distribution $P_A(t)$ when w_A and w_B are time independent (corresponding to $F = 0$). Using (5.9) gives

$$\delta P_A \sim \exp[-(w_A + w_B) t]$$

so that

$$k_2/\gamma = w_A + w_B = 2w = 2/\tau \ . \tag{5.24}$$

Substituting (5.24) into (5.22) gives

$$K(\omega) = \frac{k_1(1 - i\omega\tau/2)}{1 + a - i\omega\tau/2} \ , \tag{5.25}$$

where $a = k_1/k_2$.

5. Area of Real Contact: Elastic and Plastic Deformations

As an illustration, let us calculate the energy "dissipation" during the same indentation cycle as in the last section. We get the energy dissipation ΔE by integrating the power $P(t) = \dot{x}(t)F(t)$ over the whole cycle:

$$\Delta E = \oint dx F = \int dt\, \dot{x} F(t) .$$

Substituting the Fourier expansions of x and F in this formula, and using

$$\int dt\, e^{i(\omega+\omega')t} = 2\pi\delta(\omega+\omega')$$

gives

$$\Delta E = \frac{1}{2\pi} \int d\omega\, (i\omega) x(-\omega) F(\omega) .$$

Since $x(t)$ is real, $x(-\omega) = x^*(\omega)$, and

$$\Delta E = -\frac{1}{2\pi} \int d\omega\, \omega\, \text{Im}\left[x^*(\omega) F(\omega)\right] . \tag{5.26}$$

Substituting (5.21) into (5.26) gives

$$\Delta E = -\frac{1}{2\pi} \int d\omega\, \omega\, \text{Im}\left[K(\omega)\,|x(\omega)|^2\right] . \tag{5.27}$$

Now, consider the same indentation process as before:

$$x(t) = -vt \quad \text{for} \quad 0 < t < t_0 ,$$
$$x(t) = -vt_0 + v(t-t_0) \quad \text{for} \quad t_0 < t < 2t_0$$

and $x(t) = 0$ otherwise. It is easy to calculate the Fourier transform,

$$x(\omega) = \frac{v}{\omega^2}\left(1 - e^{i\omega t_0}\right)^2 ,$$

so that

$$|x(\omega)|^2 = \frac{4v^2}{\omega^4}\left(1 - \cos\omega t_0\right)^2 .$$

Substituting this into (5.27) and using (5.23) and

$$\text{Im}\, K(\omega) = -\frac{k_1 a\omega \tau/2}{(1+a)^2 + \omega^2 \tau^2/4}$$

gives

$$\frac{\Delta E}{NU_0} = \frac{1}{2\pi} \int d\omega\, \frac{v^2 k^2}{\omega^2} \frac{a\tau(1-\cos\omega t_0)^2}{(1+a)^2 + \omega^2 \tau^2/4} .$$

The integral is easy to calculate, giving

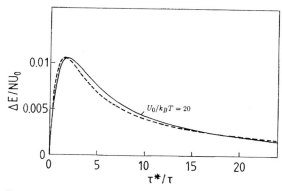

Fig. 5.26. The adhesion hysteresis as a function of τ^*/τ. The dashed line is the prediction of the rheological model in Fig. 5.25, while the solid line is the exact result from Fig. 5.24.

$$\frac{\Delta E}{NU_0} = \frac{1}{4}\frac{a}{(1+a)^3}\left(\frac{\tau}{\tau^*}\right)^2\left(\eta - \frac{3}{2} + 2e^{-\eta} - \frac{1}{2}e^{-2\eta}\right), \qquad (5.28)$$

where $\eta = 2(1+a)kx_0\tau^*/\tau$ and $\tau^* = 1/kv$. In Fig. 5.26 we compare the prediction of (5.28) (with $a = 0.53$) with the exact results from Fig. 5.24 for the case $U_0/k_BT = 20$. A similar good fit (with $a = 0.35$) is obtained for the case $U_0/k_BT = 15$ in Fig. 5.24. It is clear that the model in Fig. 5.25 constitutes a simple and useful description of the viscoelastic properties of the monolayer film.

5.3 Area of Real Contact for Polymers

The viscoelastic properties of solids enters in the description of sliding and rolling friction of polymers (see Sect. 14.8). In this section we present some basic results of the theory of viscoelasticity. The discussion focuses on rheological models of the type discussed in the last section. Recently, considerable progress has been made in deriving the viscoelastic properties of polymers from microscopic models; we refer the readers to [5.18] for an excellent introduction to this topic.

Consider a cylindrical bar of a viscoelastic material under uniaxial tension. The stress σ is related to the strain ε via the equation

$$E(\omega)\varepsilon(\omega) = \sigma(\omega). \qquad (5.29)$$

If the viscoelastic properties are described by a model of the form indicated in Fig. 5.27a, then

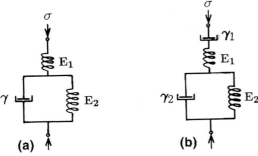

Fig. 5.27a,b. Two different rheological models for the linear viscoelastic properties of solids.

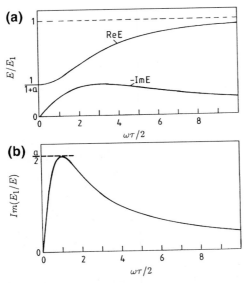

Fig. 5.28. (a) The real and imaginary part of the viscoelastic modulus $E(\omega)$ given by (5.30), corresponding to Fig. 5.27a. (b) The loss function $\mathrm{Im}[1/E(\omega)]$.

$$E(\omega) = \frac{E_1(E_2 - i\omega\gamma)}{E_1 + E_2 - i\omega\gamma} = \frac{E_1(1 - i\omega\tau/2)}{1 + a - i\omega\tau/2}, \qquad (5.30)$$

where $a = E_1/E_2$. In general, to accurately describe real materials, more springs and dashpots may be necessary. For example, the model (5.30) does not describe the continuous "flow" that occurs when polymer systems without cross links (e.g., silly putty) are exposed to external forces, and extensions, such as adding another dashpot in series with the spring E_1 (see Fig. 5.27b), are necessary in order to describe this "creep" motion. Figure 5.28 shows the real and imaginary part of $E(\omega)$ and the "loss function" $\mathrm{Im}\,[1/E(\omega)]$, as a

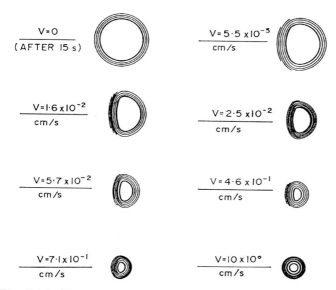

Fig. 5.29. Experimental observation of the variation of the contact area: glass ball rolling at different speeds on epoxy resin. From [5.24].

function of frequency ω as obtained from (5.30). Note that the solid behaves purely elastically for "low" and "high" frequencies, and that it is stiffer at high frequencies. Since the characteristic frequencies a sliding or rolling object exert on the substrate are proportional to the sliding (or rolling) velocity v, it is clear that the area of real contact will depend on v. To illustrate this, Fig. 5.29 shows the interference fringes surrounding the actual contact area for a glass sphere rolling on a viscoelastic epoxy resin at different speeds. At "low" speeds, the contact area is "large" and circular. With increasing v the contact area decreases and becomes first non-circular and then, at "high" rolling speeds, circular. The size effect occurs because of an effective "stiffening" of the elastomer at high rolling speeds. The asymmetry effect occurs in the transition region where the imaginary part of the complex elastic modulus is large and can be understood [5.19] based on the simple viscoelastic model shown in Fig. 5.27a.

Let us consider a cylindrical bar of a viscoelastic material. If the stress σ_0 is applied to the upper surface of the cylinder at time $t = 0$ the cylinder will deform viscoelastically. If the material is described by an elastic modulus of the form (5.30) it is easy to show that the strain at time t is given by

$$\varepsilon = \frac{\sigma_0}{E_1} \left[1 + a \left(1 - e^{-2t/\tau} \right) \right] . \tag{5.31}$$

The model in Fig. 5.27b gives an expression for $\varepsilon(t)$ with an additional term in (5.31) which increases linearly with time. Thus, if the cylindrical bar represents a surface asperity of a block squeezed against a rigid substrate, from the models above one expects the area of real contact to increase with time as $\sim A[1-\exp(-2t/\tau)]+Bt$. This time dependence of the area of real contact is not usually observed experimentally even for polymers. For example, in the study by *Dieterich* and *Kilgore* [5.6] reported on in Sect. 5.1, the area of real contact increased $\sim \ln(1+t/\tau)$ for acrylic plastic, as is indeed observed for most other materials as well. The origin for this may be that the stresses in the contact areas in most cases are so large that the linear viscoelastic theory is not valid, and the theory developed in Sects. 11.4 and 13.2 [which predict $\Delta A \sim \ln(1+t/\tau)$] must be used. However, in studies of the deformations over larger length scales (as in Fig. 5.29) the stress may be small enough for the linear viscoelastic theory to be valid.

The internal friction gives a very important contribution to the sliding or rolling friction of polymers. We will discuss this topic in detail in Sect. 14.8, but present the basic equations here. The energy dissipation in a viscoelastic media is in general given by

$$\Delta E = \int d^3x\, dt\; \sigma_{ij}\dot{\varepsilon}_{ij}\,.$$

For uniaxial deformations of a cylindrical bar this formula reduces to

$$\Delta E = \int d^3x\, dt\; \sigma\dot{\varepsilon} = -\frac{V}{2\pi}\int d\omega\, \omega\, \mathrm{Im}\,[\sigma(\omega)\varepsilon^*(\omega)]$$

or

$$\Delta E = -\frac{V}{2\pi}\int d\omega\, \omega\, \mathrm{Im}\, E(\omega)\,|\varepsilon(\omega)|^2$$
$$= \frac{V}{2\pi}\int d\omega\, \omega\, \mathrm{Im}\left(\frac{1}{E(\omega)}\right)|\sigma(\omega)|^2\,, \tag{5.32}$$

where V is the volume of the solid. Note that

$$\mathrm{Im}\left(\frac{1}{E(\omega)}\right) = \frac{a}{E_1}\frac{\omega\tau/2}{1+(\omega\tau/2)^2}\,. \tag{5.33}$$

This function has a maximum at $\omega = 2/\tau$ (Fig. 5.28b).

We can use the equations above to estimate the contribution from the internal friction to the sliding or rolling friction of elastomer. Sliding or rolling of an elastomer on a substrate results in fluctuating stresses characterized by the frequency $\omega_0 \sim v/l$, where v is the sliding (or rolling) velocity and l a length of the order of the diameter of the contact area. The main part of the energy "dissipation" will occur in a volume element $V \sim l^3$; see Fig. 5.30. If F_n is the normal force then the fluctuating stress $\sigma \sim \sigma_0 \cos\omega_0 t$, where $\sigma_0 = F_n/l^2$. Substituting this into (5.32) gives

$$\Delta E \sim l^3 \sigma_0^2 \omega_0 T\, \mathrm{Im}\left(\frac{1}{E(\omega_0)}\right)\,, \tag{5.33a}$$

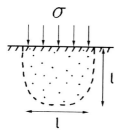

Fig. 5.30. The fluctuating stress $\sigma(t)$ acting on the surface area $A \sim l \times l$ gives rise to energy "dissipation" via the internal friction of the solid. The main part of the "dissipation" occurs in the dotted volume element $V \sim l \times l \times l$.

where T is the total time the oscillating stress has acted on the solid. The energy dissipation per unit time $\Delta E/T$ must equal the product Fv between the friction force and the sliding velocity so that

$$F \sim \frac{l^3 \sigma_0^2 \omega_0}{v} \operatorname{Im}\left(\frac{1}{E(\omega_0)}\right).$$

Since $F_n = l^2 \sigma_0$ we finally get

$$\mu = \frac{F}{F_n} \sim \sigma_0 \operatorname{Im}\left(\frac{1}{E(\omega_0)}\right). \tag{5.34}$$

We can estimate σ_0 as follows. In experiments with rubber it is usually found that the friction coefficient μ is (nearly) independent of the load L. (This is not true when the load L is very small or very large. In the former case, adhesion between the rubber and the substrate becomes important. In the latter case, the area of real contact approaches the apparent area of contact, and the area of real contact cannot increase linearly with the load.) This is the case, for example, in the experiments by Grosch and by Mori et al. (see below), and it is also expected from the model calculation of Greenwood which shows that, when two *elastic* bodies (no plastic deformation) with rough surfaces are pushed together, the area of real contact and the average pressure σ_0 in the contact areas are (nearly) independent of the load (see Sect. 5.1). Since no plastic deformations is assumed to occur, the elastic modulus E is the only quantity in the problem with the same unit as pressure, and it follows immediately that $\sigma_0 = CE$, i.e., *dimensional analysis alone predicts that the (average) pressure in the areas of real contact is proportional to the elastic modulus E of the rubber*, where C (which may be larger or smaller than unity) depends only on the nature of the surface roughness. The model of Greenwood gives an explicit expression for C, but the only relevant result here is that, for the very rough surfaces typically involved in rubber friction, C is of order unity. For sliding surfaces $E(\omega_0)$ is complex, and we will assume that $\sigma_0 \approx C \mid E(\omega_0) \mid$. Note that σ_0 depends on the sliding

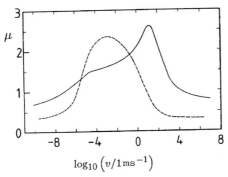

Fig. 5.31. The coefficient of friction for acrylonitrile-butadiene rubber sliding on silicon carbide paper (*solid curve*) and a smooth glass surface (*dashed curve*). The silicon carbide particles have a size of ~ 0.01 cm. The glass surface was polished by alumina powder (particle size ~ 100 Å) and is likely to have surface roughness with a wavelength ~ 100 Å. The temperature $T = 20°$C. From [5.20].

velocity v: increasing v leads to a stiffening of the elastic properties and (for a given load) to a reduced contact area (Fig. 5.29), and to an increased surface stress σ_0 in the contact areas. With this σ_0, (5.34) takes the form

$$\mu \sim -C \frac{\mathrm{Im} E(\omega_0)}{\mid E(\omega_0) \mid} \qquad (5.34\mathrm{a})$$

Since $-\mathrm{Im} E(\omega_0)/\mid E(\omega_0) \mid \sim 1$ when $\omega_0 \sim 2/\tau$ one expects the maximum of the coefficient of friction to be of order unity.

Equation (5.34a) illustrate some consequences of the internal friction, to be discussed in greater detail in Sect. 14.8. In some cases one wants the internal friction and energy "dissipation" to be as large as possible at some frequency ω_0. This can be arranged by choosing a polymer characterized by a relaxation time τ such that $\omega_0 \sim 2/\tau$. In other cases, e.g., during rolling of a wheel on the road, one wants the internal friction to be as small as possible. This can be arranged by using a material for which, for typical rolling velocities, ω_0 differs appreciably from $2/\tau$.

To illustrate the discussion above, Fig. 5.31 shows the friction coefficient at room temperature ($T = 20°$C) for acrylonitrile-butadiene rubber sliding on a silicon carbide paper (solid curve) and on a smooth glass surface (dashed curve) [5.20]. The figure shows μ up to $v \sim 10^5$ m/s but the actual measurements were performed only at low sliding velocities (up to ~ 1 cm/s), in order to avoid heating effects. However, it was found that the sliding friction depends only on the the product $v\tau$, where τ is a relaxation time with $1/\tau$ exhibiting an activated temperature dependence. Thus the curves in Fig. 5.31 have been constructed from a series of measurements at different temperatures, which were "translated" to the temperature $T = 20°$C. The strong (exponential) temperature dependence of μ for both systems was found to be the same as that of the internal friction of the rubber, strongly suggesting that μ is mainly due to the internal friction of the rubber rather than due to an interfacial dissipation process, such as shearing of a molecular thin contamination layer. To give further support for this conclusion, we now show how the peak positions of the two curves in Fig. 5.31 can be explained based on the known bulk viscoelastic properties of the rubber and the surface roughness of the two substrates.

The silicon carbide particles have a size of ~ 0.01 cm. The glass surface was polished by alumina powder (particle size ~ 100 Å) and is likely to have surface roughness with a wavelength $\sim 50 - 100$ Å. Bulk rheological experiments have shown that at room temperature the function $- \operatorname{Im} E(\omega)/|E(\omega)|$ of the rubber is maximal for $\omega = \omega_1 \sim 10^5 \, \text{s}^{-1}$. If l denotes the wavelength of the characteristic surface roughness ($l \sim 100$ Å and ~ 0.01 cm for the glass and silicon carbide surfaces, respectively), then the surface asperities will exert fluctuating forces on the rubber characterized by the frequency $\omega = \omega_0 \sim v/l$. The sliding friction coefficient will be maximal when $\omega_0 \sim \omega_1$, i.e., when $v = l\omega_1$, which gives $v \sim 10 \, \text{m/s}$ and $0.001 \, \text{m/s}$ for the silicon carbide and glass surfaces, respectively, in good agreement with the experimental data (see Fig. 5.31). [Note: the sliding friction for rubber on smooth surfaces, such as the glass surface in Fig. 5.31 (dashed line), is usually associated with an adhesional process, where small regions of the rubber at the interface perform local stick-slip motions [5.21]. This mechanism would also operate on a perfectly (atomically) smooth surface, where the contribution from surface roughness would be absent. However, such an additional mechanism is not necessary in order to explain the present experimental data.]

Adhesion of Elastomer and Its Role in Friction

Rubber exhibit unusual sliding behaviour, which is derived from the fact that it is elastically very soft. Furthermore, the area of real contact between rubber and a rough hard substrate is strongly affected by adhesion. In this section we will discuss some aspects of adhesion and sliding friction for elastically soft solids, such as rubber or gelatine.

For elastically hard materials, adhesion usually does not manifest itself on a macroscopic scale. For hard solids, the area of real contact consist of nearly randomly distributed contact areas (junctions) where surface asperities from the two surfaces "touch". The surface asperities are elastically deformed and since they have different sizes and shapes, the junctions will "pop" one after another as the block is removed, giving rise to a negligible adhesional force when removing the block.

The situation is drastically different for rubber and other elastically soft solids. In these cases, even a weak adhesive junction, e.g., resulting from the van der Waals interaction between the surfaces, may be elongated before breaking by a distance which is larger than the "hight" of the surface asperities. Thus, at one stage during the removal of the block, a large fraction of

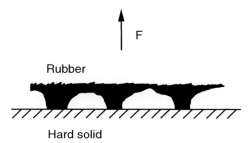

Fig. 5.32. Elastically deformed junctions during pull-off of an elastically soft solid from a hard substrate.

the junctions will be simultaneously elastically elongated and exert a force on the block in the direction towards the substrate (Fig. 5.32). Furthermore, for an elastically soft solid the area of real contact will in general be much larger than for an elastically hard solid under similar conditions. In particular, the area of real contact is nonzero even when the external load vanishes which implies that the friction coefficient $\mu \to \infty$ as $L \to 0$. The effects discussed above also explains how some adhesives, e.g., tease film, function.

Let us study the area of real contact when an elastic body with a flat surface is squeezed against a rough surface of a hard solid (Fig. 5.33a). Assume that the surface roughness is characterized by the hight h and the wavelength l. If the elastic solid completely fill out a "cavity", the elastic energy stored in the local deformation field will be of order $E_{el} \approx Glh^2$, see (9.12), where G is the shear modulus, while the adhesion energy $E_{ad} \approx \Delta\gamma l^2$ will be gained ($\Delta\gamma$ is the change in the interfacial free energy per unit area when the two solids comes in contact). Thus, if the adhesion energy is just large enough to deform the solid so that it completely fills out the cavity, then $E_{el} = E_{ad}$ or $h/l = (\Delta\gamma/Gl)^{1/2}$. For rubber at room temperature, $\rho \approx 1000\,\mathrm{kg/m^3}$ and $G \approx nk_BT \approx 10^6\,\mathrm{N/m^2}$ (n is the number of rubber molecules per unit volume) and with $\Delta\gamma = 3\mathrm{meV/\mathring{A}}^2$ and $l = 1000\mathring{A}$ this gives $h/l \approx 1$. We conclude that when rubber is slid at low speed on the polished glass surface in Fig. 5.31 (for which $l \approx h \approx 100\mathring{A}$), the rubber in the contact area will deform (because of the adhesional forces) in such a manner as to *completely follow the short-wavelength surface roughness profile* of the glass substrate (Fig. 5.33b). At high sliding velocity, the rubber becomes stiffer, and the area of real contact decreases.

Let us estimate the contribution to the friction from the adhesive interaction. Let us consider sliding of rubber on a macroscopically smooth surface, polished by small particles with the (average) diameter D. Such a surface will have a roughness with the characteristic wavelength $l \sim D$ and hight fluctuation $h \sim D$. We assume that D is so small that the rubber will completely fill out the cavities as a result of the adhesion force. As shown above, for rubber at room temperature this requires that D is smaller than $\approx 1000\mathring{A}$. Now, the

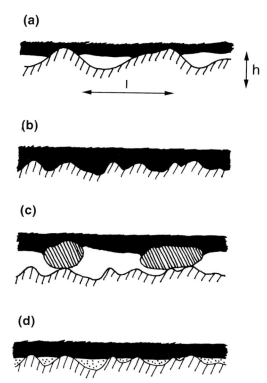

Fig. 5.33. (a) Elastic body (e.g., rubber) on a hard rough substrate. (b) Because of adhesion to the substrate, the rubber in the contact area deforms in such a manner as to completely follow the short-wavelength surface roughness profile of the substrate. (c) Rubber surface dusted by small particles sliding on a hard substrate. (d) Rubber sliding (at low velocity) on a water covered surface.

perpendicular force F_n which must act on the surface of the rubber within the area $l \times l$ in order for the rubber to fill out a "cavity" is $F_n \approx 2 |G| lh$ (note: the shear modulus G is complex at finite frequencies). During sliding (velocity v) the rubber will be deformed elastically by the adhesional forces at the characteristic frequency $\omega_0 \sim v/l$. Thus, using (5.33a), the total energy "dissipation" in the rubber during the time period T and for the surface area A (with $\sim A/l^2$ "cavities") is

$$\Delta E \approx (A/l^2)l^3(F_n/l^2)^2(v/l)T \operatorname{Im}(1/E) = 4A \mid G \mid^2 (h/l)^2 v T \operatorname{Im}(1/E).$$

This must equal vTF so that the frictional stress $\sigma = F/A$ equals

$$\sigma = 4 \mid G \mid^2 (h/l)^2 \operatorname{Im}(1/E) \approx -(h/l)^2 \operatorname{Im} G(\omega_0)$$

since $G \approx 0.3E$ for rubber. At room temperature and for frequencies in the range $0.1\,\text{s}^{-1} < \omega < 10^6\,\text{s}^{-1}$, $-Im\,G \sim 10^5 - 10^6\,\text{N/m}^2$ (Fig. 14.37), so that if $h \sim l$, $\sigma \sim 10^5 - 10^6\,\text{N/m}^2$. This is of the same order of magnitude as the observed frictional stress in most cases [5.25].

It is clear from the discussion above that if the adhesive interaction between the substrate and rubber can be reduced, the friction force will decrease. For example, if the glass surface in the experiments reported on above (Fig. 5.31) is dusted with magnesia powder, thus preventing direct contact between the rubber and the track, the friction is found to remain almost constant over the whole range of temperatures and velocities, instead of the peak-structure (dashed line in Fig. 5.31) observed for the clean glass surface and associated with the internal friction of the rubber. The magnesia particles are embedded in the rubber surface (Fig. 5.33c), leading to reduced adhesion and viscoelastic deformations of the rubber. On the other hand, the high-velocity peak in the friction coefficient for rubber sliding on the silicon carbide surface (solid line in Fig. 5.31) is not reduced for the dusted surface; this result is expected, since the long-wavelength deformations of the rubber, from which this peak is derived, are the same for the dusted and clean surfaces. Similarly, if rubber is slid on a substrate (assumed hydrophilic) covered by water, the sliding friction is reduced [5.25]. The water is trapped in the surface cavities of the substrate, leading to reduced viscoelastic deformations of the rubber (Fig. 5.33d). Another proof of the important role of adhesion on sliding friction are experiments with solids with high- and low-surface free energy. Let us illustrate this with the following example.

Mori et al. [5.26] have studied the influence on friction by the adhesional interactions between rubber and two different substrates: Teflon with the surface free energy $\gamma \approx 1.8\,\text{meV/Å}^2$ and surface roughness $h = 2100\,\text{Å}$ and aluminum with $\gamma \approx 3.8\,\text{meV/Å}^2$ and $h = 2500\,\text{Å}$. The rubber was prepared using three varions molds differing in surface free energy, resulting in rubber surfaces with different surface free energies but with identical bulk properties. [Note: molds with many polar groups give rubber surfaces with numerous nitrile groups, resulting in high surface free energy. Molds with low surface free energy (e.g., Teflon) result in rubber surfaces having many butadiene groups which are low-energy surfaces.] We denote the three rubber surfaces by **1**, **2** and **3**, with the surface free energies ≈ 2.6, 2.3 and $2.0\,\text{meV/Å}^2$, respectively. Figure 5.34 shows the relation between the kinetic friction coefficient (sliding velocity $v = 0.5\,\text{cm/s}$) and the normal load for the three rubber surfaces sliding on (a) the aluminum surface and (b) the Teflon surface. Note that, because of the molecular adhesion, $\mu \to \infty$ as $L \to 0$, and that the friction coefficient decreases with increasing load, becoming constant at a high-enough load. [Note: the constant friction coefficient at "high" load implies that the area of real contact increases proportional to the load for high load. This is also observed in the measurements by Grosch presented earlier. It implies that the substrate and/or the rubber surface have long-wavelength

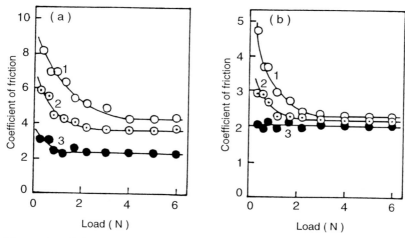

Fig. 5.34. Effect of the surface free energy of rubber on the relation between the friction coefficient and load for (**a**) an aluminum substrate and (**b**) a Teflon substrate. From Ref. [5.26].

Fig. 5.35. Rubber sliding on a hard substrate having short-ranged and long-ranged surface roughness. Because of the adhesional interaction the rubber deforms and fill out the "cavities" associated with the short-ranged surface roughness.

roughness; the argument due to Greenwood presented in Sect. 5.1 then shows that the area of real contact is proportional to the load. Note, however, that adhesion is very important in each asperity contact area associated with the long-wavelength roughness, as it will force the rubber to fill out all the small-sized (say $D < 1000\,\text{Å}$) "cavities" of the substrate in each such contact region, see Fig. 5.35.] The load at which the coefficient of friction becomes constant, increases with increasing surface free energy. Note also that, as expected, the coefficient of friction increases with increasing surface free energy of the rubber. This shows that the adhesional interaction contribute in an important manner by locally deforming the rubber in each junction as indicated in Fig. 5.35. For the Teflon substrate the adhesion to rubber is smaller than with the aluminum substrate. In particular, for the low energy rubber surface **3** the coefficient of friction for the Teflon substrate did not change with load above 0.2N.

Finally, we note that because of its low elastic modulus, rubber often exhibit elastic instabilities during sliding. The most well-known involves the compressed rubber surface in front of the contact area undergoing a buckling which produces detachment waves which propagate from the front-end to the back-end of the contact area. These so called Schallamach waves were first discovered 1971 and have been intensively studied [5.27].

Sliding Friction of Polymers

We have seen that when rubber is slid on a hard substrate, with a surface roughness characterized by a (single) length scale l, very high friction occur when $\tau \approx \tau^*$, where $1/\tau^* = v/l$ is the frequency of the fluctuating forces the substrate asperities exert on the rubber, and $1/\tau$ is the frequency for which the internal friction of the rubber is maximal. Most other polymers exhibit much lower sliding friction than rubber; one extreme case is Teflon for which the static and kinetic friction coefficients typically are smaller than 0.1. This implies that for most polymers the internal friction gives a much smaller contribution to the sliding friction than for rubber. The physical origin of this can be understood as follows.

Note first that $1/\tau$ is the rate by which some (specific) molecular segments in the polymer switches between different configurations, say between two states A and B, separated by an energy barrier of hight ϵ. Thus, $\tau \sim \exp(\epsilon/k_B T)$ depends exponentially (or faster, see Sect. 7.2) on the inverse of the temperature. Now, for most polymers the activation energy ϵ is so large that, at room temperature, τ is so large that for typical sliding velocities $\tau^* \ll \tau$ and the internal friction can be neglected. Note, however, that at high enough temperature one may have $\tau^* \approx \tau$, in which case the internal friction is important.

A measure of the activation energy ϵ is the glass transition temperature T_g, which is defined as the temperature where the viscosity of a fluid or glassy solid equals 10^{12} kg/ms (this is the highest viscosity which can be easily measured experimentally) (Sect. 7.2). For most types of rubber, $T_g \approx 200$ K. On the other hand, for many other (glassy) polymers $T_g \approx 350$ K, and in these cases the internal friction gives a negligible contribution to the sliding friction under typical sliding conditions. In this context we note that the main difference between summer tires and winter tires is that the glass transition temperature for the rubber used for winter tires is lower than that for summer tires, in such a way that the characteristic time τ for the rubber used for winter tires, at typical winter temperatures, say $\sim -10°$C, will be of similar magnitude as for the rubber used for summer tires, at typical summer temperatures, say $\sim 20°$C.

When the contribution from the internal friction is small, the friction force will be determined mainly by an interfacial process, in most cases probably involving the shearing of a molecular thin organic contamination layer (this process may also be the origin of the flat "background" contribution to the friction coefficient seen in Fig. 5.31 away from the resonance peak). Indeed, the available experimental evidence on the frictional behaviour of most polymers other than rubber indicate that the superposition principle discussed above (which implicitly assume that the sliding friction is dominated by the internal friction), is not applicable to these polymers.

Since most polymers are soft, it is unlikely that the organic contamination layer will be squeezed out by the relative low pressure which occur in the contact areas. The friction coefficient when a steel ball is slid on most polymers at "low" sliding velocities (to avoid heating effects) is in the range 0.3 ± 0.1 and probably reflect shearing of the contamination layer. At "high" sliding speeds, because of the low thermal conductivity and low melting point of most polymers, melting of the polymer usually occur, resulting in a rapid drop in the friction force with increasing sliding velocity (just as in frictional melting of ice, see Sect. 14.2), and to large wear. The spectacular low friction of Teflon is probably related to its low surface energy (the Teflon surface is not wetted by most organic liquids, see Sect. 7.5).

Rubber Wear

Let us briefly discuss wear in the context of rubber friction. When a block of rubber is exposed to low-frequency shear stresses at room temperature, the rubber response is elastic, and there is likely to be very small wear. However, strong wear may occur at low temperatures, or at very high frequencies, where the rubber behaves as a glassy brittle material. Thus, at low temperature, rubber can fracture by crack propagation (recall the Challenger catastrophe) as there is no time for the rubber molecules to deform elastically by thermal excitation over the energy barriers [5.29]. Similarly, at high enough frequency the rubber will respond in a glassy brittle manner even at room temperature. Now, when a rubber block is sliding on a rough surface with roughness on many different length scales (fractal surface), the very small surface asperities will generate very high-frequency pulsating forces on the rubber surface: $\omega_l \sim v/l$, where l is the linear size of a contact area. At room temperatures most types of rubber will be in the glassy state when $\omega > 10^8 - 10^9$ s^{-1}. If we consider the sliding velocity $v \sim 10$ m/s (typical for a wheel during braking on a road), this gives brittle or glassy response for $l < 100-1000$ Å. Thus one expects high wear in this case, with rubber particles of linear sizes $\sim 100 - 1000$ Å being removed (by brittle fracture) during braking, see Fig. 5.36. This estimate assumes that the rubber surface does not heat up to any great extent as a result of the frictional energy dissipation. This assumption may only hold at the initial phase during emergency braking (locked wheels): for sliding at high velocity over a "long" time period the rubber surface temperature may become so high that the rubber does not behave as a glassy solid even when exposed to fluctuating stresses in the frequency range $\omega_l \sim 10^9 - 10^{10}$ s^{-1}. Based on the *Williams–Landel–Ferry equation*, which state that the frequency and temperature dependence of the complex elastic modulus of rubber (and of many other glassy solids) can be (approximately) related via $E(\omega, T) \approx E(\omega a_T)$, where

$$a_T \approx \exp\left[-8.86(T - T_g - 50)/(51.5 + T - T_g)\right],$$

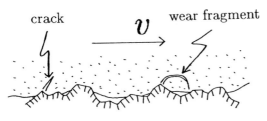

Fig. 5.36. Wear of rubber at high sliding velocity v on a hard, rough substrate. The wear results from the small-sized surface asperities which generate pulsating forces with frequencies in the glassy region of the rubber loss spectra.

one can estimate that a temperature increase from 300 K (room temperature) to 400 K gives approximately an order of magnitude shift of the glassy region to higher frequencies.

The situation is different for slowly sliding rubbers. Thus, when $v < 1$ cm/s, as is the case for the extremely important application to ABS (Anti-Lock Braking Systems) of automotive tires on dry or wet road surfaces (where $v \sim 0.01 - 1$ cm/s in the incipient part of the foot-print area), then the rubber will be able to deform and fill out the nano-scale cavities associated with the short-ranged surface roughness. This leads to an increased friction coefficient but small wear, since the relevant frequencies v/l are well below those corresponding to the glassy brittle region of most rubbers.

The discussion above is in agreement with experimental data. Figure 5.37 shows the temperature dependence of the wear and the friction when rubber is slid over a hard rough substrate at a fixed velocity. Note first that the friction is small at high temperature where, at the relevant frequencies $\omega_l = v/l$ associated with the asperity–rubber interaction, the rubber is in the rubbery region of the viscoelastic response and where the loss function $\mathrm{Im}E(\omega_l)/\mid E(\omega_l) \mid$ is small. When the temperature is reduced, the friction initially increases and reaches a maximum when the perturbing frequencies ω_l are located in the transition region between the rubbery region and the glassy region, where $\mathrm{Im}E(\omega_l)/\mid E(\omega_l) \mid$ is maximal. Finally, when the temperature is reduced below about $-40°$C, the friction coefficient decreases as ω_l now moves into the glassy region where $\mathrm{Im}E(\omega_l)/\mid E(\omega_l) \mid$ is small. Thus, from the discussion above we expect the wear to increase when T is reduced below $-40°$C as is indeed observed experimentally (see Fig. 5.37). Note, however, that the wear also increases at high temperatures. This can be understood as follows. During sliding the asperities exert (oscillating) shear stresses on the rubber, but in the rubbery region this stress by itself is usually not large enough to break the strong covalent bonds (e.g., the sulfur bonds) in the rubber. However, at high enough temperatures thermal fluctuations may break a bond, in particular if it is already stretched as a result of the external stress.

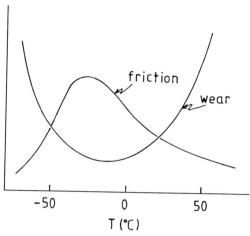

Fig. 5.37. The temperature dependence of the wear and the friction coefficient for rubber sliding on a rough, hard substrate at a fixed velocity (schematic). The figure is based on experimental results from [5.30].

This stress-aided thermally activated bond-breaking will give rise to a rapid increase in the wear at high temperatures. Thus, we expect the temperature dependence of the wear to have the U-shaped form shown in Fig. 5.37. Note also that the wear is minimal close to the point where the friction is maximal. Thus, there is no simple relation between wear and friction and, in particular, a large friction coefficient does not necessarily imply a large wear.

Other wear processes discussed in the literature involve the influence of sunlight and oxidation on the rubber surface, leading to a thin, hardened and brittle surface layer which may be removed relatively easily by the stresses the rubber surface is exposed to during its practical use.

The wear rate during rolling of an automotive tire on a road is about $1 \, \mathrm{mm}/10^4 \, \mathrm{km}$, which corresponds to $\sim 1 \, \mathrm{Å/rotation}$, or $10^5 \, \mathrm{Å/h}$ during steady driving at $100 \, \mathrm{km/h}$. This wear rate is a factor $\sim 10^4$ higher than the wear of the engine cylinders which is of the order $10 \, \mathrm{Å/h}$ or less in a good engine.

6. Sliding on Clean (Dry) Surfaces

Around 1940 Bowden and Tabor presented a simple theory for the origin of the sliding friction for *clean* surfaces [6.1]. They assumed that the friction force is the force required to shear cold-welded junctions formed between the solids.

We have seen that macroscopic bodies always have rough surfaces, at least on a microscopic level, and if one places two solid materials in contact, some regions on their surfaces will be so close together that the surface atoms of one material "touch" the surface atoms of the other material, while in other regions, the surface atoms are separated by large distances, e.g. $\sim 1\,\mu\mathrm{m}$ (Fig. 6.1). The regions of contact are referred to as *junctions*, and the sum of all the junctions is called the area of real contact ΔA. The rest of the apparent area of contact is usually much larger than the real area of contact, but plays essentially no part in determining the sliding friction. In particular, the long range van der Waals interaction gives a negligible contribution to the friction force, see Sect. 14.5.

We have also shown that the real area of contact in most practical cases can be estimated accurately by assuming that plastic deformation has occurred at each junction and that all the junctions are in a state of incipient plastic flow. This assumption gives $\Delta A = L/\sigma_\mathrm{c}$ where L is the load and σ_c (the penetration hardness) the largest compressive stress that the materials can bear without plastic yielding.

Coulomb's friction "law" (2.1) is now easy to understand: The force F necessary to shear the junctions with the total area ΔA equals $F = \tau_\mathrm{c} \Delta A$

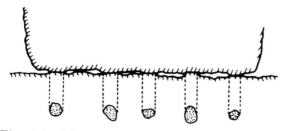

Fig. 6.1. Schematic illustration of an interface, showing the apparent and real areas of contact.

94   6. Sliding on Clean (Dry) Surfaces

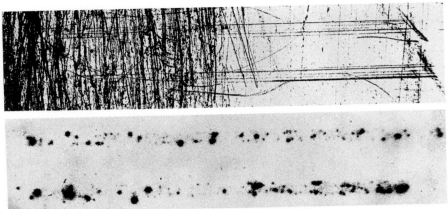

Fig. 6.2. Micrograph (*top*) and autoradiograph (*bottom*) of a copper surface, part of which has a roughness of 250 Å, the other part 5000 Å, after a radioactive copper block had been slid over it. Load 4 kg and speed 0.01 cm/s. The friction force and the wear are independent of the surface roughness. From [6.2].

where τ_c is the yield stress during shear. But $\Delta A = L/\sigma_c$ so that $F = (\tau_c/\sigma_c)L$, i.e., the friction coefficient $\mu = \tau_c/\sigma_c$.

This "derivation" not only explains why the friction force is proportional to the load but also why it is independent of the surface area A and of the nature of the surface roughness (as long as the surfaces are not too rough or too smooth). Furthermore it explains why typically $\mu \sim 1$ for the sliding of clean surfaces; this follows from the fact that the τ_c and σ_c are of similar magnitude.

Let us discuss the nature of the wear induced by the sliding of clean metal surfaces (reducing the wear is even more important in engineering applications than reducing the sliding friction but the two effects are related). During sliding of clean (unlubricated) metal surfaces, the oxide film may be locally worn away and direct metal to metal contact can occur. The surface asperities weld together momentarily and are broken apart again to produce wear debris largely composed of metallic particles. This is called adhesive wear and as discussed above it may be the main source of the friction force necessary to slide *clean* metal surfaces. This wear process has been studied experimentally using radiotracers: A radioactive material is slid over a non-radioactive surface, and the amount of radioactivity transferred to the non-radioactive surface is measured, e.g., using a piece of photographic film in contact with the surface to give the distribution of transferred particles. As an example, Fig. 6.2 shows both a micrograph (top) and the autoradiography (bottom) of a copper surface, part of which has a roughness amplitude of ~ 250 Å, the other part ~ 5000 Å, after a radioactive copper block had been slid over it. Note that the wear (and the friction force) is independent of the amplitude of the surface roughness.

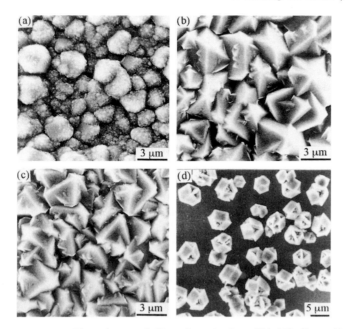

Fig. 6.3. Four diamond films deposited on Si(100). From [6.3].

For lubricated surfaces, the friction force is usually *not* determined by the mechanism discussed above, since it is known from radioactive tracer experiments of the type discussed above, that while a boundary lubricant may reduce the sliding friction by a factor of ~ 10, it may reduce the metallic transfer by a factor of $\sim 10^4$ or more [6.2]. Under these conditions the metallic junctions contribute very little to the friction force; the friction is due almost entirely to the force required to shear the lubricant film itself. Nevertheless, some wear still occurs, and this may finally lead to the failure of the sliding device. The formation of cold-welded junctions can be reduced, however, if at least one of the metals is replaced by a more inert material; one such example was given in Chap. 4 involving ball bearings with ceramic balls. Alternatively, metals can be coated by hard inert layers, e.g., titanium carbide (TiC) or nitride (TiN) or by diamond-like carbon layers, as shown in Fig. 6.3. However, unless these layers are very smooth, large wear and a high friction may result when "sharp" tips of the hard material dig in and cut the surface of the softer counterface material. Let us illustrate this with the example of diamond-like carbon films.

A Case Study: Diamond-like Carbon Overcoats

Diamond is the hardest material known to man (indentation hardness $\sim 1 \times 10^{11} \, \text{N/m}^2$), followed by cubic boron nitride (hardness $\sim 0.75 \times 10^{11} \, \text{N/m}^2$).

In diamond, every carbon atom is surrounded by 4 nearest-neighbor carbon atoms, located at tetrahedral sites relative to the carbon atom at the origin. The C–C bond results from the overlap of sp^3-hybridized atomic orbitals, which point towards the nearby carbon atoms. This gives rise to strong bonding in all three spatial directions, which is necessary for a high yield strength and indentation hardness. This should be contrasted with graphite, which is a layered material with very strong bonds between the carbon atoms within the planes (these bonds are, in fact, even stronger than those in diamond) but with very weak interactions, mainly of the van der Waals type, between the planes. As a result, the graphite planes can easily slide over one another, resulting in a low yield strength. Each carbon atom within a graphite plane is surrounded by three nearest-neighbor carbon atoms; the C–C bonds result from the overlap of sp^2-hybridized atomic orbitals of neighboring carbon atoms. Note that while diamond surfaces are usually "passivated" by hydrogen atoms, which bind to the surface dangling bonds, the clean graphite surface is already very inert and foreign atoms and molecules, if present, are mainly physisorbed.

Diamond coatings are attractive for many applications because of their extreme hardness and chemical inertness. Diamond or diamond-like carbon films can be grown on various substrates by, e.g., chemical vapor deposition or sputtering. The films have different physical properties depending on the deposition techniques and deposition conditions. Figure 6.3 shows four diamond films grown on a silicon substrate [6.3]. Note the large surface roughness, which will generate a large wear if a softer body is slid on the film surface. Furthermore, in contrast to most other engineering surfaces (see above), the friction coefficient depends on the surface roughness of the films [6.4], with the smoother films approaching the reported value for natural diamond (e.g., ~ 0.04 when silicon nitride is slid on diamond). To illustrate this, Fig. 6.4 shows the variation of the friction coefficient when 1 cm diameter silicon nitride (Si_3N_4) balls are slid on two silicon surfaces coated by diamond films with different roughnesses. Films A and B have the rms roughness 410 Å and 1250 Å, respectively. The average friction of film B is about 5 times higher than that for film A, and the wear rate (not shown) is about 9 times higher. The large fluctuations in the friction traces and the very high initial friction coefficients of rough diamond films are probably due to ploughing: sharp asperity tips of diamond crystals can dig in and cut the surface of softer counterface material, thus causing severe abrasion and ploughing. During successive sliding passes, the sharp asperity tips are progressively rounded and the valleys between asperities are filled with blanket and wear debris particles. This eventually results in a somewhat lower friction coefficient, i.e., 0.2 on the rough diamond film B.

Amorphous carbon films have the advantage over the crystalline diamond films studied above in that they have smoother surfaces, leading to lower friction and wear [6.5–8]. These films consist of a three-dimensional network of

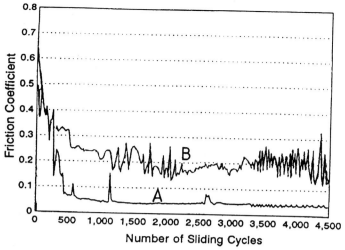

Fig. 6.4. Variation of the friction coefficient of two diamond films, A and B, during sliding against Si_3N_4 balls under a 2N load. From [6.4].

sp^3- and sp^2-bonded carbon atoms. Owing to their different bonding configurations, amorphous carbon films can have a wide range of hardness values, as well as varied friction and wear properties. Amorphous carbon films, with a thickness of a few hundred Angstroms, are widely used as overcoats on magnetic recording disks, protecting against wear even during severe sliding conditions.

A Case Study: Sliding of Metallic Surfaces with Thin Adsorbate Layers

How thick must a protective layer (say, an oxide coating) be in order to effectively reduce the formation of cold-welded junctions and hence reduce friction and wear for metals? It is plausible that a monolayer (or less) of atomic adsorbates, e.g., oxygen or sulfur, on a metallic surface will not reduce the formation of cold-welded junctions to any appreciate level, as the conduction electron wave functions "leak-out" on the vacuum side beyond the O or S atoms, leading to strong adhesional forces when the surfaces are brought in contact. This conclusion is supported by experiments on single-crystal surfaces covered by sub-monolayers of O, S, Cl and I, prepared and studied in ultrahigh vacuum in order to avoid the formation of oxide and organic contamination layers, which form rapidly in the normal atmosphere [6.9]. As an example, Fig. 6.5 shows the variation of the static friction coefficient with the sulfur coverage on Cu(111), when another copper crystal [with a slightly curved (radius ~ 2.5 cm) (111) surface] with the same sulfur coverage, is slid on it. The friction coefficient was found to be independent of the load, temperature and sliding velocity. Similar results were obtained for Cl and I on

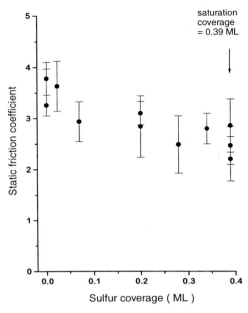

Fig. 6.5. The static friction coefficient as a function of sulfur coverage on Cu(111). Load $L \approx 40$ mN; sliding velocity $v \approx 20$ μm/s; temperature $T = 300$ K. From [6.9].

Cu(111). We conclude that, even at saturation coverage, these adsorbates have only a relative small affect on the sliding friction.

At this point we note that most lubrication fluids contain compounds known as extreme pressure additives, which are usually sulfur, phosphorus or chlorine containing organic compounds. Such additives protect surfaces from damage during sliding at a high normal force. The reaction of the additives with exposed metal surfaces result in the formation of metal sulfides, phosphides or chlorides, which are stable to high temperatures and protect the underlying metal surfaces from coming in direct contact. The thickness of the films is typically ~ 100 Å.

In the studies in [6.10], the friction coefficient at saturation coverage was reduced by $\sim 30\%$ by sulfur, $\sim 40\%$ by chlorine and not at all by iodine adsorption. The clean copper crystals showed strong adhesion, which decreased monotonically with increasing adsorbate coverage. Thus, with sulfur the adhesion coefficient (defined as the ratio between the pull-off force measured upon separation of the samples and the applied normal force) decreased monotonically from 0.48 ± 0.13 to 0.22 ± 0.12 at a saturation coverage of sulfur. The adhesive forces after sliding has occurred are much higher. In contrast, surfaces that are well-lubricated with multilayers of adsorbed alcohols or with thick contaminant films from exposure to the atmosphere exhibit negligible adhesive forces before or after sliding.

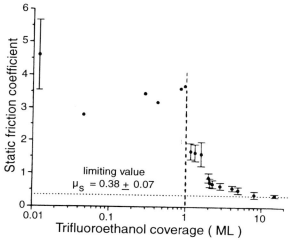

Fig. 6.6. The static friction coefficient as a function of trifluoroethanol coverage on Cu(111). Load $L \approx 40\,\text{mN}$; sliding velocity $v \approx 1\,\mu\text{m/s}$; temperature $T = 120\,\text{K}$. From [6.11].

In another experiment the Cu(111) surfaces where exposed to the ambient laboratory atmosphere for a few day [6.11]. This resulted in poorly defined carbon- and sulfur-containing films of thickness $\sim 20\,\text{Å}$. These contaminant films reduced the static friction coefficient from $\mu_s \approx 4.4$ (clean surfaces) to $\mu_s \approx 0.7$. When the contamination layer was gradually removed, a transition to higher friction occurred at a contaminant film thickness of 5–8 Å.

Atomic adsorbates such as oxygen or sulfur, bind strongly to most metallic surfaces and sit relatively "deep" into the metallic substrate; as a result the conduction electron wave functions extend beyond the atomic adsorbates on the vacuum side of the crystal. This is not the case when large molecules, e.g. hydrocarbons, bind to the surfaces. However, the adsorbate-substrate bond and, more important, the lateral corrugation of the adsorbate-substrate interaction potential is often relatively small for molecular adsorbates. As a result, the high pressure in the contact regions between the two solids may squeeze out the adsorbates from the contact areas, leading to a large wear and a high friction. We will discuss this topic in Sect. 7.4 but here we present experimental data which illustrate this effect. Figure 6.6 shows the static friction coefficient μ_s when a Cu(111) surface is slid on Cu(111), where both surfaces are covered by trifluoroethanol (CF_3CH_2OH) [6.12]. Note that below monolayer coverage the molecular adsorbates have *no influence* on the static-friction coefficient, while a sharp drop in μ_s occurs at monolayer coverage. This can be explained by assuming that below monolayer coverage the lubrication film has "holes" where the squeeze out process (layering transition) can start (Sect. 7.4). In another case a Ni(100) surface with a monolayer of S was covered by molecular films of ethanol ranging in thickness from submonolayer to multilayer [6.12]. The sulfided Ni(100) surface exhibited high sliding friction ($\mu \approx 3$) which was *continuously* reduced by adsorption of ethanol even at submonolayer coverages. This indicates that the adsorbate-substrate binding energy for the ethanol molecules on the sulfided Ni(100) surface exhibits a lateral corrugation larger than for trifluoroethanol on Cu(111), so that it takes a much longer time to squeeze out the molecules from the contact areas.

7. Sliding on Lubricated Surfaces

A lubricant is used to lower the friction and reduce the wear between two sliding solid bodies. Here we consider only liquid lubricants although some solids, e.g., graphite and molybdenum disulfide, are also used as lubricants. As pointed out before, almost all surfaces are normally covered by a layer of grease, which may have a similar influence on the sliding friction as an added lubricant.

Why is an oil usually a better lubricant than water? We all know, of course, that a drop of water between clean hands does not make the hands slip easily over each other while a drop of oil or some other high viscosity fluid does [remember trying to pick up the soap from the water when taking a bath; wet soap is covered by a layer of a high viscosity fluid (soap-water mixture)]. At first this may seem a paradox: high viscosity means large energy dissipation during shearing of the fluid and a large friction force.

A large viscosity has, however, a second effect. Suppose we push together two solid surfaces separated by a layer of fluid. The fluid will be squeezed out from the contact areas between the two solids, but this process will take a long time if the viscosity is high. Now, if the two solids are in relative lateral motion (a block sliding on a substrate) and if the velocity is high enough, there will be insufficient time for the fluid to be squeezed out from the contact areas and the two surfaces will be separated by a relatively thick fluid film (e.g., 0.01 mm), see Fig. 7.1a. This is called *hydrodynamic* or *fluid* lubrication where the sliding friction is very low. However, if the sliding velocity is small enough, or if the viscosity of the fluid is low enough, the fluid will be squeezed out leading to direct contact between the surfaces. This is the region of *boundary lubrication* where the sliding friction is much higher (typically by a factor of 100), and where the friction force is nearly independent of velocity. Figure 7.1b shows schematically the variation of the friction force with the parameter $\mu v/P$, where μ is the viscosity, v the sliding velocity and P the pressure with which the two surfaces are pushed together. Note that in hydrodynamic lubrication the friction force *increases* with increasing sliding velocity and increasing viscosity so that the viscosity of a fluid to be used in an engine, for example, should be large enough to guarantee that hydrodynamic lubrication occurs when the engine is run at its normal speed, but not larger, in order to minimize friction. During starting and stopping of an

102   7. Sliding on Lubricated Surfaces

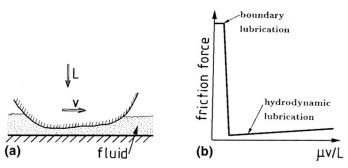

Fig. 7.1. (a) A solid body sliding on a lubricated surface. (b) The relation between the friction force and the sliding velocity. The pressure is $P = L/A$, where L is the load and A the contact area.

engine, boundary lubrication will always occur and this must be taken into account when designing lubricants (commercial lubrication oils have additives to improve their boundary lubrication properties).

Hydrodynamic or fluid lubrication is quite well understood and the friction forces can be calculated from the Navier–Stokes equations of fluid dynamics [7.1]. In contrast, boundary lubrication is not well understood. Experiments have indicated that in boundary lubrication at most a few monolayers of lubrication molecules are present between the sliding surfaces, but these have a profound effect on the sliding process. While the viscosity is the most important parameter in fluid lubrication, this parameter is completely irrelevant for boundary lubrication, where the nature of the direct interaction between the solid surfaces and the lubrication molecules is of greatest importance.

7.1 Hydrodynamic Lubrication

The principles of hydrodynamic lubrication can be understood from a simple calculation. Consider the sliding configuration in Fig. 7.1a in a reference frame where the substrate moves with the velocity $-v_0$ along the x-axis (Fig. 7.2). In most practical cases the lubrication fluid can be considered as incompressible. Hence, the continuity and the Navier–Stokes equations have the form

$$\nabla \cdot \boldsymbol{v} = 0 , \qquad (7.1)$$

$$\frac{\partial \boldsymbol{v}}{\partial t} + \boldsymbol{v} \cdot \nabla \boldsymbol{v} = -\frac{1}{\rho}\nabla P + \nu \nabla^2 \boldsymbol{v} , \qquad (7.2)$$

where ρ is the mass density and $\nu = \mu/\rho$ the kinematic viscosity of the fluid. (Note: in this section μ denotes the viscosity and f the friction coefficient.)

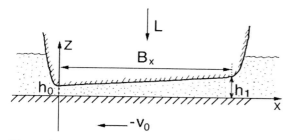

Fig. 7.2. A sliding junction in a reference frame where the substrate moves with the velocity $-v_0$ along the x-axis while the "block" is stationary.

If we measure velocity in units of v_0, length in unit of the film thickness h, time in units of h/v_0, and pressure in units of ρv_0^2, then (7.2) takes the form

$$\frac{\partial \boldsymbol{v}}{\partial t} + \boldsymbol{v} \cdot \nabla \boldsymbol{v} = -\nabla P + \frac{1}{R}\nabla^2 \boldsymbol{v}, \tag{7.3}$$

where the Reynolds number $R = v_0 h/\nu$. Laminar flow occurs if R is below a critical value R_c, while turbulent flow occur if $R > R_c$. The critical Reynolds number R_c depends on the geometry of the sliding junction, but in most cases relevant to hydrodynamic lubrication $R_c \sim 1000$. In a typical case $v_0 \sim 1 \text{m/s}$, $h \sim 10^{-5}\text{m}$ and $\nu \sim 10^{-3}\text{m}^2/\text{s}$ giving $R \sim 0.01 \ll R_c$. Hence the flow is usually laminar and during steady sliding the $\partial \boldsymbol{v}/\partial t$ term in (7.2) and (7.3) vanishes. Next, note that the ratio between the nonlinear term $\boldsymbol{v} \cdot \nabla \boldsymbol{v}$ and the viscosity term $R^{-1}\nabla^2 \boldsymbol{v}$ in (7.3) is of order $Rh/B_x \ll 1$, where $B_x \gg h$ is the width of the contact area (Fig. 7.2), and the nonlinear term can be neglected. Hence, (7.2) reduces to

$$\nabla P = \mu \nabla^2 \boldsymbol{v}. \tag{7.4}$$

If we neglect the side leakage (in the y-direction), we can approximate

$$\boldsymbol{v} \approx \hat{x} v(x, z). \tag{7.5}$$

Substituting (7.5) into (7.4) gives

$$\frac{\partial P}{\partial x} = \mu \nabla^2 v, \tag{7.6}$$

$$\frac{\partial P}{\partial z} = 0. \tag{7.7}$$

From (7.7), $P = P(x)$. Furthermore, since $B_x \gg h$,

$$\frac{\partial^2 v}{\partial z^2} \sim \frac{v_0}{h^2} \gg \frac{\partial^2 v}{\partial x^2} \sim \frac{v_0}{B_x^2}$$

and (7.6) reduces to

which can be integrated to give

$$v = P'z(z-h)/2\mu + v_0(z/h - 1) ,\tag{7.8}$$

where we have determined the two integration constants from the boundary conditions $v = -v_0$ for $z = 0$ and $v = 0$ for $z = h(x)$. Now, because of the approximations made above it is not possible to satisfy the continuity equation (7.1) locally. But we can determine the pressure distribution $P(x)$ so that the same amount of fluid flows per unit time through each cross-sectional area (normal to the direction x) of the sliding junction. Hence the quantity

$$Q = \int_0^h dz\, v_x \tag{7.9}$$

must be independent of x. From (7.8) and (7.9),

$$\frac{dP}{dx} = -6\mu v_0 \left(\frac{1}{h^2} + \frac{C}{h^3} \right) ,\tag{7.10}$$

where $C = 2Q/v_0$. Now, assume that

$$h = h_0 + ax ,$$

where $a = (h_1 - h_0)/B_x$ (Fig. 7.2). In this case it is trivial to integrate (7.10). The integration constant and the constant C in (7.10) are determined by the boundary conditions $P = P_{\text{ext}}$ for $x = 0$ and for $x = B_x$, where P_{ext} is the pressure in the fluid outside the contact area. This gives

$$P = P_{\text{ext}} + \frac{6\mu v_0}{a} \left[\left(\frac{1}{h} - \frac{1}{h_0} \right) - \frac{h_0 h_1}{h_0 + h_1} \left(\frac{1}{h^2} - \frac{1}{h_0^2} \right) \right] .\tag{7.11}$$

The load L on the "upper" body must equal the pressure difference $P - P_{\text{ext}}$ integrated over the contact area $A = B_x B_y$. This gives

$$L = \int_A dx\, dy (P - P_{\text{ext}}) = \frac{\mu A B_x v_0}{h_0^2} \alpha ,\tag{7.12}$$

where

$$\alpha = \frac{6}{(\xi - 1)^2} \left[\ln \xi - \frac{2(\xi - 1)}{\xi + 1} \right] \tag{7.13}$$

and $\xi = h_1/h_0$. The friction force F is obtained by integrating the tangential stress σ_{xz} exerted by the fluid on the surface $z = 0$ over the contact area A. Since $v_z \approx 0$ we have

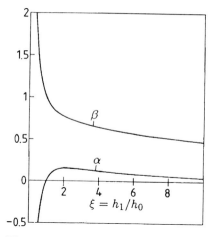

Fig. 7.3. The "pressure" and "friction" parameters α and β are shown as a function of $\xi = h_1/h_0$.

$$\sigma_{xz} = \mu \frac{\partial v_x}{\partial z}$$

and

$$F = \mu \int_A dx\, dy\, \frac{\partial v_x}{\partial z}(z=0) \,. \tag{7.14}$$

Using (7.8), (7.11), and (7.14) gives the friction force

$$F = \frac{\mu A v_0}{h_0} \beta \,, \tag{7.15}$$

where

$$\beta = \frac{1}{\xi - 1}\left[4\ln\xi - \frac{6(\xi-1)}{\xi+1}\right] \,. \tag{7.16}$$

The parameters α and β are shown in Fig. 7.3 as a function of $\xi = h_1/h_0$. If we define a friction coefficient f in the usual way, $f = F/L$, then (7.12) and (7.15) give

$$f = (h_0/B_x)(\beta/\alpha) \,. \tag{7.17}$$

Since typically $h_0/B_x \sim 10^{-4}$ and $\beta/\alpha = 10$, one obtains $f \sim 10^{-3}$, which is about a factor of 100 smaller than the the typical friction coefficient in boundary lubrication. Let us define the load per unit contact area $P = L/A$. Figure 7.4a shows the variation of h_0 with the parameter $\mu v_0/P$ as given by (7.12). For a given external load P, $h_0 \to 0$ as $v_0 \to 0$. But the theory discussed above breaks down when the separation h_0 between the surfaces

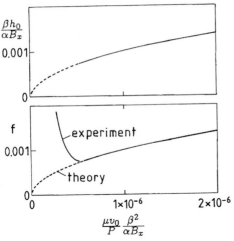

Fig. 7.4. The film thickness h_0 (in units of $\alpha B_x/\beta$) and the friction coefficient f as a function of $\mu v_0/P$ (in units of $\alpha B_x/\beta^2$). The dashed line is the result of hydrodynamic lubrication theory for perfectly flat surfaces. The solid curves are experimental observations. Below a "critical" value of $\mu v_0/P$ direct solid–solid contact occurs and the sliding friction increases rapidly with further decrease of $\mu v_0/P$. Simultaneously, plastic deformation of surface asperities occurs leading to a continuously increasing solid–solid contact area.

becomes too small. For perfectly smooth surfaces the break-down may not occur until $h_0 \sim 100$ Å or so (Sect. 7.4) but most engineering surfaces are rough on a mesoscopic scale. Assuming a surface roughness of about 1 µm implies that the theory breaks down when $h_0 \sim 1$ µm. Hence, while the theoretical hydrodynamic friction coefficient f has the dependence on $\mu v_0/P$ indicated by the dashed line in Fig. 7.4b, the observed friction has the form indicated by the solid line.

When the sliding velocity v_0 is low enough, direct solid–solid contact occurs and the friction increases rapidly with decreasing v_0. While the friction in the hydrodynamic regime depends on the lubrication fluid only via its viscosity, in the boundary lubrication regime this parameter is irrelevant and the sliding friction depends on the nature of the direct interaction between the solid walls and the lubrication molecules. This is clearly seen if the friction coefficient f is plotted as a function of $\mu v_0/P$. In the hydrodynamic region different lubrication fluids have the same value of f for the same value of $\mu v_0/P$. This is illustrated in Fig. 7.5 for mineral and lard oils (which have very different viscosities). But when v_0 has decreased so that direct solid–solid contact occurs, f depends on the nature of the detailed interaction between the oil molecules and the solid walls, e.g., on the lateral corrugation of the oil–wall interaction potential, and f is no longer a universal function of $\mu v_0/P$ (Fig. 7.5).

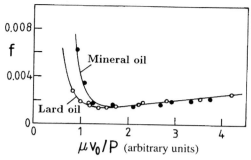

Fig. 7.5. The friction coefficient f as a function of the parameter $\mu v_0/P$ for two different lubrication oils. In the hydrodynamic sliding regime both oils fall on a universal curve. When the parameter $\mu v_0/P$ is small enough direct solid–solid contact occurs and the sliding friction increases rapidly and in a non-universal manner. The experimental data indicate that lard oil is a better boundary lubricant than mineral oil, as is indeed the case [7.1].

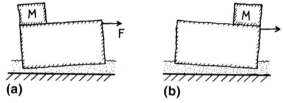

Fig. 7.6. (a) Stable sliding motion. The fluid pressure between the block and the substrate is larger than the external pressure and balances the load of the block. (b) Unstable sliding motion. The fluid pressure between the block and the substrate is below the external pressure and the two surfaces will be sucked together leading to boundary lubrication and a large friction force.

Note that for $\xi < 1$, i.e., $h_1 < h_0$, the parameter $\alpha < 0$ which implies that $P - P_{\text{ext}}$ is *negative*. In this case the system is unstable and the two surfaces will be "sucked" together during sliding, which results in boundary lubrication and a large friction force. This fact is easily observed experimentally: if a block is pulled on a lubricated surface under the influence of a perpendicular force (the "load") acting at the rear end of the block, then (if the sliding velocity is high enough to guarantee that hydrodynamic lubrication occurs) $h_1 > h_0$ as in Fig. 7.6a and the sliding friction is very low. On the other hand, if the external load acts at the front end, $h_1 < h_0$ giving $P < P_{\text{ext}}$, and the surfaces are sucked together, leading to boundary lubrication and a high sliding friction (Fig. 7.6b).

The great advantage of hydrodynamic lubrication is that there is, in the ideal case, no wear of the moving parts and the friction is extremely low. The resistance to motion arises solely from the viscosity of the lubricant. Clearly

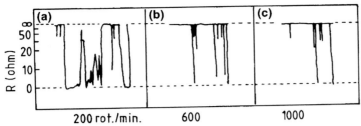

Fig. 7.7. Cathode ray traces of electrical resistance across a piston ring and cylinder wall of a running engine. (**a, b**) and (**c**) show the results for 200, 600, and 1000 rotations per minute, respectively. From [7.13].

the lower the viscosity the lower the viscous resistance. But, as pointed out above, there is a limit to this for, as the viscosity is decreased, the distance of nearest approach diminishes. If this becomes less than the height of the surface irregularities penetration of the lubrication film occurs. For this reason it is customary to use an oil of viscosity just sufficient to give a distance of nearest approach great enough to ensure the maintenance of an unbroken lubrication film.

The transition from hydrodynamic lubrication to boundary lubrication has been studied for running engines. In one investigation a small single cylinder engine was used. One of the piston rings was electrically insulated from the cylinder. The electrical resistance across the piston ring and the cylinder wall was recorded on a cathode ray oscillograph. Some typical cathode ray traces are shown in Fig. 7.7 as a function of the speed (cycles per minute). Note that the degree of breakdown of the lubricant film (as manifested by low resistivity) is reduced at higher speed. Similar measurements shows increased breakdown of the lubrication film with decreasing viscosity of the lubrication oil or increasing temperature (which, in fact, is equivalent to decreasing viscosity).

We have emphasized above that in most practical applications laminar flow occurs. This has been proven in experiments on bearings made from glass, where the flow pattern can be studied by introducing color filaments into the lubrication fluid. However, under extreme conditions the flow may become turbulent and the simple theory outlined above is no longer valid [7.2]. Turbulent flow occurs when the Reynolds number exceeds some critical value R_c which for radial bearings is on the order of 1000. As an example, Fig. 7.8 shows the result of an experiment for a water-lubricated radial bearing under small load. The friction parameter C_F is shown as a function of the Reynolds number $R = v_0 h/\nu$, where $v_0 = r\omega$ is the velocity and h is the thickness of the lubrication film. The friction parameter is defined as

$$C_F = \frac{\sigma}{\rho v_0^2/2},$$

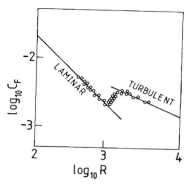

Fig. 7.8. The friction coefficient C_F as a function of the Reynolds number for a water-lubricated radial bearing [7.2].

where σ is the friction stress. Note that for $R < R_c \sim 1000$ the friction parameter $C_F \propto 1/R$ (as expected since $\sigma \propto v_0$ in the laminar region) while the friction in the turbulent region $R > R_c$ is much larger than expected for laminar flow.

Another implicit assumption in the theory above is that no slip occurs between the lubricant and the solid surfaces. This is an excellent approximation in most practical systems. Nevertheless, as will be discussed in Sect. 8.6.1, slip often occurs and may manifest itself in future applications involving very smooth surfaces and thin lubrication films.

Finally, let us discuss heating effects. The friction force F in hydrodynamic lubrication is caused by the internal friction of the oil film and results in heating of the oil. The dissipation of this heat is a complicated problem which depends on the solid bodies, the condition of the surrounding atmosphere, and other factors. For the present, let us assume that the oil film takes up all the heat generated, resulting in its temperature rising from T_1 at the inlet to $T_0 > T_1$ at the outlet. The amount of fluid flowing through the clearance space in unit time equals $QB_y = v_0 B_y h_0 \xi/(1+\xi)$, where we have used equations (7.9–11). The thermal energy transported away by the lubrication fluid per unit time is then

$$QB_y \rho C_v (T_0 - T_1),$$

where C_v is the heat capacitance of the oil. This quantity must equal the mechanical work Fv_0 so that

$$T_0 - T_1 = \frac{Fv_0}{QB_y \rho C_v} = \frac{\nu B_x v_0 \beta (1+\xi)}{h_0^2 C_v \xi}. \qquad (7.18)$$

As an example, assume that $\nu \sim 10^{-3}\,\mathrm{m^2/s}$, $B_x \sim 1\,\mathrm{cm}$, $v_0 \sim 1\,\mathrm{m/s}$, $\xi \sim \beta \sim 1$, $h_0 \sim 10^{-5}\,\mathrm{m}$ and $C_v \sim 10^3\,\mathrm{m^2/s^2K}$. In this case (7.18) gives $T_0 - T_1 \approx 100\,\mathrm{K}$.

7.2 Fluid Rheology

The viscosity of a lubrication oil is usually chosen so that when the sliding device is operated at its "normal" speed, the friction force is near the minimum in the force-velocity diagram in Fig. 7.1b. But in many applications the sliding speed is not constant, sometimes reaching very high values, where the friction, because of the viscosity of the lubrication fluid, becomes very high. But large friction also means a large heat production and the lubrication oil may heat up considerably. This in turn lowers its viscosity. Thus the friction at high sliding speeds may in fact be much lower than expected when the temperature increase of the lubrication oil is neglected. This important effect should be taken into account when designing lubrication oils for specific applications.

The viscosity of lubrication oils at *constant temperature* usually depends very little on the shear rate $\dot{\gamma}$, unless $\dot{\gamma}$ is extremely high. Fluids for which a linear relation between the shear stress σ and the shear rate $\dot{\gamma}$ holds accurately are usually said to be *Newtonian*. The reasons why many fluids σ and $\dot{\gamma}$ are linearly related over a large range of $\dot{\gamma}$-values become understandable when the molecular mechanism of internal friction is considered. The macroscopic motion of the fluid can cause only a small change in the (microscopic) statistical properties of the molecular motion when the characteristic time of the macroscopic motion (the reciprocal of the shear rate) is long compared with the characteristic time of the irregular thermal motion of the fluid molecules, which in the case of gases is given by the average time between collisions. For air at normal temperatures and pressure the average time between collisions is about 10^{-10} s and the Newtonian approximation should hold for shear rates below 10^{10} s^{-1}. However, when the shear rate becomes very high, most fluids exhibit *shear thinning*, where the effective shear viscosity decreases with increasing shear rate.

Most lubrication oils have molecular weights below 1000 so that they are relatively small organic molecules. The forces between the molecules are dominated by van der Waals interactions. The viscosity is strongly temperature dependent and usually well described by

$$\mu = \mu_0 e^{\epsilon/k_B T},$$

where the activation energy ϵ depends on the size of the molecule. Figure 7.9a shows the activation energies of linear hydrocarbons (alkanes) as a function of the chain length N. Note that ϵ saturates at $0.2 - 0.3$ eV for $N > 30$. For longer chain molecules the activation energy remains almost constant, suggesting that longer chains move by segmental motion, each segment being of order 20–30 carbons long.

For most hydrocarbon oils, the dependency of the viscosity on the hydrostatic pressure P is well approximated by

$$\mu = \mu_1 e^{\alpha P},$$

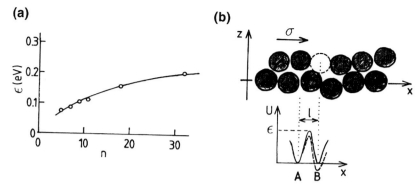

Fig. 7.9. (a) Activation energies in viscous flow of linear hydrocarbons (alkanes) as a function of the number of carbon atoms n in the chain. The activation energy reaches a steady value of $\epsilon \approx 0.23\,\text{eV}$ for a molecular length of about 30 carbon atoms. (b) The Eyring model of liquid viscosity for simple liquids. Two "layers" of molecules, parallel to the external shear plane, are shown in a reference frame where the lower layer has zero drift velocity. The upper layer is exposed to a surface stress σ from the flowing fluid above it. Thus, on the average, a fluid molecule in the upper layer experiences a tangential force $F = \sigma \delta A$, where $\delta A = A/N$, where N is the number of molecules associated with a plane and A is the area of a plane. This tangential force will lead to some drift motion of the upper layer of molecules relative to the lower layer. If the shear stress is weak, a molecule in the upper layer can move relative to the lower layer only by thermal excitation. In a semi-quantitative picture one can visualize the motion as occurring by the generation (via a thermal fluctuation) of a "cavity" of molecular size into which a nearby molecule can jump. Thus, in the figure the molecule at position A can jump to position B by overcoming a potential barrier of height ϵ: In equilibrium, jumps in both directions occur at equal frequency. When a shear stress σ is applied, the energy $l\delta A\sigma$ is gained when the atom move from position A to position B. This reduces the thermal activation energy for jumps from A to B to $\epsilon - l\delta A\sigma$ while it increases the energy for jumps from B to A to $\epsilon + l\delta A\sigma$. This provides a stress-aided thermally activated process for viscous flow.

where $\alpha \sim 10^{-8}\,\text{Pa}^{-1}$. Thus for a pressure of $1.5 \times 10^9\,\text{Pa}$, which may occur in the contact areas in, e.g., ball bearings and gears, μ may increase over a million-fold. *The harder one presses the oil the more difficult it is to extrude it.*

At very high pressures $P > 1/\alpha$, the onset of shear thinning occurs at much lower shear rates than at lower pressures. Furthermore, if the shear stress σ is changing in time, the shear rate depends not only on σ but also on $\dot{\sigma}$ as would be expected if the "fluid" were to have elastic properties. The relation between $\dot{\gamma}$ and σ under extreme pressure conditions is not well understood at present and represents a frontier field of research.

The dependency of the viscosity on the temperature and pressure described above can be understood qualitatively based on a model due to Eyring. The basic idea is simple: The liquid is envisaged as an assembly of molecules which generally have a high degree of mobility; they can change places as a result of thermal fluctuations (Fig. 7.9b). For small molecules the potential barrier ϵ that must be overcome to enable change of location to occur is about one third of the heat of vaporization. This may be considered as the energy required to open up a hole large enough for the molecule to move into. If a shear stress σ is applied to the liquid a certain amount of mechanical work is done on a molecule in the course of its movement to a neighboring site. If δA is the cross-sectional area of a molecule and the average distance it moves is l, the work done is $l\delta A\sigma$. Clearly this work favors displacements in one direction and hinders them in the opposite direction. The movement of the molecule is thus a stress-aided thermally activated process biased in the direction of the applied shear, see Fig. 7.9b. If a pressure P is now applied, an additional work $P\delta V$ is necessary in order to open up a cavity to enable movement to occur. Thus the barrier to movement is $\epsilon + P\delta V - l\delta A\sigma$ in the direction of the stress and $\epsilon + P\delta V + l\delta A\sigma$ in the opposite direction. As shown in Sect. 8.3. [see (8.15)], this gives the following expression for the shear rate

$$\dot{\gamma} = C \left(e^{l\delta A\sigma/k_B T} - e^{-l\delta A\sigma/k_B T} \right) e^{-(\epsilon + P\delta V)/k_B T}, \qquad (7.19a)$$

where the prefactor C is independent of σ. At low shear stress this formula reduces to

$$\dot{\gamma} = 2C(l\delta A\sigma/k_B T)e^{-(\epsilon + P\delta V)/k_B T}, \qquad (7.19b)$$

so that the viscosity

$$\mu = \frac{k_B T}{2Cl\delta A} e^{(\epsilon + P\delta V)/k_B T}. \qquad (7.19c)$$

Thus, at low shear stress σ, $\dot{\gamma}$ is linearly related to σ with the prefactor having the same temperature and pressure dependency as observed in experiments. For large shear rates, *shear thinning* occurs, where the viscosity decreases with increasing shear rate. From (7.19a) the onset of shear thinning occurs when $l\delta A\sigma/k_B T \sim 1$, i.e., when $\dot{\gamma} \sim C\exp[-(\epsilon + P\delta V)/k_B T]$ so that the onset approaches lower shear rates as the hydrostatic pressure increases. Using (7.19c) the onset of shear thinning occurs when $\dot{\gamma} \sim 1/\tau$, where

$$\tau = 2\mu l\delta A/k_B T \, .$$

Since $\rho l\delta A \sim m$, where m is the mass of the fluid molecule (or the mass of a segment if the chain is so long that it undergoes segmental motion),

$$\tau = 2m\mu/\rho k_B T \, . \qquad (7.20)$$

The same result can be derived from the Rouse model, see [7.3]. As an example, for n-hexadecane ($C_{16}H_{34}$) this formula gives $\tau \sim 10^{-10}$ s. Thus, n-hexadecane will behave as a Newtonian fluid, except at extremely high shear

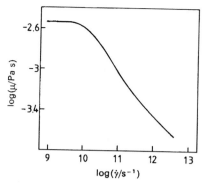

Fig. 7.10. Logarithm of shear viscosity μ (in Ns/m^2) as a function of shear rate $\dot{\gamma}$ (in s^{-1}) for n-hexadecane. From [7.4].

rates: if a lubrication layer with hexadecane has the thickness of 100 Å, sliding velocities of about 100 m/s are required before $\dot{\gamma} \sim 1/\tau$. These conditions are usually not encountered in practical applications. However, at very high pressures, μ may be a million times larger and $\tau \sim 10^{-4}$ s. In this case a 1000 Å thick film would already exhibit shear thinning for $v \sim 1$ mm/s; these conditions often occur in extreme lubrication applications (high pressure and high shear rates).

According to (7.19c) the pressure index $\alpha = \delta V/k_B T$. This expression is consistent with experimental data. Thus, one would expect the volume expansion δV to be somewhat smaller than the molecular volume. To check this for hydrocarbon oils, note that the viscosity is due to segmental motion, each segment being ~ 20 carbon atoms long. The molecular volume of C$_{20}$H$_{40}$ is $v \approx 400$ Å^3 so that at room temperature $v/k_B T \approx 10^{-7}$ Pa^{-1}, to be compared with the measured pressure index $\approx 2 \times 10^{-8}$ Pa^{-1}. The molecular volume for an H$_2$O molecule in water is $v \approx 30$ Å^3 so that $v/k_B T \approx 7 \times 10^{-9}$ Pa^{-1}, to be compared with the measured pressure index $\approx 6.7 \times 10^{-10}$ Pa^{-1}. Thus, in these cases the volume expansions involved in the flipping of molecular segments or molecules are smaller than the molecular volumes by factors of 5 and 10, respectively.

What happens if $\dot{\gamma} > 1/\tau$? Experiments and computer simulations have shown that the effective viscosity decreases with increasing $\dot{\gamma}$ when $\dot{\gamma} > 1/\tau$. As an example, Fig. 7.10 shows the logarithm of the shear viscosity of liquid n-hexadecane ($T = 477.6$ K and pressure $P = 0.69 \times 10^9$ N/m^2) as a function of the logarithm of the shear rate $\dot{\gamma}$ (in s^{-1}) [7.4]. The viscosity is constant for shear rates below $\dot{\gamma} \sim 10^{10}$ s^{-1} while it decays monotonically for $\dot{\gamma} > 10^{10}$ s^{-1}. Note that the onset of shear thinning is in good agreement with (7.20): using $\mu = 3 \times 10^{-3}$ Ns/m^2 from the computer simulations, (7.20) gives $\tau \approx 1.2 \times 10^{-10}$ s, so that $\dot{\gamma} \approx 1/\tau$ at the onset of shear thinning.

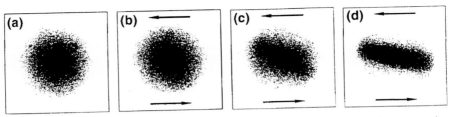

Fig. 7.11. Molecular orientational distribution function for selected shear rates (in s^{-1}): (a) $\dot\gamma = 0$, (b) $\log\dot\gamma = 9.5$, (c) $\log\dot\gamma = 10.0$, and (d) $\log\dot\gamma = 12.0$. From [7.4].

What is the physical origin of shear thinning? Experiments and computer simulations have shown that there is considerable intermolecular rearrangement when shear thinning occurs. For example, Fig. 7.11 shows the molecular orientational distribution function $f(\boldsymbol{x})$ projected onto the xz-plane. This function is obtained by *translating* all the molecules so that their centers of mass coincide with some reference point. The resulting mass density (projected onto the xz-plane) defines the function $f(\boldsymbol{x})$. At equilibrium there is no preferred direction, and a spherical distribution of molecular orientations result. The distribution $f(\boldsymbol{x})$ becomes more and more ellipsoidal as the shear rate is increased, showing that an increasing *intermolecular alignment* occurs at non-zero shear rates. The intuitive picture is that this alignment allows the molecules to slide easily past one another with a corresponding decrease in the fluid's viscosity.

An increase in the effective viscosity of fluids occurs not only at high pressure, but also when the "fluid" is confined between solid surfaces if the spacing is of the order of a few molecular diameters. This has been observed with the surface force apparatus [7.5,6], and is of importance for boundary lubrication. The same effect should also occur for fluids in narrow channels as may occur in some biological systems and in soil mechanics. The confinement leads to a large reduction in molecular mobility, and a strong increase in the effective viscosity. The effects come about because the fluid molecules cannot easily change positions, i.e., the energy barriers towards molecular rearrangements in the confined space may be strongly increased, similar to the effect of an increase in the pressure. As a result, the "fluid" may be in a glassy solid state characterized by very long relaxation times, where shear thinning occurs at very low shear rates.

As pointed out above, most lubrication oils behave as Newtonian fluids under most practical conditions. However, sometimes "additives" in the basic fluid give rise to shear thinning even at moderate shear rates. For example, the lubrication fluid in animal and human joints consist of a fluid with relatively low viscosity with the addition of a long-chain polymer with high molecular weight. This lubrication fluid is non-Newtonian even at a relatively low shear rate, with a shear viscosity which decreases almost linearly with increasing shear rate.

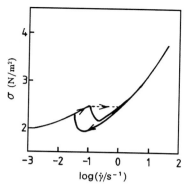

Fig. 7.12. The relation between the shear stress σ and the shear rate $\dot{\gamma}$ for a polystyrene suspension in water. *Solid lines*: steady-state stress as a function of increasing and decreasing shear rate. *Dashed line*: As a function of the applied stress the shear rate changes discontinuously from $\sim 0.1\,\text{s}^{-1}$ to $\sim 1.0\,\text{s}^{-1}$. From [7.8].

Other intensively studied systems which exhibit non-Newtonian dynamics are colloidal systems, liquid crystals [7.7], and suspensions of small particles. One particularly interesting study involves a dense suspension of charge stabilized polystyrene spheres in water [7.8]. This system exhibits long range orientational order at rest. The system evolves from a crystalline state in equilibrium to a polycrystalline state if subjected to a very low shear stress. When the shear stress is increased, an abrupt increase in the shear rate $\dot{\gamma}$ occurs at some critical shear stress σ_a to a new sliding state where partial order is reestablished with the appearance of sliding layer flow. Figure 7.12 shows the relation between shear stress σ and shear rate $\dot{\gamma}$ when the shear stress is increased from zero. For $\sigma < 0.6\,\text{N/m}^2$ the sample behaved as a perfectly elastic solid. For $0.6 < \sigma < 2\,\text{N/m}^2$, the sample deformed in a plastic manner (creep), but did not reach a steady-state rate of deformation during the period of the measurement. For $2.0 < \sigma < 2.5\,\text{N/m}^2$, the steady-state shear was achieved in 5 min. or less, and the shear rate increased smoothly with increasing σ. At $\sigma = 2.5\,\text{N/m}^2$, the system jumped from a state with $\dot{\gamma} = 0.09\,\text{s}^{-1}$ to a new steady state with $\dot{\gamma} = 0.85\,\text{s}^{-1}$.

Accompanying these rheological changes were alterations in the suspension microstructure. The equilibrium structure in the absence of a shear stress consisted of close-packed hexagonal layers oriented parallel to the rheometer wall. At any small but finite steady-state rate of deformation, the suspension broke into small crystallites. The polycrystalline state was retained up to a shear rate of $\dot{\gamma} = \dot{\gamma}_1 \approx 1.0\,\text{s}^{-1}$ at which time regions of a new microstructure appeared, suggestive of a coexistence of two microstructures. The new, lower-viscosity microstructure consisted of sliding layers, and the fraction of this structure increased monotonically with increasing shear stress so that for $\sigma > 3\,\text{N/m}^2$ the whole system consisted of the sliding-layer phase.

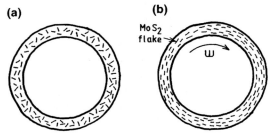

Fig. 7.13. The orientation of MoS$_2$ flakes in a lubrication fluid between two rotating cylinders. (a) Zero or low shear rate. (b) High shear rate.

Let us comment on the use of molybdenum disulfide additives in lubrication oils [7.9]. It has been found that this additive leads to several positive effects, e.g., lower friction and reduced wear. The MoS$_2$ particles consist of small flakes some thousand monolayers thick. At low shear rate, the flakes are randomly oriented and perform a tumbling motion in the fluid, but when the shear rate increases above some critical value the flakes orient themselves in a state which corresponds to the least dissipation of energy, where the flat surfaces (crystal planes) are entirely composed of streamlines of the undisturbed fluid motion (Fig. 7.13). This orientational transition occurs when the hydrodynamic torque [7.10] (which tends to align the flakes) becomes larger than the irregularly fluctuating torque associated with the interactions with the surrounding lubrication molecules (Brownian motion). Below the ordering transition, the Brownian motion increases the viscosity by bringing the flakes into more disordered positions with respect to the fluid motion. Associated with the transition from tumbling motion to flow alignment of the MoS$_2$ lamella is, therefore, a lowering of the shear viscosity of the lubrication fluid, i.e., shear thinning occurs. In addition, experiments have shown that the MoS$_2$ flakes adsorb strongly on the sliding metal surfaces forming a low-friction protection layer. This will reduce the wear and lower the friction under boundary lubrication conditions.

Undercooled Liquids and the Glass Transition*

When a liquid is cooled rapidly it will not have time to crystallize at the melting temperature T_m, but will remain "trapped" as an *undercooled* liquid. If the undercooled liquid is kept at a low-enough temperature, it will not crystallize even during a very long time period (say one year). Similarly, when a liquid is trapped between two closely spaced surfaces as in boundary lubrication, the limited space available will increase the barriers towards molecular rearrangements, and enhance the tendency to form a disordered state, rather than the thermodynamically stable state which may be crystalline.

When an undercooled liquid is kept close to the melting temperature, after a long-enough time it will make a transition to a crystalline state. The time necessary depends on the probability rate to form a critical nucleus of the crystalline solid. The nucleation rate can be very low for some systems, e.g., for long-chain molecules with side groups. If the undercooling is continued before crystallization has occurred, the liquid will note freeze, but its viscosity will rapidly increase; the resulting state is called a *glass*. The glass state is only metastable and will, after a long-enough time, make a transition to the crystalline state. However, the time necessary for the transition to the crystalline state may be very long (e.g., 100 years) if the temperature is low enough; in such cases the system may, for all practical purposes, be considered as trapped in the glass state.

A glass behaves as an elastic solid when it is exposed to weak and rapidly fluctuating forces, i.e., if the fluctuation time τ^* is much shorter than the time $\tau = \eta/G$, where η is the viscosity and G the (high-frequency) shear modulus. On the other hand, when exposed to forces which change so slowly with time that $\tau^* \gg \tau$, the glass will behave as a high-viscosity fluid. We can interpret τ as a characteristic time associated with those configurational changes in the glass which leads to viscous flow. When the shear stress σ is weak (linear response), the relation between σ and the shear rate $\dot{\gamma}$ is often approximated by the relation

$$\dot{\gamma} = \sigma/\eta + \dot{\sigma}/G,$$

which correctly describes the two limiting cases discussed above.

The glass transition temperature T_g is often defined as the temperature where the viscosity of the undercooled fluid equals 10^{12} kg/ms. Another definition, which in most cases gives nearly the same result, is that T_g is the temperature where the rate of change of the specific volume with temperature decreases from a "high" value for $T > T_g$ to a "low" value for $T < T_g$. For polymers, T_g is usually about 2/3 of the melting temperature of the crystalline state. It is observed that T_g and other properties of the glass, e.g., the density, depend on the cooling rate. Thus, the state of the glass is not uniquely defined by thermodynamic variables such as the temperature and pressure, but depends on the cooling rate.

The fundamental origin of the glassy state is the following: According to classical mechanics, a system of N particles have many mechanically stable configurations which correspond to local minima of the total potential energy $V(x_1, \ldots, x_N)$. Most of these local minima are not periodic systems, but disordered. By fast cooling from above the melting temperature to low temperatures, the system has not enough time to "find" the particle configuration which minimizes the potential energy, but the system ends up trapped in a local minima corresponding to a disordered solid. If the energy barriers for atom rearrangements are much higher than the thermal energy $k_B T$, the atoms in the glass will most of the time perform small-amplitude vibrations around the local energy minima and only very seldom long-range (diffusive) motion.

Let us discuss the temperature dependence of the viscosity. For temperatures above the melting temperature the viscosity of a fluid usually vary with temperature according to (7.19c). This relation is also a good approximation to the glassy state for liquids with strong covalent bonds between the atoms (such as SiO_2 or BeF_2). Thus, for SiO_2 the activation energy $\epsilon \approx 7.5\,\text{eV}$ is almost twice as large as the energy of a single Si-O bond, and nearly temperature independent. Thus, the viscous flow of glassy SiO_2 may occur by breaking a minimum of two of the four Si-O bonds. On the other hand, for liquids where the molecules interact weakly with each other, e.g., most organic liquids, the viscosity of the undercooled liquid can be written as

$$\eta = \eta_0 e^{\epsilon(T)/k_B T}$$

where the activation energy $\epsilon(T)$ *increases* with decreasing temperature.

We will now describe one mechanism by which the effective barrier $\epsilon(T)$ depend on the temperature. We assume that the same basic picture, as outlined above for the origin of the viscosity of fluids, is correct also for undercooled liquids. Thus, the barrier towards local rearrangements of the molecules is assumed to be associated with the "opening" up of a hole involving a local expansion. If the barrier is large compared with the thermal energy $k_B T$, the system will spend a long time in a local minima, performing vibrations around the bottom of the local potential well. However, the time τ^* the system spend in the barrier region during a successful jump over the barrier (the traversal time) is, in general, very short, and in most cases $\tau^* \ll \tau$, where $\tau = \eta/G$. Thus, when forming the "hole" the surrounding material will deform elastically. The activation energy ϵ is the sum of two terms:

$$\epsilon = \epsilon_1 + CG\Delta V^2/V, \tag{7.20a}$$

where C is a constant of order unity, and ΔV the increase in the volume (V is the volume before the local expansion has occured). The energy ϵ_1 represent the "surface" energy of the hole, i.e., the energy of the "broken" or "unsaturated" bonds at the surface of the hole. The term $CG\Delta V^2/V$ represent the elastic energy stored in the deformation field as a result of the local expansion. If, on the other hand, $\tau^* \gg \tau$ the transition over the barrier occur so slowly that the surrounding media responds adiabatically, i.e., it behaves as a fluid and no elastic energy will be stored in it during the transition over the barrier, and the activation energy will be equal to ϵ_1; this limiting case is not relevant in most situations. Now, when a liquid is cooled the density and the (high frequency) shear modulus G usually increases. According to (7.20a), this implies that the activation energy $\epsilon(T)$ increases with decreasing temperature.

The origin of the temperature dependence of the activation barrier described above is consistent with the observation of the time-dependent Stokes' shift of a nonpolar solute molecule in a glass-forming solvent. Here, an electronic excitation increases the effective size of the solute molecule at time $t = 0$, and the viscoelastic response of the surrounding solvent changes the transition energy and cause a time-dependent Stokes' shift for $t > 0$. That is, for short times $t < \tau$ the solvent respond as an elastic solid to the sudden increase in the effective size of the solute molecule, while it flow as a viscous fluid for times $t > \tau$.

7.3 Elastohydrodynamics

The theory of fluid lubrication described in Sect. 7.1 has been applied successfully to many sliding systems, e.g., journal bearings. However, in some applications extremely high local pressures occur, and under such conditions the theory above would predict fluid films thinner than the amplitude of the surface roughness, and one may expect fluid film lubrication to be replaced by boundary lubrication. Such high pressures occur, for example, in gears and ball bearings. However, experiments have shown that even in these extreme cases, fluid film lubrication usually persists under normal operation conditions.

The theory of fluid film lubrication outlined in Sect. 7.1. assumed that the sliding surfaces are rigid. In reality, all materials have a finite elasticity and in some cases this must be taken into account when studying the sliding process. This is quite obvious with soft elastic materials such as rubber, but is sometimes also the case for elastically hard materials such as steel (as in gears or ball bearings) .

Consider the sliding of an *elastic* solid on a rigid lubricated flat surface. During steady sliding, the hydrodynamic pressure distribution in the fluid and the elastic deformation of the sliding solid can be calculated in an iterative manner as follows: (1) Assume first that the solid is rigid. The resulting pressure distribution acting on the solid can be calculated from the equations of hydrodynamics, and is indicated by the dashed line in Fig. 7.14a. (2) In the next step, the elasticity of the sliding body is taken into account, and the elastic deformation of the body, resulting from the hydrodynamic pressure distribution obtained in step (1), is calculated using the equations of motion of an elastic continuum. The new surface profile is shown by the solid line in Fig. 7.14b. (3) As a result of the change in the surface profile of the solid, it is necessary to recalculate the pressure distribution in the fluid (dashed line in Fig. 7.14b). The whole process must be iterated until the changes in the pressure distribution and elastic deformations from one iteration to another are negligibly small. Although the basic idea is simple, in most practical cases the self-consistent solution must be obtained using a computer.

There are two important complications to the picture described above, which must be taken into account in applications involving very high pres-

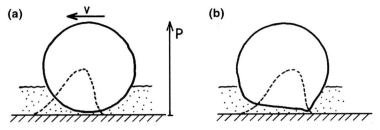

Fig. 7.14. An elastic sphere sliding on a rigid lubricated surface. (a) The dashed line shows the hydrodynamic pressure distribution in the fluid expected if the sphere were rigid. (b) The elastic deformation of the sphere in response to the pressure distribution in (a). The new pressure distribution is shown by the dashed line.

sures such as in gears and ball bearings. First, during sliding very high local temperature increases may result, and these will drastically alter the viscosity of the fluid. To obtain the temperature increase, it is necessary to study heat convection and heat diffusion. Secondly, the very high local pressures in the contact areas may increase the fluid viscosity (Sect. 7.2.). In many applications the latter effect is crucial for understanding why fluid lubrication occurs under conditions where one would otherwise expect boundary lubrication to prevail. See the Appendix at the end of this chapter for a detailed discussion of elastohydrodynamics.

To summarize, the high local stresses resulting from the load can result in two effects beneficial to fluid-film formation:

a) Local elastic deformation extends the area in which hydrodynamic pressure can be generated, thus encouraging fluid-film lubrication.
b) The lubricant viscosity may rise appreciably in regions of high pressure.

Quite generally, the lubricant in many elastohydrodynamic applications experiences rapid and very large pressure variations, a rapid transit time, possible large temperature changes and high shear rates. The severity of these conditions has called into question the normal assumption of Newtonian fluid behavior. For this reason lubricant rheology (Sect. 7.2) is a subject of great current interest.

7.4 On the Thickness of Lubrication Films

Consider a thick fluid layer between two solid surfaces (Fig. 7.15a). If the solids are pushed together with the pressure $P = L/A$, the thickness d of the fluid layer will initially decrease monotonically with time as indicated in Fig. 7.15b. However, the thickness of the fluid layer will, in general, not decrease towards zero but, if P is not too high, a "fluid" slab of thickness $d_0 \sim 10\text{--}100\,\text{Å}$ will remain trapped between the solid surfaces.

7.4 On the Thickness of Lubrication Films

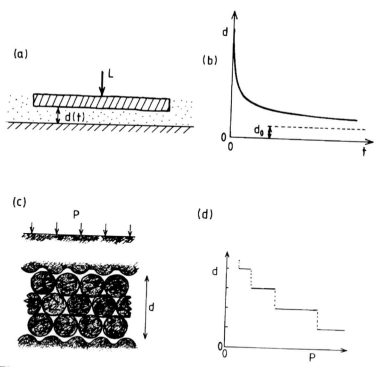

Fig. 7.15. (a) A flat circular disk is pushed towards a flat surface in a fluid. (b) The separation d between the two surfaces as a function of time t. As $t \to \infty$, $d \to d_0$, i.e., a molecularly thin "fluid" slab remains trapped between the surfaces. (c) A thin molecular slab between two plane surfaces. The molecules tend to form layers parallel to the surfaces ($n = 3$ in the present case) and the number of layers n depends on the external pressure P. (d) The variation in the thickness of the molecular film depends in a "quantized" manner on the applied pressure P.

The liquid density across such a molecular thin film is not uniform but has an oscillatory profile. The periodicity of the oscillations is close to the diameter of the liquid molecules and reflects the forced ordering of the liquid molecules into quasi-discrete layers between the two surfaces, see Fig. 7.15c. The closer the two surfaces approach each other, the sharper these density oscillations become. Furthermore, for $n \sim 5 - 10$ layers and below, the fluid slab usually solidifies. This manifests itself through the finite shear stress τ needed to slide the two surfaces relative to each other.

As a result of the layering of the molecular slab, many properties of the system exhibit "quantization". In the present case, if the normal load or pressure is varied it is found that the thickness of the slab changes in a step-like manner, see Fig. 7.15d. An increasingly high pressure P is necessary to squeeze out layer after layer of the "fluid" from the interface and, in fact, it is often not possible to squeeze out the last layer or two of trapped molecules. During the transition from $n \to n-1$ layers the film, or at least the layer being squeezed out, is believed to be in a melted or fluidized state [7.11].

Another "quantized" property is the yield stress τ_c (the tangential stress necessary to initiate the sliding of the two surfaces relative to each other), which depends on the number of molecular layers in the film [7.11]. For more than $n \sim 10$ monolayers the yield stress is usually zero as expected for a fluid but (neglecting creep) for thinner films a finite yield stress $\tau_c(n)$ is observed; furthermore $\tau_c(n)$ increases monotonically as the number of monolayers n decreases and may be very high when $n = 1$. The finite yield stress indicates that "in-plane" ordering must occur in the monolayers, where, in particular, the molecules in contact with the solid surfaces "adjust" to the corrugated substrate potentials forming pinned solid adsorbate structures. (If instead ideal incommensurate solid adsorbate structures were formed, they would experience a negligible pinning potential [7.12] resulting in zero yield stress, contrary to experimental observations.)

As stated above, it is often not possible to squeeze out the last layer or two of trapped molecules. The reason for this is related to wetting, the solid nature of the adsorbate layers, and to the fact that these are pinned by the substrate potential. But if a large enough shear stress is applied the adsorbate layer may fluidize or shear melt (in which case the two solid surfaces would move laterally relative to each other); in this case it may be possible to squeeze out the last one or two monolayers as well. Furthermore, if a given normal pressure results in, say, n monolayers when the shear stress equals zero, and the surfaces are then forced to move laterally relative to each other, one or several further monolayers may be squeezed out as a result of the fluidization of the film during sliding.

The processes discussed above are of fundamental importance in boundary lubrication [7.13,14]. Since most macroscopic bodies have a rough surface, at least on a microscopic scale, when, say, a steel block is resting on a steel substrate, very high normal pressures will result in "contact points" (junctions) between the two bodies. As discussed in Chap. 5, this will in general result in plastic deformation of the metals at the contact points in such a manner that each junction will be in a state of incipient plastic flow [7.14]. The normal pressure at a junction will therefore be close to the largest compressive stress σ_c (the penetration hardness) that the material can bear without plastic yielding. This, for steel, is $\sigma_c \sim 10^9 \, \text{N/m}^2$. Now, if a lubricant is present between the two solid surfaces, the fundamental question is whether the very high local normal pressures will squeeze out the film completely (which would lead to a high sliding friction and large wear) or whether one or more monolayers of lubricant molecules will remain in the contact areas. For many lubricant molecules the latter must be the case, since it is known that the presence of lubricant reduces the sliding friction at low sliding velocities (boundary lubrication) by a factor of 10 or so, while the wear may be reduced by several orders of magnitude.

In this section we study how the thickness of a fluid film between two surfaces changes when the surfaces are pushed together with the pressure P. We consider first the time dependence of the film thickness when the film is thick enough to behave as a fluid; in this case the squeeze-out occurs in a continuous manner (Fig. 7.15b) and can be studied using the Navier–Stokes equations. Next we study the nature of the film thickness transition $n \to n-1$ for a molecular thin film, under the assumption that the layer to be squeezed out is in a two-dimensional fluid state. We assume that the squeeze-out involves a "nucleation" of the film thickness transition and compare the theoretical results with the experimental data [7.11].

Squeezed Film

Consider a rigid flat circular (radius R) disk separated from a rigid flat substrate by a fluid film of thickness h where $h \ll R$ (Fig. 7.16). Assume that the disk is pushed towards the substrate with a force F. In this section we calculate the separation $h(t)$ as a function of time. The velocity field of the fluid is assumed to satisfy the Navier–Stokes equation for an incompressible fluid. We introduce cylindrical coordinates (ρ, ϕ, z) with the origin at the center of the lower surface and with the positive z-axis pointing towards the upper surface. Consider now the flow of the fluid through the cylindrical surface area indicated by the vertical dashed lines in Fig. 7.16, as the flat surfaces move towards each other by the amount $\Delta h = -\dot{h}\Delta t$. If ρ is the radius of the cylinder and if $\bar{v}$ denotes the radial velocity of the fluid averaged over the thickness h of the fluid film, then the amount of fluid flowing through the cylinder's surface area must equal $2\pi\rho h \bar{v} \Delta t$. But for an incompressible fluid this is just the change in volume $\pi \rho^2 \Delta h = -\pi \rho^2 \dot{h} \Delta t$, so that

$$2\pi \rho h \bar{v} = -\pi \rho^2 \dot{h},$$

or

$$\bar{v} = -\frac{\rho \dot{h}}{2h}. \tag{7.21}$$

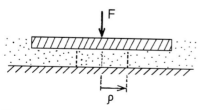

Fig. 7.16. A flat circular disk is pushed towards a flat surface in a fluid.

7. Sliding on Lubricated Surfaces

When $h \ll R$ the velocity field is accurately given by

$$\boldsymbol{v} = \hat{\rho} v(\rho, z, t) \,.$$

Substituting this into the Navier–Stokes equations and making similar approximations as in Sect. 7.1 gives

$$\frac{\partial P}{\partial z} = 0 \,, \tag{7.22}$$

$$\frac{\partial P}{\partial \rho} = \mu \frac{\partial^2 v}{\partial z^2} \,. \tag{7.23}$$

Equation (7.22) gives $P = P(\rho, t)$ so that (7.23) can be integrated to yield

$$v = \frac{z(z-h)}{2\mu} \frac{\partial P}{\partial \rho} \,.$$

Hence,

$$\bar{v} = \frac{1}{h} \int_0^h v = -\frac{h^2}{12\mu} \frac{\partial P}{\partial \rho} \,. \tag{7.24}$$

Comparing (7.21) and (7.24) gives

$$\frac{\rho \dot{h}}{2h} = \frac{h^2}{12\mu} \frac{\partial P}{\partial \rho} \,.$$

This equation is easy to integrate, using $P = P_{\text{ext}}$ (the external pressure) for $\rho = R$,

$$P = P_{\text{ext}} - (3\mu \dot{h}/h^3)\left(R^2 - \rho^2\right) \,. \tag{7.25}$$

The external force F must equal the pressure difference $(P - P_{\text{ext}})$ integrated over the contact area $\rho < R$, so that

$$F = \int_0^R d\rho \, 2\pi\rho (P - P_{\text{ext}}) = -\frac{3\pi \dot{h}}{2h^3} \mu R^4 \,. \tag{7.26}$$

If we introduce the perpendicular stress $\sigma_0 = F/\pi R^2$ and the diameter $D = 2R$, then (7.26) takes the form

$$\frac{dh}{dt} = -\frac{8h^3 \sigma_0}{3\mu D^2} \,.$$

This equation is easily integrated to obtain

$$\frac{1}{h^2(t)} - \frac{1}{h^2(0)} = \frac{16 t \sigma_0}{3\mu D^2} \,. \tag{7.27}$$

As an application, consider a 1 µm thick film of Castor oil between a flat steel surface and a steel asperity with a flat top surface of diameter

100 μm. Castor oil has the viscosity $\mu \approx 0.5 \, \text{Ns/m}^2$ and assuming that the perpendicular stress σ_0 is equal to the yield stress of steel ($\sim 10^9 \, \text{N/m}^2$), one can calculate $16\sigma_0/(3\mu D^2) \approx 1 \times 10^{18} \, \text{m}^2/\text{s}$. Hence it will take about 1 s to reduce the thickness from 1 μm to 100 Å and about 7 min to reduce the thickness to 5 Å, assuming that the Navier–Stokes theory works even for such thin films. Thus the theory predicts that if a steel block is located on a steel table lubricated by Castor oil, after a few minutes the lubrication fluid would be completely removed from the contact areas leading to direct metal contact and to a static friction force identical to that of clean surfaces. However, from experiments we know that this is not the case, proving that the continuum theory (with constant bulk viscosity) is not valid for molecularly thin lubrication films.

Equation (7.25) predicts that if two surfaces are separated rapidly (i.e., $\dot{h} > 0$) in a fluid, very low pressure or even cavity formation may occur between the surfaces. The physical reason for this is that because of the viscosity of the fluid it takes time for the fluid to flow in under the surfaces and fill the "empty" space generated by the separation of the surfaces. The same phenomena is the reason for why fingers covered with a high viscosity fluids, such as honey, "feel" sticky.

Suppose that two identical *elastic* spheres are pushed together. If no fluid is present between the spheres, a small *flat* circular contact region (radius R) would be formed where the pressure at the interface is of the form $P = (3/2)P_0[1 - (\rho/R)^2]^{1/2}$ where P_0 is the average pressure [see (5.3)]. Now, consider pushing together the spheres in a fluid. We study the influence of the fluid on the elastic deformations of the spheres. This is a problem in elastohydrodynamics but we can obtain a qualitative picture of what will happen as follows: We know that the pressure distribution $P = (3/2)P_0[1 - (\rho/R)^2]^{1/2}$ will lead to *locally flat* surfaces (Chap. 5). If a fluid layer is squeezed between two flat solid surfaces the pressure distribution is given by (7.25), i.e., $P = 2P_0[1 - (\rho/R)^2]$. This pressure is larger (by a factor 4/3) in the center of the contact area than that which would result in locally flat surfaces. Thus, the elastic spheres tend to deform as indicated in Fig. 7.17.

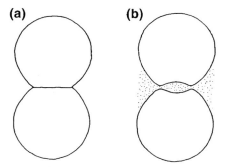

Fig. 7.17. (a) Elastic deformation of two elastic spheres squeezed together in vacuum. (b) Elastic deformation (snapshot picture) of two elastic spheres squeezed together in a fluid.

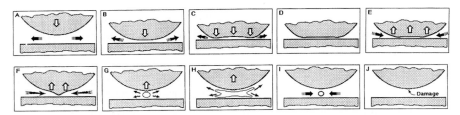

Fig. 7.18. Schematic picture of a cavitation growth and collapse mechanism revealed by visualization of two solid surfaces moving towards ($A \to D$) and away ($E \to J$) from each other. The rapid fingering-like growth of a cavity ($G \to H$) appears to be associated with an equally rapid recoil of the surfaces and the simultaneous onset of surface damage. From [7.15].

Based on the results presented above, let us briefly discuss what happens if two solid bodies are rapidly pushed together and then rapidly removed from each other. Figure 7.18 shows snapshots of the elastic deformations of the solids [7.15]. If the motion is rapid enough during separation a negative pressure may occur in the fluid between the solid bodies [see (7.25) with $\dot{h} > 0$] which can result in cavitation. The figure shows the cavitation growth and collapse mechanism revealed by visualization of two solid surfaces moving towards ($A \to D$) then away ($E \to J$) from each other. The rapid fingering-like growth of a cavity ($G \to H$) appears to be associated with an equally rapid recoil of the surfaces and the simultaneous onset of surface damage.

Layering Transition – Nucleation

The transition from liquid bulk behavior for a thick fluid film to solid-like behavior for a molecular thin film is a continuous function of the thickness d of the film. Thus, the effective viscosity μ of a fluid layer between two flat solid surfaces, deduced from shear experiments using the standard definition $\sigma = \mu \dot{\gamma}$, grows *continuously* without bound as d decreases towards a critical thickness d_0 of order $10-100$ Å. This is illustrated in Fig. 7.19 for n-dodecane [$CH_3(CH_2)_{10}CH_3$], a simple flexible chain molecule with the bulk viscosity $\mu = 0.001\,\text{Ns/m}^2$ [7.6,16]. In (a) the viscosity at low shear rate is shown as a function of the thickness d of the film. Note that the viscosity diverges for $d_0 \approx 25$ Å. In (b) the viscous shear force is shown as a function of shear velocity for a $d = 27$ Å thick film. Finally, (c) shows a log–log representation of the viscosity as a function of the strain rate $\dot{\gamma}$ for a $d = 27$ Å thick film. Note that shear thinning already starts to occur for $\dot{\gamma} \sim 10\,\text{s}^{-1}$ while for the bulk fluid the viscosity is independent of $\dot{\gamma}$ up to $10^{10}\,\text{s}^{-1}$ (compare with the results for n-hexadecane in Fig. 7.10).

We consider now a molecularly thin lubrication film in a solid state, i.e., $d < d_0$. As pointed out previously, such a film confined between two smooth solid surfaces tends to form layers of molecular thickness parallel to the sur-

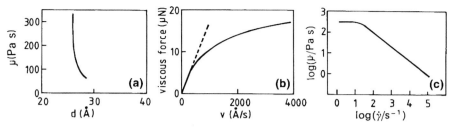

Fig. 7.19. Results of shear experiments on an ultrathin film of dodecane [$CH_3(CH_2)_{10}CH_3$], a simple flexible chain molecule. (a) Effective viscosity at low shear rate. (b) The viscous shear force as a function of shear velocity for a $d = 27$ Å thick film. (c) Log–log representation of the effective viscosity as a function of the strain rate. From [7.16].

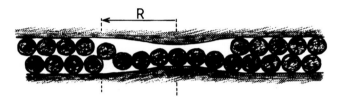

Fig. 7.20. A thermal fluctuation has created an $n = 1$ "nucleus" (radius R) in an $n = 2$ fluid layer. Because of the external pressure P, the surfaces of the elastic solids expand ("relax") into the "hole" as indicated.

faces. In this section we study the process of squeezing out of a monolayer as a function of an applied normal pressure [7.17]. We estimate the pressure necessary to "nucleate" the thickness transition ($n \to n - 1$, where n is the number of monolayers). The "growth" or "spreading" of the removal process after the nucleation has occurred will be studied in Sect. 8.6.2.

We consider the layering transition $n = 2 \to 1$ for which experimental information is available for liquid OMCTS (a silicone liquid with approximately spherical molecules of diameter ~ 8.5 Å) between two flat mica surfaces [7.11]. We assume that the layer to be squeezed out is in a fluid state, which is the case if the two mica surfaces are in relative parallel motion. A small circular region with $n = 1$ is assumed to first nucleate somewhere in the contact region followed by a "spreading" or "growth" of the $n = 1$ region and a corresponding decrease of the $n = 2$ region so that the total area $A_{n=1} + A_{n=2} = A_0$ is a constant.

We consider a system with no imperfections, e.g., steps, on the two solid surfaces. The two confined molecular monolayers are laterally in a fluid state, but in the normal direction very well defined. Assume now that due to a thermal fluctuation a small circular hole is formed in the uppermost two-dimensional fluid layer, see Fig. 7.20. If R denotes the radius of the "hole" the adiabatic work to form the hole is the sum of three terms

$$U(R) = 2\pi R \Gamma + \pi R^2 p_0 - \alpha R^3 \ . \tag{7.28}$$

The first term, $2\pi R \Gamma$, represents the free energy associated with the unsaturated bonds of the molecules at the periphery of the "hole" (Γ is a "line-tension"). The second term, $\pi R^2 p_0$, is the change in the interface free energy. If γ_s, γ_{lv} and γ_{sl} denote the solid–vacuum, liquid–vapor and solid–liquid interface free energies then $p_0 = \gamma_s - \gamma_{lv} - \gamma_{sl}$. The two-dimensional pressure p_0 is usually called the spreading pressure [7.18–20]. For organic molecules on metal oxides one typically finds $p_0 \sim 4\,\mathrm{meV}/\mathrm{Å}^2$. The third term in (7.28), $-\alpha R^3$, is an elastic relaxation energy, namely when a hole has been formed, the two solid confining materials will deform elastically as indicated in Fig. 7.20; this relaxation energy will tend to stabilize the hole. It will be shown below that the elastic relaxation energy scales with the radius of the hole as $\sim R^3$ and will, therefore, for a large enough hole, dominate over the interface energy term. Furthermore, we show below that $\alpha \propto P_0^2$ where P_0 is the pressure with which the two solid surfaces are squeezed together. The probability for a hole with the radius R to be formed by a thermal fluctuation is proportional to the Boltzmann factor

$$\mathrm{e}^{-U(R)/k_\mathrm{B}T} \ ,$$

where T is the temperature and k_B the Boltzmann constant. If R_c denotes the critical radius, i.e., the radius for which $U(R)$ is maximal, i.e., $U'(R_\mathrm{c}) = 0$, the rate w of nucleation of the $n = 1$ "phase" will be

$$w = w_0 \mathrm{e}^{-U(R_\mathrm{c})/k_\mathrm{B}T} \ . \tag{7.29}$$

From (7.28) we get

$$2\pi \Gamma + 2\pi R_\mathrm{c} p_0 - 3\alpha R_\mathrm{c}^2 = 0 \tag{7.30}$$

and

$$U(R_\mathrm{c}) = \pi R_\mathrm{c}^2 p_0 / 3 + 4\pi R_\mathrm{c} \Gamma / 3 \ . \tag{7.31}$$

The prefactor w_0 can be calculated using the theory of activated processes. Accounting only for radial fluctuations of the nucleus, it can be shown [7.17] that in the present context $w_0 \sim 10^{17}\,\mathrm{s}^{-1}$, and if we formally define "nucleation" to occur when $w \sim 0.01\,\mathrm{s}^{-1}$ (i.e., about one nucleus per minute) then it follows from (7.29) that nucleation will occur when $U(R_\mathrm{c})/k_\mathrm{B}T = \kappa$ where $\kappa \approx 44$.

Let us now evaluate the coefficient α in the elastic relaxation energy. We will treat the two solid bodies as isotropic elastic media, which is a good approximation for most "practical" cases, e.g., for most metals and metal oxides. Now, before the hole has been opened up the $n = 2$ monolayer must exert a normal pressure P_0 on the upper and lower elastic bodies. After the hole (radius R) has been opened up, the pressure on the solid bodies is assumed to be zero for $0 < r < R$, where $r = 0$ is at the center of the circular hole. For $r > R$ the normal pressure will be some function of r, $P = P(r)$, where $P(r) \to P_0$ as $r \to \infty$, and with

$$2\pi \int_R^\infty dr\, r\, [P(r) - P_0] = \pi R^2 P_0 ,$$

so that the total normal force (the "load") on the elastic body is unchanged. The function $P(r)$ (for $r > R$) can be determined from the theory of elasticity. The solution depends on the boundary conditions for $r > R$. In the literature two limiting cases have been studied. In the present context, we assume that the parallel (or tangential) stress vanishes everywhere on the elastic bodies. In this case $P(r)$ is approximately given by

$$P(r) = P_0 + \frac{1}{2} P_0 R \left[(r^2 - R^2)^{-1/2} - r^{-1} \right] \tag{7.32}$$

for $r > R$. If u_0 and $u(\boldsymbol{x})$ denote the normal displacement fields of the "upper" elastic solid at the surface contacting the $n = 2$ layer, before and after opening up the hole, respectively, then the elastic relaxation energy associated with the "upper" solid body can be written as

$$U_{\rm el} = \frac{1}{2} \int d^2x\, [P(\boldsymbol{x})u(\boldsymbol{x}) - P_0 u_0] , \tag{7.33}$$

where, for an isotropic elastic medium [7.21],

$$u(\boldsymbol{x}) = \frac{1 - \nu^2}{\pi E} \int d^2x'\, \frac{P(\boldsymbol{x}')}{|\boldsymbol{x} - \boldsymbol{x}'|} , \tag{7.34}$$

where E is Young's modulus and ν the Poisson's ratio. Substituting (7.34) into (7.33) gives

$$U_{\rm el} = -\frac{1 - \nu^2}{2\pi E} \int d^2x\, d^2x'\, \frac{1}{|\boldsymbol{x} - \boldsymbol{x}'|} \left[P_0^2 - P(\boldsymbol{x}) P(\boldsymbol{x}') \right] . \tag{7.35}$$

If we write $P(\boldsymbol{x}) = P_0 + P_0 f(r/R)$ where

$$f(\xi) = -1 \quad \text{for} \quad \xi < 1 ,$$

and

$$f(\xi) = \frac{1}{2} \left[(\xi^2 - 1)^{-1/2} - \xi^{-1} \right] \quad \text{for} \quad \xi > 1 ,$$

then it is easy to show that (7.35) reduces to

$$U_{\rm el} = (1 - \nu^2) P_0^2 I R^3 / E , \tag{7.36}$$

where I is a number,

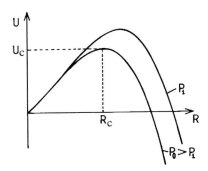

Fig. 7.21. The dependence of the energy of a "nucleus" on the radius R of the nucleus. As the pressure P increases ($P_0 > P_1$), the "activation" energy U_c, which must be overcome in order to nucleate the layering transition $n \to n-1$, decreases.

$$I = \int_0^\infty d\xi d\xi' \int_0^{2\pi} d\phi \frac{\xi\xi' f(\xi) f(\xi')}{(\xi^2 + \xi'^2 - 2\xi\xi' \cos\phi)^{1/2}} \approx 1.4 \; .$$

However, not only the upper elastic body will relax when the hole opens up, but so will the lower body. We can take this into account by doubling I; hence we will use $I = 2.8$ below. Any other reasonable choice of $f(\xi)$ [i.e., of $P(r)$] would give a small numerical change of the integral I but would not change the general form of (7.36). Using (7.36) we obtain

$$\alpha = (1 - \nu^2) P_0^2 I / E \; . \tag{7.37}$$

In Fig. 7.21 we show the general form of $U(R)$ for two different external pressures P_1 and $P_0 > P_1$. If P is increased from some low value, both R_c and $U(R_c)$ decrease. The transition $n = 2 \to n = 1$ will nucleate when $U(R_c)/k_B T = \kappa$ where $\kappa \approx 44$.

Using (7.31) we get the nucleation condition

$$\pi R_c^2 p_0 / 3 + 4\pi R_c \Gamma / 3 \approx \kappa k_B T$$

or

$$R_c \approx -\frac{2\Gamma}{p_0} \pm \left(\frac{4\Gamma^2}{p_0^2} + \frac{3\kappa k_B T}{\pi p_0} \right)^{1/2} , \tag{7.38}$$

where the + and − signs refer to $p_0 > 0$ and $p_0 < 0$, respectively. At room temperature ($k_B T \approx 25\,\text{meV}$) and for organic molecules on mica ($p_0 \approx 2\,\text{meV}/\text{Å}^2$), $\Gamma \approx 10\,\text{meV}/\text{Å}$ (we estimate Γ from the product of the liquid–vapor surface free energy, γ_{lv}, and the thickness of a molecular monolayer), we get $R_c \approx 15\,\text{Å}$, which is large enough for the continuum approximation involved in the calculation of the elastic relaxation energy to be valid. Next, using (7.30) and (7.37) we get

$$P_0 = \left[\frac{(2\pi\Gamma + 2\pi R_c p_0)E}{3(1-\nu^2)IR_c^2}\right]^{1/2}. \tag{7.39}$$

Let us tentatively apply this formula to OMCTS between two mica surfaces. The elastic properties of mica are highly anisotropic but the most relevant elastic coefficients can be semi-quantitatively described by choosing $E = 1.75 \times 10^{10}\,\mathrm{N/m^2}$ and $\nu = 0.44$. Using (7.39) we can now calculate the "critical" pressure necessary to nucleate the $n = 1$ area; we get $P_0 \approx 7 \times 10^8\,\mathrm{N/m^2}$. This should be compared with the observed critical pressure [7.11], $(2 \pm 1) \times 10^7\,\mathrm{N/m^2}$, which is a factor of ~ 30 smaller than the theoretical estimate. This indicates that the actual nucleation may occur at some "weak" point between the mica surfaces where "imperfections", e.g., adsorbed water molecules or some organic contamination, may occur. This may locally reduce the spreading pressure (which may even become negative, i.e., non-wetting) which would reduce the "critical" pressure P_0 (see below). The situation is obviously similar to that for three-dimensional systems, where the centers of formation of a new phase, e.g., solidification of an undercooled liquid, usually occur at various kinds of "impurities" (dust particles, ions, etc.).

As another application of (7.38) and (7.39), consider a metal block on a metal substrate with a lubricant fluid. Assuming that the spreading pressure $p_0 \approx 4\,\mathrm{meV/\mathring{A}^2}$ as is typical for organic fluids on metal oxides [7.20], (7.38) gives $R_c \approx 12\,\mathring{A}$. For steel, $E \sim 1 \times 10^{11}\,\mathrm{N/m^2}$, and (7.39) gives $P_0 \approx 2 \times 10^9\,\mathrm{N/m^2}$; this is similar to the pressure which occurs in the contact regions between the steel surfaces, which is of the order of the plastic yield stress (the penetration hardness), i.e., about $10^9\,\mathrm{N/m^2}$. Accounting for the imperfect nature of the real contact areas it is very likely that *if* the lubrication layer is in a fluid state it will be squeezed out from the contact areas. On the other hand, if the lubricant layer is in a solid state, even this high pressure may not be able to remove the last layer or two of lubrication molecules from a junction region. But as the temperature increases, the lubrication film will finally melt (two-dimensional fluid) and the lubrication fluid can be "squeezed out" from the contact regions, as is indeed observed in many sliding systems. For example, Fig. 7.22 shows the temperature variation of the sliding friction coefficient f and the amount of metal transfer (the "pick-up") from one metal surface to another, as a cadmium block is slid over a lubricated (palmic acid) cadmium substrate [7.22]. The metal transfer is measured by making the block radioactive and detecting the amount of radioactive metal transferred from the block to the substrate. Note that in a narrow temperature interval, from $110°$ to $130°\mathrm{C}$, there is an abrupt increase in the sliding friction (by a factor of ~ 10) and an increase in the amount of metal transfer by almost a factor 10^5. This can be explained if it is assumed that the palmic acid film melts in this temperature interval, and that the melted film is squeezed

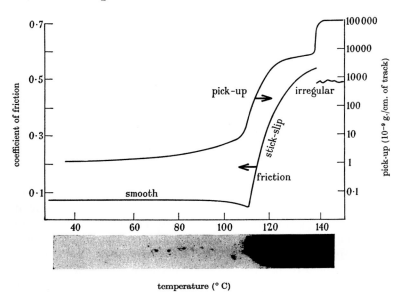

Fig. 7.22. The temperature dependence of the friction coefficient and of the metal transfer ("pick-up") for cadmium surfaces lubricated with palmic acid. The amount of pick-up is deduced from the radiograph shown at the bottom. From [7.22].

out from the contact regions between the two surfaces. In fact, since above $\sim 130°\mathrm{C}$ the sliding friction and the metal transfer are essentially identical to those for clean surfaces, this is very likely to be the correct explanation.

It must be pointed out, however, that the "squeezing out" of a lubrication film may occur very slowly (Sect. 8.6.2) so that, during sliding, if the transition $n = 1 \to n = 0$ were to nucleate somewhere in the contact region, the nucleus may not have enough time to grow before the two contact areas have passed each other. [The overall diameter of a contact area (junction) is not large, typically about 0.01 mm.] This is also probably the reason why during sliding of lubricated surfaces the metal fragments transferred from one of the sliding surfaces to the other are usually very small compared with the case of unlubricated surfaces, where the metal fragments are of similar size to the contact area itself, as shown by radioactive tracer experiments.

In Fig. 7.23 we show the critical pressure P_0 and the critical radius R_c as a function of the spreading pressure p_0. In the calculation we have taken $\Gamma = 10\,\mathrm{meV}$, $k_\mathrm{B}T = 25\,\mathrm{meV}$, $\kappa = 44$ and the elastic constant $E_* = E/(1 - \nu^2) = 1 \times 10^{11}\,\mathrm{N/m}^2$ (the results for other values of E_* can be obtained from the fact that $P_0 \propto E_*^{1/2}$ while R_c is independent of E_*). Note that P_0 vanishes for $p_0 < -3\Gamma^2/(\pi\kappa k_\mathrm{B}T)$, i.e., for $p_0 < -0.3\,\mathrm{meV}$ in the present case. Hence if the spreading pressure is negative the layering transition may nucleate spontaneously (i.e., without an external pressure). This may be the

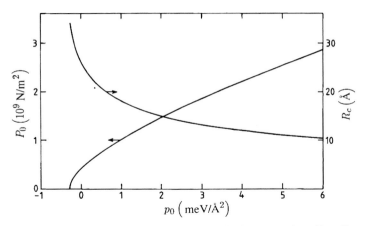

Fig. 7.23. The critical pressure P_0 and the critical radius R_c as a function of the spreading pressure p_0. In the calculation, $\Gamma = 10\,\text{meV}$, $k_BT = 25\,\text{meV}$ and $E/(1-\nu^2) = 10^{11}\,\text{N/m}^2$ are assumed as reasonable values for a generic situation.

case, for example, for most fluids between two teflon surfaces, or between two metal oxide surfaces covered by fatty acid monolayers.

A fundamental problem which we have not addressed above is how the layering transition occurs when the molecular layer between the mica surfaces is in a solid state rather than a fluid state (above we assumed that the two mica surfaces were in relative–parallel motion, so that the molecular layer was in a fluidized state). It has been found experimentally that for a solid film a larger normal pressure is necessary in order to squeeze out one molecular layer ($n \to n-1$); it has been suggested that the "critical" pressure may now be determined by the pressure necessary to induce the transition (solid layer → fluid layer). That a pressure (i.e., *normal* force per unit area) can give rise to a fluidization of the adsorbate layer can be understood as follows. Consider an elastic body in the form of a cylinder on a (rigid) flat substrate; see Fig. 7.24a. If a pressure P is applied to the upper surface of the cylinder and if the friction on the lower surface vanishes, as would be the case if a thin fluid layer were to separate the two surfaces, then the cylinder will deform elastically as indicated in Fig. 7.24b. However, if the lubrication layer is in a solid state which is pinned by the solid surfaces, then the cylinder will deform as indicated in Fig. 7.24c and a shear stress $\tau \propto P$ will develop at the interface between the cylinder and the lubrication film. When the pressure P is large enough, the shear stress τ will reach the stress necessary for fluidization of the lubrication film (i.e., the static yield stress). At this point, the lubrication film will fluidize and the elastic cylinder will evolve towards the "relaxed" configuration shown in Fig. 7.24b. It is likely that the layering transition will nucleate during the period in which the lubrication film is in the fluidized state.

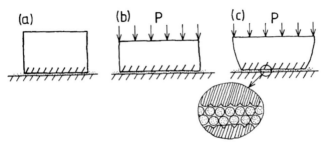

Fig. 7.24. (a) An elastic block on a (rigid) flat substrate. (b) If a pressure P is applied to the upper surface of the block and if the friction on the lower surface vanishes, the block will deform elastically as indicated in the figure. (c) If the lubrication layer is in a solid state which is pinned by the solid surfaces, the block will instead deform as indicated. When the shear stress $\tau \propto P$ at the block–substrate interface reaches the static yield stress, the lubrication film will fluidize and the elastic block will evolve towards the "relaxed" configuration in (b).

Exactly how the layering transition occurs is not well understood at present. More theoretical and experimental studies need to be performed to determine the exact nature of this important transition.

7.5 Wetting and Capillarity

We have seen in the last section that a lubrication fluid should bind or adhere so strongly to the sliding surfaces, that at least one or two layers of lubrication molecules remains trapped between the surfaces, i.e., the very high pressure P in the contact areas (for steel $P \sim 10^9 \, \text{N/m}^2 \sim 10 \, \text{meV/Å}^3$) should not be able to squeeze out the lubrication film completely. The tendency for the fluid to adhere or *wet* the solid surfaces is described by the *spreading pressure* $p_0 = \gamma_s - \gamma_{sl} - \gamma_{lv}$, where γ_s, γ_{sl} and γ_{lv} denote the interfacial free energy per unit area (surface tension) of the solid–vacuum, solid–liquid and liquid–vapor interfaces, respectively (Fig. 7.25). The importance of the spreading pressure p_0 for practical purposes was first recognized by Cooper and Nuttal in connection with the spraying of insecticides on leaves. Large positive p_0 favors the spreading of a liquid. It is important to note that p_0 is *not* a thermal equilibrium quantity, since in the definition of p_0 occurs the solid–vacuum free energy γ_s rather than the solid–vapor free energy γ_{sv}. In general, γ_s can be very different from γ_{sv}. Thus, we will give an example below where p_0 is large (and positive) but where no complete wetting occurs in the sense that in dynamical equilibrium with the fluid state (say a fluid drop on the solid surface) only a monolayer film of adsorbates is formed on the solid surface rather than a thick fluid slab. In general we may say that p_0 represents the tendency of a fluid to form a *monolayer* film on a surface. Whether the film

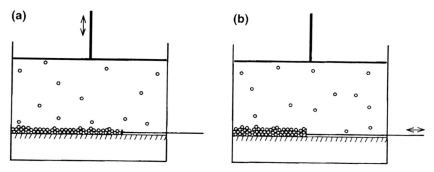

Fig. 7.25. (a) The work required to change the gas volume by δV equals $P_0 \delta V$ where P_0 is the 3D-pressure in the vessel. (b) The work required to displace the adsorbates from the area δA equals $p_0 \delta A$ where p_0 is the 2D-pressure in the adsorbate layer. p_0 is usually is called the spreading pressure. The figure shows the case where the adsorbates wet the surface. The gas phase is saturated (i.e., in equilibrium with its own fluid) and the surface is covered by a thin fluid slab. The area covered by the fluid adsorbate slab is reduced by δA which requires the work $(\gamma_s - \gamma_{sl} - \gamma_{lv})\delta A$. Thus $p_0 = \gamma_s - \gamma_{sl} - \gamma_{lv}$.

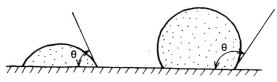

Fig. 7.26. A small liquid drop on a flat surface in thermal equilibrium with the solid and the vapor. Complete wetting occurs when the contact angle $\theta = 0$.

will grow beyond monolayer thickness depends on the interaction between the fluid and the surface covered by the monolayer film. In this section we discuss some aspects of wetting and capillarity that are relevant for sliding friction.

Incomplete Wetting: Droplet Shape and Contact Angle

Consider a small fluid drop on a flat solid surface (area A) in thermal equilibrium with the solid and vapor. The shape of the drop and the contact angle θ (Fig. 7.26) can be obtained by minimizing the free energy with the constraint that the volume of fluid is a constant V_0. It is obvious by symmetry that the height h of the drop is only a function of the distance ρ in the (x,y)-plane from the origin of the circular contact region between the drop and the substrate. Let γ_{sv} be the interfacial energy per unit area of the solid–vapor interface. The free energy of the system has the form

$$F = (A - \pi R^2)\gamma_{sv} + \pi R^2 \gamma_{sl} + 2\pi \gamma_{lv} \int_0^R d\rho\, \rho \left(1 + [h'(\rho)]^2\right)^{1/2},$$

where R is the radius of the circular base of the drop. To find the shape of the drop we must minimize F under the constraint that the volume of the fluid is a given constant V_0, i.e.,

$$2\pi \int_0^R d\rho\, \rho h(\rho) = V_0 \,.$$

This is a standard problem in variational calculus and is most easily solved by introducing the Lagrange multiplier λ and considering

$$\bar{F} = F + \lambda \left[\int_0^R d\rho\, \rho h(\rho) - V_0/2\pi \right] \,. \tag{7.40}$$

We now allow h to vary, $h \to h + \delta h$, and require that $\delta \bar{F}$ vanishes to first order in δh. This gives

$$\gamma_{\rm lv} \int_0^R d\rho\, \rho h' \left(1 + h'^2\right)^{-1/2} \delta h' + \lambda \int_0^R d\rho\, \rho\, \delta h = 0 \,. \tag{7.41}$$

Performing a partial integration of the first term in (7.41) and using $\delta h(R) = 0$ and $h'(0) = 0$ gives

$$\int_0^R d\rho \left\{ -\gamma_{\rm lv} \frac{d}{d\rho} \left[\rho h' \left(1 + h'^2\right)^{-1/2} \right] + \lambda \rho \right\} \delta h = 0 \,.$$

Since δh is arbitrary this gives

$$\frac{d}{d\rho} \left[\rho h' \left(1 + h'^2\right)^{-1/2} \right] = \frac{\lambda \rho}{\gamma_{\rm lv}} \,.$$

This equation is easy to integrate to get

$$h' \left(1 + h'^2\right)^{-1/2} = \frac{\lambda \rho}{2\gamma_{\rm lv}} \,.$$

Solving for h' and performing the ρ-integration, using $h(R) = 0$, gives

$$h(\rho) = \frac{2\gamma_{\rm lv}}{\lambda} \left\{ \left[1 - \left(\frac{\lambda R}{2\gamma_{\rm lv}}\right)^2\right]^{1/2} - \left[1 - \left(\frac{\lambda \rho}{2\gamma_{\rm lv}}\right)^2\right]^{1/2} \right\} \,. \tag{7.42}$$

Hence the droplet has the form of a spherical cup. Note that $h'(R) = -\tan\theta$ and since from (7.42)

$$h'(R) = (\lambda R/2\gamma_{\rm lv}) \left[1 - (\lambda R/2\gamma_{\rm lv})^2\right]^{-1/2} ,$$

we get

$$\sin\theta = -\lambda R/2\gamma_{\rm lv} \,. \tag{7.43}$$

Substituting (7.42) into (7.40) and using (7.43) to eliminate λ gives, after some simplification,

$$\bar{F} = A\gamma_{sv} + \frac{2\gamma_{lv}}{R}V_0 \sin\theta + \pi R^2 \left[\gamma_{sl} - \gamma_{sv} + \gamma_{lv}\frac{2}{3}\left(\frac{1}{1+\cos\theta} + \cos\theta\right)\right].$$

Now, $\bar{F}$ must be stationary as a function of R and λ or, equivalently, of R and θ. The conditions $\partial \bar{F}/\partial R = 0$ and $\partial \bar{F}/\partial \theta = 0$ give

$$\gamma_{sl} - \gamma_{sv} + \gamma_{lv}\cos\theta = 0 . \tag{7.44}$$

Since the surface tension γ_{lv} is easy to determine, from the study of the contact angle θ for small droplets it is possible from (7.44) to deduce the difference $\gamma_{sv} - \gamma_{sl}$ for surfaces where incomplete wetting occurs. If $\gamma_{sl} - \gamma_{sv} + \gamma_{lv} = 0$ then $\theta = 0$, i.e., a small drop of the fluid on the solid surface spreads out to form a uniform layer. The contact angle is a thermal equilibrium quantity. Since $\cos\theta \leq 1$ we must have $\gamma_{sl} - \gamma_{sv} + \gamma_{lv} \geq 0$ where the equals sign corresponds to complete wetting. Using the definition of the spreading pressure we get $p_0 \leq \gamma_s - \gamma_{sv}$, and for complete wetting, $p_0 = \gamma_s - \gamma_{sv}$.

High-Energy and Low-Energy Surfaces

From studies of bulk cohesive energy we know that there are two main types of solids, (a) hard solids (covalent, ionic, or metallic) and (b) soft molecular solids such as solid organic polymers. Examples of hard solids are the ordinary metals, metal oxides, nitrides, sulphides, silica, glass, ruby, and diamond. Soft organic solids are waxes, solid organic polymers, and most other solid organic compounds. Hard solids have "high energy surfaces" with $\gamma_s \sim (10-300)\,\text{meV/Å}^2$, while soft molecular solids have "low energy surfaces" with $\gamma_s \sim (1-3)\,\text{meV/Å}^2$ (Table 7.1). [Let us expand on the notion of hard and soft solids. Hard solids usually have strong bonds between *all nearest neighbor* atoms as in diamond. Thus, clean surfaces of a hard covalent solid have unsaturated bonds or *dangling bonds* (the number of which is usually reduced from the ideal value by surface reconstruction). Such surfaces are therefore very reactive and under normal atmospheric conditions they are always covered by foreign atoms and molecules which convert them into a lower (but not necessarily low) energy surfaces, e.g., diamond surfaces are usually passivated by hydrogen and oxygen atoms. Soft solids (e.g., organic molecular solids) consist typically of building blocks ("molecules") with very strong internal bonds, but with weak (typically van der Waals) bonds between the building blocks. The surface of a soft solid exposes the inert surfaces of the building blocks, and therefore has a low surface free energy and a low tendency to react with foreign atoms and molecules.]

Because of the comparatively low specific surface free energies of organic and most inorganic liquids (Table 7.2), these usually spread freely on solids of high surface energy, since this results in a large decrease in the surface free energy. But since the specific surface free energies of most liquids are

Table 7.1. Surface energies for soft low-energy surfaces (the first six materials) and hard high-energy surfaces (the last seven materials) at room temperature. From [7.23,24].

Material	Surface energy γ_s (meV/Å^2)
nylon	2.9
polyvinyl chloride	2.43
polystyrene	2.06
polyethylene	1.9
paraffin wax	1.56
PTFE (Teflon)	1.14
NaCl	10
Al$_2$O$_3$	40
Si	80
Al	70
Ag	90
Fe	150
W	280

Table 7.2. Surface energies of a number of fluids at $T = 20°$C. From [7.23,24].

Liquid	Surface energy γ_{lv} (meV/Å^2)
water	4.56
benzene	1.80
n-pentane	1.00
n-octane	1.35
n-dodecane (C$_{12}$H$_{26}$)	1.59
n-hexadecane (C$_{16}$H$_{34}$)	1.72
n-octadecane (C$_{18}$H$_{38}$)	1.75

comparable to those of low-energy solids, incomplete spreading often occurs on low-energy solids.

The spreading pressure is, loosely speaking, the difference between the fluid–substrate and the fluid–fluid binding energy. For van der Waals liquids, such as hydrocarbon or silicon fluids, the fluid–substrate and fluid–fluid binding energies are roughly determined by the polarizabilities of the solid and fluid. Since "hard" solids usually have higher polarizability (per unit volume) than most fluids, it follows that for most fluids on hard solids the spreading pressure is positive, e.g., most "hard" surfaces are wetted by hydrocarbon fluids.

Thus, most liquids having low surface free energy spread freely on a smooth, clean, high-energy surface at ordinary temperatures. But there is one exception: Some liquids can give rise to an adsorbed monolayer film that converts it into a low-energy surface which cannot be wetted by the liquid.

Table 7.3. The difference $\gamma_s - \gamma_{sv}$ in the surface energies between the solid–vacuum and the solid–saturated-vapor surfaces for a few different high-energy surfaces and fluids. From [7.20].

Solid/vapor	$\gamma_s - \gamma_{sv}$ (meV/Å^2)
TiO$_2$/water	19
TiO$_2$/n-heptane	4
SiO$_2$/water	20
SiO$_2$/n-heptane	4
Fe$_2$O$_3$/n-heptane	3
Ag/n-heptane	2

This is often the case with a liquid made up in whole or in part of molecules with a polar head such as fatty acids of fluorinated fatty acids, where a close packed layer of oriented molecules is formed at the solid–liquid interface with terminal $-CH_3$, $-CF_2H$, or $-CF_3$ groups. In this case the liquid is non-spreading because the molecules adsorbed on the solid form a film which cannot be wetted by the liquid itself. Nevertheless, the spreading pressure is high (since the monolayer film binds strongly to the substrate) and even the high local pressures in the contact regions in a sliding friction experiment, will in general not be able to remove the monolayer film. Thus, fatty acid monolayers are usually very good boundary lubrication films, which not only lower the sliding friction but also protect the surfaces and reduce the wear.

The difference $\gamma_s - \gamma_{sv}$ varies widely for different systems. With water on metal oxides, $\gamma_s - \gamma_{sv} \sim 20$ meV/Å^2 while for hydrocarbons on oxides, $\gamma_s - \gamma_{sv} \sim 4$ meV/Å^2. Molecules which bind strongly to surfaces tend to give rise to a large $\gamma_s - \gamma_{sv}$. Table 7.3 shows the quantity $\gamma_s - \gamma_{sv}$ for a few different fluids and high-energy surfaces. On low-energy surfaces such as Teflon this quantity is nearly zero, and nearly no molecules are adsorbed on the surface at normal temperatures and pressures. More generally, whenever a liquid exhibits a large contact angle on a low-energy solid there is good experimental evidence that there is negligible adsorption of the vapor.

Low-energy surfaces can give rise to partial or complete wetting, depending on the solid and the fluid. *Zisman* [7.20] has studied the wetting of n-alkanes on different surfaces. For solid polyethylene complete wetting is found for the whole series of liquid alkanes. But in other cases a finite contact angle θ occurs, which varies within the homologous series. A useful way of representing these results is to plot $\cos \theta$ as a function of the surface tension γ_{lv} of the liquid. Two examples are shown in Fig. 7.27. For n-alkanes on a polytetrafluoroethylene (Teflon) surface (Fig. 7.27a) complete wetting never occurs, i.e., $\cos \theta = 1$ is never reached. Extrapolation shows that complete wetting would require a hydrocarbon fluid with $\gamma_{lv} < \gamma_c$ where the critical surface tension $\gamma_c \approx 1$ meV/Å^2. Thus if a Teflon surface were lubricated with a hydrocarbon fluid it is likely that the fluid would be completely removed from the contact

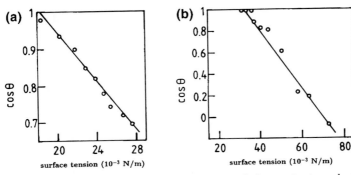

Fig. 7.27. Two Zisman plots: the cosine of the contact angle θ as a function of the surface tension γ_{lv} of the fluid. (a) Liquid alkanes on polytetrafluoroethylene (Teflon) where the critical surface tension $\gamma_c \approx 1\,\text{meV}/\text{Å}^2$. (b) Organic liquids on polyethylene where the critical surface tension equals $1.94\,\text{meV}/\text{Å}^2$. From [7.20].

Table 7.4. The critical surface tension for a few low-energy surfaces. From [7.20].

Solid	Critical surface tension γ_c (meV/Å^2)
nylon	2.87
polyvinyl chloride	2.43
polyethylene	1.94
polytetrafluoroethylene (teflon)	1.12

areas during sliding at low velocities, see Sect. 7.4. On the other hand, the critical surface tension for polyethylene is $1.94\,\text{meV}/\text{Å}^2$, which is higher than those of the liquid n-alkanes and polyethylene is wetted by these fluids.

In general, one would expect γ_c to depend not only on the solid, but also on the liquid series. However, when dealing with simple molecular fluids (where the van der Waals interaction dominates), Zisman observed that γ_c is essentially independent of the nature of the liquid, and is a characteristic of the solid alone. Typical values are listed in Table 7.4.

Some comments:

a) The system with high γ_c (nylon, PVC) are those most wettable by organic liquids. They carry relatively strong permanent dipoles.
b) Among systems that are dominated by van der Waals interactions, we note that the CF_2 groups are less wettable (i.e., less polarizable) than the CH_2 groups. In practice, many protective coatings (anti-stain, waterproofing, etc.) are based on fluorinated systems.
c) It is possible to study specifically the wetting properties of terminal groups CF_3- or CH_3- by depositing a surfactant monolayer on a polar solid surface (Fig. 7.28). For CH_3 groups, $\gamma_c \approx 1.7\,\text{meV}/\text{Å}^2$, and for CF_3, γ_c is amazingly small $\sim 0.4\,\text{meV}/\text{Å}^2$. The studies of Zisman and coworkers

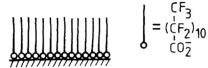

Fig. 7.28. Idealized structure of a surfactant monolayer attached to a polar solid (e.g., a metal oxide). The particular example chosen represents one of the least wettable surfaces ever found.

on the wettability of surfaces covered with highly fluorinate-substituted alkyl groups stimulated several research laboratories to apply these surface properties to polymeric coating materials for textile fibers and fabrics as a means of imparting to them non-staining, oil-, and water-resistant properties. For more details on all these fascinating questions, see the beautiful review of *Zisman* [7.20].

If we want to find a molecular liquid that completely wets a given low-energy surface, we must choose a liquid of surface tension $\gamma_{lv} < \gamma_c$. Thus γ_c may be called a "critical" surface tension and is clearly the essential parameter for many practical applications.

Dynamics of Spreading

In the course of his pioneering work on wettability, Hardy recognized that a spreading droplet is announced by a precursor film, which may be as thin as a single monolayer, which shows up ahead of the nominal contact line. The precursor film revealed itself through its lubricating effect: a small test particle can slip more easily on the solid when the film is present. Hardy believed that these films occurred only with volatile liquids, which could condense ahead of the advancing droplet. This process may very well exist, but recent studies have shown that the film is present even in the absence of any vapor fraction. In fact, in some cases the precursor film may result from the force associated with a gradient in the 2D film pressure, which is proportional to the spreading pressure; in this case the spreading velocity of the precursor film is limited only by the sliding friction which occurs when a molecularly thin adsorbate layer slides on a solid surface; see Sect. 8.6.3.

Capillary Bridges: Adhesion and Sliding Friction

Consider two small fluid drops connected by a thin pipe as shown in Fig. 7.29. In the pipe is a rigid membrane (cross-sectional area A) which separates the fluid on the right from the fluid on the left. The hydrostatic pressures in the fluid on the two sides are denoted by P_1 and P_2. If the membrane is displaced a distance δx, a volume of fluid $\delta V = A\,\delta x$ flows into the drop **1**, while the same volume of fluid flows out of drop **2**. If the displacement δx

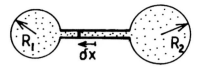

Fig. 7.29. Two small fluid drops connected via a narrow pipe. In the pipe is a rigid membrane which separates the fluids from each other.

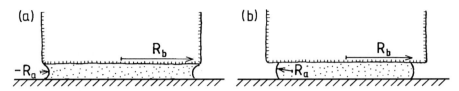

Fig. 7.30a,b. Capillary bridges between a block and a substrate.

occurs slowly so that negligible viscous energy dissipation occurs, then the work $(P_1-P_2)\delta V$ must equal the change in the surface energy of the two drops which can be written as $\gamma_{lv}(\delta A_1 + \delta A_2)$. Assume that the drops are spherical with the radii R_1 and R_2, respectively. Then $\delta V = 4\pi R_1^2 \delta R_1 = -4\pi R_2^2 \delta R_2$, $\delta A_1 = 8\pi R_1 \delta R_1$ and $\delta A_2 = 8\pi R_2 \delta R_2$. Thus $(P_1 - P_2)\delta V = \gamma_{lv}(\delta A_1 + \delta A_2)$ gives

$$P_1 - P_2 = 2\gamma_{lv}\left(\frac{1}{R_1} - \frac{1}{R_2}\right).$$

If we let $R_2 \to \infty$ then $P_2 \to P_{\text{ext}}$ (the external pressure; we neglect gravitational effects) so that the pressure increase $\Delta P = P_1 - P_{\text{ext}}$ equals

$$\Delta P = 2\gamma_{lv}/R_1 . \tag{7.45}$$

In a more general case, for non-spherical fluid surfaces, (7.45) is replaced with

$$\Delta P = \gamma_{lv}\left(\frac{1}{R_a} + \frac{1}{R_b}\right) \tag{7.46}$$

where R_a and R_b are the local radii of curvature of the fluid surface, which are either *positive* or *negative* depending on whether the origin of the radius of curvature is located inside or outside the fluid (in Fig. 7.30a, $R_a < 0$ and $R_b > 0$ while both R_a and R_b are positive in Fig. 7.30b).

Formula (7.46) is the basis for understanding some aspects of adhesion and sliding friction. Consider first the force necessary to remove a block from a substrate with a *very thin* fluid layer. If a capillary bridge of the form indicated in Fig. 7.30a is formed between the block and the substrate, a very large force might be necessary to separate the surfaces. This follows directly from (7.46) (with R_a negative and $R_b \gg |R_a|$), since the pressure in the fluid

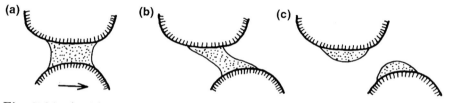

Fig. 7.31. (a, b) and (c) show the successive development of a capillary bridge during sliding.

bridge can be much smaller than the external pressure. As an example, consider a 1 μm thick water bridge between two flat surfaces which are wetted by water. In this case $R_a = -0.5\,\mu\text{m}$ so that (7.46) gives $\Delta P = 144 \times 10^3\,\text{N/m}^2$; thus the force necessary in order to separate two such surfaces with the contact area $100\,\text{cm}^2$ would be $\sim 1440\,\text{N}$! Note that this contribution to adhesion is independent of the speed with which the surfaces are separated. This is different from the other mechanism of fluid "adhesion" discussed in Sect. 7.4 related to the viscosity of the fluid, which depends on the speed with which the surfaces are separated, and, in particular, vanishes when the surfaces are separated very slowly. Furthermore, the present mechanism is particularly important for fluids such as water, which have high surface tension, but smaller for hydrocarbon oils which have much smaller surface tension. Finally, we note that adhesion resulting from capillary bridges occurs in a very wide range of phenomena which are well-known to all of us. For example, it is the reason why one can only build sand castles with moist sand, but not with dry or completely wet sand. It is also the reason why it is very hard to separate two flat glass surfaces with a drop of water in between, while it is easy to separate them when they are completely surrounded by water.

Capillary bridges will also contribute to the sliding friction. If capillary bridges are formed between the block and the substrate, as the block is displaced parallel to the surface, the bridges will first "tilt" and elongate, and exert a force on the block opposite to the sliding direction (Fig. 7.31). When a bridge is finally broken, the surface energy stored in the bridge is dissipated (via the bulk viscosity of the fluid) in the rapid fluid flow which follows the "necking". At low sliding velocity this gives rise to a velocity independent friction force. This is the reason why the friction force during skiing on slush or wet snow ($T = 0°\text{C}$) is usually larger than for dry snow (say $T = -1°\text{C}$); see Sect. 14.2.

Effect of Humidity on Friction: Moisture-Induced Aging

For many sliding systems it is observed that the static friction coefficient increases with the time t for which the two surfaces have been in stationary contact. There are many possible reasons for this behavior. Here we show that in a humid atmosphere, for surfaces which are wetted by water, the

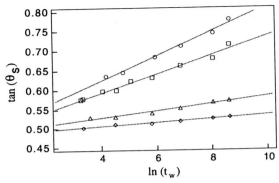

Fig. 7.32. Dependence of the static friction coefficient on the time of stationary contact for a paper–paper interface. Relative humidity (8, 20, 34 and 40%) increases from lower to upper curves. From [7.35]

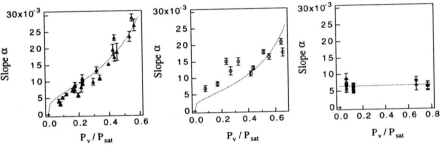

Fig. 7.33. Variation of the slope α with the relative humidity P/P_{sat} for paper (*left*), glass (*middle*) and Teflon (*right*). From [7.35]

formation of capillary bridges leads to a time dependent increase in the adhesion and static frictional forces [7.35]. The formation of liquid bridges in an under-saturated atmosphere is a thermally activated process, which results in a logarithmic dependence of the adhesion and friction forces on the time of stationary contact. This effect has been observed for stone, paper, and glass surfaces, but no influence of the humidity is observed for hydrophobic materials such as Teflon.

In one set of experiments, a block is placed on a plane, and after a waiting time t the tilting angle θ (see Fig. 1.3) is increased until the block begins to move, and the corresponding angle θ_s is measured. Figure 7.32 shows the dependence of the static friction coefficient $\mu_s = \tan\theta_s$ on the logarithm of the waiting time t, for a paper–paper interface. Note that μ_s increases logarithmically with the waiting time, $\mu_s = \mu_0 + \alpha \ln(t/\tau)$, with a prefactor α which increases with increasing humidity. In Fig. 7.33 (left) the slope α

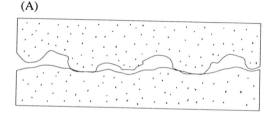

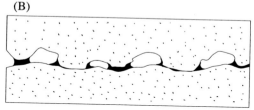

Fig. 7.34. Contact between two hydrophilic surfaces in a humid atmosphere. (**A**) immediately after contact; (**B**) after a long contact time.

of the curves in Fig. 7.32 is plotted as a function of the relative humidity. Similar results for glass and Teflon interfaces are also shown in Fig. 7.33.

Moisture-induced aging has also been observed in granular media. The static properties of granular systems can be (approximately) described in terms of the frictional properties between different layers, leading to the relation $\mu_s = \tan\theta_s$ between the angle θ_s of repose of a granular pile and the coefficient of static friction. Experiments have shown that for granular media in a moisture atmosphere the angle of repose θ_s increases continuously with the time of stationary contact, at a rate which increases with the relative humidity.

In one set of experiments, small glass particles (radius ~ 200 μm) are contained in a rotating drum. The initial state of the system is prepared by rotating the drum and hence mixing the particles. The system is then left to age for some given time period t, after which the angle θ_s is measured (note: θ_s is the rotation angle at which the first avalanche takes place). It was found that the static frictional force increases as $\sim \ln(t/\tau)$, with increasing aging time t.

The results presented above can be understood based on the (thermally activated) formation of water bridges, see Fig. 7.34. That is, in the presence of an under-saturated water vapor, liquid bridges form between two surfaces in contact, and produces an attractive capillary force between them. The curvature $1/R$ of the liquid–vapor interface of a stable water bridge is given by the *Kelvin equation* (see below). In the present case this gives R of the order of a few nanometers. Thus, only liquid bridges of nanoscopic size are stable and can be formed. Such bridges can only form in the regions where the solid surfaces are separated by at most a few nanometers. Practically all surfaces of solids are rough on the nanometer scale, and there will always be regions where the surfaces are separated by a gap h of nanometer order. Let us consider the problem in greater detail.

Liquids that wet surfaces will spontaneously condense (as bulk liquid) from an under-saturated vapor phase, into cracks, pores, or closely spaced regions between two solids in contact. At *equilibrium* the meniscus curvature, $1/R \equiv 1/R_\mathrm{a} + 1/R_\mathrm{b}$, is related to the relative vapor pressure P/P_sat by the *Kelvin equation*:

$$k_\mathrm{B} T \ln(P/P_\mathrm{sat}) = \gamma v_0 / R \qquad (7.46\mathrm{a})$$

where v_0 is the molecular volume in the liquid. We will show below, however, that it may take a very long time to reach the equilibrium state, since the formation of a water bridge is a thermally activated process. The formation of an increasing number of capillary bridges, with increasing time, leads to an increase in the adhesion force between the surfaces, and to an increase in the static friction force.

Let us prove (7.46a). The chemical potential for a molecule in the gas phase (pressure P) is $\mu(P) = k_\mathrm{B} T \ln(P/P_0)$, where P_0 is a reference pressure. The chemical potential for a molecule in the *saturated* vapor phase must be the same as in the bulk liquid (flat surface) so that $\mu_\mathrm{sat} = \mu_\mathrm{liq}$, where $\mu_\mathrm{sat} = k_\mathrm{B} T \ln(P_\mathrm{sat}/P_0)$. Since the pressure in the vapor phase is higher than in the liquid meniscuses by the amount $\Delta P = -\gamma/R$ (where R is negative), a pressure-work of magnitude $\Delta P v_0$ is necessary in order to bring a molecule from the fluid meniscuses into the vapour phase. Using the equation $\mu_\mathrm{liq} = \mu(P) + \Delta P v_0$ gives (7.46a).

Note that under ambient conditions, (7.46a) gives R of nanometer dimensions: for $P/P_\mathrm{sat} = 0.9$, 0.5 and 0.1 we get $R = -100$, -16, and -5 Å, respectively. Thus, liquid bridges are able to form only in nanometer-scale interstices. Since most surfaces are rough on the nanometer scale, there are many regions of the type indicated in Fig. 7.35 where a bridge is thermodynamically stable, but where there is a barrier that hampers formation of the bridge.

Consider a region between two surfaces where the surfaces are within nanometer distance from each other (see Figs. 7.34 and 7.35). Thin wetting films coat both surfaces. For gaps of less than the a critical distance, of order the Kelvin radius R, this state is metastable: capillary condensation should occur. However, an energy barrier has to be overcome, as the coating films have to grow and coalesce in order to fill the gap between the surfaces (see Fig. 7.35). Along this path, the free energy of the system increases to a maximum as the films are about to merge (Fig. 7.35, middle). The energy barrier, ΔE, is roughly the free energy cost of condensing the corresponding water volume from the under-saturated vapor phase: $\Delta E \approx \Delta \mu \, v_1/v_0$, where v_1 is the liquid volume needed to nucleate the liquid

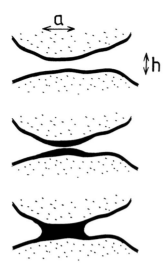

Fig. 7.35. Formation of a liquid bridge (in an under-saturated atmosphere) between two surface asperities separated by a distance h of the order of a nanometer. *Top:* immediately after contact. The surfaces are covered by thin wetting films. *Middle:* activated complex. *Bottom:* thermal equilibrium configuration.

bridge, and where $\Delta\mu = k_{\rm B}T \ln(P_{\rm sat}/P)$ is the difference in the chemical potential between the liquid and the under-saturated water vapor. If we consider two asperities separated by a distance h, the volume $v_1 \approx ha_0^2$, where a_0^2 is a typical nucleation area. Assuming an activated process, the (average) time τ to form a liquid bridge is $\tau = \tau_0 \exp(\Delta E/k_{\rm B}T)$, where τ_0 is a microscopic time. Because the surfaces are rough, many bridges can form in the contact region. It is clear that the nucleating sites will exhibit a broad distribution of gaps h, and that the activation times are accordingly widely distributed. At a given time t, only the bridges with the activation time $\tau < t$, or $\exp(\Delta E/k_{\rm B}T) < t/\tau_0$, have formed. These bridges occur at interstitial regions where $h < h_{\rm max}(t) = (k_{\rm B}T/\Delta\mu)(v_0/a_0^2)\ln(t/\tau_0)$. Once a liquid bridge has formed, it rapidly expand and locally fills the volume surrounding the nucleation site, until the Kelvin equilibrium condition (7.46a) is met. Thus, because of roughness, only a fraction $\Phi(t)$ of the total wettable area is wetted at time t. To a first approximation, $\Phi(t) = h_{\rm max}(t)\rho$, where ρdh equals the number of nucleation sites with surface separation in the interval h to $h+dh$, normalized by the total number of available sites where capillary bridges are thermodynamically stable (we have assumed ρ to be a constant). Thus we expect the adhesion force to increase with time according to

$$F_{\rm adh} \sim \frac{\rho v_0/a_0^2}{\ln(P_{\rm sat}/P)} \ln(t/\tau_0) \,. \tag{7.46b}$$

If we assume that the static friction force depends linearly on F_{adh}, then (7.46b) predicts that the frictional force increases linearly with $\ln t$, and that it depend on the relative humidity according to $[\ln(P_{\text{sat}}/P)]^{-1}$. Both these predictions are in agreement with the experimental data presented above.

The formation of liquid bridges will have two effects on the static frictional force. First, the liquid bridges will give rise to an adhesive force between the solids which will increase the area of real contact, just like an increase in the external load. This will lead to an increase in the shear force necessary to break the junctions. Secondly, as discussed before, if the two solids are slid relative to one another, the liquid bridges will elongate and tilt and exert a force on the solid bodies, opposite to the displacement direction (see Fig. 7.31): This effect will also increase the static frictional force.

7.6 Introduction to Boundary Lubrication

We have seen that at low sliding velocity the lubrication fluid will be squeezed out from the contact areas between the two sliding solids, except for a few monolayers which tend to wet the solid surfaces. This is the regime of boundary lubrication where, at low sliding velocities, the friction force is nearly velocity independent. In this section we present a short overview of boundary lubrication. We discuss the magnitude of the static and kinetic friction coefficients for a number of different sliding systems. We discuss the relation between friction and adhesion hysteresis, and show that both quantities often depend on the ambient atmosphere. We also present an introduction to sliding dynamics resulting from simple friction laws.

Boundary Lubrication

Figure 7.36a shows snapshots from a computer simulation of the sliding of a five-layer thick film of hexadecane (n-$C_{16}H_{34}$) confined between two gold substrates exposing (111) surfaces [7.25]. Both surfaces have gold asperities four atomic layers thick. The relative sliding velocity is $\sim 10\,\text{m/s}$. Note that the lubrication molecules tend to form molecular layers parallel to the gold surfaces. Furthermore, even though the hexadecane–gold surface interaction is rather weak, the lubrication fluid is never completely squeezed out from the contact region between the the asperities, even though the local stress is high enough to plastically deform the asperities. Note also that the layering of the lubrication molecules is clearly seen in the region between the asperities, starting from five layers at "large" asperity separation down to two trapped layers during the act of plastic deformation of the asperities. No "intermediate" layering configuration occurs in any of the pictures indicating that the squeeze-out process (transition from n to $n-1$ layers) is rapid compared to the time the system spends with an integer number of layers between the gold surfaces.

7.6 Introduction to Boundary Lubrication 149

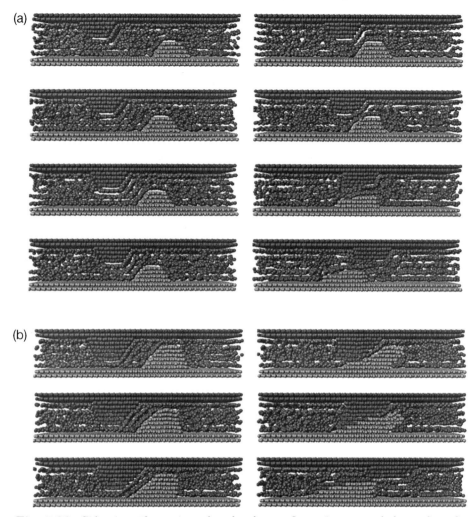

Fig. 7.36. Side view of atomic and molecular configurations recorded at selected times during molecular dynamics simulations of the sliding process of nonuniform gold Au(111) surfaces (lighter and darker grey balls describe the bottom and top surfaces, respectively), lubricated by hexadecane molecules (the molecular segments are described by the medium grey balls). The top surface moves with the velocity $v = 10$ m/s relative to the lower surface. (**a**) Surface separation 23.2 Å. (**b**) Surface separation 21.3 Å. From [7.25].

After the asperities have passed each other their heights are reduced and they are flattened out over a larger surface area (the "base" of the asperity on the lower surface has increased from 14 to 18 Au atoms). This type of "surface smoothening" as a result of plastic deformations is likely to occur, e.g., during the running-in time of a car engine. If the gold surfaces are squeezed together

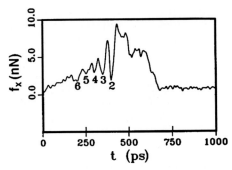

Fig. 7.37. Total force f_x in the x-direction on the gold solid surfaces from the alkane molecules, as a function of time t, for the system in Fig. 7.36a. The numbers designate the "quantized" number of layers in the interasperity zone (Fig. 7.36a). Note the correlation between the force oscillations and the structural variations in the lubricant. From [7.25].

with such a high force that less than five layers of molecules are trapped between the surfaces (Fig. 7.36b), during the collisions between asperities the lubrication layer may be completely squeezed out, leading to metal transfer and wear; the running-in of a car engine should be "gentle" enough to avoid such a complete break-down of the lubrication film.

Figure 7.37 shows the time variation of the force in the shear direction necessary for steady sliding with $v = 10\,\mathrm{m/s}$. Of particular interest are the oscillations in the force prior to the collisions between the asperities, which reflect the layering stages of the lubricating film: compare Figs. 7.36a and 7.37 and note that the marked minima in f_x correspond to a successively decreasing discrete number of molecular layers in the interasperity region. As discussed in Sect. 7.4, the number of layers in this region evolves in a "quantized" manner as successive layers are squeezed out when the asperities come closer together.

The local stress at the asperities at the onset of plastic deformation in Fig. 7.36 is $\sim 4 \times 10^9\,\mathrm{N/m^2}$, i.e., about a factor of 10 *higher* than the penetration hardness of gold. As discussed in Sect. 5.2, this is related to the fact that plastic deformation of gold in a macroscopic measurement, e.g., an indentation measurement, is determined by the motion of pre-existing dislocations which are absent in the model study presented above, as well as in real systems when the asperities are as small as those in the model study (Sect. 5.2).

It is clear from the discussion above that a good boundary lubricant should have a strong tendency to wet the sliding surfaces so that at least one or two monolayers of lubrication molecules remain trapped between the surfaces even when two asperities collide. The interaction between saturated hydrocarbons and metal or metal oxide surfaces is mainly via the relatively weak van der Waals interaction leading to a relatively low adsorbate–substrate binding energy and a low tendency to wet the surfaces. For this reason, commercial lubrication oils, the basis of which is usually long-chain hydrocarbons, have "additives" to improve their boundary lubrication properties. One common additive is long-chain fatty acids which consist of a hydrocarbon chain with a polar head which binds strongly to many metal oxides, forming a dense film with the inert hydrocarbon tail pointing away from the surface; see Fig. 2.5a. These molecules are usually not removed even when two asperities collide. However, inert metals like gold or platinum are not effectively lubricated by fatty acids. Conversely, chemically reactive metals, such as copper or cadmium, are effectively lubricated by fatty acids. We have also seen in Sect. 5.2 that n-hexadecanethiol [$CH_3(CH_2)_{14}CH_2SH$] binds strongly to gold surfaces, forming an excellent protection and boundary lubrication film. It is clear that, depending on the nature of the sliding surfaces, different boundary lubricants must be used.

As long ago as 1920, Langmuir showed that a single molecular layer of a fatty acid deposited on glass reduces the friction to the same value as observed with excess fatty acid. Nevertheless, when repeated sliding over the same film of fatty acid is performed, a single molecular film is worn away. It is clear that the best protection and lubrication will be afforded by an excess of lubricant.

Experiments have shown that for smooth sliding the (kinetic) friction force shows little change when the speed varies from $0.001\,\mathrm{cm/s}$ to $2\,\mathrm{cm/s}$. When stick-slip motion occurs, the mean coefficient of friction (which is equal to the kinetic friction force if the motion is underdamped, see below) shows little change, but the maximum coefficient of friction at stick diminishes with increasing sliding velocity (Sect. 12.2). Experiments have shown, however, that lubricants containing fatty acids are much more likely to provide smooth sliding than lubricants containing only paraffins (saturated hydrocarbons). The reason for this will be discussed below.

Under a heavy load the boundary lubricant can be completely removed from the asperity contact region, and the metal oxide layer broken, leading to direct contact between the metal surfaces and to the formation of cold-welded junctions. Chlorinated hydrocarbons (e.g., CH_2Cl_2 or CCl_4) or polysulfides [$H_3C\text{-}(S)_n\text{-}CH_3$] are commonly added to the basic lubricating fluid when the lubricant is used under extreme pressure conditions [7.26,27]. When these molecules come into contact with, e.g., a bare iron surface they decompose and form hard and inert protective coatings on the steel surface, consisting of thin layers of $FeCl_2$ or FeS. However, the chlorine and sulfur compounds are often corrosive and may destroy the sliding surfaces. For example, polysulfide lubricant additives effectively passivate and protect ferrous metals but often corrode copper-containing metal alloys such as bronze and brass. This is a serious limitation, as ferrous and non-ferrous metals are commonly used to fashion different parts of the same mechanical system. The interaction of polysulfide molecules with Cu_2O surfaces is at present being studied in order to understand the origin of the corrosion and passivation mechanisms [7.28].

Static and Kinetic Friction

As discussed earlier, most "normal" surfaces are covered by an (unknown) layer of grease or other weakly bound molecules, which may act as a boundary lubricant during sliding. As a result, many reported values of the static and kinetic friction coefficients refer to ill-defined systems.

In Table 7.5 we show the static and kinetic friction coefficients for a number of carefully prepared sliding systems. The data has been grouped into four categories.

The first category consists of organic fluids (e.g., saturated hydrocarbons) which interact relatively weakly with the sliding surfaces, and it is likely that they go from a solid pinned state at "stick" to a fluidized state during sliding. Note that while μ_s and μ_k vary strongly between the different systems (by a factor of ~ 20) the ratio μ_k/μ_s is always close to 0.5. In fact, this ratio is also nearly equal to 0.5 for many "clean" surfaces. But, as pointed out above, "clean" surfaces, unless special care is taken, are usually covered by a layer of "grease" (hydrocarbons) which may have a very similar influence on the sliding friction to the lubricants quoted in Table 7.5. It is interesting to note that there is no correlation between the viscosity of a lubrication oil and the friction coefficients μ_s and μ_k and, in particular, the ratio $\mu_k/\mu_s \approx 0.5$ is independent of the film viscosity. For example, the oils in Table 7.5 have the viscosities (in Ns/m^2):

Atlantic spindle oil: 0.03; Liberty aero oil: 0.89; Castor oil: 0.47,

Table 7.5. The static and kinetic friction coefficients, μ_s and μ_k, for a number of different sliding systems. From [7.29–31]

System	Lubricant	μ_s	μ_k	μ_k/μ_s
steel on babbitt	atlantic spindle oil	0.25	0.13	0.52
steel on babbitt	castor oil	0.12	0.06	0.50
steel on babbitt	lard oil	0.10	0.05	0.50
steel on steel	castor oil	0.15	0.08	0.53
steel on steel	diethylene glycol stearate	0.089	0.083	0.93
steel on steel	calcium stearate	0.113	0.107	0.95
steel on steel	aluminium stearate	0.113	0.107	0.95
steel on steel	lithium-12-hydroxy stearate	0.218	0.211	0.97
steel on steel	molybdenum disulfide	0.053	0.050	0.94
steel on steel	barium hydroxide	0.163	0.151	0.93
steel on steel	silver iodide	0.245	0.231	0.94
steel on steel	borox	0.226	0.210	0.93
diamond on diamond	arbitrary	~ 0.1	~ 0.1	
teflon on teflon	arbitrary	~ 0.05	~ 0.05	

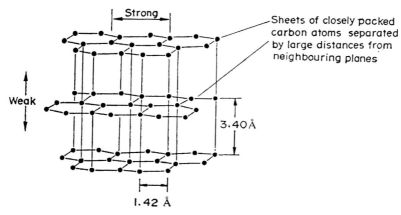

Fig. 7.38. Sliding of one graphite lamella over another.

but μ_k and μ_s are very similar in all cases. The ratio μ_k/μ_s averaged over all the 12 measurements reported by Campbell [7.29] equals 0.51 ± 0.14.

The second category in Table 7.5 are fatty acid lubricants which react so strongly with the surfaces that no fluidization of the adsorbate structures takes place and sliding is likely to occur between the ends of the hydrocarbon tails (dotted plane in Fig. 2.5b). In this case the ratio μ_k/μ_s is close to unity.

The third category in Table 7.5 are inorganic layered lattice systems (similar to graphite). For these systems the lubricity is due to the sliding of one lamella over another (Fig. 7.38), which is made possible by the strong bonding forces within the planes and the relatively weak bonding forces between the planes. Here again, the static and kinetic friction are nearly the same.

Finally, in Table 7.5 we show the friction coefficients for sliding of diamond on diamond and teflon on teflon. These systems are remarkable because the sliding friction is nearly the same with or without an organic lubrication fluid. This can be understood as follows: For diamond, because of its high penetration hardness ($\sim 1 \times 10^{11}\,\mathrm{N/m^2}$), the local pressure in the contact areas is likely to be so high that the lubrication fluid is completely squeezed out. Note that diamond surfaces are usually passivated by hydrogen or oxygen, and are relatively inert. Teflon surfaces are extremely inert and most organic fluids do not wet teflon surfaces. Hence, in this case, too, the lubrication fluid may be squeezed out from the contact areas even if the local pressure is relatively low (Sect. 7.4).

The fact that μ_k/μ_s is close to 0.5 for surfaces lubricated by many organic fluids, e.g., hydrocarbon molecules, while it is close to unity for fatty acids, indicates that two fundamentally different sliding scenarios occur for these two classes of systems. Hydrocarbons are bound to solid surfaces mainly via the van der Waals interaction and the lateral corrugation of the interaction potential is weak. As a result, the shear force in a sliding experiment is strong enough to pull the adsorbates over the lateral energy barriers in the system, i.e., the lubrication layer during motion will slide relative to the solid surfaces. In this kinetic state the lubrication layer is likely to be in a 2D-fluid state whereas during "stick" the adsorbate layer forms a solid pinned state which, in simple cases, may be a commensurate solid, but in most practical cases is probably a glassy state. Thus the transition from stick to slip may involve a *discontinuous dynamical phase transition*, and there is no reason why the friction force during sliding should be close to the static friction force. In fact, arguments will be given in Sect. 8.9 for why μ_k/μ_s may be close to 0.5 for this class of systems.

Next we consider lubrication by fatty acids. The fatty acid molecules bind with relatively strong bonds to specific sites on the solid surfaces so that usually no fluidization or shear melting of the lubrication film will occur during sliding. It is still possible for μ_s to be larger than μ_k, e.g., for the following reason: During stick the hydrocarbon tails may *interdiffuse* (Fig. 12.6). The free energy barriers for interdiffusion are often quite large and the system may need a long time to find the minimum free energy configuration for which the static friction force takes its highest value. Hence, the static friction force will, in general, increase slowly with increasing time of stationary contact. However, fatty acid monolayers are usually rather compact which inhibits interdiffusion. More importantly, however, is the very high pressure in the contact areas which tends to increase the barrier towards interdiffusion (Sect. 7.2); thus during typical stick periods very little interdiffusion of the hydrocarbon tails occurs, which could lead to a difference between the static and kinetic friction force. Similarly, for the layered material in Table 7.5, e.g., graphite, one does not expect any slow relaxation processes to occur, which may increase the friction force with increasing time of stationary contact, and as a result $\mu_k \approx \mu_s$.

Influence of Temperature and Ambient Conditions on Friction and Adhesion Hysteresis

We have pointed out that fatty acid monolayers on metal oxides usually form dense layers; this, in combination with the very high local pressures which occur in the contact areas during sliding of metals on metals, leads to large barriers towards interdiffusion of the hydrocarbon chains and to correspondingly long relaxation times τ. As explained above, this is probably the reason why the kinetic friction force nearly equals the static friction force when metal surfaces are lubricated by fatty acids (Table 7.5). However, it is also possible to prepare grafted monolayer films (Fig. 7.39a) with large separation between the chains and to perform measurements at much lower pressures, e.g., using the surface forces apparatus; in these cases the barrier towards interdiffusion may be small. One such example was given in Fig. 3.8a, where τ is so short that the sliding dynamics were fluid-like without stick-slip. More generally one may, loosely speaking, distinguish between *solid-like*,

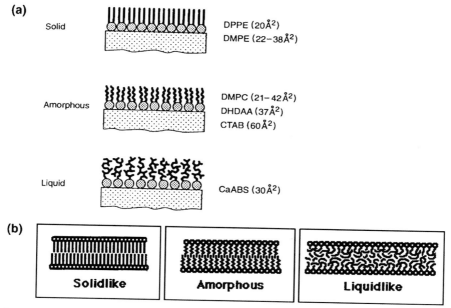

Fig. 7.39. (a) Surfactant monolayers used in friction and adhesion experiments and their typical phase states (room temperature). (b) Different states of grafted monolayers (schematic). Low friction is exhibited by solid-like ($\tau^* \ll \tau$) and liquid-like ($\tau^* \gg \tau$) layers; high friction is exhibited by amorphous ($\tau^* \sim \tau$) layers. This classification refers to a particular experiment involving the time scale τ^* and τ is a relaxation time associated with the monolayer films, e.g., related to interdiffusion of the chains from the two surfaces. From [7.32].

amorphous, and *liquid-like* monolayers (Fig. 7.39b) depending on whether $\tau^* \ll \tau$, $\tau^* \sim \tau$, or $\tau^* \gg \tau$, respectively, where τ^* is a time characterizing the speed with which the monolayer is probed, e.g., $\tau^* \sim D/v$ in a (steady) sliding friction experiment (where v is the sliding velocity and D is on the order of the chain length). This definition of the "aggregate state" of the monolayer film is convenient but does not refer to any fundamental difference in the thermal equilibrium properties in the three different cases (as would be the case if these states were separated by phase transitions) but rather to a classification which refers to *the time scale τ^* involved in a particular experiment*. Thus a film which may be liquid-like with respect to a "slow" experiment may be solid-like with respect to a "fast" experiment. This effect is, of course, not limited to monolayer films but is quite general. Thus a layer of tar, for example, behaves as a viscous fluid with respect to the slowly varying force from a plant which penetrates it from below, while it behaves as a solid when a human walks on it.

In general, with increasing temperature a monolayer film goes from solid-like → amorphous → liquid-like, even if no sharp (phase) transition occurs as

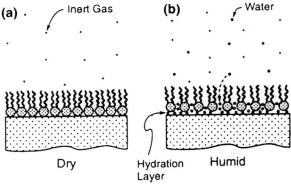

Fig. 7.40. Schematic illustration of the structure of grafted monolayers exposed to (a) inert dry atmosphere and (b) humid air. From [7.33].

a function of temperature. It has been found that the "state" of the monolayer film at a given temperature often depends on the nature of the ambient atmosphere. For example, increased vapor humidity leads to increased water penetration, mainly into the head-group–substrate interface region (Fig. 7.40). It results in the swelling of the monolayer and an increased fluidity of the film.

As discussed in Chap. 5, adhesion hysteresis and sliding friction are small when the monolayer film is in a *liquid-like* or *solid-like* state but high in the intermediate *amorphous* state. Thus for fluid chains (fast relaxation times), the system is at equilibrium at all stages of a loading–unloading cycle (or friction experiment) and the adhesion energy is reversible and equal to the thermodynamic equilibrium value. With amorphous chains, if their characteristic interdigitation or relaxation time is comparable to the inverse of the loading–unloading rate, the unloading energy (or separation energy) is much higher than the loading energy, resulting in a large adhesion hysteresis. As we approach the other extreme of solid-like monolayers (whose chains are usually close-packed), no interdigitation occurs over the time scale of the measurement, and the adhesion energy is again low and reversible.

As an illustration of the discussion above, Fig. 7.41 shows the adhesion hysteresis of two DMPE monolayers at 25°C at 0% and 100% relative humidity. When exposed to dry conditions the DMPE film is in the solid state, but when exposed to humid air, the film becomes more amorphous. This explains why the hysteresis is larger for the 100% relative humidity case. Figure 7.42 shows the kinetic friction force F for the two DMPE layers; as expected, F increases with increasing humidity.

7.6 Introduction to Boundary Lubrication   157

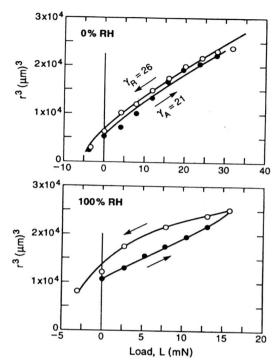

Fig. 7.41. Adhesion hysteresis at 25°C for two DMPE monolayers at 0% and 100% relative humidity. The radius of the contact area is denoted by r. From [7.34].

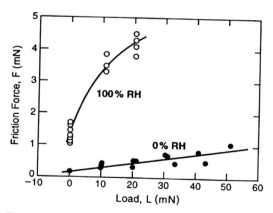

Fig. 7.42. Friction forces at 25°C for two DMPE monolayers at 0% and 100% relative humidity. From [7.34].

Sliding Dynamics Resulting from Simple Friction Laws

Consider a block of mass M on a substrate and assume that a spring with force constant k_s is connected to the block as indicated in Fig. 3.1. Assume that the block is rigid and that the free end of the spring moves with the velocity v_s. In this section we discuss the nature of the motion of the block, under the assumption that the friction force, $-F_0(t)$, only depends on the instantaneous velocity $\dot{x}(t)$ of the block. In particular, we consider the friction "laws" (a–f) indicated in Fig. 7.43. As will be shown below, using friction laws which only depend on the instantaneous velocity $\dot{x}(t)$ is, in general, not a good approximation. A more accurate study of sliding dynamics will be presented in Chaps. 12 and 13.

Let $x(t)$ be the position coordinate of the block at time t, and assume that at $t = 0$ the block is stationary relative to the substrate [$x(0) = 0$ and $\dot{x}(0) = 0$] and that the spring force vanishes. Now, for "low" spring velocities v_s the solid block is either in a pinned state relative to the substrate (i.e., $\dot{x} = 0$) in which case the friction force is $F_0 \leq F_a$ (in Fig. 7.43a the static friction force vanishes), or else in a sliding state where the friction force equals $F_0(v)$ given by Fig. 7.43. The equation of motion for the block has the form

$$M\ddot{x} = k_s(v_s t - x) - F_0 . \tag{7.47}$$

If we measure time in units of $(M/k_s)^{1/2}$, distance in units of F_a/k_s, velocity in units of $F_a(Mk_s)^{-1/2}$ and the friction force F_0 in units of F_a then (7.47) takes the form

$$\ddot{x} = v_s t - x - F_0 . \tag{7.48}$$

In Fig. 7.44 we show the spring force $F_s = v_s t - x(t)$ as a function of time. The different curves have been calculated using the friction "laws" (a–f) of Fig. 7.43 and in all cases the spring velocity increases abruptly at $t = 50$, as indicated in Fig. 7.44.

In cases (a) and (b) smooth sliding occurs for all values of v_s and k_s. In cases (c) and (d) stick-slip motion occurs for all parameter values. Finally, in cases (e) and (f) stick-slip motion occurs when v_s is below some critical value v_c^+ which depends on k_s, while smooth sliding occurs when $v_s > v_c^+$.

Let us analyze cases (d) and (f) in more detail. The result shown in Fig. 7.44d can be understood as follows: Initially, as time increases the spring will extend, but the solid block will not move until the force in the spring reaches F_a. At this point the two surfaces can start to slide relative to each other. During sliding the friction force is $F_b < F_a$ and since the spring force is F_a at the onset of sliding the block will initially accelerate to the right. Note that owing to the inertia of the block, the velocity $\dot{x}$ is initially lower than v_s and the spring will continue to extend for a while before finally $\dot{x} > v_s$. The maximum spring force will therefore not occur exactly at the onset of sliding but slightly later, where the spring force is greater than F_a. In Fig. 7.44d

7.6 Introduction to Boundary Lubrication   159

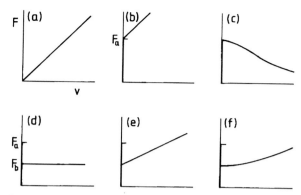

Fig. 7.43. The friction "law" is given by (a) $F_0 = F_a v/v_0$ while in cases (**b–f**), $F_0 \leq F_a$ if $v = 0$ and if $v > 0$, (b) $F_0 = F_a(1 + v/v_0)$, (c) $F_0 = F_a/\left[1 + (v/v_0)^2\right]$, (d) $F_0 = F_b, (F_b < F_a)$, (e) $F_0 = F_b(1 + v/v_0)$, (f) $F_0 = F_b\left[1 + (v/v_0)^2\right]^{1/2}$.

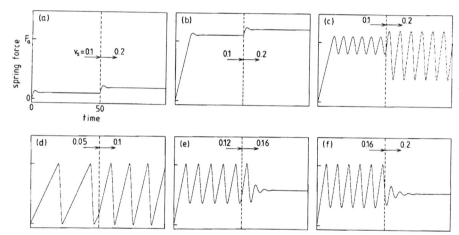

Fig. 7.44. The spring force F_s as a function of time for the friction laws (**a–f**) in Fig. 7.39. In all cases, $F_b = F_a/2$ and $v_0 = 1$ while in cases (**a–c**), $v_s = 0.1$ for $t < 50$ and 0.2 for $t > 50$ and in case (**d**), $v_s = 0.05$ for $t < 50$ and 0.1 for $t > 50$, and in case (**e**), $v_s = 0.12$ for $t < 50$ and 0.16 for $t > 50$, and in case (**f**), $v_s = 0.16$ for $t < 50$ and 0.2 for $t > 50$.

this inertia effect is very small (this is typically the case in sliding friction experiments) and the maximum spring force is nearly equal to F_a. When the sliding velocity $\dot{x} > v_s$ the spring force decreases, but the analysis presented below shows that the sliding motion does not stop when the spring force equals F_b but continues until it reaches $\approx 2F_b - F_a$ where the motion stops, and the whole cycle repeats itself. This model is simple enough that one can analytically evaluate the amplitude ΔF of the oscillations in the spring force which occur during sliding, as well as the stick and slip time periods. Assume that the sliding process starts at $t = 0$ and let t_0 denote the time when the slip starts so that $v_s t_0 = F_a$. The general solution of (7.48) valid during the first slip period is

$$x(t) = v_s t - F_b - A \sin(t + \alpha) \tag{7.49}$$

and the spring force is

$$F_s(t) = v_s t - x(t) = F_b + A \sin(t + \alpha) . \tag{7.50}$$

At $t = t_0$ both x and $\dot{x}$ vanish so that

$$x(t_0) = v_s t_0 - F_b - A \sin(t_0 + \alpha) = 0 ,$$

$$\dot{x}(t_0) = v_s - A \cos(t_0 + \alpha) = 0 ,$$

where $v_s t_0 = F_a$. Hence

$$A = \left[v_s^2 + (F_a - F_b)^2\right]^{1/2} , \tag{7.51}$$

$$t_0 + \alpha = \arcsin\left[(F_a - F_b)/A\right] \equiv \phi . \tag{7.52}$$

Now, during the first slip period $t + \alpha$ increases continuously from ϕ to $2\pi - \phi$ [note that $\cos(2\pi - \phi) = \cos\phi$]. Hence, slightly after the beginning of the sliding, for $t + \alpha = \pi/2 > \phi$, the spring force takes its largest value $F_b + A$ [see (7.50)]; this delay is caused by inertia effects. Similarly, slightly before the end of the sliding, for $t + \alpha = 3\pi/2 < 2\pi - \phi$, the spring force takes its smallest value $F_b - A$. Hence, the amplitude ΔF of the oscillations in the spring force is $\Delta F = 2A$ or, using (7.51),

$$\Delta F = 2\left[v_s^2 + (F_a - F_b)^2\right]^{1/2} .$$

The time between a maximum and the following minimum of F_s equals π. Similarly, the time between a minimum and the following maximum in F_s equals $2(F_a - F_b)/v$.

Next, let us consider case (f). Starting with steady sliding and reducing the spring velocity v_s, steady sliding is replaced by stick-slip motion when $v_s < v_c$. If one starts instead with stick-slip and increases v_s, stick-slip disappears at $v_s = v_c^+$ where $v_c^+ \geq v_c$. In Fig. 7.45 we show the (k_s, v_s) "phase diagram"; the dashed area indicates the region where stick-slip motion occurs when starting with stick-slip and increasing v_s. The boundary line $v_s = v_c^+(k_s)$ in the figure has been obtained by numerical integration of the equation of motion. On the other hand, the critical velocity v_c can be determined by a simple linear instability analysis. During steady sliding $x = x_0 + v_s t$ which satisfies (7.47) if $k_s x_0 + F_0(v_s) = 0$. To determine when steady sliding becomes unstable, let us write

$$x = x_0 + v_s t + \xi ,$$

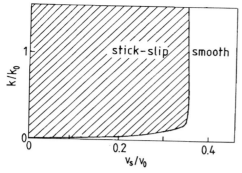

Fig. 7.45. The (k_s, v_s) "phase diagram" for the friction law shown in Fig. 7.43f. The dashed area shows the region where stick-slip motion occurs (when increasing v_s from zero), while smooth sliding occurs in the other region of the dynamical phase diagram.

where $\xi(t)$ is a small perturbation. Substituting this equation into (7.47) and expanding $F_0(v_s + \xi) \approx F_0(v_s) + F_0'(v_s)\xi$ gives

$$M\ddot{\xi} = -k_s\xi - F_0'(v_s)\dot{\xi} \ . \tag{7.53}$$

Assume that $\xi \propto \exp(\kappa t)$. Substituting this into (7.53) gives

$$\kappa^2 + [F_0'(v_s)/M]\kappa + (k_s/M) = 0 \ .$$

This equation has two roots. If the real part of the roots are negative then the perturbation ξ of the steady motion will decay with increasing time, i.e., the steady sliding state is stable with respect to *small* perturbations. On the other hand, if a root has a positive real part the steady motion is unstable. Thus the line $v_s = v_c(k_s)$ in the (k_s, v_s)-plane, separating steady sliding from stick-slip motion, is determined by $\mathrm{Re}\kappa = 0$, i.e., by $F_0'(v_s) = 0$. Note that this condition is *independent* of k_s. Thus for all models where the friction force depends only on the instantaneous velocity of the block the $v_s = v_c(k_s)$ curve will be a vertical line in the (k_s, v_s)-plane. This is contrary to experimental observations, where it is found that stick-slip can always be eliminated by using a stiff enough spring k_s. However, as discussed in Chap. 3 (see also Chaps. 9 and 12), it follows naturally in models where the static friction force increases monotonically with the time of stationary contact. There are several different physical origins of this effect, e.g, chain interdiffusion as discussed above or a slow increase in the area of real contact during stick (Chap. 5). Note that in case (f) the condition $F_0'(v_s) = 0$ gives $v_c = 0$, i.e., when lowering v_s from a high value, steady sliding occurs the whole way down to $v_s = 0$. Since $v_c^+ > 0$ (Fig. 7.45) this implies that hysteresis occurs as a function of v_s.

The condition $F_0'(v_s) = 0$ implies that for all models where F_0 increases with increasing sliding velocity (for large sliding velocities), steady sliding will be stable for large enough v_s. This is in accordance with experiments, and real sliding systems tends to have a kinetic friction force which increases linearly with $\dot{x}$ for large $\dot{x}$ as in Fig. 7.43(b, e, and f). The fact that stick-slip disappears for large sliding velocity is in accordance with everyday experience. For example, when a door is opened slowly, a loud squeak (associated with stick-slip motion) usually occurs, whereas, if the door is opened fast enough, steady sliding occurs and no noise is generated. The friction laws (a–d) cannot explain this observation.

Finally, we note that the discussion in this section has been based on a number of implicit assumptions:

1) The elastic properties of the block have been neglected. As will be shown in Sect. 9.2, it is necessary to account for the elastic properties of the block in order to understand "starting" and "stopping" of sliding.
2) We have assumed that the friction law relates the friction force $F_0(t)$ to the velocity $\dot{x}(t)$ at *the same time* t. This assumes that the important relaxation processes in the lubrication film occur rapidly compared with the typical times involved in the sliding process (e.g., the "slip-time" during a stick-slip cycle). We have already indicated that this assumption is not a good approximation, since molecules on surfaces may exhibit large rearrangement barriers which imply long relaxation times. Hence, in general, one must replace the local (in time) relation between $\dot{x}$ and F_0 with a non-local relation, where the force F_0 at time t depends on the velocity $\dot{x}(t')$ at earlier times, $t' \leq t$.
3) In the model above the block is treated as a point mass. However, if the linear size of the block is larger than a characteristic length ξ, the so-called elastic coherence length (to be defined in Sect. 10.2), this assumption is no longer true. The elastic coherence length is determined by the elastic properties of the block and depends on the nature of the block–substrate interaction potential.
4) In contrast to the models studied above, when exposed to a constant external force $F < F_a$ the block may move slowly via thermally activated processes ("creep" motion). Hence, strictly speaking, the static friction force may vanish.

We will return to all these points later when we continue our study of sliding dynamics in Chaps. 9, 12, and 13.

Appendix: Elastohydrodynamics

The main physics of elastohydrodynamics can be understood by considering a rigid cylinder (radius R and length B) sliding or rolling on a flat elastic substrate, see Fig. 7.46. Neglecting side leakage (in the y-direction), the pressure $P(x)$ satisfies (7.10):

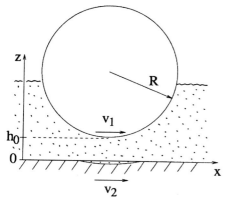

Fig. 7.46. A rigid cylinder (radius R) sliding or rolling on a lubricated substrate.

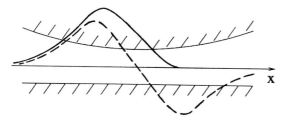

Fig. 7.47. Solid and dashed lines show the pressure distribution $P(x)$ assuming two different boundary conditions (see text).

$$\frac{dP}{dx} = 6\mu v_0 \left(\frac{h - h^*}{h^3}\right), \tag{7.54}$$

where $v_0 = v_1 + v_2$, and where h^* is the film thickness at which the pressure gradient $dP/dx = 0$. The viscosity μ is assumed to depend on P according to

$$\mu = \mu_0 e^{\alpha P}. \tag{7.55}$$

Within the region of interest, the thickness of the lubrication film is given by

$$h \approx h_0 + x^2/2R - u_z(x), \tag{7.56}$$

where $u_z(x)$ is the (elastic) normal displacement of the surface of the substrate. Assume first that $\alpha = 0$, and consider a rigid substrate so that $u_z = 0$. In this case it is possible to integrate (7.54) analytically, and if it is assumed that the pressure vanishes for $x = \pm\infty$, then the pressure distribution will take the form shown by the dashed line in Fig. 7.47. It is positive in the converging zone and equally negative in the diverging zone at exit: the total load supported by the film is clearly zero in this case. This result is unphysical since a region of large negative pressure cannot exist in normal ambient conditions. In practice the flow at exit breaks down into streamers separated by fingers of air penetrating from the rear. The pressure is ambient (i.e. nearly zero) in this region. The point of film breakdown is (approximately) determined by the condition $P = 0$ and $dP/dx = 0$. Using this boundary condition, instead of $P(\infty) = 0$, results in the pressure distribution shown by the solid curve in Fig. 7.47, and the load L supported by the film is given by

$$L = \int_{-\infty}^{\infty} dx dy\, P(x) \approx 2.45 v_0 R B \mu_0 / h_0. \tag{7.57}$$

In most applications it is the load L which is specified, and (7.57) then enables the minimum film thickness h_0 to be calculated. For effective hydrodynamic lubrication h_0 must be larger than the height of the surface irregularities.

Let us now include the elasticity of the substrate. Using the theory of elasticity it is possible to show that a normal surface stress $P(x)$ acting on a semi-infinite elastic solid gives rise to the surface displacement

$$u_z = -\frac{2}{\pi E^*} \int_{-\infty}^{\infty} dx'\, P(x') \ln \left| \frac{x - x'}{x} \right|, \tag{7.58}$$

with $E^* = E/(1-\nu^2)$ where E is the elastic modulus and ν the Poisson ratio. Let us introduce the dimensionless variables:

$$\bar{P} = \alpha P, \quad \bar{x} = (x/R)(\alpha E^*),$$

$$\bar{h} = (h/R)(\alpha E^*)^2.$$

In these variables (7.54-56) and (7.58) give

$$e^{-\bar{P}} \frac{d\bar{P}}{d\bar{x}} = \frac{6\mu_0 \alpha^4 (E^*)^3 v_0}{R} \left(\frac{\bar{h} - \bar{h}^*}{\bar{h}^3} \right), \tag{7.59}$$

$$\bar{h} = \bar{h}_0 + \frac{\bar{x}^2}{2} - \frac{2}{\pi} \int_{-\infty}^{\infty} d\bar{x}'\, \bar{P}(\bar{x}') \ln \left| \frac{\bar{x} - \bar{x}'}{\bar{x}} \right|. \tag{7.60}$$

The parameter $\bar{h}_0$ is determined by the load

$$L = \int dx dy\, P = \frac{R^2}{\alpha^3 (E^*)^2} \int d\bar{x} d\bar{y}\, \bar{P},$$

or

$$\int d\bar{x} d\bar{y}\, \bar{P} = \alpha^3 (E^*)^2 L / R^2.$$

Thus, $\bar{h}$ and $\bar{P}$ are functions of the two dimensionless parameters $\bar{L} = \alpha^3 (E^*)^2 L / R^2$ and $\bar{v}_0 = \mu_0 \alpha^4 (E^*)^3 v_0 / R$. However, it is more convenient to use two other (dimensionless) parameters, namely

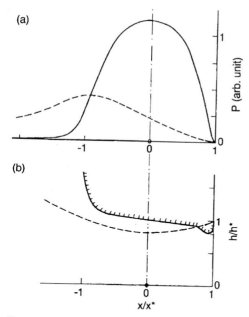

Fig. 7.48. Lubrication of elastic rollers. Pressure index $\alpha = 0$. (a) Film pressure $P(x)$; (b) film thickness $h(x)$. x^* is the x-value for which $h = h^*$. From [7.38].

$$g_\alpha = \frac{\bar{L}^3}{\bar{v}_0^2} = \frac{\alpha L^3}{v_0^2 \mu_0^2 R^4},$$

$$g_E = \frac{\bar{L}^{8/3}}{\bar{v}_0^2} = \left(\frac{L^4}{E^* \mu_0^3 R^5 v_0^3}\right)^{2/3}.$$

Note that $g_\alpha = 0$ when $\alpha = 0$, i.e. g_α vanishes for isoviscous liquids, while $g_E = 0$ when $E = \infty$, i.e. g_E vanishes for rigid solids. In general,

$$\bar{h} = f(g_\alpha, g_E, \bar{x})$$

or

$$h = R(\alpha E^*)^{-2} f(g_\alpha, g_E, \bar{x}) = R \left(\frac{\mu_0 v_0 R}{L}\right)^2 (g_E^3/g_\alpha^2) f(g_\alpha, g_E, \bar{x})$$

$$= R \left(\frac{\mu_0 v_0 R}{L}\right)^2 F(g_\alpha, g_E, \bar{x}).$$

Let us first assume that the pressure $P \ll 1/\alpha$ everywhere, so that we can neglect the pressure dependence of the viscosity: $\mu \approx \mu_0$. This is the case, for example, when a rubber block is slid on a lubricated substrate. In this case the solution to (7.59) and (7.60) depends only on the parameter g_E. Figure 7.48

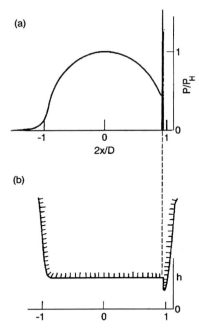

Fig. 7.49. Lubrication of elastic rollers. Pressure index $\alpha \neq 0$. (a) Film pressure $P(x)$; (b) film thickness $h(x)$. D and P_H denote, respectively, the Hertzian contact diameter and maximum contact pressure. From [7.36].

shows for a typical case the pressure distribution $P(x)$ and the film thickness $h(x)$ for a rigid substrate (dashed curves) and for an elastic substrate (solid curves). In the latter case, when the elastic flattening becomes large compared to the film thickness, the pressure distribution approaches that of Hertz for unlubricated contact. Note that close to the exit the pressure gradient dP/dx must become negative. It follows from (7.54) that this is possible only if h is smaller than h^* in this region of the contact area. This is the reason for the constriction in the film at exit which is a feature of all elastohydrodynamic film profiles. From the point of view of effective lubrication it is the minimum film thickness $h_{\min} \approx 0.8 h^*$ which is important.

Let us now assume that the viscosity index $\alpha \neq 0$. Figure 7.49 shows the typical film shape and pressure distribution in this case. Comparing Figs. 7.48 and 7.49 shows that the pressure–viscosity effect has a marked influence on the behaviour. Over an appreciable fraction of the contact area the film thickness is approximately constant. This follows from (7.59) since when the exponent $\bar{P} = \alpha P \gg 1$ the left-hand side becomes small, and hence $h - h^*$ becomes small, i.e., $h \approx h^* =$ constant. The corresponding pressure distribution is basically that of Hertz for dry contact, but a sharp pressure peak occurs on the exit side.

It is important to determine the regions in the (g_E, g_α)-plane where the elastic deformation and/or the variable viscosity effects are important, and to determine (approximate) analytical expressions for the dependence of the minimum film-thickness on the parameters g_E and g_α. This has recently been done for the important case of elliptical conjunctions. Because of its practical importance, we give here the result for a spherical object (radius R) sliding or rolling on a substrate. For this case numerical calculations have been performed for a large set of typical cases, corresponding to four different parameter regimes I–IV, and the results have been fitted to simple analytical expressions. The results can be summarized as follows [7.36]:

I: $\alpha = 0$, $E = \infty$

$$h_{\min} \approx 140R \left(\frac{\mu_0 v_0 R}{L}\right)^2 \tag{7.61}$$

II: $\alpha \neq 0$, $E = \infty$

$$h_{\min} \approx 3.4 R g_\alpha^{0.38} \left(\frac{\mu_0 v_0 R}{L}\right)^2 \tag{7.62}$$

III: $\alpha = 0$, $E \neq \infty$

$$h_{\min} \approx 3.3 R g_E^{0.67} \left(\frac{\mu_0 v_0 R}{L}\right)^2 \tag{7.63}$$

IV: $\alpha \neq 0$, $E \neq \infty$

$$h_{\min} \approx 1.7 R g_\alpha^{0.49} g_E^{0.17} \left(\frac{\mu_0 v_0 R}{L}\right)^2. \tag{7.64}$$

Regime III is relevant for soft elastic solids such as rubber, while regime IV prevails for most metal surfaces lubricated by typical oils. Based on these results it is possible to construct an approximate contour map of $h_{\min}$ in the (g_E, g_α)-plane, see Fig. 7.50. By using this figure for given values of g_E and g_α, one can determine both the prevailing fluid film lubrication regime and the approximate value of $h_{\min}$. When the lubrication regime is known, a more accurate value of $h_{\min}$ can be obtained by using the appropriate equation (7.61–64). Similar results to those presented above have also been derived for elliptical conjunctions between non-spherical surfaces [7.36].

Let us illustrate the theory above with experimental data. Figure 7.51 shows the film thickness profile parallel (a) and transverse (b) to the rolling direction, for a steel ball rolling on a glass substrate lubricated by a synthetic oil. The measurements are based on optical interferometry with 1 nm film thickness resolution. The steel ball has 12 nm rms surface roughness. The film profiles are not smooth, but show thickness fluctuations. This is due to the surface roughness variation on the steel ball, rather than the glass substrate, as the latter is much smoother. In the static contact these asperities deform

168   7. Sliding on Lubricated Surfaces

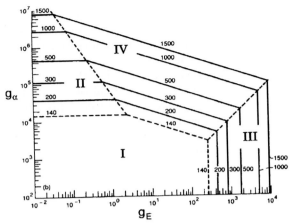

Fig. 7.50. Maps of lubrication regions. The numbers on the contour lines give $H_{\min} = (h_{\min}/R)(L/\mu_0 v_0 R)^2$. From [7.36].

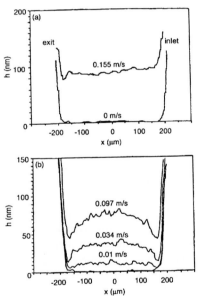

Fig. 7.51. (a) Film thickness profiles through center line parallel to rolling direction for $v_0 = 0$ and 0.155 m/s. (b) Film thickness profiles transverse to rolling direction for $v_0 = 0$, 0.01, 0.034 and 0.097 m/s. From [7.37].

elastically, giving only minor fluctuations of less than 3 nm. With increasing film thickness, the fluctuations increases from an average of 6 nm at $v_0 =$

0.034 m/s to 9 nm at 0.097 m/s. This is to be expected if the overall elasticity of the film increases as the film thickness increases. Note also the constriction in the film at exit (Fig. 7.51a). In the present case, the minimum thickness calculated from (7.64) agrees with the experimental data to within 15%.

Finally, let us give a few practical applications of the results presented above. Consider first ball bearings and gears. In these cases the local pressures in the contact areas are typically of order GPa, and the lubrication region IV should prevail. The thickness of the oil film can be estimated using (7.64), which we can write as

$$h_{\min} \approx \frac{1.7(\alpha R)^{0.49}(\mu_0 v_0)^{0.69}}{(E^*)^{0.11} L^{0.09}} .$$

Using $R \approx 0.01$ m, $v_0 \approx 10$ m/s, $L \approx 10^4$ N, $E^* \approx 2 \times 10^{11}$ Pa, $\mu_0 \approx 0.01$ Pa s and the pressure index $\alpha \approx 2 \times 10^{-8}$ Pa^{-1}, we get the minimum film thickness $\sim 0.5\,\mu$m. Thus in order to avoid direct contact between surface asperities the rms roughness of the surfaces should be below $0.1\,\mu$m. If one instead assumes $\alpha = 0$ and $E = \infty$, equation (7.61) gives $h_{\min} \approx 10^{-8}\,\mu$m; this illustrates the profound importance of the elasticity of the solids, and of the pressure dependence of the viscosity. Experiments on ball bearings have shown that large wear and a short fatigue lifetime result if $h_{\min} < r$ (where r is the rms surface roughness amplitude).

Next, let us consider railway wheels on wet or oily rails. In this case the maximum pressure in the contact areas is again of order GPa, and the lubrication region IV should prevail. If the rolling velocity $v_0 \approx 100$ m/s, equation (7.64) gives the minimum water layer thickness on a wet rail, $h_{\min} \approx 0.1\,\mu$m, i.e., much smaller than the amplitude of the surface roughness (the film thickness for other rolling velocities can be obtained by scaling, using $h_{\min} \sim v_0^{0.69}$). Thus boundary lubrication is likely to prevail in this case. However, if the rail is covered by an oil with the viscosity $\mu_0 = 0.1$ Pa s and pressure index 2×10^{-8} Pa^{-1}, then (for $v_0 = 100$ m/s) equation (7.64) gives $h_{\min} \approx 18\,\mu$m. Thus, the oil film thickness is similar to the rms surface roughness amplitude of the rail. In this case the traction will be reduced, and skidding can be expected. These results confirm the importance of clean tracks, and procedures for cleaning the rail in front of locomotive driving wheels are now being considered.

As a final application, let us consider the friction between an automobile tire and a wet road. The elastohydrodynamic lubrication of the interface will lower the friction, as indeed observed on wet roads. A road surface has roughness on many different length scales with an upper cut-off of order of a few mm. Thus, the elastohydrodynamic lubrication can occur on many different length scales. Consider, for example, an asperity with radius of curvature R. Assume that a rubber block is squeezed against the asperity with the force (or load) L, and that it slides with the velocity v_0 relative to the asperity. The thickness of the water layer between the asperity and the rubber surface can be estimated using (7.63) which we can write as

$$h_{\min} \approx \frac{3.3(\mu_0 v_0)^{0.65} R^{0.75}}{(E^*)^{0.45} L^{0.2}}.$$

If $R \approx 2$ mm, $L \approx 10$ N and $v_0 \approx 30$ m/s we get $h_{\min} \approx 1\,\mu$m. Thus $h_{\min} \ll R$ and the thin water layer will have a negligible effect on the contribution to the friction force from the asperity-induced (time-dependent) deformations of the rubber (see Sect. 5.3). The water film will, however, reduce or remove the adhesional contribution to the friction force, and may quite generally reduce the contribution from the small-scale surface roughness. In addition, water may be trapped in the substrate (road) cavities, thus inhibiting the rubber from penetrating into the cavities (see Fig. 5.33d), and reducing the time-dependent deformations of the rubber.

8. Sliding of Adsorbate Layers

In this chapter we consider the sliding of adsorbate layers on surfaces. This problem is of great interest in its own right but is also directly relevant for several sliding friction problems, as illustrated by the following two examples.

Consider an elastic block on a substrate separated by a monolayer of lubrication molecules as shown in Fig. 8.1a (left). An external force $\boldsymbol{F}_{\text{ext}}$ acts on the block parallel to the substrate. If we consider a stationary case (i.e., the bock is not accelerating) then the same force $\boldsymbol{F}_{\text{ext}}$ must act on the lubrication layer. If A is the contact area between the block and the substrate, then the block will act with the tangential stress $\sigma = F_{\text{ext}}/A$ on the lubrication layer. The actual tangential force $\boldsymbol{F}$ which acts on a particular lubrication molecule due to the block will fluctuate in time but takes the average value $\sigma \delta A$ where δA is the area occupied by a lubrication molecule. In a mean-field treatment we remove the block but assume that the force

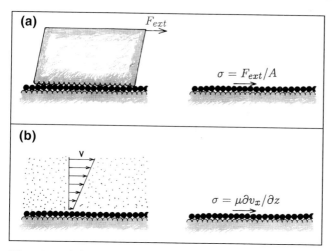

Fig. 8.1. (a) *Left:* An elastic block sliding on a substrate with a monolayer of lubrication molecules (*black dots*). An external force $\boldsymbol{F}_{\text{ext}}$ act on the block parallel to the substrate. *Right:* In a mean-field approximation the effect of the block on the lubrication layer is replaced by a uniform tangential stress $\sigma = F_{\text{ext}}/A$ acting on the adsorbed monolayer. (b) *Left:* A flowing fluid above a flat solid surface with a monolayer of adsorbed molecules (*black dots*). *Right:* Mean-field approximation.

$F = \sigma \delta A$ acts on each adsorbate and study the sliding problem shown in Fig. 8.1a (right). This mean-field theory should be a good approximation as long as the lubrication layer is homogeneous.

Next consider the flow of a fluid above a flat solid surface as in Fig. 8.1b (left). We assume that a monolayer of fluid molecules is adsorbed on the solid surface. The fluid will exert a tangential stress σ on the layer of adsorbed molecules given by $\sigma = \mu \partial v_x / \partial z$ where μ is the viscosity of the fluid. The actual tangential force $\boldsymbol{F}$ which acts on a particular adsorbed molecule due to flowing fluid will fluctuate in time but has an average value $\sigma \delta A$. In a mean-field treatment we assume that the force $F = \sigma \delta A$ acts on each adsorbate and study the sliding problem shown in Fig. 8.1b (right). The theory developed below also has direct applications to the sliding of flux line systems and charge density waves (Sect. 14.4).

8.1 Brownian Motion Model

The two examples above lead us to consider the following basic model problem. Consider an adsorbate system and assume that, in addition to the periodically corrugated adsorbate–substrate potential $U = \sum_i u(\boldsymbol{r}_i)$ and the adsorbate–adsorbate interaction potential $V = \frac{1}{2}\sum'_{ij} v(\boldsymbol{r}_i - \boldsymbol{r}_j)$, an external force $\boldsymbol{F}$ acts on each of the adsorbates. This will lead to a drift motion so that $m N_1 \bar{\eta} \langle \dot{\boldsymbol{r}} \rangle = N \boldsymbol{F}$ where $\langle ... \rangle$ stands for thermal average and where $\boldsymbol{r}$ denotes the coordinate of an arbitrary adsorbate. N_1 and N denote the number of adsorbates in the first layer in direct contact with the substrate, and the total number of adsorbates, respectively. For coverages up to one monolayer $P_1 = N_1/N = 1$ but for higher coverage $P_1 < 1$. For a weak external force $\boldsymbol{F}$, the sliding friction $\bar{\eta}$ is independent of $\boldsymbol{F}$; this is the linear response limit.

We consider adsorbates on a (100) surface of a fcc crystal, but the general results presented below should be independent of the substrate lattice structure and of the detailed form of U and V. The equation of motion for the particle coordinate $\boldsymbol{r}_i(t)$ is taken to be

$$m \ddot{\boldsymbol{r}}_i + m \eta \dot{\boldsymbol{r}}_i = -\frac{\partial U}{\partial \boldsymbol{r}_i} - \frac{\partial V}{\partial \boldsymbol{r}_i} + \boldsymbol{f}_i + \boldsymbol{F} \tag{8.1}$$

where $\boldsymbol{F}$ is the external force introduced above and $\boldsymbol{f}_i$ a stochastically fluctuating force which describes the influence on particle i of the irregular thermal motion of the substrate. η is a diagonal matrix of the form

$$\eta = \begin{pmatrix} \eta_\| & 0 & 0 \\ 0 & \eta_\| & 0 \\ 0 & 0 & \eta_\perp \end{pmatrix}. \tag{8.2}$$

The components f_i^α of $\boldsymbol{f}_i$ are related to the friction η via the fluctuation–dissipation theorem

$$\langle f_i^\alpha(t) f_j^\beta(0) \rangle = 2mk_B T \eta_{\alpha\beta} \delta_{ij} \delta(t) \ . \tag{8.3}$$

The friction $\eta_\perp$ associated with motion normal to the surface is usually much larger than the parallel friction $\eta_\parallel$. Note also that η depends on the adsorbate–substrate separation and on the position of the adsorbates along the surface. Since the adsorbate–substrate interaction is very short ranged, for multilayer adsorption, to a good approximation the friction η vanishes for all adsorbates except for those in the first layer in direct contact with the substrate.

The adsorbate–substrate interaction potential is taken to be

$$u(\mathbf{r}) = E_B \left(e^{-2\alpha(z-z_0)} - 2e^{-\alpha(z-z_0)} \right)$$
$$+ U_0 [2 - \cos(kx) - \cos(ky)] e^{-\alpha'(z-z_0)} \ , \tag{8.4}$$

so that $2U_0$ is (approximately) the activation barrier for diffusion and $k = 2\pi/a$, where a is the lattice constant of the substrate. The adsorbate binding energy E_B has been measured experimentally for many adsorbate systems and α can be deduced from the resonance frequency for the perpendicular adsorbate–substrate vibrational mode. In the present case we can consider the substrate as rigid. Writing $z = z_0 + q$, where q is the perpendicular vibrational normal mode coordinate, we get to quadratic order in q: $u \approx \alpha^2 E_B q^2$. This must equal $m\omega_\perp^2 q^2/2$, where $\omega_\perp$ is the vibrational resonance frequency of the perpendicular vibration. Hence

$$\alpha = \left(\frac{m\omega_\perp^2}{2E_B} \right)^{1/2} \ . \tag{8.5}$$

Similarly, from a knowledge of the resonance frequency $\omega_\parallel$, it is possible to estimate the barrier height $2U_0$. Using (8.4) we get

$$U_0 = m\omega_\parallel^2 / k^2 \ . \tag{8.6}$$

The adsorbate–adsorbate interaction potential V is taken as a sum of Lennard–Jones pair potentials

$$v(r) = \epsilon \left[\left(\frac{r_0}{r} \right)^{12} - 2 \left(\frac{r_0}{r} \right)^6 \right] \ , \tag{8.7}$$

where ϵ is the well depth and r_0 the particle separation at the minimum of the pair potential. In all simulations presented below we have chosen r_0/a to correspond to Xe on Ag(100); in this case $a = b/\sqrt{2} = 2.89$ Å (where b is the lattice constant of Ag) so that $r_0/a \approx 1.56$.

Equation (8.1) describes the motion of an adsorbate system on a corrugated substrate. When the external force $\mathbf{F} = 0$ the particle performs irregular motion (diffusion) with no long-term drift, i.e., $\langle \dot{\mathbf{r}}_i \rangle = 0$. For $\mathbf{F} \neq 0$, in addition to the irregular motion, the particles drift in the direction of $\mathbf{F}$ with the speed $\langle \dot{\boldsymbol{r}} \rangle = \boldsymbol{F}/m\bar{\eta}$. Note that when $U_0 = 0$, the thermal average of (8.1) gives

$$m\langle \ddot{\boldsymbol{r}} \rangle + m\eta_\parallel \langle \dot{\boldsymbol{r}} \rangle = \boldsymbol{F},$$

or, since $\boldsymbol{F}$ is constant,

$$m\eta_\parallel \langle \dot{\boldsymbol{r}} \rangle = \boldsymbol{F}$$

so that $\bar{\eta} = \eta_\parallel$ as expected in this limiting case.

On the Origin of the Friction η and the Fluctuating Force $\boldsymbol{f}_i$

The term $\boldsymbol{f}_i$ in (8.1) describes the fluctuating force acting on adsorbate i due to the thermal motion of the substrate ions and electrons. This force can "kick" the adsorbate over the potential barriers in the system and allow it to diffuse along the surface. Without this term no thermally activated motion could occur, which would correspond to zero temperature. In the absence of the driving force $\boldsymbol{F}$, one can show from (8.1-3) that $(m/2)\langle \dot{\boldsymbol{r}}^2 \rangle = 3k_\mathrm{B}T/2$ where T is the substrate temperature. For example, in the limiting case where $U = 0$ and $V = 0$ it is trivial to integrate (8.1) to get (we assume that $\eta_{\alpha\beta} = \eta \delta_{\alpha\beta}$)

$$\dot{\boldsymbol{r}}_i(t) = \dot{\boldsymbol{r}}_i(0)\mathrm{e}^{-\eta t} + \frac{1}{m}\int_0^t dt' \mathrm{e}^{-\eta(t-t')} \boldsymbol{f}_i(t')$$

and using (8.3) gives, for large times,

$$\frac{m}{2}\langle \dot{\boldsymbol{r}}_i^2(t) \rangle = \frac{1}{2m}\int_0^t dt' dt'' 2mk_\mathrm{B}T 3\eta \delta(t'-t'') \mathrm{e}^{-\eta(2t-t'-t'')} = \frac{3}{2}k_\mathrm{B}T$$

which is the standard equipartition theorem for the mean square velocity.

It is well-known that (8.1-3) cannot be microscopically correct. The friction force on an adsorbate is ultimately due to "collisions" with the substrate ions and electrons (see next section) and takes the simple form $-m\eta\dot{\boldsymbol{r}}$ only on a time scale which is long compared to the frequency and duration of individual collisions. In general, one must make the replacement

$$\eta\dot{\boldsymbol{r}} \to \int_{-\infty}^t dt' \eta(t-t') \dot{\boldsymbol{r}}(t') .$$

In order to satisfy the fluctuation–dissipation theorem, the components f_i^α of the fluctuating force $\boldsymbol{f}_i$ must satisfy

$$\langle f_i^\alpha(t) f_j^\beta(t') \rangle = mk_\mathrm{B}T\eta_{\alpha\beta}(t-t')\delta_{ij} .$$

The actual time dependence of the kernel $\eta(t)$ or frequency dependence of its Fourier transform, $\eta(\omega)$, is determined by the nature of the adsorbate–substrate coupling; if the time scale introduced by this coupling is short compared with the time scale associated with the motion of the adsorbate, one may replace $\eta(t-t')$ by $2\eta\delta(t-t')$ or $\eta(\omega) = $ const., in which case one recovers (8.1–3).

For adsorbates on surfaces the friction η and the fluctuating force $\boldsymbol{f}_i$ are due to the coupling between the adsorbate and the vibrational (phonon) and electronic (electron–hole pair) excitations of the substrate. One can show [8.1] that the coupling to the electronic excitations gives rise to a friction which, to an excellent approximation, is frequency independent and hence local $\eta(t-t') \approx 2\eta\delta(t-t')$. Coupling to the substrate phonons also gives rise to a frequency-independent friction if the characteristic frequencies involved in the adsorbate motion are well below the highest substrate phonon frequency [8.2]. For physisorption systems, which we will mainly focus on below, this is usually an excellent approximation. Thus (8.1–3) should be accurate for the applications considered below.

Finally, let us point out that (8.1–3) can be derived from first principles. The most general treatment is based on the memory function formalism and a lucid discussion is presented in D. Forster's book [8.3]. For coupling to (harmonic) substrate phonons, it is possible to eliminate (or integrate out) the infinite phonon degrees of freedom explicitly, to obtain a (nonlocal) friction force and fluctuating force in the equation of motion for the adsorbate [8.4]. As pointed out above, the result reduces to a frequency-independent friction and a $\delta(t-t')$-correlated fluctuating force in the limit where the frequencies associated with the adsorbate motion are well below the highest substrate phonon frequency.

8.2 Electronic and Phononic Friction

The equation of motion (8.1) for the adsorbate layer contains the friction matrix η. Thermal equilibrium properties of the adsorbate layer do not depend on η, but non-equilibrium properties, such as the sliding friction, do depend on this parameter. It is therefore of fundamental importance to have information about the magnitude of η and how it depends on the substrate and on the nature of the adsorbate–substrate interaction.

As stated above, the friction force which acts on an adsorbate has the simple form $-m\eta v$ only if the typical frequencies associated with the adsorbate motion are much smaller than the highest substrate phonon frequency and (for a metallic substrate) the Fermi frequency. In most cases this is equivalent to the statement that the highest adsorbate velocity v is much smaller than the sound velocity and (for a metallic substrate) the Fermi velocity of the substrate. For insulating surfaces (e.g., most metal oxides) η can only be due to phonon emission but on metallic surfaces both electronic and phononic friction occurs and $\eta = \eta_{\text{el}} + \eta_{\text{ph}}$.

Information about the friction parameter η can be deduced from infrared spectroscopy [8.5] and inelastic helium scattering [8.6] since η determines the line width of adsorbate vibrations if inhomogeneous broadening and pure dephasing processes can be neglected. Information about η can also be deduced from quartz crystal microbalance (QCM) measurements, see Sect. 8.5. Finally, as shown below, for metals the electronic contribution to the friction η can be deduced from surface resistivity measurements.

Electronic Friction

In this section we estimate the electronic contribution to η for physisorbed adsorbates on metal surfaces. Let us first consider the parallel electronic friction which can be obtained directly from surface resistivity measurements as follows [8.1,7]:

Consider a thin metallic film (thickness d) with a layer of adsorbed molecules. Assume that an electric field $\boldsymbol{E}$ acts on the electrons in the film. This induces a collective (drift) motion of the electrons, corresponding to a current $\boldsymbol{J} = ne\boldsymbol{v}$, where $\boldsymbol{v}$ is the electron drift velocity and n the number of conduction electrons per unit volume (Fig. 8.2a). Let us now move to a frame of reference where no electron current flows, but where the adsorbates move with the velocity $-\boldsymbol{v}$ relative to the electron-fluid in the metal film. In this frame there will be a friction force acting on the each adsorbate in the first layer

$$\boldsymbol{F}_{\text{fric}} = m\eta_{\text{el}}\boldsymbol{v} , \tag{8.8}$$

where m is the adsorbate mass and η_{el} the friction coefficient caused by excitation of electron–hole pairs in the metal film. The energy transfer per unit time from the adsorbates to the metal film is given by

$$N_1 \boldsymbol{F}_{\text{fric}} \cdot \boldsymbol{v} = N_1 m \eta_{\text{el}} v^2 , \tag{8.9}$$

where N_1 denotes the number of adsorbates in the first layer. But in the original reference frame, the adsorbate-induced power absorption can be related to the adsorbate-induced increase in the film resistivity $\Delta\rho$ via

$$Ad\boldsymbol{J} \cdot \boldsymbol{E} = Ad\Delta\rho J^2 ,$$

where Ad is the volume of the film (area A and thickness d). Substituting $\boldsymbol{J} = ne\boldsymbol{v}$ in this formula and comparing with (8.9) gives

$$\eta_{\text{el}} = \frac{n^2 e^2 d}{m n_a} \Delta\rho , \tag{8.10}$$

where $n_a = N_1/A$ is the number of adsorbates per unit area in direct contact with the substrate.

Figure 8.2b shows the change in resistivity [8.8] as a function of Xe coverage for the Xe/Ag adsorption system. For low Xe coverage ($n_a < 0.04\,\text{Å}^{-2}$),

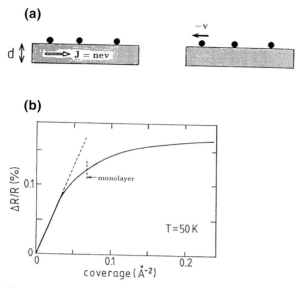

Fig. 8.2. (a) The damping (caused by electronic excitations) of the translational motion of the adsorbate layer (*right*) can be related to the adsorbate-induced increase in film resistivity by a change of reference frame (*left*). (b) The Xe-induced resistivity change ΔR of a thin silver film. From [8.11].

$\Delta \rho$ increases linearly with n_a. According to (8.10), $\partial \rho / \partial n_a \sim 1/d$ if we assume that η_{el} is independent of d as expected if the film is thick enough. This relation is well satisfied for the Xe/Ag adsorption system [8.8] (not shown). From the initial slope in Fig. 8.2b and using $m = 131$ amu (the Xe mass) gives at low Xe coverage $\eta_{el} \sim 3 \times 10^8 \, \text{s}^{-1}$. Owing to uncertainty in the metal film thickness, this friction is rather uncertain. According to Fig. 8.2b the electronic friction of a double layer of Xe is roughly 20% larger than that of the monolayer.

In Table 8.1, we have summarized results for the e–h pair friction for several different adsorption systems. In all cases we have assumed that the free electron density n corresponds to one electron per substrate atom. The first six systems in Table 8.1 are chemisorption systems characterized by an electronic friction in the range $\eta_{el} \sim 10^{10} - 10^{12} \, \text{s}^{-1}$. The last six systems are physisorption systems where $\eta_{el} \sim 10^8 - 10^9 \, \text{s}^{-1}$. In Table 8.1 we also show the cross section $\Sigma = (16/3)(m/m_e)(\eta_{el}/nv_F)$, where m_e is the electron mass, for diffusive scattering of a substrate conduction electron from an adsorbate. All the quantities in the table refers to the low-coverage limit; at saturation coverages the friction and the cross section Σ are typically 20–50% smaller than in the dilute limit.

Theories describing electronic friction have been developed for several limiting cases of the adsorption bond, namely covalent, ionic, and van der

Table 8.1. The electronic friction (at low adsorbate coverage) deduced from surface resistivity data. From [8.1].

System	$\eta_{eh}(s^{-1})$	$\Sigma(\text{Å}^2)$
H/Ni	1.0×10^{12}	8.2
O/Cu	1.4×10^{11}	18.8
CO/Ni	7.1×10^{10}	16.0
CO/Cu	2.6×10^{10}	5.8
N$_2$/Ni	2.2×10^{10}	4.8
Ag/Ag	1.0×10^{10}	13.4
CO/Ag	2.8×10^9	1.0
C$_2$H$_4$/Ag	1.4×10^9	0.5
Xe/Ag	$(3-10) \times 10^8$	$(0.6 - 2.0)$
C$_6$H$_6$/Ag	7.1×10^8	0.6
C$_6$H$_{12}$/Ag	5.9×10^8	0.6
C$_2$H$_6$/Ag	2.8×10^8	0.1

Waals bonding. Consider first the covalent contribution. This is assumed to be derived from adsorbate-induced resonance states (virtual levels) centered close to the metal Fermi energy. Such resonance states are usually derived from the highest occupied or lowest unoccupied level of the gas phase molecule (or atom). Upon adsorption on a metallic surface these levels broaden and shift and, in simple cases, they form well-defined resonances with a width $\Gamma \sim 1\,\text{eV}$ and with a (projected) density of states, $\rho_a(\epsilon)$, with a nonzero value at the Fermi energy of the substrate. The contribution to the electronic friction from a resonance state equals [8.1]

$$\eta_{el} = 2\frac{m_e}{m}\omega_F \Gamma \rho_a(\epsilon_F)\langle \sin^2\theta \rangle \tag{8.11a}$$

where $\omega_F = \epsilon_F/\hbar$ is the Fermi frequency and where $\langle \sin^2\theta \rangle$ depends mainly on the symmetry of the adsorbate orbital with $\langle \sin^2\theta \rangle \approx 0.20$ for an orbital with s or p_z symmetry and 0.33 for an orbital with p_x or p_y symmetry (the z-axis is perpendicular to the surface). As an example, consider a single Ag atom adsorbed on an Ag(111) surface. In the gas phase the Ag 5s-level contains one electron. Work function measurements show that an Ag atom adsorbed on an Ag(111) surface (at low temperatures in order to inhibit diffusion to step edges) has a small dipole moment, corresponding to a charge transfer $\sim 0.04e$ to the metal. Thus, upon adsorption the 5s-level broadens into a resonance state which must be nearly half filled so that, when accounting for both spin directions, it contains on average one electron, giving a nearly neutral Ag-adatom and a small dipole moment. Using (8.11a) and assuming that the 5s-resonance is centered right at the Fermi energy as indicated in Fig. 8.3a, gives $\eta_{el} \approx 1.1 \times 10^{10}\,\text{s}^{-1}$ and $\Sigma \approx 14.6\,\text{Å}^2$ in close

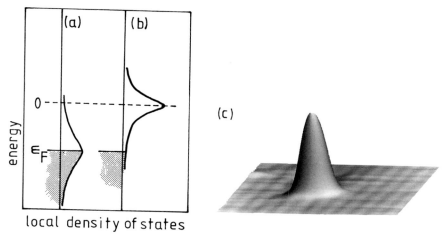

Fig. 8.3. Local (or projected) density of states induced by (**a**) an Ag adatom on Ag(111) and (**b**) a Xe or C_2H_4 molecule on Ag(111). (**c**) A constant current STM picture of Xe adsorbed on Ni(110) taken with the tip biased 0.02 V relative to the sample [8.26,27].

agreement with the result deduced from surface resistivity measurements, see Table 8.1.

Next, consider van der Waals bonding (physisorption). In this case the adsorbate–substrate interaction consists of the long-range attractive van der Waals interaction plus a short-range repulsion (Pauli repulsion) resulting from the overlap of the electron clouds of the adsorbate and the substrate. This model is relevant for, e.g., the light noble gas atoms and saturated hydrocarbons adsorbed on metal surfaces.

The electronic friction associated with the van der Waals interaction was studied in [8.9] where the metal was treated in the semi-infinite jellium model and the molecule–substrate interaction in the dipole approximation. Using second-order perturbation theory (Fig. 8.4a,b), and including screening by the substrate conduction electrons, $\eta_\parallel$ and $\eta_\perp$ both have the form

$$\eta_{\rm el} = \frac{e^2}{\hbar a_0} \frac{\left[k_F^3 \alpha(0)\right]^2}{(k_F z_0)^{10}} \frac{m_e}{m} \frac{\omega_F}{\omega_p} k_F a_0 I \; , \qquad (8.11b)$$

where a_0 is the Bohr radius, $\alpha(0)$ the static polarizability, ω_p is the plasma frequency, k_F the Fermi wave vector, and where I is a function of the electron gas density parameter r_s and of the distance z_0 between the nucleus of the adsorbate and the jellium edge. Figure 8.5a shows the dependence of I on z_0 for $r_s = 2$, 3 and 4, for both parallel and perpendicular motion of the adsorbate.

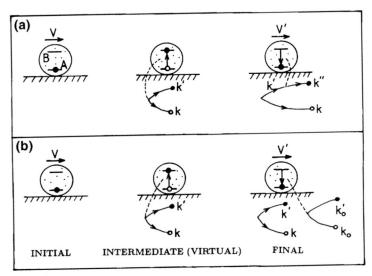

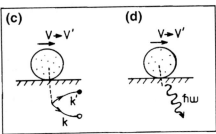

Fig. 8.4. Elementary processes involved in the electronic (**a, b, c**) and phononic (**d**) friction. Electronic friction: (**a**) A second-order process which results in one electron–hole pair. (**b**) A second-order process which results in two electron–hole pairs but which gives a vanishing contribution to the sliding friction. (**c**) A first-order process, where the effective one-particle potential of the adsorbate excites an electron–hole pair. Processes (**a**) and (**b**) are related to the van der Waals interaction, while process (**c**) is caused by the repulsive short-range interaction. (**d**) Phononic friction: Excitation of a bulk or surface phonon. From [8.9].

Note that $I \to$ const. as $z_0 \to \infty$ so that for $k_F z_0 \gg 1$ the friction (8.11b) decays as z_0^{-10} with the distance z_0 from the surface. In contrast, the contribution to the electronic friction from the short-range repulsive adsorbate–substrate interaction (the Pauli repulsion) depends on the overlap between the metal wave functions and the wave functions of the adsorbate, and therefore decays roughly exponentially with increasing z_0. Since at the equilibrium separation the attractive and repulsive adsorbate–substrate interactions forces are of identical magnitude, one would also expect their contributions to the dissipative force to be of similar magnitude.

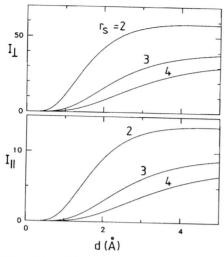

Fig. 8.5a. The friction integrals $I_\perp$ and $I_\parallel$ (defined in the text) as a function of $k_F z_0$ where z_0 is the distance of the center of the adsorbate from the jellium edge of the substrate [8.9,28].

It has been found that even for adsorbate systems where the *adsorption energy* is dominated by the van der Waals interaction (as is likely to be the case for the last six adsorption systems in Table 8.1), a large contribution to the electronic friction may stem from covalent effects. For example, C_2H_4 and C_2H_6 both bind to Ag(111) with almost the same binding energy, as expected for the van der Waals interaction, and the desorption temperatures are almost the same (96 K for C_2H_6 and 90 K for C_2H_4). Thus one would expect the van der Waals contribution to the electronic friction to be of similar magnitude for both adsorbate systems. Nevertheless, as shown in Table 8.1, the electronic friction for C_2H_4 is a factor of 5 higher than for C_2H_6. This can be explained by the observation that C_2H_4 has a π^*-resonance centered around the vacuum level (with a width $\Gamma \sim 1\,\text{eV}$), see Fig. 8.3b, with a tail extending down to the Fermi energy so that $\rho_a(\epsilon_F) \neq 0$. In this case, (8.11a) gives an electronic friction of similar magnitude to that quoted in Table 8.1 for C_2H_4. Such a π^*-resonance does not occur for C_2H_6, which has only higher-lying σ-resonances, and in this case the observed electronic friction is likely to derive mainly from the (dissipative) van der Waals interaction.

Xe has also a resonance state centered around the vacuum energy, with with a tail extending down to the Fermi energy, as in Fig. 8.3b. This resonance is derived from the Xe 6s-level, which is the lowest unoccupied level in gas phase Xe. It has been pointed out that the 6s-resonance is the reason why Xe shows up as a big bump in scanning tunneling microscopy (STM), where electron tunneling to adsorbate induced states in the vicinity of ϵ_F occurs

(Fig. 8.3c). We will return to the contribution to the electronic friction from the 6s-resonance state below, when we consider the Xe/Ag(111) system in detail.

Note that the electronic friction coefficients for physisorption systems determined from resistivity data, are of similar magnitude to the sliding friction determined for IC-monolayers using the QCM (compare Table 8.1 with Table 3.1). We will return in Sect. 8.5 to the fundamental question of the origin of the QCM sliding friction but it is clear already at this point that the electronic friction will make a very important contribution.

Phononic Friction

Consider an adsorbate sliding on a surface. If the characteristic frequencies associated with the fluctuating forces exerted on the substrate by the motion of the adsorbate are well below the maximum phonon frequency ω_c of the solid, which is usually the case for physisorption systems, then the phononic friction can be calculated using the elastic continuum model. This gives a friction force on the form $-m\eta v$ where [8.2]

$$\eta_\perp \approx \frac{3}{8\pi} \frac{m}{\rho} \left(\frac{\omega_\perp}{c_T}\right)^3 \omega_\perp , \qquad (8.12a)$$

and

$$\eta_\parallel \approx \frac{3}{8\pi} \frac{m}{\rho} \left(\frac{\omega_\parallel}{c_T}\right)^3 \omega_\parallel \qquad (8.12b)$$

where ρ and c_T are the mass density and the transverse sound velocity of the substrate and where

$$\omega_\parallel^2 = \frac{1}{m}\frac{\partial^2 U}{\partial^2 x} , \qquad \omega_\perp^2 = \frac{1}{m}\frac{\partial^2 U}{\partial^2 z} .$$

For a particle performing vibrations around a local minima of $U(\mathbf{x})$, $\omega_\parallel$ and $\omega_\perp$ are the resonance frequencies for parallel and perpendicular vibrations, respectively. Although $\partial^2 U/\partial x^2$ and $\partial^2 U/\partial z^2$ depend on the position of the molecule, for physisorption systems these quantities may be reasonably estimated from the observed frequencies of vibration around an equilibrium location. The friction coefficients (8.12a) and (8.12b) have their origin in one-phonon emission processes as indicated in Fig. 8.4d. The formulas above are valid for atoms and small molecules. For large molecules the phononic friction is smaller than predicted by (8.12). This results from the fact that the phonons emitted by, say, a linear chain molecule, treated as a rigid bar of length D, must have $|q_x| < 1/D$, where q_x is the component of the wave vector parallel to the surface in the direction of the chain. For example, for the chain molecule $C_{10}H_{22}$ the reduction amounts to roughly a factor of 3.

Here is a very simple (approximate) derivation of (8.12): In a simplified picture one regards each adsorbate oscillation as a collision with the substrate [8.68], with an energy transferred to the substrate of $\Delta E \approx E(m/m_s)$, where E is the energy of the adsorbate vibration and m_s the "effective" mass of the substrate. During the collision time $\tau \approx 1/\omega_0$ (where ω_0 stands for either $\omega_\perp$ or $\omega_\parallel$) the resulting substrate displacement field extends a distance $\sim c_T\tau = c_T/\omega_0$ into the solid. The effective mass m_s is simply the mass contained in the volume element $\Delta V \sim (c_T/\omega_0)^3$. Thus, $m_s \sim \rho(c_T/\omega_0)^3$. Since collisions with the substrate occur with the frequency $\nu = \omega_0/2\pi$, the net rate of energy transfer per unit time is $dE/dt \approx -(\omega_0/2\pi)(m/m_s)E$. This results in an exponential decay of the energy, characterized by the damping constant $(\omega_0/2\pi)(m/m_s) = m\omega_0^4/(2\pi\rho c_T^3)$ which agrees with (8.12) except for a factor of order unity.

Equations (8.12) have been derived for isolated adsorbates but should also be good approximations at finite coverages if the adsorbate system is in a 2D-fluid state so that the motions of the adsorbates are incoherent. However, for an incommensurate solid adsorbate layer the phononic friction of the sliding layer is severely reduced, and may even vanish because of destructive interference between the elastic waves emitted into the solid [8.67]. Similarly, for an ordered adsorbate overlayer, the phononic damping of the collective $\boldsymbol{q} \approx 0$ adsorbate vibrational mode, as probed by infrared spectroscopy, may differ strongly from the prediction of (8.12). Let us now consider this case in greater detail.

Phonon-Damping of the $\boldsymbol{q} = 0$ Collective Adsorbate Vibration

Many surface spectroscopies can be used to measure the damping of collective adsorbate vibrational modes in ordered adsorbate layers. With inelastic helium atom scattering and electron energy loss spectroscopy, the damping of any $\boldsymbol{q}$-phonon mode can be studied, while infrared light absorption spectroscopy only probes the $\boldsymbol{q} \approx 0$ mode.

It is easy to calculate the damping of the $\boldsymbol{q} = 0$ collective adsorbate vibration under the assumption that the adsorbate unit cell is so small that for all (non-vanishing) reciprocal lattice vectors $\boldsymbol{G}$ (associated with the adsorbate lattice), $G > \omega_0/c_T$, where ω_0 is the adsorbate vibrational resonance frequency [8.67,68]. Assume first that the adsorbates vibrate parallel to the surface, say along the x-axis, and let $s\hat{x}$ denote the adsorbate displacement vector. We have

$$m\ddot{s} = -m\omega_0^2[s - u_1(0,t)], \tag{8.12c}$$

where $u_1(0,t)$ denotes the x-component of the displacement vector of the substrate at an adsorbate on the surface $z = 0$. When the adsorbate lattice constant $a < \lambda$, where $\lambda = 2\pi c_T/\omega_0$ is the wavelength of the emitted phonons, we can, when calculating the damping of the $\boldsymbol{q} = 0$ adsorbate phonon, replace the true non-uniform surface stress distribution with an average (uniform) distribution

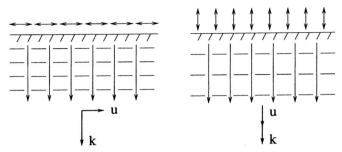

Fig. 8.5b. Elastic waves emitted from a uniform (in space) and oscillating (in time) surface stress distribution, acting parallel (*left*) and perpendicular (*right*) to the surface. The emitted elastic waves are transverse in the former case, and longitudinal in the latter case, and in both cases the waves propagate orthogonal to the surface into the elastic half-space.

$$\sigma_{13} = n_a m \omega_0^2 [u_1(0,t) - s] \, ,$$

where n_a is the number of adsorbates per unit area, and where we have used the fact that a uniform tangential surface stress gives rise to a displacement field $u_1(z,t)$, which is polarized parallel to the surface (in the direction of the vibrational displacement coordinate s) and independent of x and y (the emitted elastic waves propagate perpendicular to the surface into the elastic half space), see Fig. 8.5b (left). The elastic displacement field $u_1(z,t)$ satisfies the wave equation

$$\frac{\partial^2 u_1}{\partial t^2} - c_T^2 \frac{\partial^2 u_1}{\partial z^2} = 0 \, ,$$

with the solution

$$u_1 = u_0 e^{i(\omega z / c_T - \omega t)} \, .$$

The boundary condition

$$\rho c_T^2 \frac{\partial u_1}{\partial z} = \sigma_{13} \quad \text{for} \quad z = 0$$

gives

$$i \rho c_T \omega u_0 = n_a m \omega_0^2 (u_0 - s) \, ,$$

or

$$u_0 - s = \frac{i \rho c_T \omega s}{n_a m \omega_0^2 - i \rho c_T \omega} \, .$$

Substituting this in (8.12c) gives

$$\omega^2 s = \frac{\omega_0^2 s}{1 - n_a m \omega_0^2 / i\rho c_T \omega} , \tag{8.12d}$$

Since $n_a m \omega_0^2 / \rho c_T \omega$ is in general very small for $\omega \approx \omega_0$, we get from (8.12d) for ω close to the resonance frequency ω_0:

$$-m\omega^2 s + m\omega_0^2 \left(1 - i n_a m \omega_0 / \rho c_T\right) s = 0 .$$

Thus, the damping is

$$\eta_\| = m\omega_0^2 n_a / \rho c_T . \tag{8.12e}$$

A similar calculation to that presented above can be performed for the $q = 0$ perpendicular adsorbate vibration. In this case the displacement field in the solid is polarized perpendicular to the surface and is again independent of x and y (the emitted elastic waves propagate perpendicular to the surface into the elastic half-space), see Fig. 8.5b (right). We get

$$\eta_\perp = m\omega_0^2 n_a / \rho c_L . \tag{8.12f}$$

A Case Study: Xe on Ag(111)

Consider Xe on Ag(111) where one can distinguish between three contributions to the electronic friction: (a) the long-range attractive van der Waals interaction, (b) the short-range Pauli repulsion caused by the overlap of the electron clouds of the adsorbate and the substrate, and (c) a contribution from the 6s-resonance. Let us estimate the magnitudes of these contributions to the electronic friction. From low-energy electron diffraction studies, it is known that the separation between the Xe nucleus and the jellium edge of Ag(111) is 2.4 Å. The static polarizability of Xe is $\alpha(0) = 4.0$ Å^3. From (8.11b) the contribution to $\eta_\|$ from the van der Waals interaction is estimated to be $\approx 3.4 \times 10^8$ s^{-1}. The contribution to $\eta_\|$ from the short-range interaction has been estimated [8.1,9] as 6×10^7 s^{-1}, i.e., about a factor of 6 smaller than the contribution from the van der Waals interaction. Finally, the contribution to the electronic friction from the Xe-6s electronic resonance state, which is located around the vacuum energy [8.10] with a tail extending down to the Fermi energy, can be calculated [8.1] from (8.11a) as 1.5×10^8 s^{-1}. For the lighter noble gas atoms and for saturated hydrocarbons, chemical effects should be negligible since no electronic resonance states occur close to the Fermi energy. Assuming that the different contributions are additive, we conclude that the total electronic friction should be on the order of 6×10^8 s^{-1}. This should be compared with the electronic friction for Xe on Ag(111) deduced from surface resistivity [8.11] data (Table 8.1) $\eta_\| \approx (3 - 10) \times 10^8$ s^{-1}.

The $\sim 20\%$ increase in the electronic friction for the Xe-double layer film on Ag(111), compared to the monolayer film (Fig. 8.2b), may be due to the strong distance dependence of the van der Waals contribution. First, note that the direct contribution from the second Xe layer is negligible. However, the second layer "squeezes" the first Xe layer slightly closer to the surface (Fig. 8.17c). A decrease in d of ~ 0.1 Å is enough to increase the contribution from the van der Waals interaction by $\sim 20\%$. In addition, the 6s-resonance state of Xe is modified by the interaction with the second layer Xe atoms: the overlap between the 6s-orbitals of the first Xe layer with those of the second layer will broaden the 6s-resonance state, which increases both Γ and $\rho_a(\epsilon_F)$ in (8.11a). Furthermore, screening of the Coulomb field of a temporary Xe$^-$ by the second Xe-layer will shift the 6s-resonance state towards the Fermi energy, which increases $\rho_a(\epsilon_F)$. According to (8.11a), both effects will increase the friction of the double layer.

A Case Study: Hydrocarbons on Cu(111), Pb(111) and Diamond(111)

Many lubrication fluids consist of long-chain hydrocarbons, but very little experimental data relevant for sliding friction is available for such molecules adsorbed on metallic surfaces. Recently *Fuhrmann* et al. [8.12,67] have performed inelastic He-atom scattering from saturated hydrocarbons adsorbed on the (111) surfaces of Cu, Pb and diamond. The perpendicular vibrations of all these molecules was found to be $\hbar\omega_0 \approx 7$ meV. In the simplest interaction model between a physisorbed molecule and the substrate, the interaction strength, and the force constant, are proportional to the molecular polarizability $\alpha(0)$. Therefore the perpendicular vibrational energies should be proportional to $[\alpha(0)/m]^{1/2}$. Witte and Wöll have shown that $[\alpha(0)/m]^{1/2}$ is almost equal for all the hydrocarbons in Table 8.2, indicating that, as expected, all these molecules are physisorbed on Cu(111). In Table 8.2 we have summarized the observed damping η_0 of the perpendicular $\boldsymbol{q} = 0$ collective adsorbate vibrations, as deduced from the observed linewidth γ using $\hbar\eta = \gamma$, which holds if the contribution to the linewidth from inhomogeneous broadening and dephasing can be neglected. The large magnitudes of the friction coefficients quoted in Table 8.2 cannot be explained as resulting from the electronic contribution which gives $\eta_{el} \sim 1 \times 10^9$ s^{-1} or $\hbar\eta_{el} < 10^{-3}$ meV, i.e. roughly three orders of magnitude smaller than the observed friction. As discussed above, when $\omega_0 \ll \omega_c$ (where $\omega_c \approx 30$ meV is the highest substrate phonon frequency), the phononic friction is accurately given by (8.12f) for ordered adsorbate layers ($\boldsymbol{q} = 0$ mode). Note the good agreement between theory and experiment for Cu and diamond, but not for Pb. Also shown in Table 8.2 are the theoretical results for the damping of isolated (independent) adsorbate vibrations, η_{ind}, as given by (8.12a). This theory is clearly not applicable in the present case involving the damping of the $\boldsymbol{q} = 0$ collective phonon mode of well-ordered, densely packed overlayers, where emission of

Table. 8.2. Experimentally determined energies $\hbar\omega_0$ and damping $\hbar\eta_0$ of the $q = 0$ collective frustrated translation normal to the surface ($\perp$), for ordered layers of alkanes adsorbed on different surfaces. The values are compared with predictions for the phononic contribution to the lifetime broadening calculated from (8.12f). Also shown is the theoretically predicted damping η_{ind} (8.12a) for the case of isolated (independent) adsorbate vibrations. From [8.12,67].

System	Experiment		Theory	
	$\hbar\omega_0$ [meV]	$\hbar\eta_0$ [meV]	$\hbar\eta_0$ [meV]	$\hbar\eta_{\text{ind}}$ [meV]
Octane/Cu(111)	7.2	0.45	0.74	2.4
Hexane/Cu(111)	7.0	0.48	0.87	1.7
Nonane/Cu(111)	6.6	0.52	0.74	2.0
Decane/Cu(111)	6.6	0.52	0.74	2.0
Hexane/Pb(111)	6.7	0.34	(1.3)	(20.4)
Octane/Pb(111)	6.6	0.26	(1.1)	(25.4)
Octane/C(111)	7.0	0.38	0.35	0.03
Decane/C(111)	7.0	0.32	0.36	0.04

phonons into the substrate is strongly influenced by interference phenomena, which are taken into account in (8.12f).

For hydrocarbons on Pb(111), the phononic friction calculated from (8.12f) is much larger (by roughly a factor of 4) than the observed values. This discrepancy is due to the fact that lead is an elastically soft material, with the highest phonon frequency $\hbar\omega_c \approx 9$ meV, i.e., the resonance frequency $\omega_0 \sim \omega_c$ and the phononic friction cannot be calculated from (8.12f), since this formula is only valid if $\omega_0 \ll \omega_c$. This is illustrated in Fig. 8.6b which shows the bulk phonon density of states $g(\omega)$ of lead: the resonance frequencies of the adsorbate vibrations are indicated by the dashed lines and occur near a local minima of $g(\omega)$ rather than in the $g \sim \omega^2$ region at low frequencies (which result from the initial linear dispersion of the phonon modes). On the other hand, for copper the resonance frequencies occur in the $g \sim \omega^2$ region of the phonon spectra (see Fig. 8.6a), and (8.12f) should be accurate in this case.

Note that for diamond there is huge difference between the damping rates η_{ind} and η_0, for independent (isolated) and collective (high coverage) $q = 0$ adsorbate vibrations, respectively. This difference is related to the high sound velocities in diamond. Since η_{ind} is so small for hydrocarbons on diamond, and since it is η_{ind} which determines the solid–fluid heat transfer (see Sect. 8.8), it is clear that in sliding friction involving boundary lubrication, the lubrication film may heat up to very high temperatures even at low sliding velocities. This effect may be important in the process of polishing diamond, where the high temperature may lead to the decomposition of lubrication molecules, and to a "chemical" contribution to the polishing process.

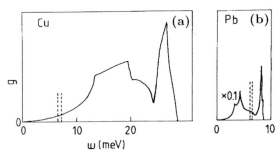

Fig. 8.6a,b. The phonon density of states $g(\omega)$ of Cu and Pb. The dashed lines indicate typical resonance frequencies of perpendicular vibrations for adsorbed (saturated) hydrocarbons.

Electronic Friction and Infrared Spectroscopy*

When a beam of infrared radiation is incident on a metal surface covered by molecules, "dips" usually occur in the reflection spectra at those frequencies which correspond to vibrational excitations of the adsorbed molecules. Since the wavelength of the incident photons is much larger than the size of the molecules, it is usually assumed that the interaction between the radiation field and an adsorbed molecule is of the dipole form. Since the perpendicular electric field component at the surface associated with a p–polarized wave is nearly twice as large as that of the incoming wave, while (due to screening) the parallel electric field component, at infrared frequencies, is several orders of magnitude smaller, the interaction energy due to the *direct* coupling between a p-polarized wave and the adsorbate dipoles will be strong only for modes with a non-zero dynamic dipole moment perpendicular to the surface. The difference in magnitude between the perpendicular and the parallel electric field components results in an (approximate) surface selection rule: *Only those vibrational modes that have a non-vanishing dynamic dipole moment perpendicular to the surface are excited in ir-spectroscopy.*

However, experiments now indicate that the surface selection rule may, in fact, not be valid [8.13,14]. In particular, formally dipole forbidden vibrations have been observed for H and D on Cu(111) and for CO on Cu(100) and Cu(111). Remarkably, the dipole forbidden vibrational modes give rise to *anti-absorption* peaks in the ir-spectra, in contrast to the absorption "dips" observed for dipole-allowed vibrational modes. This is schematically illustrated in Fig. 8.7a, which shows the ir-spectra for a layer of atomic adsorbates on a free-electron-like metal surface. The dashed line shows the reflectivity of the clean surface which is nearly equal to unity at infrared frequencies, while the solid curve is the ir-reflectivity for the adsorbate-covered surface (for clarity, the adsorbate-induced change in the reflectivity is strongly exaggerated).

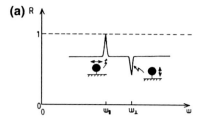

Fig. 8.7. (a) Infrared reflection spectra from an atomic adsorbate (e.g., hydrogen) on a free-electron-like substrate.

The explanation of why dipole-forbidden adsorbate vibrations can be detected with ir-spectroscopy is related to the *electronic friction or drag force* exerted on the adsorbates by the electric current induced in the surface region of the metal by the external ir-light: Scattering of the conduction electrons from the adsorbates results in a broad band absorption of the ir-light. When the ir-frequency ω coincides with the resonance frequency ω_0 of the parallel adsorbate vibrations, the molecules move in resonance with the collective drift motion of the electrons; hence the additional surface resistivity vanishes and the ir-reflectivity reaches the original value of the clean surface. This results in an anti-absorption peak which is observed at the frequency $\omega_\parallel = \omega_0$ of the molecular vibration parallel to the surface.

The following simple model calculation illustrates the basic physics of why dipole-forbidden adsorbate vibrations can be detected with ir-spectroscopy. Consider a metal surface with adsorbates. The equation of motion governing the parallel displacement $\boldsymbol{u}$ of an adsorbate is assumed to be

$$m\ddot{\boldsymbol{u}} + m\omega_0^2 \boldsymbol{u} = -m\eta_{\rm el}(\dot{\boldsymbol{u}} - \boldsymbol{v}) , \qquad (8.13)$$

where ω_0 is the resonance frequency of the frustrated translation. Note that the friction force $\boldsymbol{F}_{\rm fric} = -m\eta_{\rm el}(\dot{\boldsymbol{u}} - \boldsymbol{v})$ depends on the relative velocity between the adsorbate and the (parallel) drift velocity $\boldsymbol{v}$ of the conduction electrons at the surface, induced by the external electric field $\boldsymbol{E}$ of the incoming light beam. This is similar to the force which acts on a body in streaming water, which, of course, depends only on the *relative* velocity between the body and the water. The time–averaged adsorbate-induced power absorption is

$$P = -n_{\rm a} A \langle \boldsymbol{F}_{\rm fric} \cdot (\dot{\boldsymbol{u}} - \boldsymbol{v}) \rangle ,$$

where A is the surface area and $n_{\rm a}$ the concentration of adsorbates. If we assume that $\boldsymbol{v} = \boldsymbol{v}_0 \exp(-i\omega t) + {\rm c.c.}$, where $\boldsymbol{v}_0$ is real, then from (8.13),

$$\dot{\boldsymbol{u}} - \boldsymbol{v} = \frac{\omega^2 - \omega_0^2}{\omega_0^2 - \omega^2 - i\omega\eta_{\rm el}} \boldsymbol{v}_0 e^{-i\omega t} + {\rm c.c.}$$

and hence

$$P = 2n_\text{a} A m v_0^2 \frac{\eta_\text{el}\left(\omega^2 - \omega_0^2\right)^2}{\left(\omega^2 - \omega_0^2\right)^2 + \omega^2 \eta_\text{el}^2} \ .$$

To linear order in n_a we can calculate P by using the drift velocity v_0 obtained without adsorbates. The change in the ir-reflectivity induced by the adsorbates is given by $\Delta R = -P/I_0 A_0$ where A_0 is the cross-sectional area of the incident photon beam. The surface area A illuminated by the incident beam is $A = A_0/\cos\theta$, where θ is the angle of incidence. The intensity of the incident ir-beam is determined by the Poynting vector and is given by $I_0 = cE_0^2/8\pi$, where E_0 is the amplitude of the electric field of the incident light beam. Using these equations gives

$$\Delta R = -\frac{4}{c} \frac{n_\text{a}}{n} \frac{m}{m_\text{e}} \frac{1}{\cos\theta} \frac{\eta_\text{el}\left(\omega^2 - \omega_0^2\right)^2}{\left(\omega^2 - \omega_0^2\right)^2 + \omega^2 \eta_\text{el}^2} \ . \tag{8.13a}$$

This formula predicts an *enhanced reflectivity* at resonance, $\omega = \omega_0$ (Fig. 8.7a). This has a simple physical meaning: according to (8.13), $\dot{\boldsymbol{u}} - \boldsymbol{v} = 0$ at $\omega = \omega_0$, i.e., no relative motion occurs between the adsorbates and the collective motion of the electron fluid, resulting in zero friction and zero adsorbate-induced absorption. Away from resonance, $|\omega - \omega_0| \gg \eta_\text{el}$, the adsorbates are nearly stationary. The energy dissipation caused by the friction drag between the (flowing) electron fluid at the surface and the stationary adsorbates will generate the *background absorption* indicated in Fig. 8.7a and predicted by formula (8.13a).

The indirect coupling via the frictional drag-force does not exist for adsorbates on insulators. However, the parallel electric field at the surfaces of insulators is not screened as strongly as at metallic surfaces, so that even if an adsorbate vibrational mode has a dynamical dipole moment which vanishes in the perpendicular direction, it can couple effectively to the external electric field via the parallel dynamical dipole moment, as described by the interaction Hamiltonian $H' = -\hat{\mu}E_x = -e^* q E_x$ (we assume that the electric field component in the surface plane points along the x-axis). The dynamic dipole moment $\hat{\mu} = e^* q$, where e^* is the so-called effective charge and q the vibrational normal mode coordinate. On the other hand, for semiconductor surfaces both the electronic drag force and the direct coupling to the electric field are usually non-vanishing, so that the force (parallel to the surface) acting on an adsorbate has the form

$$F_x = -m\eta_\text{el}(\dot{u}_x - v_x) + e^* E_x \ , \tag{8.13b}$$

where $\dot{u}_x - v_x$ is the relative velocity of the adsorbate and the electron fluid at the surface. But since the concentration n of free carriers in most semiconductors is much smaller than that in metals, and since $\eta_\text{el} \sim n$, one expects in most cases the friction drag to be very small for semiconductors. In some cases, however, a very high concentration of free carriers may occur in a thin layer at a semiconductor surface. Later we consider the Si(111) system at high temperature, where this seems to be the case, and where (for some temperatures) the electronic drag force term dominates over the direct term e^*E_x.

Electromigration and the Electron Wind Force*

In the experiments described above, an (oscillating) electric current is driven in the surface region of a metal (or semiconductor) by an external electromagnetic wave. In this section we consider the $\omega \to 0$ or dc limit. This case is, of course, realized in practice by applying a potential difference between two ends of the conductor. For a rectangular metal block, this leads to a nearly uniform electric field $\boldsymbol{E}$ and current density $\boldsymbol{J} = ne\boldsymbol{v}$ in the block. For a semiconductor, the current density at the surface may be very different from that in the bulk because of band bending or conducting surface states. The relative motion between the conduction electron fluid and an "impurity" atom adsorbed on the surface will give a friction drag force $m\eta_{\mathrm{el}}\boldsymbol{v}$. In addition, there will be a contribution to the force on the impurity atom from the direct coupling to the electric field via $e^*\boldsymbol{E}$ so the total force is again given by (8.13b):

$$\boldsymbol{F} = m\eta_{\mathrm{el}}\boldsymbol{v} + e^*\boldsymbol{E} \, . \tag{8.13c}$$

For metals, the frictional drag force is usually larger than the contribution $e^*\boldsymbol{E}$, while the opposite is true for semiconductors. A force $\boldsymbol{F}$ of the form (8.13c) will also act on an "impurity" atom in the bulk of the crystal. In the latter case the frictional drag force is usually called the *electronic wind force* and the atomic motion induced by $\boldsymbol{F}$ is called *electromigration* [8.15,16]. Electromigration phenomena are of tremendous practical importance, since they affect growth and annealing of thin films as well as the performance and stability of electronic devices.

Let us discuss the physical origin of the direct force $e^*\boldsymbol{E}$. At first one may think that since an impurity atom in a metal is locally screened and neutral it should not be subjected to a direct force from a uniform electric field. This argument, however, is wrong. Thus it has been pointed out that the same argument can be applied to an electron and its accompanying exchange–correlation hole, which is also a neutral entity. One would then be led to conclude that an electron is not subject to a force in an electric field, and thus, electric current could not occur. The difficulty arises because one must solve a screening problem in a current-carrying non-equilibrium system in order to determine e^*. The basic physics is contained in a very simple argument due to Landauer [8.17]. Consider, for simplicity, the jellium model with the electron density n. Assume that an atom is embedded in the electron gas. Let Ze denote the (bare) ionic charge of the impurity atom, i.e., Ze is the total nuclear charge plus the charge (of opposite sign) contained in all the sharp, split-off electronic states lying below the metal conduction band. In general, the impurity atom will have resonance states in the conduction band and the nature (e.g., position and width) of these resonance states will determine e^*. Assume first a 1D-conductor. If we consider a stationary case the continuity equation for the electric current gives $dJ/dx = 0$, i.e., $J = $ const. In a semi-classical treatment we use a local relation between the current and the electric field so that $J(x) = \sigma(x)E(x)$, where the electronic conductivity $\sigma(x) \propto n(x)$ is proportional to the local conduction band electron density $n(x)$. Thus, if E and n denote the electric field and free electron density far away from the impurity atom and E_1 and n_1 the electric field and the free electron density at the impurity atom, then $nE = n_1 E_1$ or $E_1 = (n/n_1)E$. The force on the ion is $ZeE_1 = (n/n_1)ZeE = e^* E$. Thus $e^* = Ze(n/n_1)$, and if we write $n_1 = n + \Delta n$ we get $e^* = Ze/(1+\Delta n/n)$. Landauer has shown that essentially the same formula is valid in the 3D-case:

$$e^* \sim \frac{Ze}{1 + C\Delta n/n},$$

where C is a number of order unity. We can estimate Δn from the condition that the ion Ze must be totally screened by the metal conduction electrons. This gives $\Delta n \sim Z/\Omega$, where Ω is the "atomic volume" associated with the screening charge. Thus

$$e^* \sim \frac{Ze}{1 + CZ/n\Omega}.$$

Note that, according to this formula, $e^* \to Ze$ when $Z \to 0$, i.e., if the bare charge Ze is very weak, the direct force is given by the macroscopic field acting on the bare charge. Note also that $e^* \to 0$ as $n \to 0$ and $e^* \to Ze$ as $n \to \infty$. Both these results are physically appealing and have been confirmed by accurate calculations [8.18]. One implication of these results is that even if $e^* \sim Ze$ for an atom or molecule in the bulk of a metal, when the same atom or molecule is adsorbed on the metal surface one may have $e^* \approx 0$ because the local conduction electron density at the adsorbate may be much lower than in the bulk. Thus, Ishida [8.18] has shown that for an Si atom located in a jellium with the same electron density as aluminum $e^* \approx 3.6e$ (i.e., the macroscopic electric field "sees" almost the full bare ion charge which would result if all the four Si 3s- and 3p-electrons were removed), while when adsorbed on the surface $e^* \approx 0$. Finally, we note that for metals the electronic wind force is usually much more important than the direct contribution $e^* E$. For example, for an Ag atom on an Ag(111) surface at room temperature, the wind force is about ~ 100 times larger than the direct force $e^* E$, assuming that $e^* = e$, which is likely to be an overestimate of the effective charge.

Electromigration at Surfaces*

In this section we show how the frictional drag-force can lead to morphological instabilities on the surfaces of solids. Remarkably, this is true not only for metals but also for semiconductors, as illustrated below for the Si(111) surface [8.19,20].

One of the characteristic features of growing crystal surfaces is the presence of step systems in which the spacing between successive steps changes in some slowly varying manner. Let us consider the growth of a crystal by (random) deposition of atoms on its surface. The growth occurs by incorporation of atoms at the steps as indicated in Fig. 8.7b. Two possibilities occur: an atom may migrate to a step either from the step-up terrace (atom a in Fig. 8.7b) or from the step-down terrace (atom b). However, experiments have shown that the atoms are incorporated at the step with different probabilities γ_- and γ_+, depending on whether they arrive from the terrace above or below. In most cases $\gamma_+ > \gamma_-$ because the adsorbate–substrate potential usually has the qualitative form shown in Fig. 8.7b, with a barrier (the so called Schwoebel barrier) to incorporation experienced by atom a as a result of the decreased coordination at A. Thus the growth will occur mainly by incorporation of atoms from the lower terrace. It is easy to see that the condition $\gamma_+ > \gamma_-$ results in a nearly uniform distribution of steps on the surface. Let us first prove this in the limiting case $\gamma_- = 0$, where incorporation of atoms only occurs from the lower terrace. Assume that, due to a statistical fluctuation, terrace **1** is wider than the nearby (in the step-down direction) terrace **2**, see Fig. 8.7c. The number of atoms deposited (per unit time) on terraces **1** and **2** is proportional to the width of respective terrace. Since all atoms deposited on terrace **1** are incorporated at step **1**, the step velocity v_1 will be proportional to the width of terrace **1**. Since the same is true for terrace **2** it follows that $v_1 > v_2$ and that the width of terrace **1** will decrease with increasing time. Thus the uniform step distribution is the "stable" steady-state configuration. It is easy to show that during sublimation the opposite is true, and step bunching occurs if a Schwoebel barrier exists.

Let us analyze the situation in more detail [8.21]. Let s_n be the width of the terrace between steps n and $n+1$. We then have

$$\dot{s}_n = v_{n+1} - v_n \ .$$

The velocity v_n of step n is

$$v_n = C\left(\gamma_+ s_n + \gamma_- s_{n-1}\right) \ .$$

In this expression $C > 0$ is assumed to be a constant of the step motion, proportional to the rate of deposition of atoms on the surface. Thus

$$\dot{s}_n = C\left[\gamma_+ s_{n+1} + (\gamma_- - \gamma_+) s_n - \gamma_- s_{n-1}\right] \ . \tag{8.13d}$$

Let L be the average separation between two steps and ξ_n a small perturbation

$$s_n = L + \xi_n \ . \tag{8.13e}$$

Substituting (8.13e) into (8.13d) gives

$$\dot{\xi}_n = C\left[\gamma_+ \xi_{n+1} + (\gamma_- - \gamma_+)\xi_n - \gamma_- \xi_{n-1}\right] \ . \tag{8.13f}$$

194   8. Sliding of Adsorbate Layers

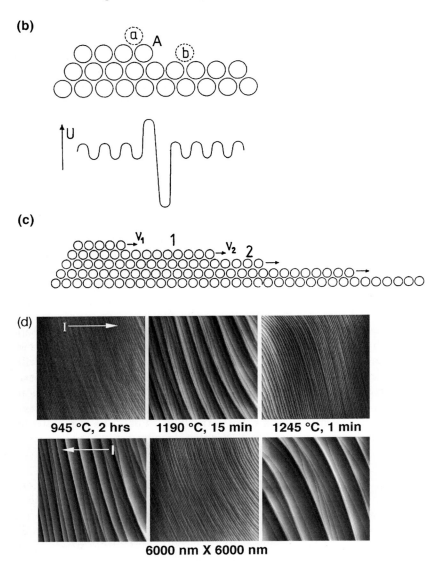

Fig. 8.7. (b) Cross section through a monatomic step in a surface, and the hypothetical potential energy surface associated with the motion of an atom over such a step. From [8.21]. (c) A cross section of the step model which initially consists of an infinite sequence of parallel steps. From [8.21]. (d) STM images of areas 6000 nm × 6000 nm showing step structures after annealing the sample with direct current in the step-up direction for the upper panel and in the step-down direction for the lower panel. The step structure depends both on the current direction and the temperature. From [8.20].

Assume now that

$$\xi_n = A e^{ikn - \lambda t} . \tag{8.13g}$$

If Re $\lambda > 0$ then ξ_n decreases exponentially with time, i.e., the uniform distribution of steps is stable. But if Re $\lambda < 0$ the perturbation ξ_n increases exponentially with time and the uniform step distribution is unstable. Substituting (8.13g) into (8.13f) gives

$$\lambda = C\left[(\gamma_+ - \gamma_-)(1 - \cos k) + \mathrm{i}(\gamma_+ + \gamma_-)\sin k\right] .$$

Thus Re $\lambda > 0$ if $\gamma_+ > \gamma_-$ which is the stability condition stated above.

Assume now that an electric field E drives an electric current J parallel to the surface but orthogonal to the steps. In this case, in addition to the corrugated substrate potential, the adsorbates will experience a force F [given by (8.13c)] parallel to the current J. In a typical case F is extremely weak, $F \sim 10^{-8}$ eV/Å, but the force nevertheless has a strong influence on the motion of the adsorbates. Thus, just as in the QCM-friction measurements, the force will tilt the potential energy surface so that the adsorbates jump (by *thermal excitation*) slightly more often in the direction of the force F than the opposite direction (Sect. 8.3). Now, if F points in the step-down direction (to the right in Fig. 8.7c), the atoms deposited on terrace **1** will drift towards step **2** and even if $\gamma_+ > \gamma_-$, the rate of incorporation of atoms at step **2** from terrace **1** may be larger than that from terrace **2**. Thus, the force F may lead to an instability resulting in step bunching.

Current-induced step bunching has been observed for Si(111). Experiments [8.22] have shown that for Si(111) at $T \sim 500°$C the Schwoebel barrier is negligible compared to $k_\mathrm{B}T$, i.e., $\gamma_+ \approx \gamma_- \approx 1$. Thus even a weak drag force F induced by a current J can "tilt" the flow of Si atoms between the terraces and the steps in such a manner that the system flips between step bunching and a uniform step distribution when the direction of the current J is reversed.

Figure 8.7d shows STM pictures of Si(111) surfaces during sublimation. For Si atoms on Si(111) the effective charge e^* turns out to be positive so that if the magnitude of the e^*E term is larger than the magnitude of the friction drag-term $m\eta_\mathrm{el}v$, then F will point in opposite direction to the current J. For Si(111) this seems to be the case for $T < T_1$ and $T > T_2$, where $T_1 \approx 1000°$C and $T_2 \approx 1175°$C, and as a result an electronic current in the step-up direction will give rise to uniformly distributed steps while reversing the current flow gives rise to step bunching. However, for $T_1 < T < T_2$ the friction drag-force dominates and F and J point in the same direction, and the dependence of step-bunching on the current direction is reversed; see Fig. 8.7d. The switching observed at T_1 and T_2 may be related to changes in the surface electronic structure [e.g., change in surface reconstruction (7×7) $\to$ (1×1) and pre-melting], and accompanying changes in the surface free carrier concentration [8.23]. Such surface transformations have been observed for Ge(111) [8.24], and are also likely to occur for Si(111).

We note finally that the "electronic wind force" may also be important for the sliding of adsorbates using the scanning tunneling microscope (STM). In STM very high local current densities occur, which locally "tilt" the adsorbate–substrate potential energy surface so that the adsorbates jump (by thermal excitation) more frequently in one direction than in the opposite direction [8.25].

Electromigration in Metallic Films*

Consider a perfect crystal at zero temperature with an electric current density J. For such a system there can be no wind force acting on the atoms in the crystal for the same reason that the resistivity is zero. However, a wind force will act on crystal defects, e.g., on foreign atoms or atoms at grain boundaries, voids or atoms adsorbed on the crystal surface, see Fig. 8.7e. This will lead to a drift motion of the "defect atoms", parallel to the direction of the electric current. It has been found that electromigration can cause device failure due to the opening of the interconnections on silicon semiconductor devices. The rapid progress in miniaturization of semiconductor devices continuously reduces the space available for the individual metallic connections, and the danger of line failures due to electromigration increases.

The wind force which acts on, e.g., an Ag atom on a silver surface or at a grain boundary is extremely small, even when the electric current density J is very high. For example, for Ag on an Ag(111) surface the electronic friction $\eta_{el} \approx 1 \times 10^{10}\,\mathrm{s}^{-1}$ (see Table 8.1) and with the current density $J = 1 \times 10^9\,\mathrm{A/m}^2$, the electronic wind force $F = m\eta_{eh}v \sim 10^{-7}\,\mathrm{eV/Å}$, and an even smaller drag force is expected for an Ag atom at a grain boundary. This force is much smaller than the barrier for lateral motion of the atom, which is $\sim 0.1\,\mathrm{eV/Å}$ for an Ag atom on Ag(111). Thus, the only role of the wind force is to slightly tilt the potential energy surface so that the Ag atoms jump (by thermal excitation) slightly more often in the direction of the electric current than in the opposite direction (compare with QCM friction measurements, where the force of inertia $\sim 1 \times 10^{-9}\,\mathrm{eV/Å}$ has the same effect). This will lead to a drift motion of the Ag atoms in the direction of the current J.

Most metallic films are surrounded by an oxide layer which acts as an elastic membrane, inhibiting the drift motion of the metal atoms. That is, the metal atoms initially drift parallel to the current J until a high enough elastic stress difference $\Delta\sigma$ has been built up between the strip ends, inhibiting further drift motion. The elastic stress gradient is equivalent to a chemical potential gradient (see below), which will tend to drive a current of metal atoms in the opposite direction to the electronic wind force. If the elastic stresses are too small to break the oxide coating, a stationary case will result where no net motion of metal atoms occurs. However, if the electric current is large enough, or if the metal strip is long enough, the elastic stresses will be sufficiently high to break the oxide coating, resulting in metal extrusion, as has been observed in many experiments. For example, Fig. 8.7f shows four

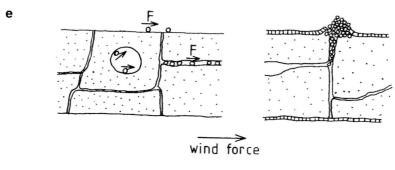

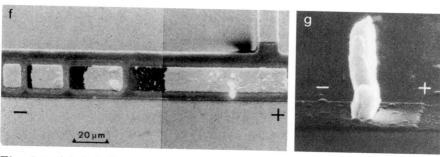

Fig. 8.7. (e) *Left*: Atoms located on a metal surface, at a grain boundary, or on the surfaces of a void, will experience a wind-force when an electric current density J is driven through a metal film. *Right*: If the current J is high enough, metal atoms which have drifted along, e.g., grain boundaries may give rise to such a high local stress "down stream" that an oxide coating is broken, resulting in hillock formation. (f) Drift of four aluminum strips with varying lengths after passage of $J = 3.7 \times 10^9 \, \text{A/m}^2$ for 15 h. (g) Aluminum extrusion formed through an etched hole in the SiN on the anode end. The current density was $1.5 \times 10^9 \, \text{A/m}^2$ for 45 h. From [8.29].

aluminum strips of various lengths after passage of $J = 3.7 \times 10^9 \, \text{A/m}^2$ for 15h. Note that for the shortest strip the elastic stresses are too small to break the oxide layer, and the metal film is virtually unaffected by the current. For the other films the stresses were large enough to break the oxide coating, resulting in hillock formation while the aluminum edge facing the negative terminal erodes away. In another experiment an aluminum strip was coated with silicon nitride (SiN). In this case it was impossible for the aluminum to break the coating and no drift motion of Al atoms could be observed. However, these effects were only seen when the entire aluminum strip was enclosed within the SiN. A partial covering of the aluminum showed no effect at all. Thus if a small hole is etched in the SiN on the anode end, an aluminum extrusion is formed (Fig. 8.7g).

The discussion above can be made quantitatively as follows. Since the driving force F is extremely weak compared to the periodic forces from the crystal atoms, one can use linear response theory where the drift velocity v

is proportional to the driving force $\boldsymbol{F}$. This gives (see Sect. 8.3)

$$m\bar{\eta}\boldsymbol{v} = \boldsymbol{F} ,$$

where the drift friction $\bar{\eta}$ has already been introduced in Chap. 3. In Sect. 8.7.2 we show that $\bar{\eta} = c^2/D_c$ where D_c is the chemical diffusivity and where $c^2 = k_B T/m$ if one assumes that the only interaction between the diffusing atoms is an on-site repulsion which forbids double occupation of the same site. The diffusivity $D_c = D_0 \exp(-\Delta E/k_B T)$, where ΔE is the activation barrier for diffusion and where the prefactor D_0 depends weakly on the temperature T. Thus

$$\boldsymbol{v} = (D_0/k_B T)e^{-\Delta E/k_B T} \boldsymbol{F} .$$

The driving force $\boldsymbol{F}$ is the sum of two contributions, $\boldsymbol{F} = \boldsymbol{F}_1 + \boldsymbol{F}_2$, where $\boldsymbol{F}_1$ is the electronic wind force and $\boldsymbol{F}_2$ the contribution from the gradient in the hydrostatic stress $\sigma(x)$. If Ω is the atomic volume, then the work done against the hydrostatic stress when bringing an atom from one end ($x = 0$) to the other end ($x = l$) of the metal strip equals $\Omega[\sigma(l) - \sigma(0)]$ so that $F_2 = -\Omega \Delta \sigma/l$, where $\Delta \sigma = \sigma(l) - \sigma(0)$. If we write the wind force as $F_1 = AJ$, it is clear that if $J < J_c$, where the critical current $J_c = \Omega \sigma_c / lA$ (σ_c being the stress necessary to break the oxide coating), then the steady state diffusion velocity v will vanish. That is, if $J < J_c$ the wind force will not give rise to a steady-state transport of metal atoms and the metal film will not erode. Note that $J_c \propto 1/l$ so that, in accordance with Fig. 8.7f, the critical current J_c is higher for a short metal strip than for a long strip.

8.3 Sliding Friction in the Low-Density Limit

8.3.1 Analytical Results for a 1D-Model

No general analytical solution, valid for arbitrary adsorbate coverage, has been obtained for the model (8.1). But for the limiting case of low adsorbate coverage, where the lateral interaction between the adsorbates can be neglected, a relatively complete analysis is possible. In this section we present a brief summary of these results because of the physical insight they supply. For simplicity we assume that $\alpha = \infty$ in (8.4); thus, the motion is strictly 2D, confined to the plane $z = z_0$.

For a single adsorbate in the potential (8.4) the motions in the x and y directions are independent. Thus, it is enough to consider a one-dimensional model, where a particle (an adsorbate) moves on a corrugated substrate. The equation of motion for the particle coordinate $x(t)$ is

$$m\ddot{x} + m\eta\dot{x} = -\frac{\partial U}{\partial x} + f + F , \qquad (8.14)$$

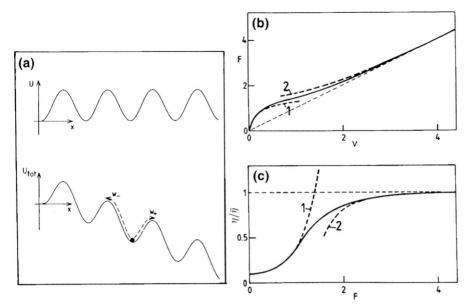

Fig. 8.8. (a) A particle moving in the potential $U = U_0(1 - \cos kx)$ under the influence of an external force F. The total potential $U_{\text{tot}} = U - Fx$. The barrier to particle diffusion is lower in the direction of F and the particle will jump over the barrier more often in this direction than in the opposite direction. (b) The relation between the driving force F and the drift velocity v. (c) The inverse of the normalized sliding friction, $\eta/\bar{\eta}$. The dashed curves 1 and 2 are the low and high force F results while the solid lines are the exact results. The quantities v and F are in natural units, $1/k\tau$ and $kk_{\text{B}}T$, respectively. In the calculation $\eta\tau = 1$ and $U_0/k_{\text{B}}T = 2.5$.

where

$$U(x) = U_0(1 - \cos kx) ,$$

so that $2U_0$ is the activation barrier for diffusion and $k = 2\pi/a$ where a is the "lattice" constant of the substrate. The stochastically fluctuating force f is related to the friction η via the fluctuation–dissipation theorem

$$\langle f(t)f(0)\rangle = 2\eta m k_{\text{B}}T\delta(t) .$$

Let us first assume that $2U_0 - Fa/2 \gg k_{\text{B}}T$, and that the friction η is so large that the particle performs jumps only to the nearest neighbor's potential wells (Fig. 8.8a). We can then use Kramer's theory of activated processes to evaluate the drift velocity (and hence the friction $\bar{\eta}$) induced by the external force F. A constant force F gives a contribution $-Fx$ to the potential, i.e., $U_{\text{tot}} = U(x) - Fx$ (Fig. 8.8a). According to Kramer's theory of thermally activated processes the particle jump rates to the right (w_+) and to the left (w_-) are given by [8.30]

$$w_+ = \nu e^{-\beta(2U_0 - Fa/2)},$$

$$w_- = \nu e^{-\beta(2U_0 + Fa/2)},$$

where $\beta = 1/k_\mathrm{B} T$ and where, in the limit of large friction,

$$\nu = \frac{2\pi U_0}{ma^2 \eta}.$$

Since the barrier to particle diffusion is lower to the right than to the left, the particle will jump more frequently to the right (i.e., $w_+ > w_-$), leading to a drift velocity

$$\langle \dot{x} \rangle = a(w_+ - w_-) = a\nu e^{-\beta 2U_0} \left(e^{\beta Fa/2} - e^{-\beta Fa/2} \right) \tag{8.15}$$

and the sliding friction $\bar{\eta} = F/m\langle \dot{x} \rangle$,

$$\bar{\eta} = \frac{\eta e^{\beta 2U_0}}{2\pi \beta U_0} \frac{\beta Fa}{e^{\beta Fa/2} - e^{-\beta Fa/2}}. \tag{8.16}$$

The dashed lines denoted by *1* in Figs. 8.8b,c show the drift velocity $v = \langle \dot{x} \rangle$, and the inverse of the normalized sliding friction $\eta/\bar{\eta}$, as a function of the driving force F, as obtained from (8.15) and (8.16).

The initial linear slope of the $F = F(v)$ curve in Fig. 8.8b is obtained by expanding $\exp(\pm \beta Fa/2) \approx 1 \pm \beta Fa/2$, which holds if $\beta Fa \ll 1$. Substituting this expansion in (8.15) gives the linear response results

$$\langle \dot{x} \rangle \approx \beta a^2 \nu e^{-\beta 2U_0} F \tag{8.17}$$

and

$$\bar{\eta} = \frac{\eta}{2\pi} \frac{k_\mathrm{B} T}{U_0} e^{\beta 2U_0}. \tag{8.18}$$

Hence, $1/\bar{\eta}$ exhibits an activated temperature dependence. One can easily extend the results above to finite adsorbate concentration if one assumes that the only adsorbate–adsorbate interaction is a "hard-core" repulsion which excludes double occupation of the lattice sites. In this case the jump rates w_+ and w_- given above must be multiplied by the probability $1 - c$ that a site is empty (c is the adsorbate "coverage"). Hence the atomic scale friction $\bar{\eta} \propto 1/(1-c)$. Note that $\bar{\eta} \to \infty$ as $c \to 1$. More generally, one expects $\bar{\eta}$ to be very large for high-coverage commensurate adsorbate structures, as indeed observed, e.g., in QCM measurements for C_6H_{12} on Ag and in molecular dynamics simulations (Sect. 8.4).

The result (8.16) is only valid as long as the effective barrier height $2U_0 - Fa/2$ is much larger than $k_\mathrm{B} T$. In this case the particle will perform a diffusive

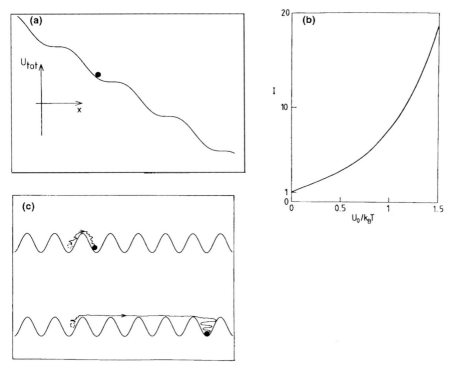

Fig. 8.9. (a) The total potential when $F > kU_0$. (b) The friction integral I, defined in the text. (c) For high friction η the particle motion is strongly damped and the particle only jumps between nearby potential wells. For low friction the particle performs long (in units of the substrate lattice constant a) jumps.

motion with a slow drift in the direction of the external force. On the other hand when $Fa/2 \gg 2U_0$ the particle will propagate rapidly in the x-direction (Fig. 8.9a). Because of its finite mass (inertia), the adsorbate will move very little in response to the rapidly fluctuating force from the substrate and one therefore expects the drift velocity to be nearly the same as that of an energetically flat surface (i.e., $U_0 = 0$). Thus, in this limit the drift velocity is limited only by the friction η and $\bar{\eta} \to \eta$ as $F \to \infty$. Let us analyze the large F limit in greater detail.

We write the coordinate of the particle as

$$x = vt + u , \tag{8.19}$$

where v is the drift velocity and u the fluctuation in the particle position caused by the substrate potential and the stochastic force f. Let us assume that $u(0) = 0$ (the final result does not depend on this assumption). Since $\langle \dot{x} \rangle = v$ by definition, we have $\langle \dot{u} \rangle = 0$ and since $u(0) = 0$ this gives $\langle u(t) \rangle = 0$ for all times. Now, when the drift velocity v is very high the particle will move a long distance before the contribution to u from the fluctuating force f reaches the order of one lattice spacing a. Let us consider the system in the time period where $|u| \ll a$. Substituting (8.19) in the equation of motion (8.14) and expanding U to linear order in u gives

$$m\ddot{u} + m\eta\dot{u} = f - kU_0 \sin\omega t - k^2 U_0 u \cos\omega t + m(\bar{\eta} - \eta)v , \qquad (8.20)$$

where $\omega = kv$. Since $\langle u(t) \rangle = 0$ the time average of (8.20) gives

$$m(\bar{\eta} - \eta)v = k^2 U_0 \langle u \cos\omega t \rangle$$

or

$$\bar{\eta} = \eta + (k^2 U_0/mv)\langle u \cos\omega t \rangle . \qquad (8.21)$$

The leading $1/F$ contribution to u can be obtained from (8.20) by neglecting the last two terms so that

$$m\ddot{u} + m\eta\dot{u} = f - kU_0 \sin\omega t . \qquad (8.22)$$

Thus

$$u = u_\mathrm{T} - \frac{kU_0}{m} \frac{\sin(\omega t + \phi)}{\omega (\omega^2 + \eta^2)^{1/2}} , \qquad (8.23)$$

where $\tan\phi = \eta/\omega$ and where u_T is the thermal contribution to u derived from the random force f. Substituting (8.23) into (8.21) gives

$$\bar{\eta} = \eta + \frac{k^4 U_0^2}{2m^2} \frac{\eta}{\omega^2 (\omega^2 + \eta^2)} \to \eta \left(1 + \frac{U_0^2}{2m^2 v^4}\right) \qquad (8.24)$$

as $\omega = kv \to \infty$. Using the definition $m\bar{\eta}v = F$ we get from (8.24), to leading order in $1/F$,

$$\bar{\eta} = \eta \left(1 + \frac{U_0^2 m^2 \eta^4}{2F^4}\right) . \qquad (8.25)$$

The dashed curves denoted by 2 in Fig. 8.8b,c shows the result for the drift velocity $v = F/m\bar{\eta}$ and for $\eta/\bar{\eta}$ as a function of the driving force F, as obtained from (8.25). The solid lines are the exact results.

Note that the drift velocity depends linearly on the driving force F both at very low and at very high F. However, the slope of the $F = F(v)$ curve is much higher at low F than at high F, see Fig. 8.8b. This is analogous to shear thinning of fluids and has a similar physical origin. In fact, the particle system can be considered as a (dilute) 2D-fluid under the influence of a shear force. At low shear stress the force F is small and the particles can move along the surface only because of *thermal excitation* over the energy barriers of $U(x)$. Since these barriers are lower in the direction of the external force F, than in the opposite direction (Fig. 8.8a), the particles will jump more often in the direction of F. This leads to a slow drift motion in the direction of F. The drift velocity v is proportional to $\exp(-2U_0/k_BT)$ [see (8.17)], where $2U_0$ is the diffusion barrier. Thus as the temperature $T \to 0$, the drift velocity $v \to 0$, and the effective shear viscosity [which is proportional to $\bar{\eta} \propto \exp(2U_0/k_BT)$] increases towards infinity. On the other hand when $F > kU_0$ the external force is large enough to "pull" the adsorbates over the barriers and the drift velocity is non-vanishing even at zero temperature. In particular, for very large F, the drift velocity $v = F/m\eta$ does not depend on the temperature.

In order to study (8.14) in a general case, it is convenient to introduce dimensionless coordinates

$$\tilde{x} = kx, \quad \tilde{t} = t/\tau,$$

where τ will be defined below. In the new coordinates (8.14) becomes

$$\frac{d^2\tilde{x}}{d\tilde{t}^2} + \tilde{\eta}\frac{d\tilde{x}}{d\tilde{t}} = \tilde{U}_0 \sin \tilde{x} + \tilde{f} + \tilde{F}, \qquad (8.26)$$

where

$$\tilde{\eta} = \eta\tau, \quad \tilde{U}_0 = k^2\tau^2 U_0/m, \quad \tilde{f} = k\tau^2 f/m, \quad \tilde{F} = k\tau^2 F/m.$$

Note that

$$\langle \tilde{f}(\tilde{t})\tilde{f}(0)\rangle = (k\tau^2/m)^2 \langle f(t)f(0)\rangle = 2\eta\delta(t)k_B T k^2\tau^4/m.$$

Hence using that $\delta(t) = \delta(\tilde{t})/\tau$ and choosing

$$\tau = \left(\frac{m}{k_B T k^2}\right)^{1/2}$$

gives

$$\langle \tilde{f}(\tilde{t})\tilde{f}(0)\rangle = 2\tilde{\eta}\delta(\tilde{t})$$

and

$$\tilde{U}_0 = U_0/k_B T, \quad \tilde{F} = F/k_B T k.$$

Equation (8.26) depends on the dimensionless parameter $\tilde{\eta}$. Let us therefore estimate $\tilde{\eta}$ in a typical QCM-friction experiment (Chap. 3). Let us write $k = 2\pi/a$ so that

$$\tau = a/(2\pi v_{\text{th}}),$$

where the thermal velocity $v_{\text{th}} = (k_B T/m)^{1/2} \sim 100\,\text{m/s}$ in a typical case (we have used $T = 200\,\text{K}$ and $m = 131$ amu). If $a \sim 2\,\text{Å}$ then $\tau \sim 10^{-13}\,\text{s}$ and $\tilde{\eta} = \eta\tau \sim 10^{-4}$ if $\eta \sim 10^9\,\text{s}^{-1}$. Hence, $\tilde{\eta} \ll 1$ and the low friction limit should prevail in most QCM-friction experiments.

The low friction limit is more complicated to analyze than the high friction limit considered above. *Risken* and *Vollmer* have presented a general solution for the sliding friction $\bar\eta$ which is valid for arbitrary magnitude of η and F. Following *Kramers* [8.30] classical treatment of thermally activated processes, the Langevin equation (8.14) is first reformulated as a Fokker–Planck equation for the particle probability distribution function $P(x,\dot x)$. For the Fokker–Planck equation effective mathematical tools have been developed which were invoked by *Risken* and *Vollmer* [8.31] in their derivation of an expression for the sliding friction. Here we consider only the linear response limit, which is directly relevant for QCM-friction measurements. In this case Risken and Vollmer have shown that in the low friction case

$$\bar\eta = \eta I(U_0/k_B T) \tag{8.27}$$

where

$$I(\xi) = \frac{1}{4\pi^{3/2}} \int_0^{2\pi} dx e^{\xi(1+\cos x)} \Big/ \left(\int_0^\infty d\epsilon \frac{e^{-\epsilon}}{\int_0^{2\pi} dx\, [\epsilon + \xi(1+\cos x)]^{1/2}} \right). \tag{8.28}$$

This function is shown in Fig. 8.9b. Note that $I \to 1$ and hence $\bar\eta \to \eta$ as $U_0 \to 0$ while $I \to (4/\pi) \exp(2U_0/k_B T)$ and hence $\bar\eta \to (4\eta/\pi)\exp(2U_0/k_B T)$ as $U_0 \to \infty$. In the latter case, $1/\bar\eta$ exhibits an activated temperature dependence but the prefactor is different from that found earlier, (8.18), in the case of large friction.

Note that the drift velocity scales as $1/\eta$ for *both* high friction (8.17) and low friction (8.27). In the high friction case this follows directly from the fact that the rate of thermally exited jumps over a barrier is proportional to $1/\eta$. However, in the low fiction limit the jump rate is proportional to η. But once thermally exited over a barrier, because of the low friction η, the particle will on average propagate for a long distance before finally falling into a new potential well (Fig. 8.9c). The net result is that in both cases the drift velocity scales as $1/\eta$.

It is interesting to note that the sliding friction (8.27) is a non-analytical function of U_0 at $U_0 = 0$. This implies that it is impossible to perform an expansion of the sliding friction in powers of U_0. This is likely to be true not only in the dilute limit (non-interacting particles) but more generally when the adsorbate layer is in a 2D-fluid state, and an analytical calculation of the sliding friction for the fluid state is therefore likely to be extremely complicated even when U_0 is very small. However, when the adsorbate layer is in an IC-solid state, then $\bar\eta$ is an analytical function of U_0 for $U_0 = 0$ and, for "small" U_0, it can be expanded in a power series in U_0. The leading U_0-contribution in such an expansion is $\propto U_0^2$; it is clear from symmetry that no linear U_0 term can occur since translating the substrate by $a/2$ in the x and y-directions, which is equivalent to replacing $U_0 \to -U_0$ in the original potential, will leave the sliding friction unchanged.

8.3.2 A Case Study: Chaotic Motion for a 3D-Model

Suppose we set the microscopic friction to zero, $\eta = 0$, in (8.14) which, according to the fluctuation–dissipation theorem, implies that the fluctuating force $f(t)$ also vanishes. In this case, in the absence of the driving force F, (8.14) has only periodic solutions corresponding to either localized vibrations in a potential well or, if the kinetic energy is large enough, propagating solutions where the speed of the particles is periodically modulated by the force from the substrate. This type of solution is, in fact, common to all 1D-Hamiltonian systems with periodic potentials. However, a particle moving in two or more spatial dimensions may exhibit chaotic motion, where the trajectory of the particle is extremely sensitive to the initial conditions. Thus a particle moving above a surface may, even in the absence of a fluctuating force $f(t)$, exhibit a quasi-diffusive motion rather than ballistic motion as in the 1D-model (8.14). More generally, whenever the phase space (that is, the space spanned by the coordinates and momenta of all the particles) of a dynamical system has a dimension larger than 2, chaotic motion is possible and, as a function of the initial conditions, the motion may be regular in some region of phase space and chaotic in other regions.

In this section we discuss the motion of "hot" atoms above surfaces. We show that if the kinetic energy (parallel to the surface) of the atom is of order of the lateral barrier height of the adsorbate–substrate potential energy surface, the distance traveled by the atom will, in general, be limited by elastic scattering from the substrate atoms rather than by the direct energy transfer to the substrate via the electronic or phononic friction. The discussion is illustrated with molecular dynamics calculations for high energy oxygen atoms injected parallel to an aluminum surface following the dissociation of an O_2 molecule. It will be shown that the distance traveled by an oxygen atom is limited mainly by the rapid randomization of the adsorbate motion by quasi-elastic scattering from the substrate potential, and not primarily by the direct energy transfer to the substrate degrees of freedom.

The simulations start with an oxygen atom close to an Al(111) surface, with the oxygen velocity vector parallel to the surface, corresponding to the initial kinetic energy 3.5 eV. The diffusion barrier of an adsorbed oxygen atom is about 0.4 eV (and the overall corrugation about 1.5 eV), i.e., much larger than the thermal energy, and diffusion of thermalized oxygen atoms is negligible on the time-scale of the simulations. The motion of the substrate atoms is treated explicitly and all curves in Fig. 8.10c–e have been obtained by averaging over 64 trajectories obtained with different initial conditions for the substrate atoms generated from separate simulations of a clean Al substrate at a temperature of 300 K.

Figure 8.10a shows 64 oxygen atom trajectories, obtained with different initial conditions for the substrate atoms as described above. The black dots indicate the equilibrium lattice positions of the Al atoms in the top layer. The time dependence of the distance of the oxygen atom from the first Al-layer

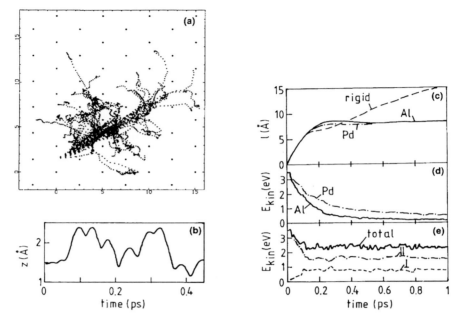

Fig. 8.10. (a) The lateral positions of the oxygen atom for 64 trajectories at consecutive times with the time step of 9fs. The equilibrium positions of the surface atom are indicated by dots. (b) The distance from the first surface atom layer, z, as a function of time for one of the trajectories (randomly chosen). (c) Time dependence of the average lateral distance l that the oxygen atom has moved. *Solid curve*: Al substrate. *Dash-dotted curve*: with the mass altered to that of Pd. *Dashed curve*: frozen substrate. (d) Time dependence of the average kinetic energy. *Solid curve*: Al substrate. *Dash-dotted curve*: with the mass altered to that of Pd. (e) Time dependence of the average kinetic energy when the substrate atoms are fixed at their equilibrium positions (no energy transfer to the substrate). The total kinetic energy and the kinetic energy corresponding to motion parallel and perpendicular to the surface are displayed. From [8.32].

for a randomly chosen trajectory is shown in Fig. 8.10b. From this figure it is immediately clear that the notion of ballistic motion is only applicable for a very short time period.

The average displacement along the surface, $l = \langle |\boldsymbol{x}_\parallel(t) - \boldsymbol{x}_\parallel(0)| \rangle$, is shown in Fig. 8.10c for an oxygen atom on an Al(111) surface (solid curve), and for the case where the substrate atoms are kept fixed at their equilibrium positions (in the absence of the oxygen atom), i.e., for a rigid substrate (dashed curve). Also shown is the case of a "heavy substrate" where the mass of the substrate atoms equals the atomic mass of Pd, but where the adsorbate–substrate interaction potential is the same as for aluminum (dash-dotted curve). Note that, although the effective phononic friction η is roughly twice as large for an Al substrate as for the "heavy" substrate, l is virtually the same for the two systems. For a rigid substrate (infinite substrate mass) no energy transfer to the substrate is possible, and $l \to \infty$ as $t \to \infty$. Note, however, that there is a decrease in the slope at about 0.2 ps, i.e., beyond $t = 0.2$ ps the displacement l increases more slowly than in the initial phase.

From the simulation with a frozen substrate is is clear that two distinct phases with different types of adsorbate motion can be identified and that the transition between the two is not connected to any energy transfer to the substrate.

Figure 8.10d shows the time dependence of the total kinetic energy of the oxygen atom for the Al substrate and for the "heavy" substrate. At about $t = 0.2$ ps, where the saturation in l takes place (Fig. 8.10c), the kinetic energy is roughly 1 eV, which is still high compared to the diffusion barrier which is about 0.4 eV. Note also that for the "heavy" substrate the decay of the oxygen kinetic energy is much slower than for the Al substrate, but the average distance an oxygen atom moves before stopping is almost the same in the two cases (Fig. 8.10c). We conclude that the distance l is determined mainly by randomization of the adsorbate velocity by scattering from substrate atoms, rather than by energy transfer to the substrate.

The time dependency of the kinetic energy of the oxygen atom on the frozen substrate is shown in Fig. 8.10e. Here the total kinetic energy is also shown decomposed into parts corresponding to adsorbate motion parallel and perpendicular to the surface. The decay in the kinetic energy in the first 0.1 ps is compensated for by an increase in the potential energy (the total energy is conserved). Initially, the oxygen atom is moving in the lateral direction, but after 0.2 ps a substantial amount of energy has been transferred to the normal degree of freedom. For times $t > 0.2$ ps equal amounts of kinetic energy are, on average, distributed over the three translational degrees of freedom.

The two regions in the time dependence of l are seen to be connected with the equipartition of the kinetic energy into the different translational degrees of freedom. This is achieved without any energy transfer to the substrate and it is established surprisingly fast. The efficient coupling between the x, y, and z degrees of freedom is caused by the corrugation of the potential energy surface. The net displacement along the surface takes place exclusively in the first, ballistic phase. After the equipartition has taken place, the atom still has enough energy to perform jumps between chemisorption sites, but the directions of the jumps are randomized making this type of motion very ineffective for lateral transport. On the time scale 10 ps no increase in l can be detected except for the case with a frozen substrate where the initial total energy is intact.

An understanding of the factors governing the mobility of adsorbates in a transient phase is a vital ingredient in a more general picture of such phenomena as epitaxial growth and chemical reactions. The distance traveled by the "hot" atoms when an oxygen molecule dissociates in the vicinity of a metal surface has recently been studied experimentally using the scanning tunneling microscope.

208   8. Sliding of Adsorbate Layers

Let us now assume that in addition to the adsorbate–substrate interaction, an external force $\boldsymbol{F}$ acts on the adsorbate parallel to the substrate. If F is very large, the adsorbate will propagate with high velocity essentially ballistically, and the sliding friction is determined mainly by the direct energy transfer to the substrate. However, when the kinetic energy of the particle is of the order of the corrugation amplitude of the potential energy surface, strong scattering from the substrate potential will occur, and the sliding friction $\bar{\eta}$ will be very high and mainly determined by these scattering processes rather than by the direct energy transfer to the substrate. More generally, one may ask what is the primary source of friction when an adsorbate layer slides on a surface: is it the *direct* transfer of energy to the substrate via the friction η, or is it the *indirect* channel where energy is *first* transferred into internal degrees of freedom of the adsorbate layer and then transferred to the substrate via the friction η? In the latter case the sliding friction could be quite insensitive to η. We will address this question in Sect. 8.5.

8.4 Linear Sliding Friction

At finite adsorbate coverage, where the adsorbate–adsorbate interaction is important, it is, in general, not possible to study (8.1) analytically. But the sliding friction can be obtained from computer simulations [8.33-35]. In this section we consider the linear response limit where the driving force $\boldsymbol{F}$ in (8.1) is so weak that the drift velocity $\langle \boldsymbol{v} \rangle$ depends linearly on $\boldsymbol{F}$. This limit is directly relevant for QCM-friction studies. In Sect. 8.9 we consider the non-linear sliding friction which is relevant for boundary lubrication.

Let us first prove that the QCM-friction measurements probe the linear sliding friction. The force of inertia which acts on the adsorbate slab in a QCM measurement (which is the reason why the slab will slide relative to the substrate) is extremely small, at least for thin adsorbate layers. The force of inertia acting on an adsorbate is $F \sim mA\omega_0^2$, where A is the vibration amplitude and ω_0 the vibration frequency of the quartz crystal. Typically $A \sim 100\,\text{Å}$ and $\omega_0 \sim 10^8\,\text{s}^{-1}$ which gives $F \sim 10^{-8}\,\text{eV/Å}$ which is extremely small compared with the force due to the corrugated substrate potential, which is of order U_0/a where $a \sim 1\,\text{Å}$ is the substrate lattice constant and where U_0 is the amplitude of the corrugation of the adsorbate–substrate potential energy surface. For physisorption systems, $U_0 \sim 1\,\text{meV}$. Thus $U_0/a \sim 10^{-3}\,\text{eV/Å} \gg F$ and the linear response approximation is very accurate. Recent QCM-friction experiments have indeed shown that the friction force is directly proportional to the sliding velocity.

The sliding friction $\bar{\eta}$ can be obtained from computer simulations based on the Langevin equations (8.1) and (8.3). In this section we assume again

$2U_0/\epsilon = 2$ $k_B T/\epsilon = 1$

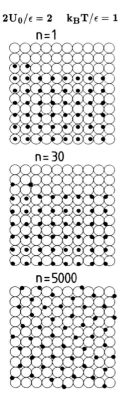

Fig. 8.11. Snapshots (after $n = 1, 30$ and 5000 time steps) from Langevin dynamic simulations.

that $\alpha = \infty$ in (8.4), i.e., we consider a strict 2D-model. The random forces f_i are assumed to be Gaussian random variables, generated by adding many random numbers which are equally distributed in the interval [0,1]. The time variable is discretized with the step length $\Delta = 0.01\tau$. The basic unit is chosen as a square containing $M \times M$ substrate atoms, where typically $M = 10$. In the snapshot pictures of adsorbate structures shown below, it is assumed that the hollow sites have the largest adsorbate–substrate binding energy, i.e., these sites correspond to the local minima of $u(\mathbf{r})$ given by (8.4). If N denotes the number of adsorbates in the basic unit, then the coverage $\theta = N/(M \times M)$. In all simulations periodic boundary conditions were used. The system was "thermalized" by $\sim 10^6$ time steps which corresponds to the actual "preparation" time $10^4 \tau$; this was enough in all cases to reach thermal equilibrium. Figure 8.11 shows an example for $2U_0/\epsilon = 2$ and $k_B T/\epsilon = 2$, and with $N = 50$ (i.e., the coverage $\theta = 0.5$) where the particles are initially located in the lower part of the basic unit. After ~ 1000 time steps a $c(2 \times 2)$ structure is formed which is the thermal equilibrium structure.

210   8. Sliding of Adsorbate Layers

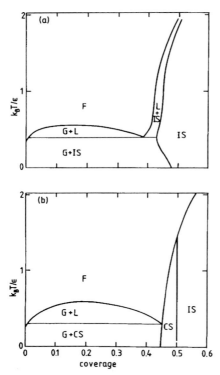

Fig. 8.12. The adsorbate phase diagram for (**a**) $2U_0/\epsilon = 0.5$ and (**b**) $2U_0/\epsilon = 2.0$. The phase boundaries are only roughly correct. In the figure F, G, L, IS, and CS denote fluid, gas, liquid, incommensurate solid and commensurate solid, respectively.

The sliding friction $\bar{\eta}$ was obtained from the simulations using the definition $\bar{\eta} = F/m\langle v \rangle$ where F is a weak external force acting in the x-direction on each adsorbate and which results in the adsorbate drift velocity $\langle v \rangle$, obtained from the simulations by averaging over all the particles in the basic unit and over many time steps corresponding to the time $\sim 10^4 \tau$.

Figure 8.12a, b shows the adsorbate phase diagrams for $2U_0/\epsilon$ equal to 0.5 and 2, respectively. In the former case, the substrate corrugation is so weak that the phase diagram is essentially that of a two-dimensional Lennard–Jones system on a *flat* surface [8.36]. In the figure F, IC, G, and L denote the fluid phase, the incommensurate solid phase, the gas, and the liquid phases, respectively; the gas and liquid phases are the fluid phase at low and high density, respectively. Similarly, G+L and G+IC denote the gas–liquid and gas–incommensurate solid coexistence regions, respectively. The phase diagram in Fig. 8.12b differs most importantly from that in (a) by a new phase around the coverage $\theta = 0.5$. This phase, a $c(2 \times 2)$ commensurate solid (CS) phase, is shown in Fig. 8.13b. It is obvious that this structure is stable if $2U_0/\epsilon$

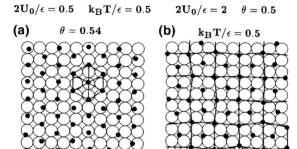

Fig. 8.13. Snapshots of the adsorbate structure for (a) $2U_0/\epsilon = 0.5$ and coverage $\theta = 0.54$ where a triangular incommensurate structure occurs and (b) $2U_0/\epsilon = 2.0$ and $\theta = 0.5$ where a commensurate c(2 × 2) structure occur. In both cases $k_B T = 0.5\epsilon$.

is large enough because all adsorbates bind at hollow sites, i.e., the adsorbate–substrate binding energy is maximized while the separation between nearby adsorbates is close to the equilibrium separation in the pair potential $v(r)$. But in Fig. 8.12a the corrugation of the substrate potential is so weak that the adsorbates prefer to minimize the nearest neighbor repulsive adsorbate–adsorbate interaction energy by forming a triangular incommensurate solid phase; see Fig. 8.13a.

In Fig. 8.14 we show the result for $\eta/\bar{\eta}$ as a function of coverage and for a few different temperatures. Let us first consider the case of a "weakly"-corrugated adsorbate–substrate interaction potential, $2U_0/\epsilon = 0.5$ (Fig. 8.14a). In this case, as the coverage increases, the sliding velocity $\langle v \rangle$ increases monotonically towards v_0, where v_0 is the sliding velocity which would occur for an energetically flat surface (i.e., if $U_0 = 0$). For large coverage the adsorbate system is dominated by the adsorbate–adsorbate interaction and a practically perfect triangular adsorbate structure is formed, as shown in Fig. 8.13a (for $\theta = 0.54$ and $k_B T/\epsilon = 0.5$). This structure can slide with practically no activation barriers on the substrate, since, as some adsorbates move uphill during sliding, other adsorbates moves downhill. Hence, the *pinning* potential induced by the substrate is negligible in this case and $\langle v \rangle \approx v_0$. This also explains why $\langle v \rangle$ is almost temperature independent for $\theta > 0.55$ – the barriers which must be overcome during sliding are very small and can be neglected.

On the other hand, for low adsorbate coverage, $\langle v \rangle$ is much smaller that v_0 and strongly temperature dependent. This is easy to understand in the limit of very low adsorbate coverage where a dilute lattice gas occurs (Sect. 8.3.1). Here each adsorbate atom performs an independent random-walk type of motion with a slight drift in the direction of the weak external force, F. The particles have to "jump" over barriers of height $\sim 2U_0$, which separate the local minima of the ground state potential energy surface; at low

212   8. Sliding of Adsorbate Layers

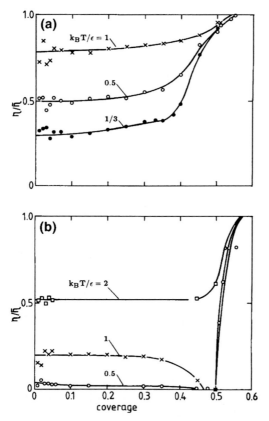

Fig. 8.14. The inverse of the normalized drift friction, $\eta/\bar{\eta}$ (equal to the normalized drift velocity $\langle v \rangle/v_0$), as a function of coverage for several different temperatures. (a) $2U_0/\epsilon = 0.5$ and (b) $2U_0/\epsilon = 2.0$.

temperature the rate of this thermally activated process is proportional to $\exp(-2U_0/k_B T)$. Note that the external force F by itself is much too weak to "move" adsorbates over the barriers in the system (except for incommensurate structures).

Let us now consider the case of a "strongly" corrugated substrate, $2U_0/\epsilon = 2$, where the adsorbate phase diagram is shown in Fig. 8.12b. Again, at high coverage $\langle v \rangle \to v_0$. That is, at high enough coverage the adsorbate–adsorbate interaction dominates over the adsorbate–substrate interaction and the pinning potential is negligible. Note, however, that in the present case it is necessary to go to slightly higher coverage than for $2U_0/\epsilon = 0.5$, before one enters into the incommensurate solid phase region; this is exactly what one would expect when the corrugation of the adsorbate–substrate interaction potential increases. (In practice it may be impossible to reach this high coverage where $\langle v \rangle \approx v_0$ before desorption or multilayer absorption occur.)

The drift velocity for very low adsorbate coverage, where a dilute lattice gas exists, is of exactly the same nature in the present case as in the earlier case where $2U_0/\epsilon = 0.5$ (see above). But if $k_B T < 1.5\epsilon$, the drift velocity *decreases* continuously with increasing coverage up to $\theta = 0.5$. At $\theta = 0.5$ the drift velocity is extremely low, in particular at low temperatures; the reason is obvious if one considers a snapshot of the adsorbate system at this coverage (Fig. 8.13b). Obviously, except for thermal displacements of the adsorbates away from the hollow sites, a perfect $c(2 \times 2)$ structure occurs. This structure is strongly pinned by the adsorbate–substrate interaction. In principle, elementary excitations can be thermally induced in the $c(2 \times 2)$ structure and drift in the direction of the applied force $\boldsymbol{F}$, but in the present system these excitations have such a high energy that they occur in a negligible concentration. The point is that moving an adsorbate from a hollow site to any of the nearby empty hollow sites leads to a very large increase in the adsorbate–adsorbate repulsion energy. This will effectively block the movement of the adsorbates and lead to a very low drift velocity for the $c(2 \times 2)$ system. But for coverage just slightly higher than 0.5, the adsorbate system undergoes a phase transition to the incommensurate triangular structure shown in Fig. 8.13a for which the pinning potential is small. Note also that for $k_B T > 1.5\epsilon$ no $c(2 \times 2)$ structure occurs even in the present case and the sliding friction decreases monotonically with increasing coverage.

Influence of Defects (Pinning Centers)

For $2U_0/\epsilon = 2$ and $k_B T = 0.5\epsilon$ at the coverage $\theta = 0.5$ the adsorbate system forms a commensurate solid structure (Fig. 8.13b) which is pinned by the substrate. According to linear response theory this structure will not slide on the surface, i.e., the drift velocity vanishes and the sliding friction is infinite (neglecting the contribution from thermally exited imperfections). However, an adsorbate structure can also be pinned by surface defects such as steps or immobile impurity atoms. At "low" adsorbate coverages or "high" temperatures, where the adsorbate layer is in a 2D-fluid state, surface imperfections will increase the sliding friction but have no drastic effect on the sliding process. At "high" coverages the adsorbate layer may form an incommensurate solid state. If the lateral corrugation U_0 of the adsorbate–substrate potential is weak compared with the 2D-elastic properties of the adsorbate layer, then the periodic substrate potential cannot pin the adsorbate layer (Sect. 10.1). However, it is easy to show that a finite concentration of *randomly* distributed defects will *always* give rise to a finite pinning barrier. We will discuss this problem in detail in Sect. 14.4, but the result is so important that we will already present the basic idea here [8.37,38].

Assume that the interaction between the IC-solid adsorbate lattice and the defects is "weak". In a certain area $A_c = \xi^2$ there is short-range order and the adsorbate system forms an almost perfect hexagonal lattice as in Fig. 8.13a. Here ξ is the elastic coherence length which will be calculated in Sect. 14.4. Since inside the area A_c the lattice is almost regular and the defects randomly distributed, the pinning forces acting on the lattice nearly compensate for each other. From random walk arguments, the total pinning force acting on the area A_c is of order $g_0 N_d^{1/2}$ where g_0 is the root-mean-square force of interaction of an individual defect with the lattice, and N_d the number of pinning centers in the area A_c, i.e., $N_d = n_d A_c$ where n_d is the concentration of defects. The pinning force *per unit area* is therefore

$$F_p = g_0 N_d^{1/2}/A_c = g_0 (n_d/A_c)^{1/2} \ .$$

Thus in the linear response limit, where the driving force F is arbitrarily weak, at zero temperature the adsorbate lattice will always be pinned. [Note, however, that $F_p \to 0$ as $A_c \to \infty$, i.e., the *pinning force per unit area vanishes for a rigid lattice* (for which $\xi = \infty$).] However, for nonzero temperatures thermally activated motion will occur and the adsorbate layer will drift in the direction of the applied force F. If the thermal energy $k_B T$ is much smaller than the barrier height

$$U_1 \approx F_p A_c a = g_0 (n_d A_c)^{1/2} a \ ,$$

where a is the lattice constant, the thermally activated drift motion involves units of size $\xi \times \xi$ displacing coherently (creep motion). If $k_B T \gg U_1$ the defects have a negligible influence on the sliding dynamics and the whole adsorbate lattice drifts coherently with a speed limited mainly by the finite friction coupling to the substrate so that $\bar{\eta} \approx \eta$.

If the adsorbate–defect interaction is "strong", the IC solid monolayer may be pinned by the defects even at relatively high temperature, resulting in a nearly infinite sliding friction. As an example of the latter case, let us assume that one of the adsorbates in Fig. 8.13a is fixed (i.e., immobile), but that it interacts with all the other adsorbates by the same pair-potential that acts between the other adsorbates. Note that because of the periodic boundary condition we actually have a finite concentration of regularly spaced impurity atoms. In this case the IC-solid layer is able to slide on the surface only if F is so large that *local fluidization* of the adsorbate layer occurs in the vicinity of the localized adsorbates; we will give an example of this in Sect. 8.9 when we study the non-linear sliding friction. However, such a local fluidization cannot occur in the linear response case, where F is arbitrarily weak, and as a result the adsorbate layer is pinned. This is illustrated in Fig. 8.15 for the same parameters as in Fig. 8.14a (i.e., $U_0/\epsilon = 0.5$), but now with one of the atoms fixed to a hollow site. At low and intermediate coverages, where the layer is in a 2D-fluid state, the sliding friction is very similar to that without any pinning centers (dashed lines, from Fig. 8.14a). However, when the IC-solid monolayer is formed the sliding velocity vanishes.

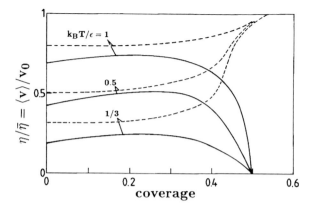

Fig. 8.15. The inverse of the normalized drift friction, $\eta/\bar{\eta}$ (equal to the normalized drift velocity $\langle v \rangle/v_0$), as a function of coverage for several different temperatures and for $2U_0/\epsilon = 0.5$. The dashed lines are from Fig. 8.14a, while the solid lines have been calculated with one of the adsorbates in the basic unit fixed to a hollow site.

8.5 A Case Study: Xe on Silver

In this section we study the forces required to slide thin layers of xenon on a silver surface [8.39]. We will show that for an IC-monolayer film, on a surface free from defects, the sliding friction $\bar{\eta}$ is nearly identical to the parallel microscopic friction $\eta_\parallel$ which acts on the individual adsorbates, which is of mainly *electronic origin*. For the bilayer film a small fraction of the sliding friction arises from internal excitations in the film, and the rest from the *direct* energy transfer to the substrate via the electronic friction. We also show that for monolayer coverage and below, the motion of the adsorbates perpendicular to the surface is irrelevant and a 2D-model gives nearly the same result for the sliding friction as a 3D-model. In the light of the theoretical results we discuss the recent QCM-friction measurements for monolayers and bilayers of Xe on Ag(111).

Suppose that an adsorbate layer is sliding on a substrate. In the steady state the energy pumped into the adsorbate layer by the external force must ultimately be transferred to the substrate. This can occur in two different ways as shown in Fig. 8.16. First [case (a)], there is a direct energy transfer from the center of mass motion of the adsorbate layer to the substrate via the processes discussed in Sect. 8.2. Secondly [case (b)], due to the lateral corrugation of the adsorbate–substrate interaction potential and due to interaction with surface defects (e.g. steps), during sliding energy will be transferred into internal degrees of freedom (lattice vibrations) of the adsorbate layer. If the frictional coupling between the adsorbate layer and the substrate is weak, the latter energy may be completely thermalized in the overlayer before it

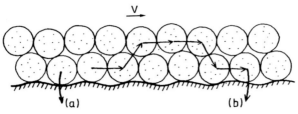

Fig. 8.16. A Xe bilayer sliding on a surface. Two processes contribute to the friction force acting on the bilayer. In (*a*) a direct process occurs where the translational kinetic energy of the bilayer is directly transferred to the substrate via the microscopic friction acting on the first layer of adsorbates in direct contact with the substrate. This energy transfer channel also occurs for an energetically perfectly smooth substrate surface (i.e., semi-infinite jellium model). Process (*b*) is an *indirect* process where the *corrugated* substrate potential excites vibrations (virtual phonons) in the bilayer. These vibrational modes may thermalize before the energy is finally being transmitted to the substrate via the same microscopic friction which determines the direct adsorbate–substrate energy transfer [channel (*a*)].

is finally transferred to the substrate. It is important to note that only very low frequency phonons (frequency $\omega_{\rm ph}$) can be excited in the sliding film: If v denotes the sliding velocity and a the lattice constant of the substrate, or the spatial extent of the defect potential, then $\omega_{\rm ph} \sim kv$, where $k = 2\pi/a$. In a typical case, $v \sim 1\,{\rm cm/s}$ and $a \sim 1\,{\rm \AA}$ giving $\omega_{\rm ph} \sim 1 \times 10^9\,{\rm s}^{-1}$, corresponding to the wave number $\sim 0.01\,{\rm cm}^{-1}$. Since both energy and momentum must be conserved in the excitation process, and since the corrugated substrate potential is characterized by the wave vector $k = 2\pi/a$, it is clear that the periodic substrate potential cannot (directly) excite any *real* phonons in the sliding film. However, "higher-order" effects such as phonon anharmonicity and defects in the film, "break" the momentum conservation and allow the virtual phonons to become real. Similarly, the finite lifetime of the phonons caused by the friction coupling to the substrate [via process (*a*)], will make channel (*b*) possible if the lifetime of a (virtual) phonon (with wave vector $k = 2\pi/a$) is short enough that Heisenberg's uncertainty relation is satisfied. Due to these "higher-order" effects, the substrate potential will give a contribution to the sliding friction via process (*b*). Since the potential from surface defects contains all possible wave vectors, both energy and momentum can be directly conserved when surface defects (e.g., steps) are involved in the phonon excitation processes.

Model Parameters

For Xe on silver, $\omega_\perp \approx 22\,{\rm cm}^{-1}$ and $E_{\rm B} \approx 0.23\,{\rm eV}$ and (8.5) gives $\alpha = 0.72\,{\rm \AA}^{-1}$. In the simulations presented below we have also taken $\alpha' = 0.72\,{\rm \AA}^{-1}$. The frequency $\omega_\parallel$ has not been measured for Xe on Ag(100) [or Ag(111)], but it has been measured [8.40] using inelastic helium scattering for the $(\sqrt{3} \times \sqrt{3})$Xe structure on Cu(111) where, at the zone center,

8.5 A Case Study: Xe on Silver

Table 8.3. The parameter values used in the simulations unless otherwise stated.

$\omega_\perp = 22\,\mathrm{cm}^{-1}$	$\omega_\parallel = 2.7\,\mathrm{cm}^{-1}$
$E_0 = 0.23\,\mathrm{eV}$	$\alpha = 0.72\,\mathrm{\AA}^{-1}$
$U_0 = 0.95\,\mathrm{meV}$	$\alpha' = 0.72\,\mathrm{\AA}^{-1}$
$\epsilon = 19\,\mathrm{meV}$	$r_0 = 4.54\,\mathrm{\AA}$
$a = 2.89\,\mathrm{\AA}$	$\eta_\perp = 2.5 \times 10^{11}\,\mathrm{s}^{-1}$

$\omega_\parallel = 3 \pm 1\,\mathrm{cm}^{-1}$. In the calculations reported we have used $\omega_\parallel = 2.7\,\mathrm{cm}^{-1}$ and from (8.6) $2U_0 = 1.9\,\mathrm{meV}$.

The friction parameters $\eta_\parallel$ and $\eta_\perp$ can be estimated as follows. Consider first the phonon contribution to the friction. As discussed in Sect. 8.2, if the characteristic frequency of the forces exerted on the substrate by the motion of the adsorbate are well below the maximum phonon frequency of the solid, then the phononic friction can be calculated using the elastic continuum model. For Xe on silver this condition is satisfied and with $\omega_\perp \approx 22\,\mathrm{cm}^{-1}$ and $\omega_\parallel \approx 2.7\,\mathrm{cm}^{-1}$, (8.12) and (8.13) give $\eta_\perp = 2.5 \times 10^{11}\,\mathrm{s}^{-1}$ and $\eta_\parallel = 5.5 \times 10^7\,\mathrm{s}^{-1}$.

The phononic frictions $\eta_\perp$ and $\eta_\parallel$ calculated above for isolated adsorbates should also be good approximations for fluid adsorbate layers. However, the phononic friction associated with the coherent sliding of incommensurate solid adsorbate layers, is expected to vanish [8.41,42]. In the applications below we find that the electronic contribution to $\eta_\parallel$ is much more important than the phononic contribution, and since the electronic friction varies relatively little with the coverage (in particular, it does not vanish at incommensurate coverage) the uncertainty in the value for the phonon friction is of no practical importance.

The parallel electronic friction has been estimated from surface resistivity measurements as well as from theoretical calculations; both theory and measurements give an electronic contribution to $\eta_\parallel$ of order $\sim (3-10) \times 10^8\,\mathrm{s}^{-1}$ which is about one order of magnitude larger than the estimated (low coverage) phonon contribution to $\eta_\parallel$. On the other hand, for fluid adsorbate layers the electronic contribution to $\eta_\perp$ is negligible compared with the phonon contribution. Table 8.3 summarizes the parameter values used below unless otherwise stated.

Results

Figure 8.17a shows the number of adsorbates N_1, N_2, and N_3 in the basic unit cell in the first, second, and third layers, respectively, as a function of the total number N of adsorbates. Note that all Xe atoms occupy the first layer for N up to 68 atoms in the 12×12 basic unit; for $N = 69$ one Xe atom occupies the second layer. Thus the monolayer Xe coverage in the simulations correspond to $n_a = 68/(144 a^2) \approx 0.0565\,\mathrm{\AA}^{-2}$. Close to the completion of the

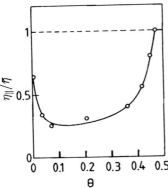

Fig. 8.19. The dependency of $\eta_\parallel/\bar{\eta}$ on the coverage up to completion of the first Xe monolayer. In the calculation $\eta_\parallel = 2.5 \times 10^9 \text{ s}^{-1}$ and the other parameters are as in Table 8.3.

Table 8.4. The inverse of the normalized sliding friction, $\eta_\parallel/\bar{\eta}$, for a Xe monolayer and bilayer. In the calculation $\eta_\parallel = 6.2 \times 10^8 \text{ s}^{-1}$.

System	$\eta_\parallel/\bar{\eta}$
monolayer ($N = 68$)	0.98 ± 0.04
bilayer ($N = 132$)	0.83 ± 0.02

Table 8.4 shows $\eta_\parallel/\bar{\eta}$ for the IC-solid monolayer and the bilayer when $\eta_\parallel = 6.2 \times 10^8 \text{ s}^{-1}$. Note that for the monolayer film $\bar{\eta} = \eta_\parallel$ within the accuracy of the simulation, while for the bilayer film the sliding friction $\bar{\eta}$ is about 20% larger than $\eta_\parallel$.

Comparison with Experiments

Figure 3.22 shows the sliding friction $\bar{\eta}$ as a function of coverage for Xe on Ag(111), where the monolayer coverage is defined as 0.0597 Xe atoms per Å^2 (this is the highest possible density of Xe atoms in the first layer in direct contact with the substrate). The sliding friction of the monolayer and bilayer Xe films are $\bar{\eta} = (8.4 \pm 0.3) \times 10^8 \text{ s}^{-1}$ and $(10.7 \pm 0.4) \times 10^8 \text{ s}^{-1}$, respectively. The slip times at low adsorbate coverage tend to vary from one system to another indicating that surface defects may have an influence on the sliding dynamics at low coverage. However, the sliding friction for the monolayer and the bilayer films are very reproducible indicating that the surface defects have a negligible influence on the sliding dynamics of these layers, and we mainly focus on these limiting cases below.

The experimental data (Fig. 3.22) is for the Xe/Ag(111) system while our simulation is for Xe on Ag(100). We have chosen to study the latter system because it is easier to analyze analytically. Nevertheless, the quali-

8.5 A Case Study: Xe on Silver

Table 8.3. The parameter values used in the simulations unless otherwise stated.

$\omega_\perp = 22\,\mathrm{cm}^{-1}$	$\omega_\parallel = 2.7\,\mathrm{cm}^{-1}$
$E_0 = 0.23\,\mathrm{eV}$	$\alpha = 0.72\,\mathrm{Å}^{-1}$
$U_0 = 0.95\,\mathrm{meV}$	$\alpha' = 0.72\,\mathrm{Å}^{-1}$
$\epsilon = 19\,\mathrm{meV}$	$r_0 = 4.54\,\mathrm{Å}$
$a = 2.89\,\mathrm{Å}$	$\eta_\perp = 2.5 \times 10^{11}\,\mathrm{s}^{-1}$

$\omega_\parallel = 3 \pm 1\,\mathrm{cm}^{-1}$. In the calculations reported we have used $\omega_\parallel = 2.7\,\mathrm{cm}^{-1}$ and from (8.6) $2U_0 = 1.9\,\mathrm{meV}$.

The friction parameters $\eta_\parallel$ and $\eta_\perp$ can be estimated as follows. Consider first the phonon contribution to the friction. As discussed in Sect. 8.2, if the characteristic frequency of the forces exerted on the substrate by the motion of the adsorbate are well below the maximum phonon frequency of the solid, then the phononic friction can be calculated using the elastic continuum model. For Xe on silver this condition is satisfied and with $\omega_\perp \approx 22\,\mathrm{cm}^{-1}$ and $\omega_\parallel \approx 2.7\,\mathrm{cm}^{-1}$, (8.12) and (8.13) give $\eta_\perp = 2.5 \times 10^{11}\,\mathrm{s}^{-1}$ and $\eta_\parallel = 5.5 \times 10^7\,\mathrm{s}^{-1}$.

The phononic frictions $\eta_\perp$ and $\eta_\parallel$ calculated above for isolated adsorbates should also be good approximations for fluid adsorbate layers. However, the phononic friction associated with the coherent sliding of incommensurate solid adsorbate layers, is expected to vanish [8.41,42]. In the applications below we find that the electronic contribution to $\eta_\parallel$ is much more important than the phononic contribution, and since the electronic friction varies relatively little with the coverage (in particular, it does not vanish at incommensurate coverage) the uncertainty in the value for the phonon friction is of no practical importance.

The parallel electronic friction has been estimated from surface resistivity measurements as well as from theoretical calculations; both theory and measurements give an electronic contribution to $\eta_\parallel$ of order $\sim (3-10) \times 10^8\,\mathrm{s}^{-1}$ which is about one order of magnitude larger than the estimated (low coverage) phonon contribution to $\eta_\parallel$. On the other hand, for fluid adsorbate layers the electronic contribution to $\eta_\perp$ is negligible compared with the phonon contribution. Table 8.3 summarizes the parameter values used below unless otherwise stated.

Results

Figure 8.17a shows the number of adsorbates N_1, N_2, and N_3 in the basic unit cell in the first, second, and third layers, respectively, as a function of the total number N of adsorbates. Note that all Xe atoms occupy the first layer for N up to 68 atoms in the 12×12 basic unit; for $N = 69$ one Xe atom occupies the second layer. Thus the monolayer Xe coverage in the simulations correspond to $n_\mathrm{a} = 68/(144a^2) \approx 0.0565\,\mathrm{Å}^{-2}$. Close to the completion of the

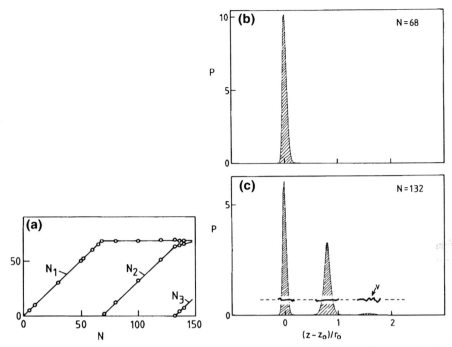

Fig. 8.17. (a) Variation of the number of adsorbates in the first, N_1, second, N_2, and third, N_3, Xe layer as a function of increasing numbers of adsorbates in the basic unit cell. Note that the third layer starts to become populated before the second layer is completed. (b) The probability distribution $P(z)$ of particles in the direction normal to the surface for one complete monolayer of Xe on Ag(100) (coverage $\theta = 68/144$). $P(z)$ is normalized so that the integral over all z equals unity. (c) The same quantity for two monolayers [132 particles in the substrate (12×12) unit cell]. The horizontal wavy line shows the drift velocity v (arbitrary units). In the calculation $\eta_\parallel = 6.2 \times 10^8 \, \mathrm{s}^{-1}$ and the other parameters are as in Table 8.3.

bilayer one additional Xe atom is transferred to the first monolayer giving a compressed layer with $n_\mathrm{a} = 0.0574 \, \text{Å}^{-2}$. Experimentally, at T = 77.4 K, Xe condenses onto Ag(111) as an "uncompressed" solid monolayer with 0.05624 atoms/Å^2. The monolayer accommodates further atoms by compressing, until it reaches 0.0597 atoms/Å^2.

Figure 8.17b,c shows the Xe probability distribution $P(z)$ and the drift velocity $\langle v \rangle$ (arbitrary units) as a function of the distance z from the surface for one complete monolayer of Xe (coverage $\theta = 68/144$) and for the bilayer (132 particles in the basic unit). In both cases $P(z)$ is normalized so that

$$\int d(z/r_0) P(z) = 1 \, .$$

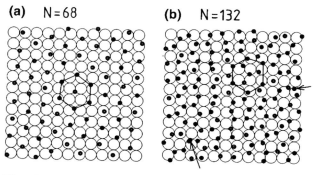

Fig. 8.18. Snapshots of the adsorbate layer in cases (b) and (c) of Fig. 8.17.

In the calculation $\eta_\parallel = 6.2 \times 10^8\,\text{s}^{-1}$. Note that the Xe atoms form well-defined layers and that the probability distribution $P(z)$ for the first layer is narrower than that of the second layer. This is, of course, related to the stronger Xe–substrate interaction as compared with the Xe–Xe interaction, which results in a higher frequency of the perpendicular Xe–substrate vibration as compared with the perpendicular Xe vibration of the second layer Xe atoms. Note also that there is a small occupation of the third layer even before the second layer is fully occupied (Fig. 8.17a).

The horizontal wavy lines in Fig. 8.17c show the variation of the drift velocity with the center of mass position of the adsorbates. The fluctuations in the curve come from the limited time period over which the velocity has been averaged and are particularly large in the regions where the probability of finding a particle is low. It is clear that the drift velocity of the second (and higher) layer, within the accuracy of the simulations, is identical to those of the first layer. This is true even when the second (or third) layer is incomplete and simply reflects the fact that the lateral corrugation experienced by an Xe atom in the second (or third) layer is much higher than the (small) corrugation of the adsorbate–substrate interaction potential.

Figure 8.18 shows snapshots of the adsorbate layer of parts (b) and (c) of Fig. 8.17. As expected, the Xe monolayer forms a hexagonal structure which is incommensurate (IC) with respect to the substrate. In the bilayer case the second layer atoms (filled circles) occupy the hollow sites in the first layer (the two arrows point at Xe atoms in the third layer occupying two hollow sites of the second layer).

Figure 8.19 shows the dependency of $\eta_\parallel/\bar{\eta}$ on the coverage up to one monolayer. In the calculation $\eta_\parallel = 2.5 \times 10^9\,\text{s}^{-1}$. Note the initial drop in $1/\bar{\eta}$ when going from the dilute lattice gas state to the formation of small solid islands and the increase of $1/\bar{\eta}$ close to the formation of the full monolayer phase. At monolayer coverage (68 Xe atoms in the unit cell) the sliding friction equals $\eta_\parallel$.

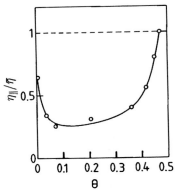

Fig. 8.19. The dependency of $\eta_\parallel/\bar{\eta}$ on the coverage up to completion of the first Xe monolayer. In the calculation $\eta_\parallel = 2.5 \times 10^9 \,\text{s}^{-1}$ and the other parameters are as in Table 8.3.

Table 8.4. The inverse of the normalized sliding friction, $\eta_\parallel/\bar{\eta}$, for a Xe monolayer and bilayer. In the calculation $\eta_\parallel = 6.2 \times 10^8 \,\text{s}^{-1}$.

System	$\eta_\parallel/\bar{\eta}$
monolayer ($N = 68$)	0.98 ± 0.04
bilayer ($N = 132$)	0.83 ± 0.02

Table 8.4 shows $\eta_\parallel/\bar{\eta}$ for the IC-solid monolayer and the bilayer when $\eta_\parallel = 6.2 \times 10^8 \,\text{s}^{-1}$. Note that for the monolayer film $\bar{\eta} = \eta_\parallel$ within the accuracy of the simulation, while for the bilayer film the sliding friction $\bar{\eta}$ is about 20% larger than $\eta_\parallel$.

Comparison with Experiments

Figure 3.22 shows the sliding friction $\bar{\eta}$ as a function of coverage for Xe on Ag(111), where the monolayer coverage is defined as 0.0597 Xe atoms per Å^2 (this is the highest possible density of Xe atoms in the first layer in direct contact with the substrate). The sliding friction of the monolayer and bilayer Xe films are $\bar{\eta} = (8.4 \pm 0.3) \times 10^8 \,\text{s}^{-1}$ and $(10.7 \pm 0.4) \times 10^8 \,\text{s}^{-1}$, respectively. The slip times at low adsorbate coverage tend to vary from one system to another indicating that surface defects may have an influence on the sliding dynamics at low coverage. However, the sliding friction for the monolayer and the bilayer films are very reproducible indicating that the surface defects have a negligible influence on the sliding dynamics of these layers, and we mainly focus on these limiting cases below.

The experimental data (Fig. 3.22) is for the Xe/Ag(111) system while our simulation is for Xe on Ag(100). We have chosen to study the latter system because it is easier to analyze analytically. Nevertheless, the qualitative results obtained above should also be valid for the Ag(111) surface. In particular, the sliding friction of the IC-solid monolayer system should be nearly equal to $\eta_\parallel$, i.e., a negligible contribution to the sliding friction from internal excitations in the film is expected. This conclusion is supported by the fact that the experimental sliding friction for the compressed monolayer film, $\bar\eta = (8.4 \pm 0.3) \times 10^8 \, \text{s}^{-1}$, is similar to the electronic friction deduced from surface resistivity data $[(3-10) \times 10^8 \, \text{s}^{-1}$ from Table 8.1], and is also in accordance with theoretical estimates of the electronic friction (Sect. 8.2). For a Xe bilayer on Ag(100) we have found that a small fraction (about 20%) of the sliding friction is due to internal excitations in the film. The observed sliding friction for the Xe/Ag(111) system is approximately 27% larger for the bilayer than for the monolayer film, i.e., of the same order of magnitude as we obtain in the simulations. Nevertheless, surface resistivity measurements (Fig. 8.2b) have shown that the electronic friction of a bilayer film may be about 20% larger than for the monolayer film; the observed increase in the sliding friction for the bilayer film may therefore result partly from internal excitations in the film, and partly from a slight increase in the electronic friction.

Analysis of the Results of the Computer Simulation

In this section we address the following fundamental questions related to the simulations presented above: (A) Why is the sliding friction for the IC-solid monolayer essentially equal to the parallel friction $\eta_\parallel$, i.e., why is the contribution from the perpendicular motion unimportant? (B) Why is the contribution from internal excitations [process (b) in Fig. 8.16] negligible for the compressed IC-monolayer film?

A) Perpendicular Motion

Let us first discuss the relative importance of the "perpendicular" and "parallel" friction. We first derive a formally exact expression for $\bar\eta$. The power absorption (per adsorbate) P induced by the force $\boldsymbol{F}$ can be written in two different ways. First, from the definition of $\bar\eta$

$$P = \langle \boldsymbol{v} \cdot \boldsymbol{F} \rangle = \langle \boldsymbol{v} \rangle \cdot \boldsymbol{F} = m\bar\eta \langle \boldsymbol{v} \rangle^2 \ . \tag{8.29}$$

Next, let us write the velocity of an adsorbate as

$$\boldsymbol{v} = \langle \boldsymbol{v} \rangle + \delta \boldsymbol{v} \ ,$$

where $\langle \delta \boldsymbol{v} \rangle = 0$. Now in the steady state the power P that the external force $\boldsymbol{F}$ "pumps" into the adsorbate layer must equal the energy transfer per unit time from the adsorbate layer to the substrate. The latter quantity is given by

$$P = -(\langle \boldsymbol{v} \cdot \boldsymbol{F}_\text{f} \rangle - \langle \boldsymbol{v} \cdot \boldsymbol{f} \rangle)$$
$$= m\eta_\perp \left(\langle \delta v_\perp^2 \rangle - \langle \delta v_\perp^2 \rangle_0 \right) + m\eta_\parallel \left(\langle \delta v_\parallel^2 \rangle - \langle \delta v_\parallel^2 \rangle_0 \right) + m\eta_\parallel \langle \boldsymbol{v} \rangle^2 \ , \tag{8.30}$$

where $\boldsymbol{f}$ is the fluctuating force of (8.1), $\langle ... \rangle_0$ stands for thermal average when $\boldsymbol{F} = 0$, and where

$$\boldsymbol{F}_{\mathrm{f}} = -m\left(\eta_\perp \boldsymbol{v}_\perp + \eta_\parallel \boldsymbol{v}_\parallel\right) .$$

Equation (8.30) can also be derived directly from (8.1) if the latter is multiplied by $\dot{\boldsymbol{r}}_i$ and summed over i. Combining (8.29) and (8.30) gives

$$\bar{\eta} = \eta_\perp \left(\langle \delta v_\perp^2 \rangle - \langle \delta v_\perp^2 \rangle_0\right)/\langle v \rangle^2 + \eta_\parallel \left[\left(\langle \delta v_\parallel^2 \rangle - \langle \delta v_\parallel^2 \rangle_0\right)/\langle v \rangle^2 + 1\right] . \quad (8.31)$$

The relative importance of the contributions to $\bar{\eta}$ from the "perpendicular" and "parallel" friction for the IC-solid monolayer film can be estimated as follows. Consider first the perpendicular term. As an adsorbate drifts with the velocity $\langle v \rangle$ parallel to the surface its distance from the surface will fluctuate by the amount δz. We can obtain δz from the potential (8.4) by assuming that the adsorbate, as it translates along the surface, always occupies the position normal to the surface which minimizes the adsorbate–substrate interaction energy, i.e., by requiring $\partial u/\partial z = 0$. This gives $z = z_0 + \delta z$ where

$$\delta z = (U_0/2\alpha E_{\mathrm{B}})(2 - \cos kx - \cos ky) .$$

The contribution to the perpendicular velocity from the drift motion parallel to the surface is therefore $\delta v_\perp = \delta \dot{z}$ or

$$\delta v_\perp = (kU_0 v/2\alpha E_{\mathrm{B}}) \sin kx ,$$

where we have assumed that the drift motion occurs along the x-axis with the speed v (i.e., $x = vt$). Thus

$$\left(\langle \delta v_\perp^2 \rangle - \langle \delta v_\perp^2 \rangle_0\right)/\langle v \rangle^2 \sim \frac{1}{2}\left(\frac{kU_0}{2\alpha E_{\mathrm{B}}}\right)^2 . \quad (8.32)$$

For the Xe/Ag(100) system the prefactor of $\eta_\perp$ in the expression (8.31) for $\bar{\eta}$ becomes $\sim 3 \times 10^{-5}$. The friction parameter $\eta_\perp$ has been estimated above using the elastic continuum model $\eta_\perp \sim 2.5 \times 10^{11}\,\mathrm{s}^{-1}$. Hence, the perpendicular contribution to $\bar{\eta}$ for Xe on Ag(100) is of order $7 \times 10^6\,\mathrm{s}^{-1}$. The contribution from the parallel friction to (8.31) for Xe on Ag(100) is of order (see above) $10^9\,\mathrm{s}^{-1}$. Thus, the parallel contribution to $\bar{\eta}$ is a factor $\sim 10^2$ larger than the perpendicular contribution.

We have shown that the perpendicular adsorbate motion is unimportant at monolayer coverage. Our simulation shows that this is true also for coverage below monolayer. For example, in Table 8.5 we show the sliding friction for $\theta = 52/144$ for two different values of the decay constant α occurring in the potential $u(\boldsymbol{r})$ in (8.4). A large α implies that the particles cannot displace in the direction normal to the surface, i.e., the motion is nearly two-dimensional. We note that increasing α from $0.72\,\text{Å}^{-1}$ to $1.98\,\text{Å}^{-1}$ leads to a negligible

Table 8.5. The inverse of the normalized sliding friction, $\eta_\parallel/\bar\eta$, for two different decay constants α. In the calculation $N = 52$ and $\eta_\parallel = 2.5 \times 10^9 \, \text{s}^{-1}$.

$\alpha \, (\text{Å}^{-1})$	$\eta_\parallel/\bar\eta$
0.72	0.38 ± 0.03
1.98	0.40 ± 0.02

change in the sliding friction, i.e., a strict 2D-model gives nearly the same result for the sliding friction as the 3D-model.

B) Internal Excitations in the Film

Next, let us discuss why, for the compressed IC-monolayer film, there is a negligible contribution to the sliding friction from internal excitations in the film. We use a 2D-model since we have just proved that the motion in the z-direction is unimportant. Let us write the coordinates for the particles in the IC-solid sliding state as

$$\boldsymbol{r}_i = \boldsymbol{v}t + \boldsymbol{x}_i + \boldsymbol{u}_i , \tag{8.33}$$

where $\boldsymbol{v}$ is the drift velocity, $\boldsymbol{x}_i = (x_i, y_i)$ the perfect lattice sites (in a reference frame moving with the velocity $\boldsymbol{v}$) of the hexagonal structure, and $\boldsymbol{u}_i$ the (fluctuating) displacements away from these sites. Because of the weak corrugation of the substrate potential, $|\boldsymbol{u}_i|$ is small. Substituting (8.33) into the equation of motion (8.1) and expanding U and V to linear order in $\boldsymbol{u}_i$ gives

$$m\ddot{\boldsymbol{u}}_i + m\eta_\parallel \dot{\boldsymbol{u}}_i + \sum_j K_{ij} \cdot \boldsymbol{u}_j$$
$$= \boldsymbol{f}_i - kU_0 \left[\hat{x} \sin k(v_x t + x_i) + \hat{y} \sin k(v_y t + y_i)\right]$$
$$- k^2 U_0 \left[\hat{x} u_{xi} \cos k(v_x t + x_i) + \hat{y} u_{yi} \cos k(v_y t + y_i)\right] + m(\bar\eta - \eta_\parallel)\boldsymbol{v} , \tag{8.34}$$

where the force constant matrix K_{ij} has the components $K_{ij}^{\alpha\beta} = \partial^2 V/\partial u_i^\alpha \partial u_j^\beta$. The time average of (8.34) gives (note $\langle \boldsymbol{u}_i \rangle = 0$)

$$m(\bar\eta - \eta_\parallel)\boldsymbol{v} = k^2 U_0 \left[\hat{x}\langle u_{xi} \cos k(v_x t + x_i)\rangle + \hat{y}\langle u_{yi} \cos k(v_y t + y_i)\rangle\right]$$

or, if the drift motion occurs along the x-axis,

$$\bar\eta = \eta_\parallel + (k^2 U_0/mv)\langle u_{xi} \cos k(v_x t + x_i)\rangle . \tag{8.35}$$

To linear order in U_0 (8.34) gives

$$m\ddot{\boldsymbol{u}}_i + m\eta_\parallel \dot{\boldsymbol{u}}_i + \sum_j K_{ij} \cdot \boldsymbol{u}_j = \boldsymbol{f}_i - kU_0 \left[\hat{x} \sin k(v_x t + x_i)\right.$$
$$\left. + \hat{y} \sin k(v_y t + y_i)\right] . \tag{8.36}$$

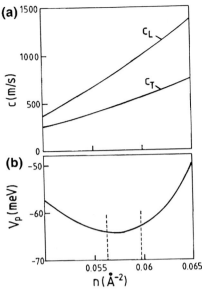

Fig. 8.20. (a) The transverse and the longitudinal sound velocity and (b) the lateral interaction energy, as a function of the Xe coverage.

Let us introduce the matrix $K(q)$ with components

$$K^{\alpha\beta}(q) = \sum_j K^{\alpha\beta}_{ij} e^{-i q \cdot (x_i - x_j)},$$

or

$$K^{\alpha\beta}(q) = \frac{6\epsilon}{r_0^2} \sum_{n \neq 0} \left(\left(\xi_n^4 - \xi_n^7\right) \delta_{\alpha\beta} + \left(14\xi_n^8 - 8\xi_n^5\right) \frac{x_n^\alpha x_n^\beta}{r_0^2} \right) \left(1 - e^{i q \cdot x_n}\right),$$

where $\xi_n = r_0^2/x_n^2$ and where the sum is over all the sites x_n of the hexagonal lattice of the adsorbate layer (excluding the site at the origin). Later we will need the long-wavelength limit of $K(q)$. To quadratic order in q we get

$$K^{\alpha\beta} = m c_T^2 q^2 \delta_{\alpha\beta} + m \left(c_L^2 - c_T^2\right) q_\alpha q_\beta.$$

Figure 8.20a shows the transverse c_T and the longitudinal c_L sound velocities as a function of Xe coverage. Figure 8.20b shows (for $U_0 = 0$) the Xe–Xe interaction potential, $V_p = \sum'_j v_{0j}/2$, as a function of coverage. All quantities refer, of course, to zero temperature. The two vertical dashed lines correspond to the coverage of the uncompressed Xe monolayer film ($n_a = 0.05624 \, \text{Å}^{-2}$) and the compressed film ($0.0597 \, \text{Å}^{-2}$) at the temperature $T = 77.4 \, \text{K}$.

Assume that the sliding velocity $\boldsymbol{v} = v \hat{x}$. Let $\boldsymbol{e}_1$ and $\boldsymbol{e}_2$ ($\boldsymbol{e}_i \cdot \boldsymbol{e}_j = \delta_{ij}$) denote the longitudinal and transverse eigenvectors of $K(q)$ for $q = (k, 0)$, corresponding to the eigenvalues $m\omega_1^2$ and $m\omega_2^2$, respectively. If we expand $\hat{x} = \alpha_1 e_1 + \alpha_2 e_2$, the solution to (8.36) can be written as

$$\boldsymbol{u}_i = (\boldsymbol{u}_i)_\mathrm{T} - \frac{kU_0}{m}\left(e_1 \frac{\alpha_1 \sin(\omega t + x_i + \phi_1)}{\left[(\omega^2 - \omega_1^2)^2 + \omega^2 \eta_\parallel^2\right]^{1/2}}\right.$$

$$\left. + e_2 \frac{\alpha_2 \sin(\omega t + x_i + \phi_2)}{\left[(\omega^2 - \omega_2^2)^2 + \omega^2 \eta_\parallel^2\right]^{1/2}}\right), \qquad (8.37)$$

where $\tan\phi_1 = \omega\eta_\parallel/(\omega^2 - \omega_1^2)$ and $\omega = kv$ and where $(\boldsymbol{u}_i)_\mathrm{T}$ is the thermal contribution to $\boldsymbol{u}_i$ derived from the random forces $\boldsymbol{f}_i$. Substituting (8.37) into (8.35) gives

$$\bar{\eta} = \eta_\parallel + \frac{k^4 U_0^2 \eta_\parallel}{2m^2}\left(\frac{\alpha_1 e_{x1}}{(\omega^2 - \omega_1^2)^2 + \omega^2 \eta_\parallel^2} + \frac{\alpha_2 e_{x2}}{(\omega^2 - \omega_2^2)^2 + \omega^2 \eta_\parallel^2}\right). \qquad (8.38)$$

In the present case $v \to 0$ so that $\omega = kv \ll \omega_1, \omega_2$ and (8.38) reduces to

$$\bar{\eta} = \eta_\parallel + \frac{k^4 U_0^2 \eta_\parallel}{2m^2}\left(\frac{\alpha_1 e_{x1}}{\omega_1^4} + \frac{\alpha_2 e_{x2}}{\omega_2^4}\right). \qquad (8.39)$$

It is important to note the physical mechanism behind the two terms in (8.39). The first term describes the direct energy transfer to the substrate due to the center of mass motion of the adsorbate system [process (a) in Fig. 8.16]. The second term describes the transfer of energy from the center of mass motion into sound waves in the adsorbate system [process (b) in Fig. 8.16]. This is caused by the corrugated substrate potential which exerts an oscillating force on the (sliding) elastic 2D solid. In the analytical calculation above the sound waves are only damped via the friction $\eta_\parallel$ which is very small. However, in real systems (and in the computer simulations) additional damping comes from phonon–phonon collisions derived from the anharmonic terms in the Xe–Xe interaction potential and from imperfections in the Xe film, e.g., vacancies in the monolayer film or a low concentration of Xe atoms in the second layer. These processes can be accounted for phenomenologically if the damping $\eta_\parallel$ in the second term in (8.39) is replaced with a new damping $\eta_\parallel' = \eta_\parallel + \eta_\parallel''$ where $\eta_\parallel''$ has its origin in the scattering processes described above:

$$\bar{\eta} = \eta_\parallel + \frac{k^4 U_0^2 \eta_\parallel'}{2m^2}\left(\frac{\alpha_1 e_{x1}}{\omega_1^4} + \frac{\alpha_2 e_{x2}}{\omega_2^4}\right). \qquad (8.40)$$

Physically, the reason for the decrease of the second term in (8.40) with increasing ω_1 and ω_2 is that the larger these phonon frequencies are, the shorter the time for which the corresponding virtual phonon excitations can

Table 8.6. The frequency factors f_1 and f_2 for four different coverages.

$N = 60$ ($n = 0.0499\,\text{Å}^{-2}$)	$f_1 = 1.9$	$f_2 = 1.4$
$N = 65$ ($n = 0.0540\,\text{Å}^{-2}$)	$f_1 = 4.2$	$f_2 = 3.1$
$N = 68$ ($n = 0.0565\,\text{Å}^{-2}$)	$f_1 = 5.7$	$f_2 = 4.4$
$N = 69$ ($n = 0.0574\,\text{Å}^{-2}$)	$f_1 = 6.2$	$f_2 = 4.8$

exist without violating Heisenberg's uncertainty relation, and the less likely that they will be able to undergo scattering, which could convert them to real excitations and contribute to the friction. Equation (8.40) explains the sharp decrease in $\bar{\eta}$ as the monolayer film compresses (coverage range $0.36 < \theta < 0.47$ in Fig. 8.19 or $0.057\,\text{Å}^{-2} < n_\text{a} < 0.07\,\text{Å}^{-2}$ in Fig. 3.22): During compression the phonon frequencies ω_1 and ω_2 increase and as a result the last term in (8.40) decreases. For the compressed monolayer film the contribution from the last term in (8.40) is negligible and $\bar{\eta} \approx \eta_\|$ (see below).

It is convenient to write

$$\omega_1 = \left(\frac{6\epsilon}{mr_0^2}\right)^{1/2} f_1(n_\text{a}), \quad \omega_2 = \left(\frac{6\epsilon}{mr_0^2}\right)^{1/2} f_2(n_\text{a}), \tag{8.41}$$

where f_1 and f_2 are dimensionless functions of the adsorbate coverage n_a. For $N = 68$ and 69 Xe atoms in the basic unit we get $(f_1, f_2) = (5.7, 4.4)$ and $(6.2, 4.8)$, respectively. Substituting (8.41) into (8.40) gives

$$\bar{\eta} = \eta_\| + \frac{4}{9}\left(\frac{\pi r_0}{a}\right)^4 \left(\frac{U_0}{\epsilon}\right)^2 \left(\frac{\alpha_1 e_{x1}}{f_1^4} + \frac{\alpha_2 e_{x2}}{f_2^4}\right)\eta'_\| = \eta_\| + A\eta'_\|. \tag{8.42}$$

If the sliding occurs along the e_2 direction we obtain $A = 9 \times 10^{-4}$ and 6×10^{-4}, for $N = 68$ and 69, respectively. Note that if $\eta'_\| \approx \eta_\|$ then, since $A \ll 1$, the contribution from the corrugated substrate potential (proportional to U_0^2) to the sliding friction is negligibly small and $\bar{\eta} \approx \eta_\|$, as observed in the simulations. Table 8.6 gives f_1 and f_2 for some different coverages.

Role of Defects

The study above has not considered the contribution to the sliding friction from surface defects, e.g., steps or foreign chemisorbed atoms. Very little is known about this subject and in this section we only present an elementary introduction (see also Sects. 8.10 and 10.4).

Assume beginning with zero temperature. A 2D elastic solid on a periodically corrugated substrate can either be pinned by the substrate potential, in which case a finite force (per particle) is necessary in order to start sliding, or else, if the amplitude of the substrate potential is small enough, no pinning occurs and the static friction force vanishes. On the other hand, if a *random* distribution of defects occurs on the surface, the 2D solid will *always* be pinned by the defects. If the concentration of defects is very low one can neglect the interaction between the defects and the total pinning force is the sum of the pinning forces from the (independent) pinning centers. However, the surfaces used in the QCM measurements have a relatively high concentration of defects. In these cases, as discussed in Sect. 8.4, if the interaction between the defects and the elastic solid is weak, pinning will occur via the formation of domains of linear size ξ; each domain can be considered as an "effective" particle which is pinned individually. The effective particles experience a potential from the defects which is periodic and in the simplest case takes the form $U_1 \cos kx$ as a function of the coordinate x of the effective particle. The strength U_1 of the pinning potential and the linear size ξ of the domains is determined in Sect. 14.4.1 (see also [8.37]). For a 2D solid of linear size L one has [see (14.15)]

$$\xi \approx \frac{mc^2 l}{U_\mathrm{d} \left[4\pi \ln(L/\xi)\right]^{1/2}},$$

where c is the sound velocity in the film and where U_d now denotes the strength of the defect potential (i.e., the strength of the interaction potential between a *single* defect and the elastic 2D solid) and where $l = n_\mathrm{d}^{-1/2}$ is of the order of the average distance between two neighboring defects. In the present case, if $L \sim 1\,\mathrm{cm}$, $l \sim 100\,\mathrm{\AA}$ and $U_\mathrm{d} \sim 10\,\mathrm{meV}$ and using $mc^2 = 0.3\,\mathrm{eV}$ as calculated for the compressed Xe monolayer on silver and $c = 430\,\mathrm{m/s}$ from Fig. 8.20a, we get $\xi \approx 300\,\mathrm{\AA}$. A $300\,\mathrm{\AA} \times 300\,\mathrm{\AA}$ area contains approximately $N = 10^4$ Xe atoms which the force of inertia $NF_\mathrm{ext} \sim 10^{-5}\,\mathrm{eV/\AA}$ acts upon. This is smaller than the pinning force which acts on a domain, which is of order kU_1, where (see Sect. 14.4.1)

$$U_1 \approx \xi U_\mathrm{d}/l = mc^2/[4\pi \ln(L/\xi)]^{1/2}.$$

Using $mc^2 = 0.3\,\mathrm{eV}$ gives $U_1 \approx 25\,\mathrm{meV}$ and $kU_1 \sim 10\,\mathrm{meV/\AA} \gg NF_\mathrm{ext}$. Thus no sliding motion is possible at zero temperature. However, for nonzero temperatures thermally activated motion will occur. Thermal excitation over the barrier U_1 depends on the ratio $U_1/k_\mathrm{B}T$; for $U_1/k_\mathrm{B}T \ll 1$ sliding occurs as if no barrier were present at all, in which case $\bar\eta \approx \eta$. But at low temperature $\bar\eta \propto \exp(2U_1/k_\mathrm{B}T)$ and it is clear that by studying the temperature dependent of the sliding friction it is possible to deduce information about the pinning barrier. We urge our experimental colleagues to perform such temperature dependence measurements, in particular since the pinning barrier deduced above seems to be too large: if $U_1 \sim k_\mathrm{B}T$ the sliding friction should be strongly influenced by the defects while the observed sliding friction for the compressed monolayer film seems insensitive to the defect concentration. More experimental and theoretical work is needed to resolve this problem.

8.6 Applications

The linear sliding friction $\bar{\eta}$ studied above is of direct relevance for several topics, in addition to the QCM measurements discussed earlier. In this section we present three such applications, namely (1) boundary conditions in hydrodynamics, (2) layering transition, and (3) spreading of wetting liquid drops.

8.6.1 Boundary Conditions in Hydrodynamics

The flow of a fluid is usually described by the Navier–Stokes equations of hydrodynamics. For an incompressible viscous fluid these equations take the form

$$\frac{\partial \bm{v}}{\partial t} + \bm{v} \cdot \nabla \bm{v} = -\frac{1}{\rho} \nabla P + \nu \nabla^2 \bm{v} \,, \tag{8.43}$$

$$\nabla \cdot \bm{v} = 0 \,, \tag{8.44}$$

where P is the pressure and ν the kinematic viscosity. When solving these equations it is necessary to specify boundary conditions. At a solid–fluid interface it is usually assumed that not only the normal component of the fluid velocity vanishes but also the tangential component. For example, in the book on hydrodynamics by Landau and Lifshitz one can read: "There are always forces of molecular attraction between a viscous fluid and the surface of a solid body, and these forces have the result that the layer of fluid immediately adjacent to the surface is brought completely to rest, and adheres to the surface." Similarly, in Feynman's Lectures on Physics: "It turns out – although it is not at all self-evident – that in all circumstances where it has been experimentally checked, the *velocity of a fluid is exactly zero at the surface of a solid.*" From quartz-crystal microbalance (QCM) measurements we now know that *these statements are incorrect*, or at least not very accurate. In this section we discuss exactly under which experimental conditions slip boundary conditions manifest themselves [8.33].

Flow Between two Parallel Surfaces

Consider a fluid enclosed between two parallel planes moving with a constant relative velocity $\bm{v_0}$. We take one of these planes as the xy-plane, with the x-axis in the direction of $\bm{v_0}$, see Fig. 8.21. It is clear that all quantities depend only on z, and that the fluid flow is everywhere in the x-direction, i.e., $\bm{v} = \hat{x} v(z)$. We have from (8.43) for steady flow

$$dP/dz = 0, \quad d^2 v/dz^2 = 0 \,.$$

Hence $P = \mathrm{const.}$, and $v = az + b$. Now, assume that the fluid sticks to one of the surfaces, e.g. the surface $z = d$ (where d is the distance between the planes). This gives

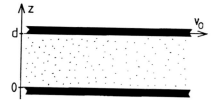

Fig. 8.21. A fluid between two surfaces in relative motion.

$$ad + b = v_0 .\tag{8.45}$$

On the surface $z = 0$ we assume that slip occurs. More precisely, if F is the force (in the x-direction) which acts on a fluid molecule adsorbed on this surface, due to the fluid above, then the drift velocity is

$$m\bar{\eta}v(0) = F .$$

But if n_a is the concentration of adsorbed molecules, the force F is related to the stress tensor component $\sigma_{xz}(z)$ via $n_\mathrm{a} F = \sigma_{xz}(0)$. Since

$$\sigma_{xz} = \rho\nu dv/dz = \rho\nu a$$

we get

$$m\bar{\eta}b = \rho\nu a/n_\mathrm{a} .\tag{8.46}$$

Equations (8.45) and (8.46) are easily solved to yield

$$\sigma_{xz} = \frac{\nu\rho}{d}\frac{v_0}{1+l/d} ,$$

with

$$l = \frac{\nu n}{\bar{\eta} n_\mathrm{a}} ,$$

where $n = \rho/m$ is the number of molecules per unit volume of fluid.

For "stick" boundary conditions $\bar{\eta} \to \infty$ and $l \to 0$ and we recover the standard result for the stress $\sigma_{xz} = \nu\rho v_0/d$. The deviation from this no-slip limit is determined by the parameter l/d, i.e., by the ratio between the "slip length" l and the separation d between the two planes.

Flow Through a Pipe

The result above is general. For example, consider the flow of a fluid through a pipe of circular cross-section. If the pipe has the length L and the radius R and if the pressure difference between the ends of the pipe is ΔP, then the mass Q of fluid passing per unit time through any cross-section of the pipe is given by

$$Q = \frac{\pi \Delta P R^4}{8\nu L}\left(1 + \frac{4l}{R}\right).$$

In the limit $\bar{\eta} \to \infty$ we recover a standard result (Poiseuille's formula), and the deviation from this limit is again determined by the ratio between the slip length l and a characteristic length (the radius R) of the system.

Slip Length l

The slip length l enters in both cases above. Let us evaluate l for a typical case. Consider the water–silver interface, which has been studied by *Krim* et al. [8.43]. In this case, $n_a/n \sim r \sim 3\,\text{Å}$ where r is a typical nearest neighbor separation. Furthermore, from the measurements by Krim et al. at H_2O monolayer coverage (see Table 3.1), $\bar{\eta} \approx 2.5 \times 10^8\,\text{s}^{-1}$ and with the observed kinematic bulk viscosity of water (for $T = 300\,\text{K}$), $\nu \approx 0.01\,\text{cm}^2/\text{s}$, one gets $l \sim 2 \times 10^5\,\text{Å}$. Hence, the "standard" no-slip boundary condition of hydrodynamics is not valid when shearing a water film between two *flat* silver surfaces if the water layer is thinner than, say, $100\,\mu\text{m}$.

Influence of Surface Roughness

The discussion above has assumed perfectly flat surfaces. But real surfaces are (almost) never perfectly flat, having roughness on various length scales depending on the material and the preparation method. Microscopic roughness, such as adatoms and steps, will act as pinning or scattering centers for the first layer of adsorbates and will affect (renormalize) the sliding friction $\bar{\eta}$. Roughness on a mesoscopic length scale, i.e., on a length scale beyond a few lattice spacings but below the (usually macroscopic) length determined by the overall dimension of the systems under study (i.e., the radius R of the pipe in Poiseuille flow) will give rise to a renormalization of the slip length l. That is, we can replace the rough surfaces, characterized by the *local* slip length l, with smooth surfaces with an effective slip length l_{eff} (Fig. 8.22). It is possible to derive an expression for l_{eff} by solving the Navier Stokes equations for a rough surface with the local slip length l and identifying the flow pattern far away from the surface with that obtained for a smooth surface with the effective slip length l_{eff}. For example, if the surface roughness profile is given by

$$h(x) = h_0 \cos kx,$$

then, to quadratic order in kh_0, this procedure gives [8.44]

$$l_{\text{eff}} = \frac{l\left[1 - 5(kh_0)^2/4\right] - kh_0^2\left(1 + 2kl\right)}{1 + (kh_0)^2\left(1/2 + kl\right)}. \tag{8.47}$$

Note that

$$l_{\text{eff}} \approx -(kh_0)^2/k \tag{8.48}$$

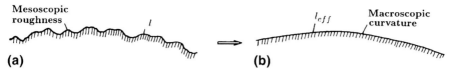

Fig. 8.22. With respect to the hydrodynamic flow of a fluid above a rough surface, on a length scale large compared with the amplitude and periodicity of the surface roughness, one can replace the rough surfaces of the real system (**a**), with smooth surfaces (**b**) having the effective slip length l_{eff}. Note that l_{eff} is a function of the local slip length l of the rough surfaces and of the amplitude and the wavelength of the surface roughness.

if $kl \ll (kh_0)^{-2}$ and

$$l_{\text{eff}} \approx (kh_0)^{-2}/k \tag{8.49}$$

if $kl \gg (kh_0)^{-2}$.

Let us consider two applications of this formula.

Consider first a thin silver film evaporated at room temperature onto a smooth glass substrate. The surface topography of such a film has been studied by STM [8.7] and one has $h_0 \approx 10\,\text{Å}$ and the wavelength of the roughness $\lambda \approx 300\,\text{Å}$ so that (note $k = 2\pi/\lambda$) $(kh_0)^2 \approx 0.04$. Hence, with $l = 2 \times 10^5\,\text{Å}$ (see above) one get $kl \gg (kh_0)^{-2}$ and from (8.49), $l_{\text{eff}} \approx (kh_0)^{-2}/k \approx 1000\,\text{Å}$, i.e., l_{eff} is much smaller than l, and *independent* of l.

Next consider liquid ^{3}He in a vessel with walls covered by ^{4}He. Here, the surface is very slippery and, without any roughness on a mesoscopic scale, the ^{3}He would slide past it with little friction. In the absence of ^{4}He on the walls, Ritchie et al. [8.45] found that surface impedance measurements are consistent with l_{eff} equal to zero (or, more accurately, $l_{\text{eff}} \ll \delta$ where $\delta \sim 15\,\mu\text{m}$ is the viscous penetration depth defined below) as expected from the standard no-slip boundary condition. On the other hand, for surfaces covered with ^{4}He the experimental data implied $l_{\text{eff}} \approx 5\,\mu\text{m}$. The latter result is expected if $kl \gg (kh_0)^{-2}$ in which case (8.49) gives $l_{\text{eff}}' \approx 5\,\mu\text{m}$ if $h_0 \sim 1\,\mu\text{m}$, as was measured experimentally, and $1/k \sim 1.8\,\mu\text{m}$. In the former case, if $kl \ll (kh_0)^{-2}$ the limit (8.48) is relevant and l_{eff} is a factor ~ 10 smaller than for surfaces without ^{4}He, and the ratio between the real and the imaginary parts of the transverse surface impedance would be close to unity as observed experimentally.

Flow Above a Plane Performing In-Plane Oscillations with Applications to QCM-Friction Measurements

If a fluid layer on a surface is thinner than the amplitude of the surface roughness, the equations given in the last section are not valid. In fact, a monolayer film on a surface may slip while for a thick film the "stick" boundary condition may be an excellent approximation. One way to study how the

232   8. Sliding of Adsorbate Layers

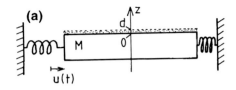

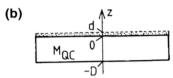

Fig. 8.23. (a) A solid slab between two elastic springs performing in-plane oscillations. (b) A quartz-crystal performing in-plane deformation vibrations.

slip boundary condition depends on the film thickness is to use the quartz-crystal microbalance (QCM). In this section we describe the basis for this, and present experimental results which illustrate the "transition" from slip for very thin films to stick for thick films. The general formulas derived below for the QCM frequency shift and damping are also of considerable practical importance because of the increased use of the QCM in liquid environments, e.g., in electrochemistry and immunology.

Consider first a solid slab (mass M) connected by elastic springs (effective spring constant k) to a rigid frame as illustrated in Fig. 8.23a. If the slab is displaced parallel to itself and then released, it will perform in-plane vibrations [displacement $u(t)$] with a frequency $\omega_0 = (k/M)^{1/2}$ and damping γ_0:

$$u = u_0 \cos \omega_0 t \, e^{-\gamma_0 t/2} .$$

Assume now that a fluid layer of thickness d is present on the upper surface of the slab. We study the influence of this fluid layer on the resonance frequency and the damping of the oscillator.

If u denotes the displacement of the solid slab from its equilibrium position, then

$$M\ddot{u} + M\gamma_0 \dot{u} + M\omega_0^2 u = A\sigma , \tag{8.50}$$

where A is the surface area and σ is the stress the fluid layer exerts on the solid slab. The velocity field of the fluid is of the form $\boldsymbol{v}(\boldsymbol{x},t) = \hat{x}v(z,t)$ and (8.43) therefore reduces to

$$\frac{\partial v}{\partial t} = \nu \frac{\partial^2 v}{\partial z^2} . \tag{8.51}$$

We now assume that

$$u = u_0 e^{-i\omega t}, \qquad v = v_0(z) e^{-i\omega t} .$$

Substituting this into (8.51) gives an equation for $v_0(z)$ which is easily solved to give

$$v_0 = a e^{\kappa z} + b e^{-\kappa z} , \tag{8.52}$$

where $\kappa = (-i\omega/\nu)^{1/2} = (i-1)/\delta$ with $\delta = (2\nu/\omega)^{1/2}$ is the viscous penetration depth. But the stress on the upper surface of the fluid must vanish, i.e.,

$$\sigma_{xz} = \rho\nu\frac{\partial v}{\partial z} = 0 \quad (z = d) ,$$

while the stress on the fluid at the solid–fluid interface is given by $-\sigma$, where

$$\sigma = n_a m \bar{\eta}[v(0,t) - \dot{u}] . \tag{8.53}$$

Using these two boundary conditions, (8.52) gives

$$v_0 = a\left(e^{\kappa z} + e^{2\kappa d - \kappa z}\right) , \tag{8.54}$$

where

$$a = \frac{-i\omega u_0}{(1 + e^{2\kappa d}) - \kappa l(1 - e^{2\kappa d})} . \tag{8.55}$$

Substituting (8.54) and (8.55) into (8.53) gives

$$\sigma = \frac{n_a m i \omega \bar{\eta} u}{1 - \frac{1}{\kappa l}\frac{1+e^{2\kappa d}}{1-e^{2\kappa d}}} . \tag{8.56}$$

Substituting (8.56) into (8.50) gives, after some simplifications,

$$(-\omega^2 - i\omega\gamma_{\text{eff}} + \omega_0^2)u_0 = 0 , \tag{8.57}$$

where the effective (complex) damping γ_{eff} is given by

$$\gamma_{\text{eff}} = \gamma_0 + \frac{M_{\text{ml}}}{M}\frac{\bar{\eta}}{1 - \frac{1}{\kappa l}\frac{1+e^{2\kappa d}}{1-e^{2\kappa d}}} \tag{8.58}$$

where $M_{\text{ml}} = n_a A m$ is the mass of a monolayer of adsorbed fluid molecules. From (8.57) and (8.58) one obtains the frequency shift $\Delta\omega$ and the increase in the damping $\Delta\gamma$, induced by the fluid:

$$\Delta\omega = (1/2)\,\text{Im}\,\gamma_{\text{eff}} = \frac{M_{\text{ml}}}{2M}\,\text{Im}\left(\frac{\bar{\eta}}{1 - \frac{1}{\kappa l}\frac{1+e^{2\kappa d}}{1-e^{2\kappa d}}}\right) , \tag{8.59}$$

$$\Delta\gamma = \text{Re}\,\gamma_{\text{eff}} - \gamma_0 = \frac{M_{\text{ml}}}{M}\,\text{Re}\left(\frac{\bar{\eta}}{1 - \frac{1}{\kappa l}\frac{1+e^{2\kappa d}}{1-e^{2\kappa d}}}\right) , \tag{8.60}$$

where the RHS is evaluated for $\omega = \omega_0$. For thick fluid films, $d \gg \delta$, and assuming the no-slip boundary condition, (8.59) and (8.60) reduce to $\Delta\omega = -\Delta\gamma/2 = -(M_{\text{film}}/2M)(\nu/\delta d)$ where M_{film} is the mass of the fluid layer. Finally, for a solid slab adsorbed on the QCM surface, the frequency shift and the damping is obtained from (8.59) and (8.60) with $\nu \to \infty$. This gives

$$\Delta\omega = -\frac{M_{\text{ml}}}{2M} \frac{\omega \bar{\eta}^2 P_1}{\omega^2 + (\bar{\eta} P_1)^2}, \tag{8.61}$$

and

$$\Delta\gamma = \frac{M_{\text{ml}}}{M} \frac{\omega^2 \bar{\eta}}{\omega^2 + (\bar{\eta} P_1)^2}, \tag{8.62}$$

where $P_1 = n_a/nd$ is the fraction of the total number of adsorbates in the first layer in direct contact with the substrate.

Let us now consider a quartz crystal (mass M_{QC}) performing in-plane deformation vibrations as indicated in Fig. 8.23b. The elastic displacement field of the vibrations is of the form $\boldsymbol{u} = \hat{x} u(z,t)$ and satisfies the wave equation,

$$\frac{\partial^2 u}{\partial t^2} - c^2 \frac{\partial^2 u}{\partial z^2} = 0$$

where c is the transverse sound velocity. The general solution is of the form

$$u = \left(A e^{ikz} + B e^{-ikz}\right) e^{-i\omega t}, \tag{8.63}$$

where $k = \omega/c$. Since the surface stress must vanish on the lower surface ($z = -D$) we get $\partial u/\partial z = 0$ for $z = -D$ which gives $A = B \exp(2ikD)$. Substituting this into (8.63) and using the condition that the stress $\rho_{\text{QC}} c^2 \partial u/\partial z$ at the upper solid surface must equal σ [given by (8.56) with $u = u(0,t)$], gives an equation for the complex resonance frequency ω. The solution to this equation gives the same general expression for the frequency shift and damping as for the rigid block in Fig. 8.23a, but with the mass M in (8.59) and (8.60) replaced by $M_{\text{QC}}/2$. This result is easy to understand physically. Assume that the blocks in Figs. 8.23a,b are of equal size and mass and consider a case where both the resonance frequency and the amplitude of the vibration of the $z = 0$ surface are exactly the same. This will result in equal adsorbate induced energy dissipation rates for the two different cases. But, as shown below, the vibrational energy of the spring-block system in Fig. 8.23a is twice that associated with the deformation mode in Fig. 8.23b; hence the decay of the amplitude of vibration in the latter case is twice as fast as for the rigid block. Since the real and imaginary parts of the function $\gamma_{\text{eff}}(\omega)$ are related via a Kramers–Kronig relation, the same mass difference also occurs for the frequency shift.

Let us prove that the vibrational energy of the spring-block system in Fig. 8.23a is twice that associated with the deformation mode in Fig. 8.23b.

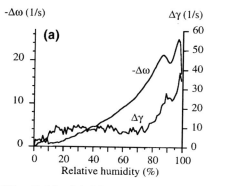
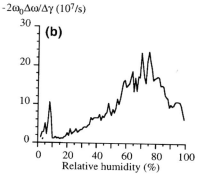

Fig. 8.24. (a) The change in the resonance frequency $\Delta\omega$ and damping $\Delta\gamma$ as a function of relative humidity for a QCM with sputter-deposited gold electrodes. (b) The ratio $-2\omega_0\Delta\omega/\Delta\gamma$ as a function of the relative humidity. The temperature was held constant at 18°C. From [8.46].

The velocity field of the quartz crystal is, to a good approximation, of the form $v = v_0 \cos k(z + D)$, where $k = \pi/D$. Hence the energy of vibration equals

$$\frac{1}{D}\int_{-D}^{0} dz \frac{1}{2} M_{\text{QC}} v^2 = \frac{1}{2} M_{\text{QC}} v_0^2 \int_{-D}^{0} \frac{dz}{D} \cos^2 \frac{\pi}{D}(z+D) = \frac{1}{4} M_{\text{QC}} v_0^2 ,$$

which is just half of the vibrational energy for the rigid block in Fig. 8.23a.

Let us now present some applications of (8.59) and (8.60). *Rodahl* and *Kasemo* [8.46] have performed QCM studies with sputter-deposited gold electrodes. The gold surfaces were mirror like, but scanning tunneling microscopy showed that they consisted of grains approximately 300 Å in diameter, with an average roughness of approximately 7 Å.

The frequency shift $\Delta\omega$ and damping $\Delta\gamma$ caused by water adsorption from humid air on the QCM gold electrode at room temperature is shown in Fig. 8.24a. To obtain an estimate of the coefficient of friction between the gold electrode and the water film, the ratio $-2\omega_0\Delta\omega/\Delta\gamma$ has been plotted as a function of relative humidity in Fig. 8.24b. As long as the water film does not form multilayers, this ratio equals the sliding friction $\bar{\eta}$. The film coverage can now be estimated using (8.59) or (8.60). Assuming that the water does not form 3D-droplets before a full monolayer is formed, a coverage of approximately one monolayer is obtained at $\sim 80\%$ relative humidity. The reason for the increase in $\bar{\eta}$ as the relative humidity increases up to 80% is likely to be the following: It is known that water molecules readily form 2D-clusters as they adsorb onto metal surfaces. As these clusters grow larger it is possible that they become ever more strongly pinned by surface defects such as steps and grain boundaries, which thereby increases the effective coefficient of sliding friction, as indeed observed in computer simulations (Sect. 8.4). The sliding friction at monolayer coverage (from Fig. 8.24b) is about $\sim 2 \times 10^8 \, \text{s}^{-1}$, in good agreement with the observations of *Krim* et al. for water on silver surfaces (Table 8.1).

If the gold electrode were perfectly flat, it would be easy to calculate the frequency shift $\Delta\omega$ and damping $\Delta\gamma$ which would result if the QCM is submerged in bulk water. Using the sliding friction $\bar{\eta} \sim 1 \times 10^8 \, \text{s}^{-1}$ which is obtained for a monolayer water film on gold (Fig. 8.24b) gives the slip length $l = 6 \times 10^5$ Å and from (8.59) and (8.60) (with $d = \infty$) one can calculate $\Delta\omega$ and $\Delta\gamma$, which turns out to be 83,000 and 210 times smaller, respectively, than observed experimentally. However, the experimentally observed $\Delta\omega$ and $\Delta\gamma$ agree to within 1% and 2%, respectively, with those calculated using the no-slip boundary condition. This is most likely due to the mesoscopic surface roughness. As shown above, only if a sufficiently smooth QCM surface can be made, should it be possible to detect liquid slippage for thick fluid films.

To close this section, we note that for most "normal" engineering applications the standard no-slip boundary condition is an excellent approximation. However, the present drive towards smaller dimensions and smoother surfaces, and new engineering applications (e.g., micromechanical devices) may make it necessary to account for slip at solid–fluid interfaces. Slip may also be important in some biological applications, e.g., in fluid flow in narrow channels, and for seepage in soils.

8.6.2 The Layering Transition

In Sect. 7.4 we studied the nucleation of the $n = 2 \to 1$ layering transition. We now study how the $n = 1$ region will spread after the nucleation has occurred [8.47]. Let $\boldsymbol{v}(\boldsymbol{x}, t)$ be the two-dimensional velocity field of the monolayer being squeezed out. Assuming an incompressible two-dimensional fluid, the continuity equation and the (generalized) Navier–Stokes equations take the form

$$\nabla \cdot \boldsymbol{v} = 0 \,, \tag{8.64}$$

$$\frac{\partial \boldsymbol{v}}{\partial t} + \boldsymbol{v} \cdot \nabla \boldsymbol{v} = -\frac{1}{mn_\text{a}} \nabla p + \nu \nabla^2 \boldsymbol{v} - \bar{\eta} \boldsymbol{v} \,, \tag{8.65}$$

where p is the two-dimensional pressure and ν the kinematic viscosity. The last term in (8.65) describes the "drag-force" of the substrate acting on the fluid. We assume that the contact area between the two solid surfaces has a circular shape with radius r_0. Assume first that the initial nucleation occurs at the center of the contact area. Hence, by symmetry the $n = 1$ region has a circular shape with the radius $r_1(t)$, see Fig. 8.25. Let p_1 be the two-dimensional pressure for $r = r_1$. At the boundary $r = r_0$ the two-dimensional pressure takes the value $p_0 < p_1$ which is a constant (the spreading pressure) determined by local equilibrium between the 3D-fluid and the adsorbed molecules. Because of symmetry

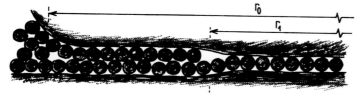

Fig. 8.25. Schematic picture of the layering transition $n = 2 \to 1$. The lubrication film is assumed to be in a fluid state.

$$v = \hat{r}v(r,t) ,$$

$$p = p(r,t) .$$

Substituting these equations into (8.64) and (8.65) gives

$$\frac{\partial v}{\partial r} + \frac{v}{r} = 0 , \tag{8.66}$$

$$\frac{\partial v}{\partial t} = -\frac{1}{mn_\mathrm{a}} \frac{\partial p}{\partial r} + \nu \frac{\partial}{\partial r}\left(\frac{\partial v}{\partial r} + \frac{v}{r}\right) - \bar{\eta}v , \tag{8.67}$$

where we have dropped the non-linear term in (8.65) since the velocity v is very small in the present applications. Note that because of the continuity equation (8.65) the viscosity term in (8.67) vanishes, i.e., the spreading does not depend on the two-dimensional viscosity. The relevant solution of (8.66) is of the form

$$v(r,t) = B(t)/r . \tag{8.68}$$

Inserting this into (8.67) gives

$$\partial p/\partial r = -mn_\mathrm{a}\left[\dot{B}(t) + \bar{\eta}B(t)\right]/r ,$$

where $\dot{B} = dB/dt$. Using the boundary condition $p(r_0) = p_0$, this equation has the solution

$$p(r) = p_0 - mn_\mathrm{a}\left[\dot{B}(t) + \bar{\eta}B(t)\right]\ln(r/r_0) . \tag{8.69}$$

Now, at the leading edge $r = r_1(t)$ of the $n = 1$ region the pressure is $p = p_1$ so that from (8.69) we get

$$\left[\dot{B}(t) + \bar{\eta}B(t)\right]\ln(r_1/r_0) = -(p_1 - p_0)/mn_\mathrm{a} . \tag{8.70}$$

But we must also have

$$\dot{r}_1 = v(r_1,t) = B(t)/r_1 ,$$

so that (8.70) can be written as

$$\frac{d}{dt}\left(\frac{dr_1^2}{dt} + \bar{\eta}r_1^2\right)\ln\left(\frac{r_1}{r_0}\right) = -\frac{2(p_1 - p_0)}{mn_a}. \tag{8.71}$$

For "large" times, r_1 varies slowly with time (see below), and we can neglect the second-order time derivative term in the expression above. Furthermore, if we introduce the area $A(t) = \pi r_1^2(t)$, (8.71) becomes

$$\frac{dA}{dt}\ln\left(\frac{A(t)}{A_0}\right) = -\frac{4\pi(p_1 - p_0)}{\bar{\eta} mn_a}. \tag{8.72}$$

If we assume that p_1 is independent of r_1, i.e., of $A(t)$, then (8.72) can be integrated to yield

$$\frac{A(t)}{A_0}\left[\ln\left(\frac{A(t)}{A_0}\right) - 1\right] = -\frac{4\pi(p_1 - p_0)t}{mn_a\bar{\eta} A_0}. \tag{8.73}$$

The time t^* taken to completely squeeze out the layer is obtained from (8.73) by putting $A(t^*) = A_0$ which gives

$$t^* = \frac{mn_a\bar{\eta} A_0}{4\pi(p_1 - p_0)}. \tag{8.74}$$

Let us now discuss the two-dimensional pressure p_1. We will show that $p_1 = p_0 + aP_0$, where a is the change of distance between the confining solid surfaces as one monolayer is squeezed out, i.e., roughly the diameter of a molecule. The easiest way to obtain p_1 is to evaluate the adiabatic work to squeeze out one monolayer, which must equal $(p_1 - p_0)A$ where A is the surface area. But the adiabatic work is also equal to the change in free energy resulting from the squeezing out of the monolayer, which is given by $[(2\gamma_{sl} + V_{ll}) - 2\gamma_{sl} + P_0 a] A - V_{ll}A$. In this expression the term [...] is the change in the free energy in the contact region while the last term is the binding energy of the squeezed out layer material to the bulk of fluid outside the contact region ($V_{ll}A$ describes the attractive van der Waals interaction which occurs if we bring two fluid slabs into contact). Hence we get $p_1 = p_0 + P_0 a$. Since in a typical case $P_0 \sim 10^8 \text{ N/m}^2$ and $a \sim 10\text{ Å}$, the term $P_0 a \sim 6 \text{ meV/Å}^2$, i.e., this term is of similar order of magnitude to p_0.

We can now use (8.74) to estimate the sliding friction $\bar{\eta}$ from the observed squeeze out time t^*. Using $p_1 - p_0 = P_0 a$ gives

$$\bar{\eta} = \frac{4\pi t^* P_0 a}{mn_a A_0}. \tag{8.75}$$

From the measurements of *Gee* et al. [8.48] for OMCTS between mica surfaces, $P_0 = 2 \times 10^7 \text{ N/m}^2$, $t^* \approx 120\text{ s}$, $a \approx 10\text{ Å}$ and $A_0 \approx 7 \times 10^{-9}\text{ m}^2$ giving $\bar{\eta} = 6 \times 10^{15}\text{ s}^{-1}$. This value is not very accurate for the following reasons:

a) The three-dimensional pressure in the contact area is not a constant but will instead take its highest value in the middle and vanish at the periphery. If we assume [8.49] that the pressure depends on r as $P(r) = (2P_0/3)\left[1 - (r/r_0)^2\right]^{1/2}$, where P_0 is the average pressure, then (8.72) can no longer be solved analytically, but a numerical integration shows that the expression (8.74) for t^* is increased by a factor 1.84 and $\bar{\eta}$ now takes the value $3 \times 10^{15}\,\text{s}^{-1}$.

b) It is not obvious that the film to be squeezed out remains in a fluid configuration during the whole time period but may oscillate between a pinned solid state and the fluid state; this possibility is suggested by the stick-slip oscillations exhibited by the experimental friction force [8.48] even in the transition period from $n = 2$ to $n = 1$.

c) In the calculation above it was assumed that the nucleation of the $n = 1$ region occurred at the center of the contact region. If the pressure $P(r)$ were constant in the contact region, one would expect nucleation to occur with equal probability anywhere within the contact region. But even if $P(r)$ varies with r as indicated above, the nucleation will, in general, occur some distance away from the center; in fact a simple analysis indicates that $\langle r \rangle \approx 0.2 r_0$ where $\langle ... \rangle$ stands for averaging over the spatial probability distribution of nucleation. Taking this into account is unlikely to change t^* by more than a factor of 2 or so.

8.6.3 Spreading of Wetting Liquid Drops

Many practical processes require the spreading of a liquid on a solid, e.g., a paint or a lubricant on a solid surface. Perhaps the first recognition of the importance of understanding the "driving force" behind spreading was the work of *Cooper* and *Nuttal* [8.50] in connection with the spraying of insecticides on leaves. Another important application is in the development of new and improved treatments for chemical-, oil-, and water-repellent finishes for clothing.

There are two possible situations for a droplet resting on a solid surface [8.51]. For partial wetting, there exists a macroscopically observable contact angle θ between the liquid droplet and the surface; see Sect. 7.5. The contact angle is determined by the solid-liquid γ_{sl}, solid–vapor γ_{sv}, and liquid–vapor γ_{lv} interfacial free energies, via (7.44), $\cos\theta = (\gamma_{\text{sv}} - \gamma_{\text{sl}})/\gamma_{\text{lv}}$. This corresponds to an equilibrium situation in which the droplet does not move with time. For complete wetting, the contact angle is zero, so that the liquid droplet tends to spread when deposited on the surface, the initial non-zero contact angle tending towards its equilibrium value with increasing time.

In a recent series of papers *Heslot* et al. [8.52–54] have shown that for complete wetting the spreading process may be driven by interfacial phenomena at the molecular scale. Using space- and time-resolved ellipsometry they studied the spreading of small drops of fluid on a very smooth silicon surface. In Fig. 8.26 we show the thickness profile of a tetrakis(2-ethylhexoxy)-silane drop [8.54]. Note the extreme distortion of the horizontal to vertical scale (a factor of 10^7). The volumes of the drops were found to be conserved over the time scale of the experiment (one week). This shows that there is negligible evaporation of the liquid.

Figure 8.26 shows that a monolayer of molecules extends for macroscopic distances out from the central macroscopic droplet; the thickness of this "foot" (10Å) is comparable to the lateral thickness of the molecules as measured with an atomic force microscope. Note that as the drop spreads and flattens, a second molecular "step" (again of thickness 10Å) appears, along with some indication of further layers. As time proceeds, the first layer continues to extend at the expense of the upper layers which progressively shrink and disappear. In these transient spreading stages, the fluid is strongly structured, with up to four distinct molecular layers forming near the solid wall.

De Genes and Cazabat [8.55] have presented a model where the spreading velocity for the monolayer in contact with the substrate is proportional $1/\bar{\eta}$. The analysis is, in fact, almost identical to that presented in Sect. 8.6.2 in the context of the squeezing out of a monolayer. The most important change is new boundary conditions, namely the two-dimensional pressure in the present case satisfies $p = p_0$ for $r < r_0$, where r_0 now denotes the radius of the three-dimensional liquid drop (we assume local equilibrium inside the drop) and $p = 0$ at the edge $r = r_1(t)$ of the spreading monolayer (the "foot") since the molecular density is assumed to vanish for $r > r_1$ (the 3D evaporation of the fluid drop is negligible and consequently no condensation of molecules from the gas phase can occur for $r > r_1$). Since the present problem is mathematically identical to that presented in Sect. 8.6.2, (8.71) must hold also in the present case but with $p_1 = 0$,

$$\frac{d}{dt}\left(\frac{dr_1^2}{dt} + \bar{\eta}r_1^2\right)\ln\left(\frac{r_1}{r_0}\right) = \frac{2p_0}{mn_a}.$$

For "large" times, r_1 varies slowly with time (see below), and we can neglect the second-order time derivative term in the expression above. This gives

$$\left(\frac{r_1}{r_0}\right)^2\left[2\ln\left(\frac{r_1}{r_0}\right) - 1\right] + 1 = \frac{4p_0 t}{mn_a\bar{\eta}r_0^2}. \qquad (8.76)$$

This relation between r_1 and time is shown in Fig. 8.27 (where $\bar{t} = t/t_0$, with $t_0 = mn_a\bar{\eta}r_0^2/4p_0$). Note that to a very good approximation $r_1 = r_0 + \text{const} \times t^{1/2}$. This result is consistent with the experimental data presented in Fig. 8.26 and in [8.52–54]: the length of the first layer (the "foot") is observed to grow with the square root of time as long as the central reservoir is present. Using (8.76), the experimental data [8.53] for PDMS (methyl terminated polydimethylsiloxane, linear long-chains of polydispersity 1.6 and of an average molecular weight 2400) on oxidized (14 Å) silicon wafers, gives $\bar{\eta} \approx 1 \times 10^{15}\,\text{s}^{-1}$.

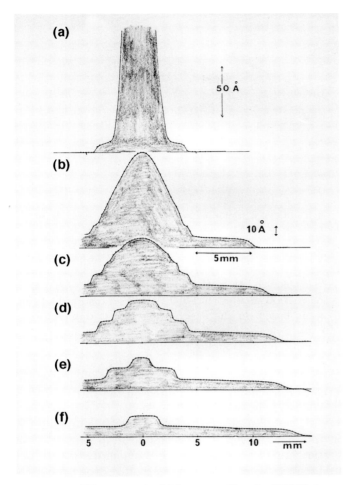

Fig. 8.26. Ellipsometric thickness profile of a PDMS drop spreading on a silicon oxide surface. Time t after deposition of droplet: (**a**) $t = 8\,\text{h}$; (**b**) $t = 71\,\text{h}$; (**c**) $t = 83\,\text{h}$; (**d**) $t = 118\,\text{h}$; (**e**) $t = 144\,\text{h}$; (**f**) $t = 168\,\text{h}$. From [8.52].

Heslot et al. [8.52] have also studied the spreading of a liquid PDMS drop on an oxidized silicon surface covered by a fatty acid monolayer, deposited using the Langmuir–Blodgett technique. In this case the spreading occurs even faster; this is expected, since the fatty acid monolayer covered surface is very inert and the sliding friction $\bar{\eta}$ may be smaller than for the clean oxidized silicon surface. However, the spreading pressure p_0 is smaller for the fatty acid case, i.e., the driving force for the spreading is smaller, but it is obvious from the experimental data that this effect, which reduces the spreading velocity, is less important than the reduction in the sliding friction. For PDMS on

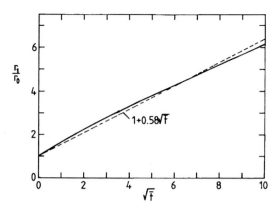

Fig. 8.27. *Solid line*: The radius of a spreading wetting drop as a function of the square root of time calculated from (8.76). *Dashed curve*: a linear "fit" to the exact result.

the fatty acid covered surface we estimate (with $p_0 \approx 0.3\,\text{meV}/\text{Å}^2$) $\bar{\eta} = 1 \times 10^{12}\,\text{s}^{-1}$, i.e., a factor ~ 1000 smaller than for PDMS on oxidized silicon.

The discussion above is based on the assumption of a two-dimensional incompressible fluid. In this case the "final" state of a spreading liquid drop will be a circular disk (radius R) of monolayer thickness, i.e., $n(r) = n_0$ for $r < R$ and $n = 0$ for $r > R$. But this cannot be the thermodynamically stable state for any real liquid, since for a finite drop on a large surface, entropy will be gained if the adsorbates are randomly distributed on the surface. But if strong lateral attraction occurs between the liquid molecules (which manifests itself in a large energy of vaporization and a high boiling temperature) it will take a very long time for the liquid molecules to evaporate from an island (island $\rightarrow$ 2D-lattice gas) and in this case one can identify two distinct steps in the spreading process: (a) a "fast" step where the drop spreads as an incompressible fluid via the mechanism discussed above, resulting in a circular disk of monolayer thickness, and (b) a "slow" process where the disk "evaporates" giving rise to a dilute 2D lattice gas as the final state.

The latter process (b) was studied in [8.52] for the spreading of a liquid drop of squalane, which assumes a quasi-spherical shape in the bulk with a diameter of about 9 Å. The spreading of this fluid on an oxidized silicon surface showed that the adsorbate density, for large times (where the local coverage was everywhere below monolayer coverage), had a Gaussian shape as expected for a "normal" diffusion process. The "effective radius" of the spreading drop again increases roughly proportionally to the square root of time, but the density profile is very different from that in Fig. 8.26. Note that some diffusion occurs also during step (a) in the spreading process – this is manifested in Fig. 8.26 by the fact that the leading edge of the "foot" is not a step function, but has a Gaussian tail which increases with time.

Note the structuring of the fluid in molecular layers parallel to the solid surface shown in Fig. 8.26. As pointed out before, this is a very general phenomenon and is even stronger when the fluid is confined between *two* flat surfaces.

Finally, note that the magnitude of $\bar{\eta}$ found above for PDMS on silicon is roughly a factor of 10^6 larger than found for inert atoms and small inert molecules (e.g., Ar, Xe, and CH_4) on smooth silver surfaces, as deduced from quartz–crystal-microbalance measurements [8.33,56]. This result can be explained by assuming that the amplitude of the lateral corrugation of the adsorbate–substrate interaction potential is considerably larger for big organic molecules on oxidized silicon surfaces, than for inert atoms and small inert molecules on smooth silver surfaces.

8.7 Two-Dimensional Fluids*

In this section we consider some general aspects of 2D fluids. We first discuss the thermodynamic foundations of 2D hydrodynamics and point out that the Navier–Stokes equations for a strict 2D fluid are ill-defined, e.g., the physical viscosity diverges in the thermodynamic limit (i.e., for an infinite system). However, for an adsorbate system, the friction coupling to the substrate (as manifested by the sliding friction $\bar{\eta}$) will regulate the theory and, in particular, lead to a finite 2D viscosity. We also show how the linear sliding friction $\bar{\eta}$ is related to the diffusivity.

8.7.1 2D Hydrodynamics

Thermal (and quantum) fluctuations become of increasing importance the lower the dimension of a system. This is, at least in part, related to the fact that the number of nearest neighbors in general decreases as the dimension of the system decreases. Thus, because of thermal fluctuations no strict 2D solid exists for $T > 0$ in the sense that $\langle \boldsymbol{u}^2 \rangle$ diverges logarithmically with the size of the system [8.57]. Similarly, for a strict 2D fluid the physical viscosity diverges logarithmically with the size of the system making hydrodynamics impossible [8.58]. However, even a weak coupling to the third dimension will in general lead to a finite viscosity and well-defined theory (see below) [8.59].

2D fluids are very common since all adsorbate systems will be in a 2D-fluid state in some parts of the adsorbate (θ, T) phase diagram [8.57]. In this section we briefly discuss the dynamical properties of fluid adsorbate systems, where the adsorbate–substrate coupling leads to well-defined hydrodynamics. In hydrodynamics one does not consider the motion of individual fluid atoms but only the motion of large enough "patches" of fluid. This gives a reasonable description of the long-time and large-scale properties of the system. Now, in addition to the usual terms which occur in the the three-dimensional Navier–Stokes equations, in the equations of motion describing the hydrodynamic properties of a fluid adsorbate layer a substrate friction force $-\rho\bar{\eta}\boldsymbol{v}$ occurs, where $\rho = n_a m$ is the adsorbate mass per unit area and $\bar{\eta}$ the sliding friction introduced earlier. We note that implicit in using such a friction force is the assumption that the substrate degrees of freedom are fast compared to the hydrodynamic processes of interest for the adsorbates. The equations of motion for the fluid velocity field $\boldsymbol{v}(\boldsymbol{x},t)$ and the density $\rho(\boldsymbol{x},t)$ are

$$\frac{\partial \rho}{\partial t} + \nabla \cdot (\rho \boldsymbol{v}) = 0 \,, \tag{8.77}$$

$$\frac{\partial \boldsymbol{v}}{\partial t} + \boldsymbol{v} \cdot \nabla \boldsymbol{v} = -\frac{1}{\rho}\nabla p + \nu \nabla^2 \boldsymbol{v} + \nu' \nabla \nabla \cdot \boldsymbol{v} - \bar{\eta}\boldsymbol{v} + \boldsymbol{f} \,, \tag{8.78}$$

where the components f_α of the fluctuating force $\boldsymbol{f}$ satisfy

$$\begin{aligned}\langle f_\alpha(\boldsymbol{x},t) f_\beta(\boldsymbol{x}',t')\rangle &= 2(k_\mathrm{B}T/m)\delta(t-t') \\ &\quad \times \left[\delta_{\alpha\beta}(\bar{\eta} + \nu\nabla^2) + \nu'\nabla_\alpha \nabla_\beta\right]\delta(\boldsymbol{x}-\boldsymbol{x}') \,.\end{aligned} \tag{8.79}$$

It should be noted that (8.77–79) are assumed to hold only down to some short-distance cutoff length Λ^{-1}. The presence of non-linear terms in the equations of motions means that the *observed* viscosities are no longer the bare parameters ν and ν' in (8.78), but are "renormalized" quantities which depend on the wave vector $\boldsymbol{q}$ and the frequency ω. The renormalized viscosities are obtained by integrating out all thermal fluctuations as follows: Let us write $\boldsymbol{v} = \langle \boldsymbol{v}\rangle + \delta\boldsymbol{v}$ and $n = \langle n\rangle + \delta n$ where $\langle ..\rangle$ stands for averaging over the thermal fluctuations induced by the stochastic force $\boldsymbol{f}$. Using (8.77) and (8.78) one can derive a set of coupled equations of motion for $\langle \boldsymbol{v}\rangle$, $\delta\boldsymbol{v}$, $\langle n\rangle$, and δn. The equations of motion for $\delta\boldsymbol{v}$ and δn can be solved by iteration. The solution, because of the non-linearities, depends on $\langle \boldsymbol{v}\rangle$ and $\langle n\rangle$. These formal solutions are then substituted into the equations of motion for $\langle \boldsymbol{v}\rangle$ and $\langle n\rangle$. The resulting equations have a similar form to the original equations (8.77) and (8.78) but without the fluctuating force $\boldsymbol{f}$ and with renormalized viscosities. For a 3D system, the renormalization adds a small term to the bare parameters but for a strict 2D system the coefficients increases logarithmically with the size of the system. But if the sliding friction $\bar{\eta}$ is finite, the renormalization is finite. For example, for an incompressible 2D fluid [8.59]

$$\nu_\mathrm{ren} - \nu \propto k_\mathrm{B}T \ln \frac{\nu\Lambda^2}{\nu q^2 + 4\bar{\eta} - 2\mathrm{i}\omega} \,,$$

which, if $\bar{\eta} > 0$, remains finite as $q \to 0$ and $\omega \to 0$.

Most experimental techniques do not probe the viscosity directly but rather some correlation function. For example, neutron scattering and inelastic helium scattering probe the density–density correlation function. If the non-linear terms in (8.77) and (8.78) are neglected, and if we assume the equation of state $p = mn_a c^2$ (where n_a is the adsorbate coverage), (8.77) and (8.78) reduce to

$$\frac{\partial n}{\partial t} + n_0 \nabla \cdot \boldsymbol{v} = 0 \, ,$$

$$\frac{\partial \boldsymbol{v}}{\partial t} = -c^2 \nabla n_a + \nu \nabla^2 \boldsymbol{v} + \nu' \nabla \nabla \cdot \boldsymbol{v} - \bar{\eta} \boldsymbol{v} + \boldsymbol{f} \, ,$$

where n_0 is the average of n_a. Using these equations it is easy to calculate various correlation functions. For example, the density–density correlation function is given by

$$C(q, \omega) = \frac{2q^2(\bar{\eta} + \nu_L q^2)}{(\omega^2 - c^2 q^2)^2 + \omega^2(\bar{\eta} + \nu_L q^2)} \tag{8.80}$$

where $\nu_L = \nu + \nu'$. If $\bar{\eta}$ is "small" the poles of $C(q, \omega)$ correspond to weakly damped phonon modes with the dispersion relation $\omega = cq$ for small q, while for "large" friction $\bar{\eta}$ the density–density correlation function has a diffusive peak centered at $\omega = 0$. The parameters which enter in correlation functions such as (8.80) are, however, renormalized and for 2D systems they exhibit logarithmic corrections. To illustrate this, let us consider the transverse sound-attenuation coefficient Γ, which, to second-order in perturbation theory, is given by [8.59]

$$\Gamma_{\mathrm{ren}} = (\bar{\eta} + \nu q^2) + \frac{k_B T}{4\pi m n_0} q^2 \left[\frac{3}{4\nu} \ln\left(\frac{\nu \Lambda^2}{\bar{\eta} + \nu q^2}\right) + \frac{2\bar{\eta}}{c^2} \ln\left(\frac{c\Lambda}{\bar{\eta} + \nu q^2}\right) \right] , \tag{8.81}$$

where we have assumed $q \ll \Lambda$, $\omega \ll \nu \Lambda^2$, and $\bar{\eta} \ll \nu \Lambda^2$. Similar logarithmic corrections occur for other transport coefficients, e.g., for the diffusion coefficient and the longitudinal sound-attenuation coefficient. Note that for $\bar{\eta} = 0$, $\Gamma_{\mathrm{ren}} \to \infty$ as $q \to 0$ (note: the smallest q are of order $1/L$, where L is the size of the system). However, if $\bar{\eta} > 0$, Γ_{ren} remains finite as $q \to 0$. Using neutron scattering or inelastic helium scattering it should be possible to probe these logarithmic corrections, which reflect the breakdown of conventional hydrodynamics for strict 2D systems. The corrections are particularly important for systems where the sliding friction $\bar{\eta}$ is small, and we suggest that physisorption systems, such as Kr or CH_4 on noble metal surfaces (where $\bar{\eta}$ is very small, of order $1 \times 10^8 \, \mathrm{s}^{-1}$), may be ideal cases on which to test the theory.

8.7.2 Relation Between Sliding Friction and Diffusion

In Sects. 8.4 and 8.5, using Langevin dynamics simulations, we have evaluated the linear sliding friction $\bar{\eta}$ directly from the definition $m\bar{\eta}\langle \dot{\boldsymbol{r}} \rangle = \boldsymbol{F}$. In this section we show that when F is weak it is easy to derive a formal expression for $\bar{\eta}$ using standard linear response theory. We also discuss the relation between $\bar{\eta}$ and the diffusivity.

Let us define the *mobility* χ via

$$\langle \dot{r} \rangle = \chi F ,$$

so that $\bar{\eta} = 1/m\chi$. The Hamiltonian for the adsorbate system is written

$$H = H_0 - \sum_i r_i \cdot F = H_0 + H' ,$$

where H_0 depends on the adsorbate–adsorbate and adsorbate–substrate interactions. Assume that the external force F is turned on at time $t = 0$. If we introduce

$$R = \sum_i r_i ,$$

then, within linear response, the drift velocity $v = \langle \dot{r} \rangle$ for $t > 0$ is given by the standard result [8.60]

$$v_\alpha = \frac{i}{N} \int_0^t dt' \left\langle \left[\dot{R}_\alpha(t), R_\beta(t') \right] \right\rangle F_\beta , \tag{8.82}$$

where the sum over repeated indices is implicitly understood and where N is the number of adsorbates. The drift mobility χ is determined from (8.82) in the limit $t \to \infty$. Using

$$R_\beta(t) = \int_0^t dt' \dot{R}_\beta(t') + R_\beta(0) ,$$

and the fact that

$$\left\langle \left[\dot{R}_\alpha(t), R_\beta(0) \right] \right\rangle \to 0$$

as $t \to \infty$, it follows that for large times

$$v_\alpha = \chi_{\alpha\beta} F_\beta$$

where

$$\chi_{\alpha\beta} = \frac{i}{N} \int_0^t dt' \int_0^{t'} dt'' \left\langle \left[\dot{R}_\alpha(t), \dot{R}_\beta(t'') \right] \right\rangle .$$

This integral is easily rewritten (as $t \to \infty$) as

$$\chi_{\alpha\beta} = \frac{i}{N} \int_0^\infty dt \ t \left\langle \left[\dot{R}_\alpha(t), \dot{R}_\beta(0) \right] \right\rangle , \tag{8.83}$$

where we have taken the limit $t \to \infty$. But the fluctuation–dissipation theorem [8.60]

$$\int_0^\infty dt \langle \dot{R}_\alpha(t)\dot{R}_\beta(0)\rangle e^{i\omega}$$
$$= (1-e^{-\beta\omega})^{-1} \mathrm{Re}\left(\int_0^\infty dt \langle [\dot{R}_\alpha(t), \dot{R}_\beta(0)]\rangle e^{i\omega t}\right)$$

gives, as $\omega \to 0$,

$$\int_0^\infty dt \langle \dot{R}_\alpha(t)\dot{R}_\beta(0)\rangle = ik_BT \int_0^\infty dt\ t \langle [\dot{R}_\alpha(t), \dot{R}_\beta(0)]\rangle .$$

Hence, (8.83) can be written as

$$\chi_{\alpha\beta} = \frac{1}{Nk_BT}\int_0^\infty dt \langle \dot{R}_\alpha(t)\dot{R}_\beta(0)\rangle .$$

For a (100) or a (111) surface of a fcc crystal, $\chi_{\alpha\beta} = \chi \delta_{\alpha\beta}$, where

$$\chi = \frac{1}{2Nk_BT}\int_0^\infty dt \langle \dot{\boldsymbol{R}}(t)\cdot \dot{\boldsymbol{R}}(0)\rangle . \tag{8.84}$$

For a single adsorbate (8.84) becomes

$$\chi = \frac{1}{2k_BT}\int_0^\infty dt \langle \dot{\boldsymbol{r}}(t)\cdot \dot{\boldsymbol{r}}(0)\rangle .$$

This relation for the mobility also holds for finite adsorbate coverage if the adsorbate–adsorbate interaction can be neglected. In this limit χ is directly related to the tracer diffusion constant D via the Einstein equation $D = \chi k_B T$. This equation follows immediately by comparing the formal expression for the diffusivity [8.61],

$$D = \frac{1}{2}\int_0^\infty dt \langle \dot{\boldsymbol{r}}(t)\cdot \dot{\boldsymbol{r}}(0)\rangle ,$$

with the expression given above for the mobility. For higher concentration of adsorbates, where the adsorbate-adsorbate interaction is important, χ cannot be related to D but rather to the chemical diffusivity D_c. The relation between D_c and χ or $\bar{\eta}$ can be derived as follows: Let us consider the adsorbates on a surface as a two-dimensional "fluid" characterized by the density $n_a(\boldsymbol{x},t)$ (the number of adsorbates per unit area) and the velocity field $\boldsymbol{v}(\boldsymbol{x},t)$. These "hydrodynamic" variables satisfy the continuity equation:

$$\frac{\partial n_a(\boldsymbol{x},t)}{\partial t} + \nabla\cdot[n_a(\boldsymbol{x},t)\boldsymbol{v}(\boldsymbol{x},t)] = 0 , \tag{8.85}$$

and the equation of motion:

$$\frac{\partial \boldsymbol{v}(\boldsymbol{x},t)}{\partial t} = -\frac{1}{mn_0}\nabla p(\boldsymbol{x},t) - \bar{\eta}\boldsymbol{v}(\boldsymbol{x},t) , \tag{8.86}$$

where n_0 is the average particle density. In (8.86) we have neglected the non-linear term $\boldsymbol{v} \cdot \nabla \boldsymbol{v}$ as well as the viscosity terms which occur in the full Navier–Stokes equations (Sect. 8.7.1). These terms are of no importance in the present context where $\boldsymbol{F}$ is assumed to be very weak (linear response) and where we have assumed that the velocity field changes very slowly in space and time. We assume that the two-dimensional pressure is $p = n_\mathrm{a} m c^2$ where c is the sound velocity. Let us linearize the continuity equation

$$\frac{\partial n_\mathrm{a}(\boldsymbol{x},t)}{\partial t} + n_0 \nabla \cdot \boldsymbol{v}(\boldsymbol{x},t) = 0 \ . \tag{8.87}$$

Taking the divergence of (8.86) and using (8.87) then gives

$$\frac{\partial^2 n_\mathrm{a}}{\partial t^2} - c^2 \nabla^2 n_\mathrm{a} + \bar{\eta}\frac{\partial n_\mathrm{a}}{\partial t} = 0 \ . \tag{8.88}$$

For very slowly varying time dependence we can neglect the second-order time derivative term in (8.88) to get

$$\frac{\partial n_\mathrm{a}}{\partial t} - \frac{c^2}{\bar{\eta}} \nabla^2 n_\mathrm{a} = 0 \ .$$

Hence

$$D_\mathrm{c} = c^2/\bar{\eta} \ .$$

Note that for non-interacting particles $p = n_a k_\mathrm{B} T$ so that $c^2 = k_\mathrm{B} T/m$ and $D_\mathrm{c} = k_\mathrm{B} T/m\bar{\eta} = \chi k_\mathrm{B} T = D$. Note also that if $\bar{\eta} = 0$ then (8.88) reduces to

$$\frac{\partial^2 n_\mathrm{a}}{\partial t^2} - c^2 \nabla^2 n_\mathrm{a} = 0 \ ,$$

which is the standard wave equation. In general, depending on the frequency ω and wave vector $\boldsymbol{k}$ characterizing the temporal and spatial dependence of $\boldsymbol{v}(\boldsymbol{x},t)$, the adsorbate system may exhibit diffusive or sound-like behavior [8.62].

8.8 Solid–Fluid Heat Transfer

Heat production and heat flow are, of course, central processes in sliding friction. In this section we discuss some simple but fundamental aspects of this topic.

Heat Transfer Coefficient α

Consider a solid–fluid interface and assume that the solid and fluid have different temperatures. In order to describe heat flow at the interface, a heat transfer coefficient α is defined via

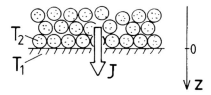

Fig. 8.28. The heat flow J at a fluid–solid interface. The layer of physisorbed fluid molecules has the temperature T_2 and the surface atoms of the solid ($z = 0$) the temperature T_1.

$$J = \alpha \left(T_2 - T_1\right) . \tag{8.89}$$

Here J is the normal component of the heat current vector at the interface, and T_2 and T_1 are the *local* temperatures *at the interface* in the fluid and in the solid, respectively (Fig. 8.28). Implicit in the definition is the assumption that the local temperature T changes discontinuously on an atomic scale when passing from the fluid to the solid. To discuss this assumption in more detail, let us first consider an "inert" fluid, e.g., a noble gas fluid interacting with a solid. In the fluid the heat transfer occurs by "collisions" with a characteristic time $\tau \sim 1/v_{\text{th}}a$ where $v_{\text{th}} \sim (k_B T/m)^{1/2}$ is the thermal velocity and $a \sim 1$ Å is on the order of the mean free path or an average interatomic separation. Note that in a fluid the atomic collisions are very effective in transferring kinetic energy owing to the equal mass of the atoms. At the solid–fluid interface a layer of adsorbed fluid atoms is assumed to occur. We further assume that the adsorbates interact effectively with the fluid atoms so that energy transfer frequently occurs between the fluid and the adsorbate layer; the latter will therefore have practically the same temperature as the fluid immediately above. But the energy transfer rate between the adsorbed layer and the substrate may be much slower than the energy transfer rate characterizing the $fluid \leftrightarrow adsorbate$ interaction. For an insulator, the reason for this is contained in the formula for the damping rate of an adsorbate vibrational mode due to one-phonon emission, namely [see (8.12)]

$$\eta_{\text{ph}} \approx \frac{3}{8\pi} \frac{m}{\rho} \left(\frac{\omega_0}{c_T}\right)^3 \omega_0 \sim \frac{m}{m_s} \left(\frac{\omega_0}{\omega_s}\right)^3 \omega_0 . \tag{8.90}$$

Hence, if the adsorbates (mass m) are much lighter than the substrate atoms (mass m_s) or, more importantly, if the substrate is hard so that the adsorbate–substrate frequency ω_0 is much smaller than the maximum substrate frequency $\sim \omega_s \sim 2\pi c_T/a$ then the one-phonon decay process is very slow. For a metal substrate the coupling to e–h pair excitations is always resonant, but we have seen in Sect. 8.2 that in most cases this process is also "slow" (see Table 8.1), at least compared with the time scale associated with the $fluid \leftrightarrow adsorbate$ energy transfer. The heat transfer in the solid itself is usually very fast and the temperature at the solid surface will be nearly the same as slightly below.

Consider now a non-equilibrium situation but assume that local thermal equilibrium occurs everywhere except at the fluid–solid interface; the energy flow through the interface is given by (8.89). At high temperature the heat transfer coefficient α can be obtained directly from a classical "friction" model. That is, consider an atomic adsorbate moving or vibrating on a surface with the velocity $\bm{v} = (\bm{v}_\|, v_\perp)$. A friction force will act on the adsorbate

$$\bm{F}_\mathrm{f} = -m\eta_\| \bm{v}_\| - m\eta_\perp \bm{v}_\perp , \qquad (8.91)$$

and the power transfer to the substrate will be

$$P = -\langle \bm{F}_\mathrm{f} \cdot \bm{v} \rangle = m\eta_\| \langle v_\|^2 \rangle + m\eta_\perp \langle v_\perp^2 \rangle . \qquad (8.92)$$

But if the adsorbate has the same temperature as the fluid above, then

$$m\langle v_\|^2 \rangle = 2k_\mathrm{B} T_2 , \qquad (8.93)$$

$$m\langle v_\perp^2 \rangle = k_\mathrm{B} T_2 , \qquad (8.94)$$

Hence

$$P = \left(2\eta_\| + \eta_\perp\right) k_\mathrm{B} T_2 . \qquad (8.95)$$

A similar formula (with T_2 replaced by the substrate temperature T_1) gives the power transfer to the adsorbate from the thermal motion of the substrate. Hence the net energy flow per unit surface area from the adsorbates to the substrate is given by

$$J = n_\mathrm{a} \left(2\eta_\| + \eta_\perp\right) k_\mathrm{B} (T_2 - T_1) \qquad (8.96)$$

and, from (8.89),

$$\alpha = n_\mathrm{a} k_\mathrm{B} \left(2\eta_\| + \eta_\perp\right) . \qquad (8.97)$$

The heat transfer coefficient at an arbitrary temperature for a molecular fluid can be related to the adsorbate resonance frequencies ω_ν and the corresponding damping rates $\eta_\nu = 1/\tau_\nu$ as discussed in [8.5,63]. In particular, at very low temperature only long-wavelength phonons are thermally excited and the simple formula (8.97) is then no longer valid.

Equation (8.97) is only valid if the (relevant) adsorbate–substrate interactions are weak so that the *fluid* ↔ *adsorbate* energy exchange is effective. If the fluid atoms (or molecules) chemisorb on the substrate surface then this condition is, in general, not satisfied; in this case we consider the chemisorbed layer as part of the solid, while the next layer of fluid atoms, which are likely to be physisorbed on the surface, constitutes the layer of adsorbates involved in the *fluid* ↔ *solid* energy transfer discussed above.

Heat Flow During Sliding

Consider two semi-infinite solids sliding along a planar surface separated by a few monolayers of lubrication fluid. Assume that the heat current J flows into the lower solid (Fig. 8.28). If the temperature of the solid equals T_0 at the beginning of sliding, the temperature $T(z,t)$ will increase with time t according to the heat diffusion law

$$\frac{\partial T}{\partial t} - \kappa \frac{\partial^2 T}{\partial z^2} = 0 \;,$$

with

$$T(z,0) = T_0, \quad -\lambda \partial T(0,t)/\partial z = J \;.$$

The thermal diffusion coefficient is $\kappa = \lambda/\rho C_\mathrm{v}$ where λ is the thermal conductivity, ρ the mass density and C_v the heat capacitance. If J is time independent for $t>0$, the solution is

$$T = T_0 + \frac{2J}{\lambda}\left(\frac{\kappa t}{\pi}\right)^{1/2} f(q) \;,$$

where $q = z/(4\kappa t)^{1/2}$ and

$$f = e^{-q^2} - 2q^2 \int_1^\infty d\eta\, e^{-\eta^2 q^2} \;.$$

The function $f(q)$ is shown in Fig. 8.29. Note that $f(0) = 1$ so that the temperature at the surface $z=0$ of the lower solid at time t equals $T_0 + \Delta T$ where

$$\Delta T = \frac{2J}{\lambda}\left(\frac{\kappa t}{\pi}\right)^{1/2} \;. \tag{8.98}$$

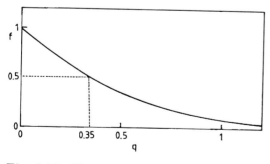

Fig. 8.29. The heat diffusion integral f as a function of $q = z(4\kappa t)^{-1/2}$.

Table 8.7. The temperature increase ΔT and the penetration depth d (see text for definitions) at contact areas (junctions) of diameter $D = 100\,\mu$m and at a sliding velocity of $1\,$m/s. The mass density ρ, the heat capacitance C_v, the heat conductivity λ, and the Brinell hardness σ_c are all for room temperature except for ice where these quantities are for $T = -5°$C. The kinetic friction coefficients μ_k are typical of boundary lubricated surfaces. For ice two values for μ_k are given, 0.02 and 0.3, corresponding to sliding with a water film and without a water film, respectively. In all cases the sliding bodies are made from the same material.

Solid	ρ (kg/m^3)	C_v (J/kgK)	λ (W/mK)	σ_c (N/m^2)	μ_k	ΔT (K)	d (μm)
steel	7900	450	75	10×10^8	0.1	35	32
granite	2700	800	2.2	20×10^8	0.1	518	7
teflon	2200	1050	2.3	0.5×10^8	0.02	2	7
diamond	3500	520	2000	800×10^8	0.1	720	232
ice	1000	2100	2.3	0.4×10^8	0.02	2	7
ice	1000	2100	2.3	0.4×10^8	0.3	30	7

We define the heat penetration depth d as the distance into the solid at which the temperature increase equals $\Delta T/2$. From Fig. 8.29 it follows that $f(q) = 1/2$ for $q \approx 0.35$. Hence, $d \approx 0.7\,(\kappa t)^{1/2}$.

Assume that σ_k is the shear stress necessary to slide two surfaces with the relative speed v. During steady sliding this requires an energy input which (per unit surface area and unit time) equals $v\sigma_k$. If this energy is completely converted into heat motion at the interfaces, then the heat currents J_1 and J_2, which flow into the lower and upper solids, respectively, must satisfy $J_1 + J_2 = \sigma_k v$. If the upper and lower solids are made from identical material then, by symmetry, $J_1 = J_2 = \sigma_k v/2$. If the two solids are made from different materials, then the ratio J_1/J_2 depends on the heat transfer coefficients α_1 and α_2.

We now give a few examples to illustrate the typical flash-temperature increase ΔT expected during sliding. First note that the contact areas (junctions) between two solids usually have a diameter of $D \sim 1-100\,\mu$m. During steady sliding at the speed v a junction will survive for a time of order $\tau = D/v$. Using $v = 1\,$m/s and $D = 100\,\mu$m this gives $\tau = 10^{-4}\,$s and the maximum temperature increase ΔT will be given approximately by (8.98) with $t = \tau$. In Table 8.7 we give ΔT and d for a few different systems, namely steel, granite, teflon, diamond, and ice. In all cases the block and the substrate are made from the same material. I assume that the kinetic friction stress can be written as $\sigma_k = \mu_k \sigma_0$ where I have used the values of the friction coefficients μ_k indicated in the table; these are typical for boundary lubricated surfaces. The perpendicular stress σ_0 in a contact area is assumed to be equal to the penetration hardness of the materials. Note the following: For steel, which is a hard material with a high thermal conductivity, the temperature rise at the surface is about 35 K and the penetration depth $d \sim 30\,\mu$m. For granite, which has a similar hardness as steel but a much lower thermal conductivity, the temperature rise is much higher, $\Delta T \sim 500\,\text{K}$, and the penetration depth much smaller, $d \sim 7\,\mu\text{m}$. Teflon has a thermal conductivity similar to granite but is much softer, and σ_k is consequently much smaller, leading to a very small temperature rise at the surface, $\Delta T \sim 2\,\text{K}$, but the penetration depth $d \sim 7\,\mu\text{m}$ is the same as that of granite. Ice has a similar hardness and heat conductivity as teflon. When ice slides on ice at low speed the friction coefficient is very high, $\mu_k \approx 0.3$, and the temperature rise (for $v = 1\,\text{m/s}$) about 30 K. When a thin water film occurs at the interface the friction is much lower, of order $\mu_k = 0.02$, which gives $\Delta T \sim 2\,\text{K}$. In this context I note that Bowden and coworkers have shown that the water film which occurs at high sliding velocities during skiing is usually produced by friction heating, rather than pressure melting (Sect. 14.2). Finally, for diamond the temperature increase is $\sim 720\,\text{K}$. Hence even though diamond has a very high thermal conductivity, its great hardness leads to very high heat flow J and a large ΔT. It is very likely, however, that because of its hardness the contact areas between two diamond crystals will, in most cases, have diameters much smaller than the $100\,\mu\text{m}$ used above. Note that $\Delta T \propto D^{1/2}$ and $d \propto D^{1/2}$ and using these scaling laws it is easy to derive the ΔT and d values for the systems in Table 8.7 for arbitrary D. For example, if $D = 1\,\mu\text{m}$ for diamond, then $\Delta T = 72\,\text{K}$ and $d = 0.7\,\mu\text{m}$.

The temperature increases estimated above are the maximum flash-temperatures at the contact areas (junctions) between the sliding bodies. After a junction is broken the local temperature rise at the surface decreases with increasing time as $t^{-1/2}$. During continuous sliding the bodies will gradually heat up. The time dependence of the temperature field depends on the size and shape of the bodies and on the external environment, but does not interest us here.

The temperature in the boundary lubrication "fluid" during sliding will be higher than the surface temperature of the surrounding solid. Let us estimate this temperature difference δT in a few different cases, namely for gold, copper, diamond, glass, and nylon. Again, we assume that both the block and the substrate are made from the same material. Assume that the sliding velocity $v = 1\,\text{m/s}$ and $\mu_k = 0.1$ and that the friction energy is initially deposited as thermal energy in the thin lubrication layer. At steady state the temperature difference δT is determined from the heat current $J = \mu_k \sigma_0 v$ via $\alpha \delta T = J/2$. Using this equation we have calculated δT and the result is presented in Table 8.8. In all cases we have assumed that $\alpha = n_a \eta_\perp k_B$ with $n_a = 0.02\,\text{Å}^{-2}$ and with $\eta_\perp$ determined by phonon emission via (8.12) with $\omega_\perp = 5\,\text{meV}$ as is typical for physisorbed molecules. The calculated $\eta_\perp$ are also given in Table 8.8. Note the following: For gold the temperature difference is only $\sim 0.2\,\text{K}$. For copper it is about ten time larger. This results from the fact that a gold atom is much heavier than a copper atom, giving a much lower sound velocity (and consequently a larger η_ph) for gold than for copper. Diamond is a very hard material, which has two effects. First σ_0

Table 8.8. The temperature difference δT between the lubrication fluid and the surfaces of the surrounding solids. We have assumed $\mu_k = 0.1$ in all cases. The perpendicular friction $\eta_\perp$ is calculated from (8.12) and the heat transfer coefficient is $\alpha = n_a k_B \eta_\perp$ with $n_a = 0.02\,\text{Å}^{-2}$.

Solid	ρ (kg/m^3)	c_T (m/s)	$\eta_\perp$ (s^{-1})	σ_0 (N/m^2)	δT (K)
gold	19000	1200	3×10^{12}	3×10^8	0.2
copper	8900	3000	3×10^{11}	4×10^8	2
diamond	3500	13000	1×10^{10}	800×10^8	14500
glass	2300	3300	1×10^{12}	60×10^8	11
nylon	1100	1070	6×10^{13}	1×10^8	0.003

and hence the heat current J is very large. Secondly, the high sound velocity of diamond results in a very low phononic friction $\eta_\perp \sim 1 \times 10^{10}\,\text{s}^{-1}$. As a result of these two effects the temperature increase is huge: $\delta T \approx 15000\,\text{K}$. Of course, such a large temperature increase will not occur in reality. First, at high temperatures not only will the adsorbate–substrate modes be thermally excited but also the high frequency internal modes of the lubricant molecules and these may be more effective than the low frequency adsorbate–substrate vibrational modes in transferring heat to the substrate. Secondly, the very high pressures which occur in the contact areas for diamond may completely remove the lubrication film. The latter conclusion is supported by the observation that for sliding of diamond on diamond, the friction of lubricated surfaces is usually nearly the same as for unlubricated surfaces [8.64]. Finally, nylon is a soft material with a low sound velocity. This gives a strong damping of the adsorbate–substrate modes (overdamped motion). Hence the energy transfer between the lubrication fluid and the substrate is very good, resulting in a negligible temperature difference between the lubrication fluid and the substrate.

8.9 Non-linear Sliding Friction

In Sects. 8.3–6 we studied the sliding of adsorbate layers on surfaces in the limit where the driving force is very weak. In this case a non-zero drift velocity occurs if the adsorbate layer is in a fluid or in an incommensurate solid state; if instead a pinned state is formed (e.g., a commensurate solid state) the sliding velocity vanishes (neglecting the contribution to sliding from thermally generated imperfections). We have shown that in QCM friction measurements, the linear response assumption is very well satisfied since the force of inertia acting on an adsorbate is much smaller than the forces due to the lateral corrugation of the adsorbate–substrate interaction potential.

In this section we study the sliding of adsorbate layers when the external force F is large enough to pull the adsorbates over the barriers [8.34,35]. This topic is directly relevant for boundary lubrication. We consider two very different microscopic models, namely the Lennard–Jones model and the Frenkel–Kontorova model. Both models give very similar results, suggesting a universal sliding scenario. This is supported by the fact that the basic results can be understood on the basis of very simple and general physical arguments.

8.9.1 Dynamical Phase Transitions in Adsorbate Layers: Lennard–Jones Model

We consider the model described in Sect. 8.1 where the adsorbates interact with each other via a Lennard–Jones pair potential of strength ϵ and where the adsorbate–substrate interaction potential is assumed to be a simple cosine corrugation in both the x and y directions with the amplitude $2U_0$. In the simulations below we have used $2U_0/\epsilon = 2, 3$, or 4, for which the adsorbate phase diagram is of the form shown in Fig. 8.12b, with a pinned commensurate c(2×2) structure centered at $\theta = 0.5$. As in Sect. 8.3, we assume $\alpha = \infty$ in (8.4), i.e., the motion is strictly 2D and we denote the parallel friction by η. Unless otherwise stated, all quantities are measured in the natural units introduced in Sect. 8.3.

The relation $\langle v \rangle = f(F)$ between the drift velocity $\langle v \rangle$ and the external force F can have two qualitatively different forms. If, when $F = 0$, the adsorbate layer is in a 2D-fluid state, which is always the case in some parts of the (θ, T) phase diagram, the drift velocity will be non-zero for arbitrarily small F. This is, of course, exactly what one expects for a fluid: an arbitrarily weak external force can shear a fluid. Furthermore, no hysteresis is observed, i.e., the relation between F and $\langle v \rangle$ does not depend on whether F decreases from a high value or increases from zero. On the other hand, if the adsorbate layer forms a solid pinned structure when $F = 0$, a minimum external force ($F = F_a$) is necessary in order to initiate sliding, and the sliding friction exhibits hysteresis as a function of F.

A) Fluid Adsorbate Structures

Let us first discuss the relation between F and v for an adsorbate system that is in a fluid state when $F = 0$. In Fig. 8.30, we show $\langle v \rangle$ as a function of F when the coverage $\theta = 0.25$. Results are presented for two temperatures, $k_B T/\epsilon = 0.5$ (circles) and $k_B T/\epsilon = 1.0$ (squares). In both cases the overlayer is in a fluid state even for $F = 0$. Note that for $F > 0.6$ (F is measured in units of $kk_B T$) the adsorbate layer slides as if there were no barriers to overcome on the surface. The sliding friction shown in Fig. 8.30 has the same general form as in the dilute limit; see Fig. 8.8c. This result is expected since both systems are in the fluid part of the adsorbate phase diagram. Note also

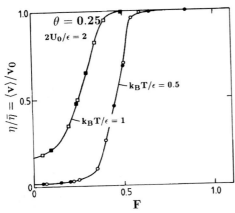

Fig. 8.30. The drift velocity v as a function of the external force F acting on each adsorbate for $\theta = 0.25$ where the adsorbate system is in a fluid state. The open and filled circles are obtained by increasing the force F from zero and by reducing F from a high value, respectively. In the calculations $2U_0/\epsilon = 2$ and $k_B T/\epsilon = 1$ and 0.5 for the two curves.

that when the temperature is increased, the sliding velocity $\langle v \rangle$ increases; this result is obvious since at the higher temperature the thermal motion of the substrate ions and electrons [as manifested by the fluctuating force $\boldsymbol{f}_i$ in (8.1)] is stronger and they will more frequently kick the adsorbates over the barriers in the system. The open and filled circles (and squares) are obtained by increasing the force F from zero and by reducing the force F from a high value, respectively. But in the present case both procedures give identical results, i.e., no hysteresis occurs. More generally, whenever the adsorbate system (for $F = 0$) is in a fluid state, no hysteresis is observed.

B) Commensurate Solid Adsorbate Structures

We consider now coverages θ and temperatures T where the adsorbate layer (when $F = 0$) is in a solid state which is commensurate with, and pinned by the substrate. We discuss how the relation $\langle v \rangle = f(F)$ depends on the microscopic friction η and the coverage θ. We also consider the influence of pinning centers on the sliding dynamics, and discuss the nature of the three types of dynamical phase transitions that occur in the simulations.

In the simulations reported below, $\epsilon/k_B T = 2$ and $2U_0/\epsilon = 2$, 3, or 4. For these parameters the c(2 × 2) structure shown in Fig. 8.32b prevails when $\theta = 0.5$ and $F = 0$. Figure 8.31 shows the inverse of the normalized sliding friction, $\eta/\bar{\eta}$, as a function of temperature, when the external force $F = 0.05$ is so weak that it has a negligible influence on the thermodynamic state of the adsorbate layer. The quantity $\eta/\bar{\eta}$ equals the ratio $\langle v \rangle/v_0$ between the actual drift velocity and the drift velocity $v_0 = F/m\eta$ which would occur if

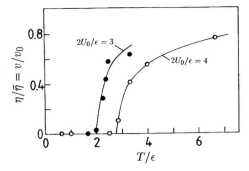

Fig. 8.31. The inverse of the normalized drift friction, $\eta/\bar{\eta}$ (equal to the normalized drift velocity $\langle v \rangle/v_0$), as a function of temperature for $\theta = 0.5$ ($M = 12$) and for $2U_0/\epsilon = 3$ (*black dots*) and $2U_0/\epsilon = 4$ (*circles*). The melting temperature of the c(2 × 2) structure is $k_B T_c = 2.0 \pm 0.25$ and $k_B T_c = 2.75 \pm 0.25$, respectively. The external force $F = 0.05$ is very weak and has a negligible influence on T_c.

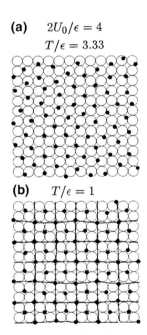

Fig. 8.32. Snapshots of the adsorbate layers in the melted state ($k_B T/\epsilon = 3.33$) and in the solid pinned state where a c(2 × 2) structure is formed ($k_B T/\epsilon = 1$). $2U_0/\epsilon = 4$ and $F = 0.05$.

the surface corrugation were to vanish (i.e., $U_0 = 0$). Note that the c(2 × 2) structure "melts" at $k_B T_m = (2.0 \pm 0.25)\epsilon$ if $2U_0/\epsilon = 3$ and at $k_B T_m =$

$(2.75 \pm 0.25)\epsilon$ if $2U_0/\epsilon = 4$. Figure 8.32 shows snapshots of the adsorbate layer in (a) the melted state and (b) in the commensurate solid (C-solid) state.

Dependence of $\bar{\eta}$ on η

Thermal equilibrium properties are independent of the friction η occurring in (8.1). But non-equilibrium properties do depend on η. In this section we study the dependence of the sliding friction $\bar{\eta}$ on η, and of the effective temperature T^* in the adsorbate layer.

Figures 8.33 and 8.34 show the variation of $\eta/\bar{\eta}$ and T^*/T with the driving force F when $\eta = 0.1$ and 0.5, respectively, and with $2U_0/\epsilon = 3$. Note that the sliding friction exhibits hysteresis: As the force F increases from zero, the ordered c(2 × 2) structure prevails and the sliding velocity vanishes, $\langle v \rangle = 0$, until F reaches $F_a \approx 1.5$. At $F = F_a$, the sliding velocity $\langle v \rangle$ increases abruptly and inspection of the adsorbate structure shows that it has changed from the c(2×2) structure which occurs for $F < F_a$ to a fluid-like (for $\eta = 0.5$) or incommensurate (IC) solid-like (for $\eta = 0.1$) state for $F > F_a$. But when the system is first "thermalized" with a large force, $F \sim 2$, and then F reduced, it does not flip back to the c(2 × 2) structure until $F = F_b \approx 0.75$ (for $\eta = 0.5$) and 0.4 (for $\eta = 0.1$). Again, inspection of adsorbate structures shows that for $F > F_b$ the system is fluid-like while for $F < F_b$ the c(2 × 2) structure prevails.

The hysteresis shown in Figs. 8.33 and 8.34 can have two different origins. The first follows from the fact that the temperature T^* in the adsorbate layer during sliding is higher than the substrate temperature T, and might be so high that the fluid configuration rather than the solid pinned state is stable for $F > F_b$. This is always the case if the adlayer–substrate heat transfer coefficient $\alpha = 2k_B\eta$ is small enough. But for large η the return to the pinned state occurs as follows. First, it will be shown below that the return to the pinned solid state at $F = F_b$ is a nucleation process. However, a drag force will act on a pinned island due to the surrounding flowing 2D-fluid. Assuming a circular pinned island, and that the drag force is uniformly distributed on the adsorbates in the island, it is shown below that the drag force is so large that the island will fluidize if $F > F_b \approx F_a/2$.

The effective temperature T^* in the adsorbate system during sliding is higher than the substrate temperature T. Figures 8.33b and 8.34b show $T^*(F)$, obtained directly from the width of the velocity distributions $P_x(v_x)$ and $P_y(v_y)$ (both P_x and P_y were found to be well approximated by Maxwellian distributions of equal width, which is necessary in order for the temperature to be well-defined).

Note that in the fluid state a "small" η gives a higher effective temperature T^* in the adsorbate layer than a "large" η. This is easy to understand since the heat transfer coefficient [see (8.97)] $\alpha = 2k_B\eta$ is proportional to η. In the steady state the power P that the external force F "pumps" into the

8.9 Non-linear Sliding Friction   259

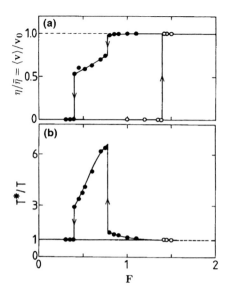

Fig. 8.33. (a) The inverse of the normalized sliding friction, $\eta/\bar{\eta}$, as a function of the external force F. (b) The effective temperature T^* in the adsorbate layer in units of the substrate temperature T, as a function of the external force F. For $\eta = 0.1$, $2U_0/\epsilon = 3$, $\epsilon/k_B T = 2$, and $M = 12$.

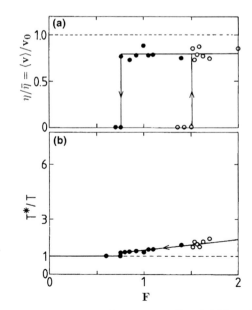

Fig. 8.34a,b. The same as Fig. 8.33 but for $\eta = 0.5$.

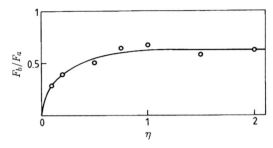

Fig. 8.35. The ratio F_b/F_a as a function of η for several simulations of the type shown in Figs. 8.33, 34. For $2U_0/\epsilon = 3$, $\epsilon/k_BT = 2$ and $M = 12$.

adsorbate layer must equal the energy transfer rate from the adsorbate layer to the substrate; this has a contribution which depends on the temperature difference $\alpha(T^* - T)$ between the adlayer and the substrate. Hence, a small heat transfer coefficient α implies $T^* \gg T$ in order to maintain a given heat current from the adsorbate layer to the substrate.

From Fig. 8.33b it is seen that the transition from the fluid state to the pinned c(2 × 2) structure occurs when the temperature T^* in the overlayer reaches $\sim 3T$ or $k_BT^* = 1.5\epsilon$, since $k_BT/\epsilon = 0.5$ in the present simulations. Note that $T^* = 1.5\epsilon/k_B$ is about 25% below the melting temperature T_c of the c(2 × 2) structure when $2U_0/\epsilon = 3$. Similarly, (see Fig. 8.36a below) for $\eta = 0.2$ and $2U_0/\epsilon = 4$ the transition from the fluid state to the pinned c(2 × 2) structure occurs when the temperature T^* in the overlayer is about 27% below the melting temperature T_c of the c(2 × 2) structure. If T^* in the overlayer exceeds T_c when $F = F_0/2$ then one would expect the return to the pinned structure to occur when the temperature T^* in the overlayer reaches the melting temperature T_c. The present result indicates that the return, in fact, occurs when the temperature is slightly below T_c but this may be related to the finite rate of nucleation (undercooling of the fluid may be necessary in order for the nucleation rate to be high enough for it to be observed during the short time of the computer simulation).

When η (and hence the heat transfer coefficient α) is large enough, the temperature in the overlayer during sliding is well below T_c for any $F < F_a$. This is the case, e.g., when $\eta = 0.5$ and $2U_0/\epsilon = 3$ as in Fig. 8.34. In this case, the return to the pinned phase occurs when $F \approx F_a/2$.

Figure 8.35 shows the variation of F_b/F_a as a function of the friction η. For large η, $F_b \approx 0.6F_a$ while $F_b \to 0$ as $\eta \to 0$. The origin of the latter was explained above: For small η the heat transfer from the adsorbate layer to the substrate is very "slow" and the temperature in the adsorbate layer will be above the melting temperature of the c(2 × 2) structure unless F is very small. On the contrary, for large η the temperature in the adsorbate layer is very close to the substrate temperature (i.e., below T_c) and the return to the pinned state occurs now at a higher critical force F_b, determined by the drag force which the surrounding flowing 2D-fluid exerts on a pinned island.

We close this section with a derivation of the drag force on a stationary circular rigid and pinned disk (radius R) in a two-dimensional fluid. The external force $\boldsymbol{F}$ acts on each fluid atom and, far away from the disk, gives rise to the drift velocity $\boldsymbol{v}_0$ where $m\bar{\eta}\boldsymbol{v}_0 = \boldsymbol{F}$. The basic equations are

$$\nabla \cdot \boldsymbol{v} = 0, \tag{8.99}$$

$$\frac{\partial \boldsymbol{v}}{\partial t} + \boldsymbol{v} \cdot \nabla \boldsymbol{v} = -\frac{1}{mn_a}\nabla p + \nu \nabla^2 \boldsymbol{v} - \bar{\eta}(\boldsymbol{v} - \boldsymbol{v}_0). \tag{8.100}$$

If we measure velocity in units of v_0 and distance in units of R then the terms

$$\boldsymbol{v} \cdot \nabla \boldsymbol{v} \;:\; \nu \nabla^2 \boldsymbol{v} \;:\; \bar{\eta}\boldsymbol{v}$$

will scale as

$$v_0^2/R \;:\; \nu v_0/R^2 \;:\; \bar{\eta}v_0 \quad \text{or} \quad v_0 R/\nu \;:\; 1 \;:\; R^2\bar{\eta}/\nu.$$

Hence, in the large-R limit we can drop the viscosity term $\nu \nabla^2 \boldsymbol{v}$ and the non-linear term $\boldsymbol{v} \cdot \nabla \boldsymbol{v}$. If, in addition, we consider stationary flow, $\partial \boldsymbol{v}/\partial t = 0$, then (8.100) reduces to

$$\nabla p + mn_a\bar{\eta}(\boldsymbol{v} - \boldsymbol{v}_0) = \boldsymbol{0}. \tag{8.101}$$

The drag force is given by

$$\boldsymbol{F}_{\text{drag}} = -\oint d\varphi\, x p, \tag{8.102}$$

where the integral is over the periphery of the disk, i.e., $|\boldsymbol{x}| = R$ and $0 \leq \varphi < 2\pi$. Since from (8.101) $\nabla \times \boldsymbol{v} = 0$ we can write $\boldsymbol{v} = \nabla \phi$ where from (8.99), $\nabla^2 \phi = 0$. The relevant solution to this equation is of the form

$$\phi = Ax/r^2 + Bx,$$

where we have assumed that the external force $\boldsymbol{F}$ points in the x-direction. Since $\boldsymbol{v} \to \boldsymbol{v}_0$ as $r \to \infty$ we get $B = v_0$. Next, since the normal component of the fluid velocity must vanish on the periphery of the rigid disk we get $\partial \phi/\partial r = 0$ for $r = R$ and hence $B = A/R^2$ or $A = R^2 v_0$. Thus

$$\phi = v_0 x \left(R^2/r^2 + 1\right). \tag{8.103}$$

Next, from (8.101),

$$\nabla[p + \bar{\eta}(\phi - xv_0)] = 0$$

or

$$p = \text{const} + \bar{\eta}(xv_0 - \phi). \tag{8.104}$$

Using (8.102–104) finally gives

$$\boldsymbol{F}_{\rm drag} = m n_{\rm a} \pi R^2 \bar{\eta} v_0 = n_{\rm a} \pi R^2 \boldsymbol{F}\ .$$

Since $n_{\rm a}\pi R^2$ is the number of adsorbates in the island, the drag force per adsorbate equals $\boldsymbol{F}$. A more general study of the drag force was presented in [8.35].

Dependence of $\bar{\eta}$ on θ

At thermal equilibrium, for $\theta = 1/2$ and $F = 0$, the ordered c(2×2) structure shown in Fig. 8.32b occurs. At slightly lower coverage the adsorbate system still has c(2×2) long-range symmetry but the structure has imperfections. In this section we discuss the nature of the non-linear sliding friction for $\theta = 0.5$, $71/144 \approx 0.493$ and $70/144 \approx 0.486$ and for $2U_0/\epsilon = 4$ and $\eta = 0.2$. Figure 8.36 shows $\eta/\bar{\eta} = v/v_0$ and T^*/T for these three cases. The major difference between the three cases is that $F_{\rm a}$ is very sensitive to the perfection of the adsorbate layer. For the perfect c(2×2) layer ($\theta = 1/2$), $F_{\rm a} \approx 2.6$, while $F_{\rm a} \approx 2.05$ for $\theta = 71/144$ and $F_{\rm a} \approx 1.25$ for $\theta = 70/144$. On the other hand, both the (IC-solid $\leftrightarrow$ fluid) and (fluid $\rightarrow$ C-solid) transitions occur at practically the same external force F in all three cases, as is indeed expected because of the small changes in the adsorbate coverage.

Influence of Pinning Centers on the Sliding Dynamics

The results presented above show that if η is small enough, at some critical force F_1 a dynamical phase transition occurs between two different sliding states, namely an IC-solid (for $F > F_1$) and a 2D-fluid (for $F < F_1$). In the IC-solid state the sliding friction $\bar{\eta}$ is nearly equal to η, i.e., the sliding occurs just as if the surface had no corrugation ($U_0 = 0$). Furthermore, the temperature in the overlayer is essentially identical to the substrate temperature. Snapshots of the system for $F > F_1$ show a hexagonal adsorbate lattice which is incommensurate with the substrate.

Now, if the sliding state for $F > F_1$ really is an IC-solid as postulated above, then if a low concentration of static (i.e., immobile) impurity atoms occurs on the surface the IC-solid sliding state should disappear and be replaced by the 2D-fluid state. We have performed simulations which show that this is indeed the case.

Consider the system in Fig. 8.36a but assume that one of the 72 adsorbates in the basic unit cell has been fixed to a hollow site. This fixed "impurity" atom interacts with all the other adsorbates with identical pair potentials to those between the mobile atoms and given by (8.7). Hence, we consider a $\theta = 0.5$ adsorbate layer with a low concentration $1/72 \approx 0.014$ of static impurities. For this case Fig. 8.37 shows $\eta/\bar{\eta}$ and T^*/T as a function of F. At thermal equilibrium and for $F = 0$ the perfect c(2×2) layer shown in Fig. 8.32b is formed where out of the two possible c(2×2) sublattices the sublattices which contains the "impurity" atoms is occupied. For this sliding

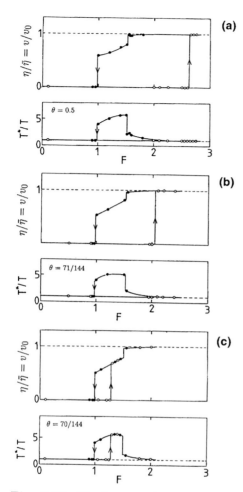

Fig. 8.36. The inverse of the normalized sliding friction, $\eta/\bar{\eta}$, and the effective temperature T^* in the adsorbate layer in units of the substrate temperature T, as a function of the external force F. For $\eta = 0.2$, $2U_0/\epsilon = 4$, $\epsilon/k_\mathrm{B}T = 2$ and $M = 12$. In (**a**, **b**) and (**c**) the coverage equals $\theta = 0.5$, $71/144$, and $70/144$, respectively.

system the IC-solid state has disappeared and the effective temperature in the overlayer increases monotonically with increasing external force F. But note that $F_\mathrm{b} \approx 1$ is very similar to the case without pinning centers (Fig. 8.36a) and for external forces above but close to F_b both the sliding friction and the effective temperature in the adsorbate layer are quite close to those without any pinning centers.

Figure 8.38a and b shows the inverse of the normalized sliding friction, $\eta/\bar{\eta}$, and the effective temperature T^* in the adsorbate layer in units of the

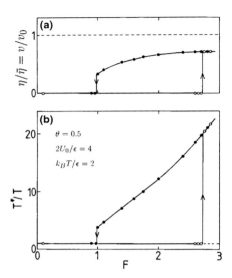

Fig. 8.37. (a) The inverse of the normalized sliding friction, $\eta/\bar{\eta}$, as a function of the external force F. (b) The effective temperature T^* in the adsorbate layer in units of the substrate temperature T, as a function of the external force F. In the simulations one out of the 72 particles is kept fixed in a hollow site. $\eta = 0.2$, $2U_0/\epsilon = 4$, $\epsilon/k_\mathrm{B}T = 2$, and $\theta = 0.5$ ($N = 72$, $M = 12$).

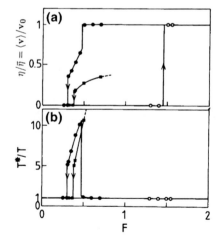

Fig. 8.38. (a) The inverse of the normalized sliding friction, $\eta/\bar{\eta}$, as a function of the external force F. (b) The effective temperature T^* in the adsorbate layer in units of the substrate temperature T, as a function of the external force F. *Black dots*: all 72 particles are allowed to move. *Black squares*: One out of the 72 particles is kept fixed in a hollow site. $\eta = 0.05$, $2U_0/\epsilon = 3$, $\epsilon/k_\mathrm{B}T = 2$, and $\theta = 0.5$ ($N = 72$, $M = 12$).

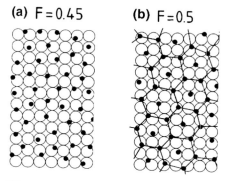

Fig. 8.39a,b. Snapshots of the adsorbate layers in the fluid state ($F = 0.45$) and in the IC-solid state ($F = 0.5$), for the case indicated by black dots in Fig. 8.38.

substrate temperature T, as a function of the external force F when $2U_0/\epsilon = 3$ and $\eta = 0.05$. The dots are results of simulations when all 72 particles in the basic unit are allowed to move while the squares correspond to the case where one of the 72 particles is kept fixed in a hollow site. Note that in *both* cases the return to the pinned state occurs when $T^* \approx 5T$. This is consistent with the criteria that for "small" η the return to the pinned state occurs when the effective temperature in the adlayer reaches the melting temperature of the $c(2 \times 2)$ structure (with $F = 0$); from Fig. 8.31 the melting temperature is $T^* \approx 2.5\,\epsilon/k_B$ or $T^* \approx 5\,T$ since $\epsilon/k_B T = 2$.

Figure 8.39 shows snapshots of the adsorbate layers in the fluid state ($F = 0.45$) and in the IC-solid state ($F = 0.5$) for the case indicated by dots in Fig. 8.38.

Fluidization or Shear-Melting Transition

Let us consider the transition from the C-solid state to the fluid state in Fig. 8.34 which occurs at $F = F_a \approx 1.5$. In Fig. 8.40 we show the instantaneous sliding velocity of the center of mass of the adsorbate layer as a function of time when, at time $t = 1000$, the force F is abruptly changed from a value below F_a to a value above F_a. For $t < 1000$ the external force $F = 0.05$ and the pinned $c(2 \times 2)$ structure is stable and the sliding velocity vanishes. For $t > 1000$ the external force is equal to 1.55 and the fluid state is the stable sliding state. Note that the system remains in the metastable $c(2 \times 2)$ structure for about 250 time units before making the transition to the fluid state. This indicates that the transition occurs by nucleation of the fluid phase in the C-solid state.

Figure 8.41 shows snapshots of the adsorbate layer during the transition from the pinned $c(2 \times 2)$ state to the fluid state of Fig. 8.40. It is obvious from these pictures that the fluidization transition occurs by nucleation.

266   8. Sliding of Adsorbate Layers

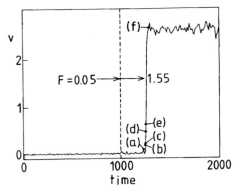

Fig. 8.40. The instantaneous sliding velocity of the center of mass of the adsorbate layer as a function of time. For $t < 1000$ the external force $F = 0.05$ and the pinned $c(2 \times 2)$ structure is stable and the sliding velocity vanishes. For $t > 1000$ the external force $F = 1.55$ and the fluid state is the stable sliding state. Note that the system remains in the metastable $c(2 \times 2)$ structure for about 250 time units before making the transition to the fluid state. The parameters are the same as in Fig. 8.34.

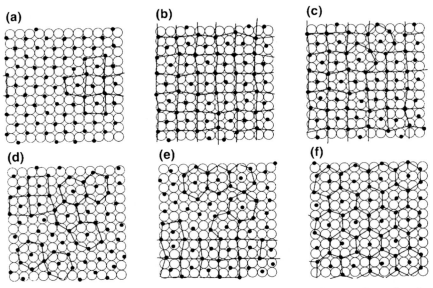

Fig. 8.41. Snapshots of the adsorbate layer during the transition from the pinned $c(2 \times 2)$ state to the fluid state in Fig. 8.40. The parts (**a–f**) correspond to the times 1230, 1250, 1251, 1255, 1258, and 1290, and are indicated in Fig. 8.40.

Freezing Transition

In this section we discuss the nature of the (fluid → C-solid) transition at $F = F_b$. We will show that this transition involves a nucleation process. Initially, a pinned island of c(2 × 2) structure surrounded by a flowing 2D-fluid occurs. If F is above F_b such a nucleus is unstable and disappears. But for F below F_b the nucleus is stable and will grow and finally fill out the whole system. In this section we present results which prove that this is the correct picture of the (fluid → C-solid) transition.

Let us consider the system in Fig. 8.36a. Figure 8.42 shows the instantaneous sliding velocity of the center of mass of the adsorbate layer as a function of time when the external force is abruptly reduced from a value above F_b to a value below F_b. For $t < 1500$ the external force $F = 1.0$ and the fluid state is stable. For $t > 1500$ the external force is equal to 0.6 and the pinned c(2 × 2) state is the stable state. However, the system does not immediately return to this state but continues to slide for a while in a well-defined metastable fluid state. Not until $t \approx 2900$ does the system abruptly return to the c(2 × 2) state. As shown in [8.35], the amount of time spent in the metastable fluid state is a stochastic quantity which depends in a random manner (but with a certain average) on the time at which F is reduced from above F_b to below F_b. This is the expected behavior of a nucleation process.

Figure 8.43 shows snapshots of the adsorbate layer at the time points (a–f) indicated by circles in Fig. 8.42. At time point (c) a large c(2 × 2) island exists, surrounded by flowing 2D fluid. At time point (f) the nucleus has grown and fills out the whole system.

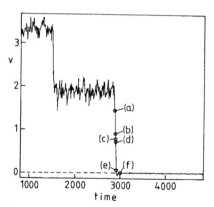

Fig. 8.42. The instantaneous sliding velocity of the center of mass of the adsorbate layer as a function of time. For $t < 1500$ the external force $F = 1.0$ and the fluid state is stable. For $t > 1500$ the external force is equal to 0.6 and the pinned c(2×2) state is stable. Note that the system remains in the metastable fluid state for about 1400 time units before making the transition to the pinned state. The parameters are the same as in Fig. 8.36a.

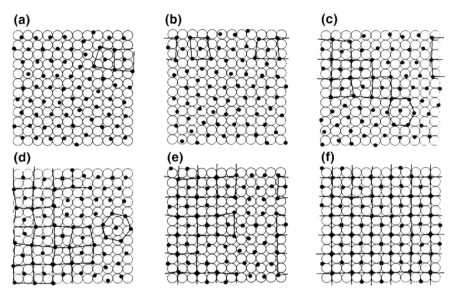

Fig. 8.43. Snapshots of the adsorbate layer during the transition from the fluid state to the pinned c(2 × 2) state in Fig. 8.42. The parts (**a–f**) correspond to the times 2870, 2885, 2890, 2895, 2910, and 3000, and are indicated in Fig. 8.42.

Fluid–Incommensurate Solid Transition

Let us now discuss the nature of the (fluid ↔ IC-solid) transition shown in Figs. 8.33, 8.36, and 8.38. As pointed out before, this dynamical phase transition only occurs if η is below some critical value (which depends on U_0, ϵ, and the substrate temperature T).

Let us consider the system in Fig. 8.36a. Figure 8.44a shows the instantaneous sliding velocity of the center of mass of the adsorbate layer as a function of time when the external force is abruptly changed at time $t = 2500$ from a value below F_1 to a value above F_1, where $F_1 \approx 1.5$ is the "critical" force associated with the (fluid ↔ IC-solid) transition (Fig. 8.36a). For $t < 2500$ the fluid state is stable while the incommensurate solid is the stable state for $t > 2500$. Note that the system remains in a metastable fluid state for about $\tau^* \approx 700$ time units before making the transition to the IC-solid state. This indicates that the transition (fluid → IC-solid)·involves a nucleation process. On the other hand, if F is abruptly changed from a value above F_1 to a value below F_1, the system immediately goes from the IC-solid state to the fluid state. This is illustrated in Fig. 8.44b where, for $t < 2500$, the external force is above F_1 and the IC-solid state is the stable state, while for $t > 2500$ the external force is below F_1 and fluid state is stable.

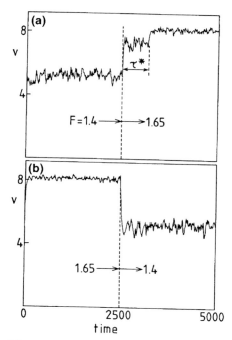

Fig. 8.44. The instantaneous sliding velocity of the center of mass of the adsorbate layer as a function of time. (a) For $t < 2500$ the external force $F = 1.4$ and the fluid state is stable. For $t > 2500$ the external force is equal to 1.65 and the IC-solid state is stable. Note that the system remains in a metastable fluid state for about 700 time units before making the transition to the IC-solid. (b) For $t < 2500$ the external force $F = 1.65$ and the IC-solid state is stable. For $t > 2500$ the external force $F = 1.4$ and the fluid state is stable. Parameters as in Fig. 8.36a.

Figure 8.45a and b shows the trajectories of 5 adsorbates when the adsorbate system is in a fluid state ($F = 1.5$) and in an IC-solid state ($F = 1.6$), respectively (note: the adsorbates drift to the right). Again the parameters are the same as in Fig. 8.36a. Note that the trajectories are much smoother in the IC-solid state than in the fluid state. The slight curvature of the orbits in the IC-solid state just reflect the finite substrate temperature: In a reference frame which moves with the drift velocity of the adsorbate system, the atoms in the IC-solid perform vibrational motion around their equilibrium positions corresponding to the temperature T of the substrate. In the fluid state the effective temperature T^* in the adsorbate layer is much higher than the substrate temperature ($T^* \approx 5T$, see Fig. 8.36), giving rise to much more curved trajectories. In addition, while in the fluid state a given adsorbate can diffuse around in the adlayer, in the IC-solid state a given adsorbate is always surrounded by the same 6 nearest neighboring atoms.

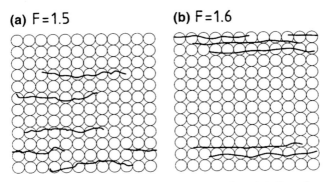

Fig. 8.45. The orbits of 5 adsorbates when (**a**) the adsorbate system is in a fluid state, $F = 1.5$, and (**b**) when the adsorbate layer is in an IC-solid state, $F = 1.6$. Parameters as in Fig. 8.36a.

The physical origin of the (IC-solid → fluid) phase transition can be understood as follows: First, it is obvious that the IC-solid state is the stable state when F is very large. This follows from the fact that, for large F, the sliding velocity of the adsorbate layer is very high, and the adsorbates (because of their finite mass) have no time to adjust to the rapidly fluctuating (rate $\omega \sim v/a$) substrate forces – hence the particle trajectories will be nearly straight lines. Due to the particle–particle interactions the adsorbates will therefore form a hexagonal structure as on an energetically flat surface (note: in the present case the substrate temperature and adsorbate coverage are such that the IC-solid structure is stable for the flat surface). Now, as F decreases the particle velocities will decrease and the adsorbates will, in response to the substrate forces, oscillate with increasing amplitude around the straight trajectories and, in analogy with the Lindemann melting criterion, the IC-solid lattice will melt when the displacements become large enough. The qualitative discussion above can be made more precise by a simple calculation. Let us write the coordinates for the particles in the IC-solid sliding state $(F > F_1)$ as

$$\boldsymbol{r}_i = \boldsymbol{v}t + \boldsymbol{x}_i + \boldsymbol{u}_i , \qquad (8.105)$$

where $\boldsymbol{v}$ is the drift velocity, $\boldsymbol{x}_i = (x_i, y_i)$ the perfect lattice sites (in a reference frame moving with the velocity $\boldsymbol{v}$) of the hexagonal structure, and $\boldsymbol{u}_i$ the (fluctuating) displacements away from these sites. For a large force F, $|\boldsymbol{u}_i|$ is small. Substituting (8.105) into the equation of motion (8.1) and expanding U and V to linear order in $\boldsymbol{u}_i$ gives (8.34). The displacement field $\boldsymbol{u}_i$ and the sliding friction $\bar{\eta}$ are given by (8.37) and (8.38):

$$\boldsymbol{u}_i = (\boldsymbol{u}_i)_T - \frac{kU_0}{m}\left(\boldsymbol{e}_1 \frac{\alpha_1 \sin(\omega t + x_i + \phi_1)}{[(\omega^2 - \omega_1^2)^2 + \omega^2\eta^2]^{1/2}}\right.$$
$$\left. + \boldsymbol{e}_2 \frac{\alpha_2 \sin(\omega t + x_i + \phi_2)}{[(\omega^2 - \omega_2^2)^2 + \omega^2\eta^2]^{1/2}}\right), \qquad (8.106)$$

$$\bar{\eta} = \eta + \frac{k^4 U_0^2 \eta}{2m^2}\left(\frac{\alpha_1 e_{x1}}{(\omega^2 - \omega_1^2)^2 + \omega^2\eta^2} + \frac{\alpha_2 e_{x2}}{(\omega^2 - \omega_2^2)^2 + \omega^2\eta^2}\right). \qquad (8.107)$$

At high sliding velocity ($\omega \gg \omega_1, \omega_2$) this formula reduces to (note: $\alpha_1 e_{x1} + \alpha_2 e_{x2} = 1$)

$$\bar{\eta} = \eta + \frac{U_0^2 \eta}{2m^2 v^4}. \qquad (8.108)$$

As F is reduced from a high value towards F_1, the differences $\omega^2 - \omega_1^2$ and $\omega^2 - \omega_2^2$ decrease and, according to (8.106), the magnitude of $\langle \boldsymbol{u}^2 \rangle$ increases. When this quantity becomes large enough the IC-solid state will melt. This is essentially the Lindemann melting criteria. [Technical comment: It is well known that, for a two-dimensional solid, $\langle \boldsymbol{u}_T^2 \rangle$ increases without bound as the size of the system increases (see, e.g., [8.57]). This divergence is due to the long-wavelength fluctuations in the system, which should have no influence on the melting process. The Lindemann melting criterion for a two-dimensional solid can be formulated in terms of the correlation function $\langle (\boldsymbol{u}_i - \boldsymbol{u}_j)^2 \rangle$, where i and j are the nearest neighbor sites.] It is important to note the physical mechanism behind the increase in $\langle \boldsymbol{u}^2 \rangle$ with decreasing $\omega = vk$: when $\omega = \omega_1$ (or $\omega = \omega_2$) the corrugated substrate potential exerts an oscillating force on the (sliding) elastic 2D-solid, which is resonant with one phonon mode [frequency ω_1, wave vector $\boldsymbol{q} = (k, 0)$] of the solid – hence the sliding motion can effectively pump energy into this lattice mode.

Effective Temperature During Sliding

Since the sliding state is a non-equilibrium state, it is not obvious that the concept of temperature is well-defined during sliding. However, we now show that, at high adsorbate coverages, the adsorbate–adsorbate interaction is so strong that the velocity distribution for the adsorbates, in a reference frame moving with the drift velocity of the adsorbate system, is approximately Maxwellian with nearly the same width (i.e., the same temperature) in the x and y-directions.

It is well-known that a stationary solid object embedded in a flowing fluid gives rise to energy dissipation. That is, collective translational energy of the fluid is converted into irregular motion finally leading to a heating of the fluid. Now, the same thing happens for a fluid adsorbate layer under the influence of an external force $\boldsymbol{F}$. The external force $\boldsymbol{F} = F\hat{\boldsymbol{x}}$ accelerates the adsorbates in the x-direction but due to "scattering" from the corrugated substrate potential $u(\boldsymbol{r})$ and due to the adsorbate–adsorbate interaction (as manifested in the viscosity of the fluid), "drift-momentum" is transferred into irregular motion (we saw an example of this already in Sect. 8.3.2). In the absent of thermal contact to a heat bath [i.e., with $\eta = 0$ in (8.1)] this "scrambling" of momentum would lead to a continuous increase of the temperature of the two-dimensional fluid. But in the present case, owing to the thermal contact to the substrate (at temperature T), energy will flow to the substrate at a rate which, when steady state has been reached, must equal the power transferred to the adsorbate system from the external force F. Using this fact, one can derive an expression for the effective temperature T^* of the adsorbate system as follows: The power (per adsorbate) transferred to the adsorbate system from the external force is

$$P = \boldsymbol{F} \cdot \langle \boldsymbol{v} \rangle = F^2/m\bar{\eta}, \tag{8.109}$$

where the latter equality follows from the definition $m\bar{\eta}\langle \boldsymbol{v} \rangle = \boldsymbol{F}$. But, at steady state, this power must equal that transferred from the adsorbate system to the substrate which has two contributions, namely a term $\alpha(T^* - T)$ proportional to the difference in temperature between the film and the substrate and another term, $m\eta\langle \boldsymbol{v}\rangle^2$, which describes the direct energy transfer to the substrate (via the friction η) from the drift motion of the adsorbate system. Hence,

$$P = \alpha(T^* - T) + m\eta\langle \boldsymbol{v}\rangle^2 . \tag{8.110}$$

The heat transfer coefficient α was derived in Sect. 8.8, $\alpha = 2k_B\eta$. Substituting this result into (8.110) and comparing with (8.109) gives

$$k_B T^* = k_B T + \frac{F^2}{2m\eta\bar{\eta}}\left(1 - \frac{\eta}{\bar{\eta}}\right) . \tag{8.111}$$

In the dimensionless variables introduced in Sect. 8.3 this equation takes the form

$$\frac{T^*}{T} = 1 + \frac{F^2}{2\eta^2}\left(1 - \frac{\eta}{\bar{\eta}}\right)\frac{\eta}{\bar{\eta}} . \tag{8.112}$$

But the temperature T^* can also be obtained directly from the simulations as follows: Let us write the velocity of an adsorbate as

$$\boldsymbol{v} = \langle \boldsymbol{v} \rangle + \delta\boldsymbol{v} ,$$

where $\langle \delta\boldsymbol{v} \rangle = \boldsymbol{0}$. Hence,

$$\langle v^2 \rangle = \langle \boldsymbol{v}\rangle^2 + \langle (\delta\boldsymbol{v})^2 \rangle .$$

Now, *if* the motion $\delta\boldsymbol{v}$ is to correspond to the film temperature T^* then $k_B T^* = m\langle(\delta\boldsymbol{v})^2\rangle/2$ so that

$$k_B T^* = m\left(\langle v^2 \rangle - \langle \boldsymbol{v}\rangle^2\right)/2 .$$

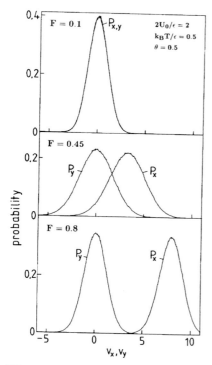

Fig. 8.46. The probability distributions P_x and P_y of adsorbate velocities for $F = 0.1$, 0.45, and 0.8 and for $\theta = 0.5$, $2U_0/\epsilon = 2$, and $k_B T/\epsilon = 0.5$. Both $P_x(v_x)$ and $P_y(v_y)$ are Maxwellian and are of equal width.

The temperature deduced using this equation agrees to within the "noise" of the simulations with that deduced from (8.111). For example, for $F = 0.42$ equation (8.111) gives $T^*/T = 2.952$ while (8.112) gives $T^*/T = 2.942$. In order for the motion $\delta \boldsymbol{v}(t)$ to really correspond to a temperature it is necessary to show that both δv_y and δv_x have Maxwellian probability distributions of equal width, i.e.,

$$P_x(v_x) = \left(\frac{m}{2\pi k_B T^*}\right)^{1/2} e^{-m(v_x - \langle v_x \rangle)^2 / 2k_B T^*},$$

$$P_y(v_y) = \left(\frac{m}{2\pi k_B T^*}\right)^{1/2} e^{-m v_y^2 / 2k_B T^*}.$$

That this indeed is the case is shown in Fig. 8.46 for $F = 0.45$. Equally good Maxwellian distributions are found for all other values of F and the effective temperature T^* deduced from the width of the probability distributions P_x and P_y never deviates by more than 3%.

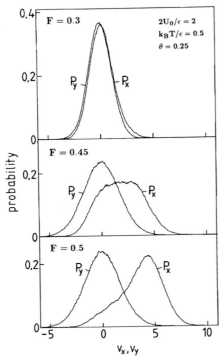

Fig. 8.47. The probability distributions P_x and P_y of adsorbate velocities for $F = 0.3$, 0.45, and 0.5 and for $\theta = 0.25$, $2U_0/\epsilon = 2$, and $k_B T/\epsilon = 0.5$.

We have shown above that for $\theta = 0.5$, when the adsorbate system is in a fluidized state, the probability distributions for $\delta v_x = v_x - \langle v_x \rangle$ and v_y are Maxwellian with equal width. Nevertheless, the adsorbate–substrate system is not in thermal equilibrium but rather in a steady state with the temperature T^* of the adsorbate system being much higher than the substrate temperature T. The reason why thermal equilibrium occurs within the adsorbate system is related to the high adsorbate coverage and concordant strong adsorbate–adsorbate interaction which tend to "randomize" the adsorbate velocities. But at low adsorbate coverage thermal equilibrium does not occur within the adsorbate layer. This is illustrated in Fig. 8.47 which shows the velocity distributions P_x and P_y for $\theta = 0.25$. In this case P_y is well described by a Maxwellian distribution. But P_x is strongly non-Maxwellian, i.e., the adsorbate–adsorbate interaction is not strong enough to "thermalize" the energy input from the external force $\boldsymbol{F}$.

8.9.2 Dynamical Phase Transitions in Adsorbate Layers: Frenkel–Kontorova Model

In the model studied above, the adsorbates interacted with each other via Lennard–Jones pair potentials. However, it was argued that the observed sliding scenario should be very general, and not depend on the details of the model. For example, analytical arguments were presented showing that the nature of the (IC-solid ↔ fluid) and (fluid → C-solid) transitions does not depend on the exact form of the adsorbate–adsorbate and adsorbate–substrate interactions. Similarly, the physical origin of the hysteresis in Figs. 8.33 and 8.34 is well understood and the mechanism is very general. In this section we present results of computer simulations [8.65] for a drastically different model, namely a 2D Frenkel–Kontorova model. We show that this model exhibits very similar dynamical phase transitions to the Lennard–Jones model, lending support for the existence of a universal sliding scenario.

Model

In the Frenkel–Kontorova model, the particles are coupled to each other by elastic springs and move on a periodic substrate potential. The displacements of the individual atoms is limited to the x-direction. Thus the equation of motion is

$$m\ddot{x}_n + m\eta\dot{x}_n = -\frac{\partial U}{\partial x_n} - \frac{\partial V}{\partial x_n} + F + f_n , \qquad (8.113)$$

where

$$U = U_0 \sum_n (1 - \cos k x_n) \qquad (8.114)$$

$$V = \sum_n \left[\frac{1}{2} K_x (x_{n+\hat{x}} - x_n - b)^2 + \frac{1}{2} K_y (x_{n+\hat{y}} - x_n)^2 \right] \qquad (8.115)$$

with $k = 2\pi/a$. The fluctuating force f_n in (8.113) is related to the microscopic friction η and to the substrate temperature T via the fluctuation–dissipation theorem

$$\langle f_n(t) f_{n'}(t') \rangle = 2\eta m k_\mathrm{B} T \delta_{nn'} \delta(t - t') . \qquad (8.116)$$

The index $n = (i, j)$ labels the particle ($i = 1, \ldots, N_x; j = 1, \ldots, N_y$), so that the system consists of $N_x \times N_y$ adsorbates. In the simulations $K_x = K_y = K$, $b = 2a$, $N_x = N_y = 10$ and periodic boundary conditions are used.

Results and Discussion

At "low" temperature and for $F = 0$ the ordered commensurate structure shown in Fig. 8.48a is formed when $b = 2a$ (Fig. 8.48b is a schematic representation of Fig. 8.48a). When the temperature increases above the critical

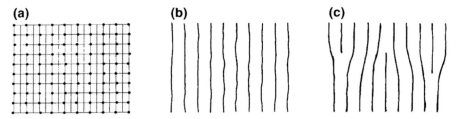

Fig. 8.48. (a) Snapshot of adsorbate layer in the commensurate solid phase, for $T = U_0/k_B$ and $b = 2a$. (b) Schematic representation of (a). (c) Snapshot of the adsorbate layer in the fluid state. From [8.65].

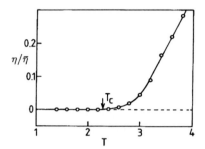

Fig. 8.49. Inverse of the normalized sliding friction $\bar{\eta}$ as a function of temperature T for $\eta = 0.5$ and $b = 2a$. From [8.65].

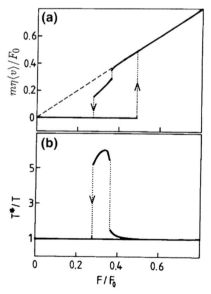

Fig. 8.50. (a) Drift velocity $\langle v \rangle$ and (b) effective temperature T^* as a function of the external force F for $\eta = 0.6$ and $T = 0.5\, U_0/k_B$. From [8.65].

temperature the adsorbate layer melts. The melting occurs by the formation of free dislocations as shown in the snapshot, Fig. 8.48c. The melting temperature T_c can be determined by studying the drift velocity $\langle v \rangle$ as a function of temperature when a weak external force F acts on the particle system. Figure 8.49 shows that in the present case $T_c \approx 2.3\,U_0/k_B$.

At zero temperature, the drift velocity $\langle v \rangle$ remains zero for any applied force less than $F_0 = kU_0$. Figure 8.50a shows the normalized drift velocity $\langle v \rangle/v_0$ as function of the external force F when $T = 0.5\,U_0/k_B$. Initially, the adlayer is in a pinned state and the drift velocity vanishes. When $F = F_a \approx F_0/2$ the drift velocity increases abruptly and, with increasing F, follows the dashed line in Fig. 8.50a, which describes sliding on an energetically uncorrugated surface (i.e., $U_0 = 0$). As in the Lennard–Jones model the sliding friction exhibits hysteresis and a (fluid ↔ solid) dynamical phase transition occurs for $F = F_1 \approx 0.36\,F_0$. The effective temperature T^* in units of the substrate temperature is shown in Fig. 8.50b. Note that the initial sliding phase ($F > F_a$) has almost the same temperature as the substrate and the substrate potential provides no additional resistance. This is a solid sliding phase and the arguments given in Sect. 8.9.1 as to why $T^* \approx T$ are still valid. As F is decreased to F_1 there is a jump in T^* from $T^* \approx 1.4\,T$ to $5.5\,T$, accompanied by a simultaneous decrease in $\langle v \rangle$. This corresponds to a dynamically induced melting transition. Finally, at $F = F_b \approx 0.29\,F_0$ the fluid phase condenses into the commensurate solid phase and the temperatures of the adlayer and the substrate become the same again. Note that the condensation occurs when the temperature in the fluid adlayer reaches the melting temperature of the pinned commensurate phase.

The results above are very similar to those of the Lennard–Jones model, lending further support to the idea that the qualitative nature of the non-linear response is universal and not valid only for specific models. One difference between the present model and the Lennard–Jones model is the dependence of F_b/F_a on the microscopic friction η. In the present case $F_b/F_a = 1$ for "large" η (Fig. 8.51), while in the Lennard–Jones model $F_b/F_a \approx 0.6$ for large η. This must be due to the fact that the nature of the fluid phase in the two models is very different. The hydrodynamics arguments given in Sect. 8.9.1 are based on the existence of a drag force acting on a nucleating domain and are not valid in the present case.

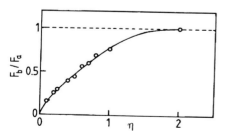

Fig. 8.51. Ratio F_b/F_a as a function of η for $T = 0.5\,U_0/k_B$. From [8.65].

8.10 Role of Defects at High Sliding Velocity and for Small Amplitude Vibrations*

In this section we study the influence of defects on the sliding friction for 1D, 2D and 3D elastic solids [8.66]. The model is presented in Sect. 8.10.1 and in Sect. 8.10.2 we calculate the leading contribution to the sliding friction at high sliding velocities. We show that for 1D and 2D solids perturbation theory breaks down at low sliding velocities. In Sect. 8.10.3 we study the sliding friction for small amplitude vibrations.

8.10.1 Model

We develop the model for incommensurate monolayers of Xe on silver surfaces. We consider the same model as studied earlier, where, in addition to the adsorbate–substrate interaction potential $U = \sum_i u(\mathbf{r}_i)$ and the adsorbate–adsorbate interaction potential $V = \frac{1}{2}\sum'_{ij} v(\mathbf{r}_i - \mathbf{r}_j)$, an external force $\mathbf{F}$ acts on each of the adsorbates. The equation of motion for the particle coordinate $\mathbf{r}_i(t)$ is taken to be

$$m\ddot{\mathbf{r}}_i + m\eta\dot{\mathbf{r}}_i = -\frac{\partial U}{\partial \mathbf{r}_i} - \frac{\partial V}{\partial \mathbf{r}_i} + \mathbf{f}_i + \mathbf{F} \; . \tag{8.117}$$

We have shown earlier that for a compressed IC-monolayer film of Xe on silver the substrate corrugation may be unimportant and we therefore take U to correspond to the interaction with surface defects only. Let us write

$$U = \sum_{jn} U_\mathrm{d} f(\mathbf{r}_j - \mathbf{R}_n) \; , \tag{8.118}$$

where $\mathbf{R}_n$ denotes the position vectors of the defects. The strength of the defect potential is denoted by U_d and the "form factor" by $f(\mathbf{x})$. For later use we will need $f(\mathbf{x})$ expanded in plane waves:

$$f(\mathbf{x}) = \int d^2q \, f(q) \mathrm{e}^{i\mathbf{q}\cdot\mathbf{x}} \; . \tag{8.119}$$

For example, if $f(\mathbf{x}) = \exp(-\alpha x^2)$ then

$$f(q) = \frac{1}{4\pi\alpha} \mathrm{e}^{-q^2/4\alpha} \; . \tag{8.120}$$

Using (8.118) and (8.119) we get

$$U = U_\mathrm{d} \sum_{jn} \int d^2q \, f(q) \mathrm{e}^{i\mathbf{q}\cdot(\mathbf{r}_j - \mathbf{R}_n)} \; . \tag{8.121}$$

8.10 Role of Defects

Let us now consider a solid adsorbate layer sliding with the velocity v on a substrate. Let us write the coordinates for the particles in the sliding state as

$$r_i = vt + x_i + u_i ,\qquad(8.122)$$

where v is the drift velocity, $x_i = (x_i, y_i)$ the perfect lattice sites (in a reference frame moving with the velocity v) of the hexagonal structure, and u_i the (fluctuating) displacements away from these sites. Substituting (8.122) into the equation of motion (8.117) and expanding V to quadratic order in u_i gives

$$m\ddot{u}_i + m\eta\dot{u}_i + \sum_j K_{ij} \cdot u_j$$
$$= f_i + F - m\eta v - \frac{\partial U}{\partial x_i}(x_i + vt + u_i) ,\qquad(8.123)$$

where the force-constant matrix K_{ij} has the components $K_{ij}^{\alpha\beta} = \partial^2 V/\partial u_i^\alpha \partial u_j^\beta$. Let us introduce the matrix $K(q)$ with the components

$$K^{\alpha\beta}(q) = \sum_j K_{ij}^{\alpha\beta} e^{-i q \cdot (x_i - x_j)} .$$

For particles interacting via Lennard–Jones potentials, an explicit expression for $K(q)$ was given in Sect. 8.5. Here we will only need the fact that, in the long-wavelength limit to quadratic order in q,

$$K^{\alpha\beta} = mc_T^2 q^2 \delta_{\alpha\beta} + m\left(c_L^2 - c_T^2\right) q_\alpha q_\beta .\qquad(8.124)$$

For steady sliding at small velocities v, only vibrations with large wavelength $\lambda \sim (c/v)a$, where a is the lattice constant and c is the sound velocity, can be exited in the adsorbed film. If $\lambda \gg a$ (i.e., $v \ll c$) one can neglect the discrete nature of the film and use the elastic continuum model. In the continuum limit $u_i(t) \to u(x,t)$, where x refers to a frame in which the defects are stationary. Using (8.124) the left hand side of (8.123) takes the form

$$m\left(D^2 u + \eta D u - c_T^2 \nabla^2 u - \left(c_L^2 - c_T^2\right)\nabla\nabla\cdot u\right)$$

where

$$D = \partial/\partial t + v\cdot\nabla .$$

The two terms in D scales as $v/a : v/\lambda$, i.e., as $1 : v/c$, and for $v \ll c$ we can neglect the gradient term so that the left-hand side of (8.123) takes the form

$$m\left(\frac{\partial^2 u}{\partial t^2} + \eta\frac{\partial u}{\partial t} - c_T^2 \nabla^2 u - \left(c_L^2 - c_T^2\right)\nabla\nabla\cdot u\right) .$$

The fluctuating force $\boldsymbol{f}_i(t) \to \boldsymbol{f}(\boldsymbol{x},t)/n_\mathrm{a}$ where $\boldsymbol{f}$ is the fluctuating force per unit area (n_a is the number of adsorbates per unit area). In the continuum limit, (8.3) takes the form

$$\langle f_\alpha(\boldsymbol{x},t) f_\beta(\boldsymbol{x}',t') \rangle = 2\rho k_\mathrm{B} T \eta \delta_{\alpha\beta} \delta(\boldsymbol{x}-\boldsymbol{x}') \delta(t-t') , \qquad (8.125)$$

where $\rho = m n_\mathrm{a}$. Finally, note that in the continuum limit the potential

$$U \to \sum_{jn} U_\mathrm{d} f(\boldsymbol{x}_j + \boldsymbol{v}t + \boldsymbol{u}(\boldsymbol{R}_n,t) - \boldsymbol{R}_n)$$
$$= n_\mathrm{a} U_\mathrm{d} (2\pi)^2 \sum_{nG} f(\boldsymbol{G}) e^{i\boldsymbol{G}\cdot[\boldsymbol{v}t + \boldsymbol{u}(\boldsymbol{R}_n,t) - \boldsymbol{R}_n]} .$$

In the simplest case one includes only the contribution from the lowest reciprocal lattice vectors $\{\boldsymbol{G}\}$ in this expansion. Thus, for example, in the 1D case, which we consider in detail in Sect. 10.4,

$$U = \mathrm{const} + 2 U_\mathrm{d} k f(k) \sum_n \cos k\left[vt + u(R_n,t) - R_n\right] ,$$
$$= \mathrm{const} + 2 U_\mathrm{d} k f(k) \sum_n \int dx \cos k[vt + u(x,t) - R_n] \delta(x - R_n), \quad (8.126)$$

where $k = 2\pi/a$, and a is the lattice constant. Thus the force

$$-\frac{\delta U}{\delta u(x)} = 2 U_0 k^2 f(k) \sum_n \sin(k[vt + u(R_n,t) - R_n]) \delta(x - R_n) .$$

Multiplying (8.123) by $n_\mathrm{a} = 1/a$ gives in the 1D case the continuum equation

$$\rho \left(\frac{\partial^2 u}{\partial t^2} + \eta \frac{\partial u}{\partial t} - c^2 \frac{\partial^2 u}{\partial x^2} \right) = f(x,t) + n_\mathrm{a} F - \rho \eta v$$
$$- \lambda \sum_n \sin(k[vt + u(R_n,t) - R_n]) \delta(x - R_n) , \qquad (8.127)$$

where

$$\lambda = 2 U_\mathrm{d} k^2 f(k) .$$

8.10.2 High Sliding Velocities

Let us calculate the leading contribution to the sliding friction for high sliding velocities and zero temperature. When the sliding velocity v is very high, the particles have no time to adjust to the rapidly fluctuating forces from the pinning centers – hence the particle trajectories will be nearly straight lines. Due to the particle–particle interactions the system will therefore form a nearly perfect hexagonal structure as expected in the absence of pinning centers. Now, as v decreases the particles will, in response to the forces from the pinning centers, oscillate with increasing amplitude around the perfect lattice sites. The treatment presented below is only valid as long as the displacement of the atoms away from the lattice sites is small compared to the lattice constant.

For large v, $|\boldsymbol{u}_i|$ is small. Substituting (8.121) into (8.123) and expanding to linear order in $\boldsymbol{u}_i$ gives

$$m\ddot{\boldsymbol{u}}_i + m\eta\dot{\boldsymbol{u}}_i + \sum_j K_{ij} \cdot \boldsymbol{u}_j = m(\bar{\eta} - \eta)\boldsymbol{v} - U_\mathrm{d} \int d^2q G(\boldsymbol{q}) f(q) i\boldsymbol{q} e^{i\boldsymbol{q}\cdot(\boldsymbol{x}_i + \boldsymbol{v}t)}$$

$$+ U_\mathrm{d} \int d^2q G(\boldsymbol{q}) f(q) \boldsymbol{q}\boldsymbol{q} \cdot \boldsymbol{u}_i e^{i\boldsymbol{q}\cdot(\boldsymbol{x}_i + \boldsymbol{v}t)}, \quad (8.128)$$

where

$$G = \sum_m e^{-i\boldsymbol{q}\cdot\boldsymbol{R}_m} .$$

Let $\langle \ldots \rangle$ stand for time averaging and for averaging over the random distributions of impurity centers. We have $\langle G \rangle = 0$ and

$$\langle G(\boldsymbol{q}) G(\boldsymbol{q}') \rangle = N_\mathrm{d} \delta_{\boldsymbol{q}+\boldsymbol{q}',0} \to (2\pi)^2 n_\mathrm{d} \delta(\boldsymbol{q}+\boldsymbol{q}') , \quad (8.129)$$

where N_d and n_d are the total number of defects and the number of defects per unit area, respectively. Furthermore since $\langle \boldsymbol{u}_i \rangle = 0$, we get from (8.128)

$$0 = m(\bar{\eta} - \eta)\boldsymbol{v} - U_\mathrm{d} \int d^2q f(q) \boldsymbol{q}\boldsymbol{q} \cdot \langle G(\boldsymbol{q}) \boldsymbol{u}_i \rangle e^{i\boldsymbol{q}\cdot(\boldsymbol{x}_i + \boldsymbol{v}t)} . \quad (8.130)$$

To linear order in U_d we obtain from (8.128)

$$\boldsymbol{u}_i = -\frac{iU_0}{m} \int d^2q \frac{G(\boldsymbol{q}) f(q) e^{i\boldsymbol{q}\cdot(\boldsymbol{x}_i + \boldsymbol{v}t)}}{(\boldsymbol{q}\cdot\boldsymbol{v})^2 + i\eta\boldsymbol{q}\cdot\boldsymbol{v} - \omega^2(\boldsymbol{q})} \cdot \boldsymbol{q} , \quad (8.131)$$

where $\omega^2(\boldsymbol{q})$ is a matrix with the components $K^{\alpha\beta}(\boldsymbol{q})/m$, which, in the long wavelength limit, takes the form [see (8.124)]

$$c_\mathrm{T}^2 q^2 \delta_{\alpha\beta} + (c_\mathrm{L}^2 - c_\mathrm{T}^2) q_\alpha q_\beta .$$

Let $\boldsymbol{e}_s(\boldsymbol{q})$ denote the eigenvectors of the matrix $\omega^2(\boldsymbol{q})$, corresponding to the eigenvalues $\omega_s^2(\boldsymbol{q})$. Expanding $\boldsymbol{q} = \sum_s \boldsymbol{q}\cdot\boldsymbol{e}_s\boldsymbol{e}_s$ and substituting into (8.131) gives

$$\boldsymbol{u}_i = -\frac{iU_\mathrm{d}}{m} \sum_s \int d^2q \frac{G(\boldsymbol{q}) f(q) e^{i\boldsymbol{q}\cdot(\boldsymbol{x}_i + \boldsymbol{v}t)}}{(\boldsymbol{q}\cdot\boldsymbol{v})^2 + i\eta\boldsymbol{q}\cdot\boldsymbol{v} - \omega_s^2(\boldsymbol{q})} \boldsymbol{q}\cdot\boldsymbol{e}_s\boldsymbol{e}_s . \quad (8.132)$$

Using (8.129) and (8.132) gives

$$\langle u^2 \rangle = \frac{(2\pi)^2 U_\mathrm{d}^2 n_\mathrm{d}}{m^2} \sum_s \int d^2q \frac{(\boldsymbol{q}\cdot\boldsymbol{e}_s)^2 |f(q)|^2}{[(\boldsymbol{q}\cdot\boldsymbol{v})^2 - \omega_s^2(\boldsymbol{q})]^2 + (\eta\boldsymbol{q}\cdot\boldsymbol{v})^2} . \quad (8.133)$$

Substituting (8.132) into (8.130), using (8.129), and assuming $\boldsymbol{v} = \hat{x}v$ gives, after some simplifications,

$$\bar{\eta} = \eta + \frac{(2\pi)^2 U_{\mathrm{d}}^2 n_{\mathrm{d}}}{m^2} \sum_s \int d^2q \, \frac{\eta q_x^2 (\boldsymbol{q}\cdot\boldsymbol{e}_s)^2 \mid f(q) \mid^2}{[q_x^2 v^2 - \omega_s^2(\boldsymbol{q})]^2 + \eta^2 q_x^2 v^2} \, . \tag{8.134}$$

We assume that η is "small", and take the limit $\eta \to 0$ when evaluating the integral in (8.134). Using

$$\frac{\Gamma}{x^2 + \Gamma^2} \to \pi\delta(x) \quad \text{as} \quad \Gamma \to 0 \, ,$$

we obtain

$$\bar{\eta} = \eta + \frac{(2\pi)^2 U_{\mathrm{d}}^2 n_{\mathrm{d}}}{m^2 v} \sum_s \int d^2q \mid q_x \mid (\boldsymbol{q}\cdot\boldsymbol{e}_s)^2 \mid f(q) \mid^2 \pi\delta[q_x^2 v^2 - \omega_s^2(\boldsymbol{q})] \, .$$

Now, note that $\omega_s(\boldsymbol{q}+\boldsymbol{G}) = \omega_s(\boldsymbol{q})$. If we assume $v \ll c_{\mathrm{L}}, c_{\mathrm{T}}$, the contributions to the integral in (8.134) will occur only for $\boldsymbol{q} \approx \boldsymbol{G}$ with $G_x \neq 0$. We find

$$\bar{\eta} = \eta + \frac{(2\pi^2)^2 U_{\mathrm{d}}^2 n_{\mathrm{d}}}{m^2 v} {\sum_{s\boldsymbol{G}}}' \mid G_x \mid (\boldsymbol{G}\cdot\boldsymbol{e}_s)^2 \mid f(G) \mid^2 /c_s^2 \, , \tag{8.135}$$

where the prime on the summation symbol indicates that in the sum over $\boldsymbol{G}$, the reciprocal lattice vectors with $G_x = 0$ are excluded. Similarly using (8.133) we obtain

$$\langle \boldsymbol{u}^2 \rangle = \frac{(2\pi^2)^2 U_{\mathrm{d}}^2 n_{\mathrm{d}}}{m^2 \eta v} {\sum_{s\boldsymbol{G}}}' [(\boldsymbol{G}\cdot\boldsymbol{e}_s)^2/\mid G_x \mid] \mid f(G) \mid^2 /c_s^2 \, . \tag{8.136}$$

In a similar way one obtains for a 1D-solid:

$$\bar{\eta} = \eta + \frac{2\pi^2 U_{\mathrm{d}}^2 n_{\mathrm{d}}}{m^2 c v^2} {\sum_G}' G^2 \mid f(G) \mid^2 \tag{8.137}$$

and

$$\langle u^2 \rangle = \frac{2\pi^2 U_{\mathrm{d}}^2 n_{\mathrm{d}}}{m^2 \eta c v^2} {\sum_G}' (G^2/G_x^2) \mid f(G) \mid^2 \, , \tag{8.138}$$

and for a 3D-solid:

$$\bar{\eta} = \eta + \frac{(2\pi)^5 U_{\mathrm{d}}^2 n_{\mathrm{d}}}{2m^2} {\sum_{s\boldsymbol{G}}}' G_x^2 (\boldsymbol{G}\cdot\boldsymbol{e}_s)^2 \mid f(G) \mid^2 /c_s^3 \tag{8.139}$$

and

$$\langle \boldsymbol{u}^2 \rangle = \frac{(2\pi)^5 U_{\mathrm{d}}^2 n_{\mathrm{d}}}{2m^2 \eta} {\sum_{s\boldsymbol{G}}}' (\boldsymbol{G}\cdot\boldsymbol{e}_s)^2 \mid f(G) \mid^2 /c_s^3 \, . \tag{8.140}$$

The most important result obtained above is the velocity dependence of $\bar{\eta} - \eta$ and of $\langle u^2 \rangle$ which we summarize:

$$1\text{D}: \propto 1/v^2, \quad 2\text{D}: \propto 1/v, \quad 3\text{D}: \propto \text{const}.$$

Now, the expansion of U in $\boldsymbol{u}_i$, on which the derivation of (8.135–140) is based, is only valid if $\langle \boldsymbol{u}^2 \rangle \ll a^2$ where a is the lattice constant. Hence it is clear that in the 3D-case, if U_d is small enough or η large enough, since $\langle u^2 \rangle$ is independent of v the results (8.139) and (8.140) are *valid for all sliding velocities* and the friction force $F_\text{f} = m\bar{\eta} v \sim v$ for all v (but $v \ll c$). [This statement is actually not quite true: In the calculation above we have neglected the modifications in the adsorbate–substrate interaction due to the motion of the substrate atoms. This is an excellent approximation at high sliding velocity but breaks down at low sliding velocity. An analysis shows that at very low sliding velocity the friction force is again proportional to the velocity but the proportionality factor is different to that in the high velocity region due to (adiabatic) renormalization of the adsorbate–substrate coupling, see next section.] However for the 1D and 2D-systems, $\langle u^2 \rangle$ diverges as $v \to 0$, and it is clear that *independent of how weak the defect potential U_d is, at low enough velocities v the expansion on which (8.135–138) are based will break down*. Physically, this difference between 3D systems and 1D and 2D systems is related to the "softness" of the elastic properties of low-dimensional solids: in 1D and 2D-solids a force applied at some point in the solid gives rise to an infinitely large elastic displacement (for a finite system of linear size L the elastic displacement scales as $u \propto L$ and $\propto \ln L$ for 1D and 2D-solids, respectively).

The condition given above for the validity of the high-velocity expansion, namely that $\langle u^2 \rangle \ll a^2$, is the correct one only if the defect concentration n_d is large enough. The point is that $\langle u^2 \rangle$ is the *average* over all the atoms in the sliding lattice, while the condition for the validity of the high velocity expansion is that the displacement u_i of *all* atoms in the lattice must be small compared to the lattice constant a. Since the displacement u_i of an atom which is (temporary) close to a defect is likely to be larger than for an atom far away, it is clear that the condition $\langle u^2 \rangle \ll a^2$ may not always guarantee the validity of the high-v expansion. To illustrate this, let us first consider the 1D-case. For a single defect it is easy to show from the 1D-version of (20b) that the maximum displacement u_i of an atom when it is close to a defect is given by

$$u_\text{max} = U_\text{d} \frac{\pi}{2mcv} \sum_G f(G).$$

Thus $\langle u^2 \rangle \sim u_\text{max}^2$ gives $n_\text{d} \sim \eta/4c$, so that for $n_\text{d} > \eta/4c$ the condition $\langle u^2 \rangle \ll a^2$ guarantee that the high-v expansion holds. Similarly, for a 2D-system one can show that if the defect concentration $n_\text{d} > 4\pi \eta v/c^2 a$, then the condition $\langle u^2 \rangle \gg a^2$ will guarantee the validity of the high-velocity expansion.

For Xe on a silver surface $\eta \sim 3 \times 10^8 \, \text{s}^{-1}$, the sound velocity $c \sim 500/\text{s}$ and with the typical sliding velocity $v \sim 1 \, \text{cm/s}$, the inequality above reduces to $n_a > 10^{-8} \, \text{Å}^{-2}$, which usually is satisfied in QCM-friction measurements.

8.10.3 Small Amplitude Vibrations

Another limiting case which can be solved exactly is the friction associated with small amplitude vibrations. We assume again a low concentration of randomly distributed defects. In this case the defects will contribute independently of each other and it is sufficient to calculate the friction force associated with a single defect. This is most conveniently done as follows. Let ξ denote the coordinate of a defect (we consider first the 1D-case so that ξ is a scalar), see Fig. 8.52. Assume that the defect is connected to the 1D elastic solid at point $x = 0$ via a spring with force constant K which is determined by expanding $U(\boldsymbol{x})$ to second-order in $\boldsymbol{x}$. We assume that the defect performs vibrations relative to the elastic solid and calculate the force acting on the defect due to the solid. This force should be the same as the force acting on a stationary defect if the elastic solid performs (translational) vibrations (with the same amplitude and frequency) relative to the defects.

The force on the defect from the elastic solid is

$$F_0(t) = -K[\xi - u(0, t)] \, .$$

The defect exerts a force $K[\xi - u(0, t)]$ on the elastic solid at point $x = 0$ so that the equation of motion of the elastic solid becomes

$$\rho \left(\frac{\partial^2 u}{\partial t^2} + \eta \frac{\partial u}{\partial t} - c^2 \frac{\partial^2 u}{\partial x^2} \right) = K[\xi - u(0, t)] \delta(x) \, . \tag{8.141}$$

This equation is easy to solve:

$$\begin{aligned} u(0, \omega) &= K[\xi(\omega) - u(0, \omega)] \frac{1}{2\pi} \int dq \frac{1}{\rho(-\omega^2 - i\eta\omega + c^2 q^2)} \\ &= A[\xi(\omega) - u(0, \omega)] \, , \end{aligned} \tag{8.142}$$

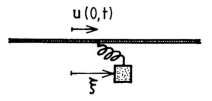

Fig. 8.52. A defect connected to a 1D elastic solid via a spring (force constant K).

where, in the limit $\eta \to 0$,

$$A = \mathrm{i}K/2c\rho\omega \ .$$

It is easy to solve (8.142) for $u(0,\omega)$:

$$u(0,\omega) = \frac{A\xi(\omega)}{1+A} \ ,$$

so that the force acting on the defect

$$F_0(\omega) = -K[\xi(\omega) - u(0,\omega)] = -\frac{K\xi(\omega)}{1+A} \to \mathrm{i}\omega\xi 2\rho c \tag{8.143}$$

as $\omega \to 0$. Thus

$$F_0(t) = -2\rho c \dot{\xi} \ . \tag{8.144}$$

It is remarkable that the force F which acts on the defect is *independent* of the original coupling strength (spring constant K) and that it is purely dissipative. That is, no *elastic* restoring force occurs for a 1D-solid. The physical origin of this is related to the fact that an arbitrarily weak force applied to a point of a 1D elastic solid gives rise to an infinite displacement (for a finite 1D solid the displacement is proportional to the linear size L of the solid). The physical origin of the friction force, $-2\rho c\dot{\xi}$, will be discussed in Sect. 10.4. If we have N_d defects the total friction force $N_\mathrm{d} 2\rho cv = Nm\bar{\eta}v$, or

$$\bar{\eta} = \eta + 2n_\mathrm{d} c \ , \tag{8.145}$$

where we have added the contribution η from the direct coupling to the substrate and used the fact that $N_\mathrm{d}\rho/Nm = n_\mathrm{d}$.

In a similar way as above it is easy to calculate the friction force for small amplitude vibrations for 2D and 3D-solids and here we only quote the results for low frequencies. For a 3D-system

$$F_0(\omega) = -\frac{K\xi(\omega)}{1+A(\omega)}$$

where

$$\begin{aligned}A &= \frac{K}{2\pi^2 \rho c_\mathrm{T}^2} \int_0^{q_c} dq \frac{q^2}{q^2 - \omega^2/c_\mathrm{T}^2 - \mathrm{i}0^+} \left(1 + \frac{q^2(c_\mathrm{T}^2/c_\mathrm{L}^2 - 1)/3}{q^2 - \omega^2/c_\mathrm{L}^2 - \mathrm{i}0^+}\right) \\ &= B(\omega) + \mathrm{i}\frac{K\omega}{6\pi \rho c_\mathrm{T}^3}\left[1 + \frac{1}{2}\left(\frac{c_\mathrm{T}}{c_\mathrm{L}}\right)^3\right] \ , \end{aligned} \tag{8.146}$$

Note that $B = \mathrm{Re}A$ and $B(\omega) \to \mathrm{const}$ as $\omega \to 0$. Thus

$$F_0(t) = -\bar{K}\xi - \frac{\bar{K}^2}{6\pi\rho c_\mathrm{T}^3}\left[1 + \frac{1}{2}\left(\frac{c_\mathrm{T}}{c_\mathrm{L}}\right)^3\right]\dot{\xi} \ , \tag{8.147}$$

where the renormalized spring constant $\bar{K} = K/[1 - B(0)]$. We note that a very similar expression for the friction force has been derived for adsorbates performing parallel or perpendicular vibrations on surfaces of semi-infinite elastic solids (see (8.12) and Ref. [8.2]).

For a 2D-system

$$A = \frac{K}{2\pi \rho c_T^2} \int_0^{q_c} dq \frac{q}{q^2 - \omega^2/c_T^2 - i0^+} \left(1 + \frac{q^2(c_T^2/c_L^2 - 1)/2}{q^2 - \omega^2/c_L^2 - i0^+}\right)$$

$$= \frac{K}{4\pi \rho c_T^2} \left[\ln\left(\frac{q_c c_T^2}{\omega c_L}\right) + \frac{c_T^2}{c_L^2} \ln\left(\frac{q_c c_T}{\omega}\right)\right] + i\frac{K}{8\rho c_T^2} \left(1 + \frac{c_T^2}{c_L^2}\right). \quad (8.148)$$

Note that the real term in this expression diverges logarithmically when $\omega \to 0$. This implies that the elastic restoring force vanishes as $\omega \to 0$. The physical origin of this is again the "softness" of 2D elastic solids where the displacement of a point P, in response to a weak force applied at P, diverges logarithmically with the linear size L of the solid. However, the logarithmic divergence in (8.148) is of no practical importance because any real measurement involves finite frequencies ω (the external force has to be turned on and turned off and this alone generates finite frequencies), and also because any real system has a finite size L which implies that the q-integral in (8.148) must be cut-off at $q_{\min} \sim 1/L$, rather than extended right down to zero, and this removes the $\omega \to 0$ divergence of (8.148). Thus, for a 2D-solid the small-amplitude friction force has the form

$$F_0(\omega) = -K\xi/(1 + A) = -(K_1 + iK_2)\xi, \quad (8.149)$$

where both the real part (elastic spring constant) K_1 and the imaginary part K_2 are finite as $\omega \to 0$.

8.11 Friction and Superconductivity: a Puzzle*

The frictional forces experienced by thin physisorbed layers of inert atoms or molecules, such as Kr, Xe or N_2, while sliding on metallic surfaces, can be measured using the Quartz-Crystal Microbalance (QCM); see Sect. 3.3. The frequency shift and the change in the damping of the QCM oscillator upon adsorption of the gas on the metal film provide direct information on the molecule–surface frictional processes. In this manner, both the phononic and electronic contribution to friction can be accessed. In particular, when the substrate is a metal that can be cooled down below the superconducting T_c, one can gauge the importance of the electronic contribution, as this alone should presumably change below T_c. This is precisely what was done in a very recent experiment, where about 1.6 ML of N_2 was adsorbed on a lead film, which can be cooled below the lead film superconducting temperature $T_c \approx 7K$ (see Fig. 3.25) [8.69]. In this section we present a discussion of sliding

8.11 Friction and Superconductivity: A Puzzle*

friction on superconductors. We first consider the frictional force acting on atoms or molecules with permanent charges or dipole moments, sliding on a metal in its normal state. Next we discuss how this electronic friction is modified as the metal becomes superconducting. We show that no mechanism presented so far is able to explain the origin of the very sharp drop in the friction observed at T_c (see Fig. 3.25).

Electronic Friction for Point Charges and Dipoles

As discussed in Sect. 8.2, for a molecule sliding above a metal surface there are many sources of the electronic friction. In principle, these contributions are not additive but will interfere. But in many cases one of the limiting mechanisms will dominate. In this section we consider the most long-range contribution which arises if the molecules have net charges or dipole moments. We note that, because of the broken inversion symmetry which occurs in the vicinity of the surface, molecules (or atoms) close to a metal surface may have net charges or dipole moments, even if these quantities vanish for the isolated molecules.

Surface Response Function $g(q,\omega)$

Let us first define the surface response function $g(q,\omega)$ which is the central quantity for most processes at a surface involving a time-dependent electric field [8.70]. Assume that a metal occupies the region $z < 0$, and that a (time-dependent) charge distribution exists for $z > d$ outside the metal, see Fig. 8.53. This charge distribution gives rise to an electric potential which, in the half-space $z < d$, can be written as a superposition of evanescent plane waves of the form

$$\phi_{ext} = \phi_0 \, e^{i(q \cdot x - \omega t) + qz} \, , \tag{8.150}$$

where $q = (q_x, q_y)$ is a two-dimensional wave vector. Note that this potential satisfies the Laplace equation, as it should, since there are no external charges for $z < d$.

The external potential (8.150) will polarize the metal substrate and give rise to a reflected electric potential which must satisfy the Laplace equation

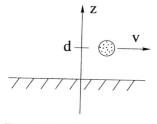

Fig. 8.53. An atom or molecule sliding at a distance d above a metal surface.

for $z > 0$. Thus, the electric potential for $0 < z < d$ is of the form

$$\phi = \phi_0 \, e^{i(q \cdot x - \omega t)} \left(e^{qz} - g(q, \omega) \, e^{-qz} \right) . \tag{8.151}$$

Note that the induced potential is evanescent in the direction *away* from the surface $z = 0$. One can show that the *loss-function* $\mathrm{Im}g(q,\omega)$ is proportional to the power absorption in the metal induced by the external potential (8.150). The reflection factor $g(q,\omega)$ is easy to calculate within local optics. If the metal substrate is described by a local dielectric function $\epsilon(\omega)$, then the electric potential in the metal will also satisfy the Laplace equation (since all the induced charges are located on the metal surface $z = 0$), so that in the metal

$$\phi = t\phi_0 \, e^{i(q \cdot x - \omega t) + qz},$$

where t is the transmission factor. The parameters t and g are determined by the standard boundary conditions that ϕ and $\epsilon \partial \phi / \partial z$ (where $\epsilon = 1$ on the vacuum side of the metal) are continuous for $z = 0$, which gives

$$1 - g = t, \qquad 1 + g = t\epsilon,$$

so that

$$g = \frac{\epsilon(\omega) - 1}{\epsilon(\omega) + 1} . \tag{8.152}$$

This is the g-function within local optics, but (8.152) can be shown to be exact in the limit $q \to 0$. Note that $g(0,0) = 1$ since $\epsilon(\omega) \to \infty$ as $\omega \to 0$. For finite q the calculation of the g-function is much more complicated. Here we only need the result to linear order in ω:

$$g(q,\omega) \approx g(q,0) + A(q)i\omega , \tag{8.153}$$

where

$$A(q) = \lim_{\omega \to 0} \frac{\mathrm{Im}g(q,\omega)}{\omega}$$

is real. In the limit of small q one can distinguish between two different contributions to $\mathrm{Im}g$, corresponding to processes where the source of the momentum necessary for the excitation is supplied by either defects in the bulk, or by the surface potential. If we assume a (bulk) Drude dielectric function

$$\epsilon = 1 - \frac{\omega_p^2}{\omega(\omega + i/\tau)},$$

where τ is a relaxation time due to electron–impurity and electron–phonon scattering, then it is easy to show from (8.152) that for $\omega \ll \omega_p$ the bulk contribution to the loss function

8.11 Friction and Superconductivity: A Puzzle*

$$(\text{Im}g)_{\text{bulk}} = \text{Im}\frac{\epsilon(\omega)-1}{\epsilon(\omega)+1} \approx 4\frac{\omega_F}{\omega_p}\frac{1}{k_F l}\frac{\omega}{\omega_p} \equiv a\,\omega/l, \tag{8.154}$$

where $l = v_F\tau$ is the mean free path due to the bulk scattering processes mentioned above, and ω_F the Fermi frequency. The q-dependent contribution is due to the surface potential experienced by the metal electrons, and is given by

$$(\text{Im}g)_{\text{surf}} = 2\xi(q)\frac{q}{k_F}\frac{\omega}{\omega_p} \equiv b\,\omega q, \tag{8.155}$$

where the dimensionless parameter $\xi(q)$ depends on the electron gas density parameter r_s. Here we will only consider the leading q-contribution to $\text{Im}g$, so that we can approximate $\xi(q) \approx \xi(0) \equiv \xi_0 \approx 1$. Since typically $a \approx b$, the surface contribution (8.155) to the loss function will dominate over the bulk contribution (8.154) when $ql > 1$. Since the mean free path in a good metallic conductor is typically 100 Å or more, it is clear that for processes occurring close to the metal surface the surface contribution will dominate.

Force on Moving Point Charge

The electric potential from a point charge Q, moving with the velocity $\boldsymbol{v}$ parallel to the metal surface at $z = d$ (see Fig. 8.53), satisfies

$$\nabla^2 \phi = -4\pi Q\,\delta(\boldsymbol{x} - \boldsymbol{v}t)\delta(z-d).$$

Using the standard formula

$$\delta(z) = \frac{1}{2\pi}\int dk\,e^{ikz},$$

we get

$$\phi_{\text{ext}} = \frac{4\pi Q}{(2\pi)^3}\int d^2q\,dk\,\frac{1}{q^2+k^2}\,e^{i\boldsymbol{q}\cdot(\boldsymbol{x}-\boldsymbol{v}t)+ik(z-d)}.$$

For $z < d$ the k-integral can be performed by closing the integration contour in the lower part of the complex k-plane. If we write $k^2+q^2 = (k+iq)(k-iq)$ only the $k = -iq$ pole will contribute. Thus,

$$\phi_{\text{ext}} = \frac{Q}{2\pi}\int d^2q\,\frac{1}{q}\,e^{i\boldsymbol{q}\cdot(\boldsymbol{x}-\boldsymbol{v}t)+q(z-d)}.$$

The induced potential (for $z > 0$) takes the form

$$\phi_{\text{ind}} = -\frac{Q}{2\pi}\int d^2q\,\frac{g(q,\omega)}{q}\,e^{i\boldsymbol{q}\cdot(\boldsymbol{x}-\boldsymbol{v}t)-q(z+d)}, \tag{8.156}$$

where $\omega = \boldsymbol{v}\cdot\boldsymbol{q}$. The induced potential will give rise to a force on the external charge given by $-Q\nabla\phi_{\text{ind}}$ evaluated at the position of the ion, i.e., for $z=d$ and $\boldsymbol{x} = \boldsymbol{v}t$:

$$\boldsymbol{F} = \frac{Q^2}{2\pi} \int d^2q \, (i\boldsymbol{q}, -q) \, \frac{g(q,\omega)}{q} \, e^{-2qd} \, . \tag{8.157}$$

Substituting (8.153) in (8.157) gives

$$\boldsymbol{F} = \boldsymbol{F}_1 + \boldsymbol{F}_2 \, ,$$

where $\boldsymbol{F}_1$ is independent of the velocity $\boldsymbol{v}$:

$$\boldsymbol{F}_1 = \frac{Q^2}{2\pi} \int d^2q \, (i\boldsymbol{q}, -q) \, \frac{g(q,0)}{q} \, e^{-2qd} \approx -\frac{Q^2}{(2d)^2} \hat{\boldsymbol{z}} \, , \tag{8.158}$$

where we have used the fact that $g(q,0) \approx 1$ for small q. Equation (8.158) is, of course, the well known image force. The force $\boldsymbol{F}_2$ is proportional to $\boldsymbol{v}$:

$$\boldsymbol{F}_2 = -m\eta\boldsymbol{v} \, , \tag{8.159}$$

where [8.71]

$$\eta = \frac{Q^2}{2m} \int dq \, q^2 \, \lim_{\omega \to 0} \left(\frac{\mathrm{Im} g(q,\omega)}{\omega} \right) e^{-2qd} \, . \tag{8.160a}$$

Using (8.154) and (8.155) this formula gives

$$\eta = \frac{Q^2}{16m} \left(\frac{a}{d^3 l} + \frac{3b}{2d^4} \right) \, . \tag{8.160b}$$

Since typically $a \approx b$, the surface contribution will dominate for $d < l$.

Force on Moving Point Dipole

We can generate a dipole, with the dipole moment $\boldsymbol{p} = p\hat{\boldsymbol{z}}$ perpendicular to the metal surface, by placing a charge Q at $(\boldsymbol{v}t, d)$ and the charge $-Q$ at $(\boldsymbol{v}t, d+\delta)$ with $p = Q\delta$. Because of the linear dependence of the electric potential on the charge density, we get from (8.156)

$$\phi_{\mathrm{ind}} = -\frac{Q}{2\pi} \int d^2q \, \frac{g(q,\omega)}{q} \, e^{i\boldsymbol{q}\cdot(\boldsymbol{x}-\boldsymbol{v}t)} \left(e^{-q(z+d)} - e^{-q(z+d+\delta)} \right)$$

$$\to \frac{p}{2\pi} \int d^2q \, g(q,\omega) \, e^{i\boldsymbol{q}\cdot(\boldsymbol{x}-\boldsymbol{v}t)+q(z-d)},$$

as $\delta \to 0$ and $Q \to \infty$ with $Q\delta = p$. The force acting on the dipole from the induced charge density is $\boldsymbol{F} = -\boldsymbol{p} \cdot \nabla \nabla \phi_{\mathrm{ind}}$ evaluated for $\boldsymbol{x} = \boldsymbol{v}t$ and $z = -d$. This gives a friction force on the form (8.159) with

$$\eta = \frac{p^2}{2m} \int dq \, q^4 \, \lim_{\omega \to 0} \left(\frac{\mathrm{Im} g(q,\omega)}{\omega} \right) e^{-2qd} \, . \tag{8.161a}$$

Using (8.154) and (8.155) this formula gives

$$\eta = \frac{p^2}{8m}\left(\frac{3a}{d^5 l} + \frac{15b}{2d^6}\right). \tag{8.161b}$$

Since typically $a \approx b$, the surface contribution will dominate for $d < l$.

Electronic Friction on a Superconductor Surface

The frictional force acting on physisorbed layers sliding on a superconducting metal films has been measured by Krim and coworkers. The result is striking: Dissipation due to friction drops at T_c to about half of its normal state value, see Fig. 3.25. The drop is clearly connected with superconductivity of the metal substrate, and is very abrupt. While the phenomenon confirms predictions about the importance of electronic friction, it is puzzling why the transition is so abrupt.

At first one may think that the explanation for the abruptness is trivial. When the metal film is in the normal state, at $T > T_c$, we have shown in Sect. 8.2 that the electronic contribution to the sliding friction is directly proportional to the resistivity change of the metal film induced by the adsorbate layer. If this argument were to remain valid when the metal film is in the superconducting state, then the electronic contribution to the sliding friction would correctly drop abruptly to zero at $T = T_c$, since the film resistivity vanishes abruptly at T_c.

However, this argument is incorrect when the metal film is in the superconducting state. First we note that the superconductivity transition is continuous, so that the fraction of the electrons in the superconducting condensate increases *continuously* from zero to one as the temperature is reduced from T_c to zero. The DC resistivity of the metal film nonetheless drops discontinuously from its normal-state value above T_c to zero at $T = T_c$. The reason is, of course, that the electrons in the condensate short circuit the metal film. Thus, no drift motion (current) occurs in the "normal" electron fluid of thermal excitations, even though just below T_c almost all the electrons belong to this "normal" fluid.

Let us now consider the system in a reference frame where the adsorbate layer is stationary. In this reference frame all electrons in the metal film move (collectively) with the speed $-v$ relative to the adsorbate layer (we will show below that this statement is actually not quite correct since, in the superconducting state, a super-current back-flow occurs). The "normal" electrons in the system will scatter from the adsorbates and give rise to energy dissipation, just as in the normal state (i.e., $T > T_c$). Thus, this argument suggests that the electronic sliding friction should decrease continuously as the system is cooled below T_c in a way that correlates with the fraction of electrons in the condensate, in sharp contrast to what is observed experimentally.

Recently, two attempts have been presented to resolve this paradox [8.72]. First, note that the sliding layer gives rise to a surface drag flow and to a bulk back-flow. The dissipative properties of the back-flow change abruptly at the superconducting transition. Secondly, both the bulk and surface contributions to η will decrease as T is lowered below T_c. However, we will show that neither of these mechanisms can explain the strong temperature dependence shown in Fig. 3.25.

Back-Flow Model

Consider first the system for $T > T_c$, and assume that there are no scattering centers for the conduction electrons inside the metal film. The sliding adsorbate layer will exert a drag force on the electrons, giving rise to an electronic current. For an open circuit, this will result in an accumulation of charges at the two ends of the metal slab, see Fig. 8.54a. The drift velocity of the electrons will vanish when the drag force from the sliding layer equals the electric force from the build up of charges. Thus in the normal state there will be a finite energy dissipation due to the frictional coupling between the adsorbates and the conduction electrons. Let us now consider the system for $T < T_c$. In this case there can be no transverse electric field in the metal film, i.e., the potential difference between the two ends of the metal film must vanish. Thus, the following scenario is valid in this case: The sliding adsorbate layer will drag the normal electrons, and since in the superconducting state there is no electric field that can inhibit this drift motion, for a perfect system (i.e., no scattering centers in the bulk), the normal electrons will at steady state drift with the *same* velocity as the sliding layer. Simultaneously, a superfluid current will flow in the opposite direction in such a manner that no net electric current occurs (see Fig. 8.54b). Thus, since the relative velocity between the

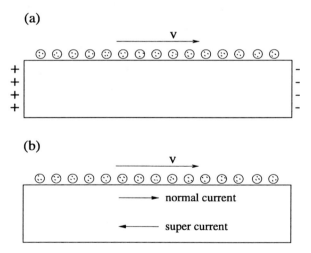

Fig. 8.54. (a) When $T > T_c$ the sliding layer will induce positive and negative charges at the end points of the metal slab. The resulting electric force acting on the conduction electrons will (at steady state) balance the drag force from the sliding layer. (b) In the superconducting state ($T < T_c$) no transverse electric field can occur in the bulk of the metal film but a supercurrent will flow to compensate for the drag of the normal fluid component by the sliding layer.

8.11 Friction and Superconductivity: A Puzzle*

adsorbate layer and the drift motion of the normal electrons vanish, there is no electronic frictional force acting on the sliding layer. Thus, the sliding friction will drop abruptly at $T = T_c$.

The real system is not perfect, but has defects in the bulk which can scatter the metal electrons. I now present a very simple model calculation (based on the same idea as above) for this more complex case. The present calculation is only valid when the electron mean free path $l = v_F \tau$ in the bulk is larger than, or of the order of, the thickness W of the metal slab. From the observed resistivity of the lead film at $T = 7K$, we estimate $l \sim 1000 \text{Å}$, which is similar to the thickness W, so that this condition is approximately satisfied. When $l < W$ the electron drift velocity v will depend on the distance from the top film surface, being highest right at the surface. The theory we present can be generalized to treat also the case of a non-uniform velocity profile, but the simple treatment presented below is enough to demonstrate the basic physics, and to estimate whether the jump in the sliding friction at T_c is of the same order of magnitude as observed experimentally.

We use a two-fluid model of the superconductor [8.73]. Let $\dot{u}$ denote the velocity of the adsorbate layer relative to the metal slab. If v denotes the drift velocity for the normal fluid component, then we have

$$m\ddot{u} + m\eta(\dot{u} - v) = F, \quad (8.162)$$

$$m_e \dot{v} - (n_a/n_N W) m\eta(\dot{u} - v) + m_e v/\tau = eE, \quad (8.163)$$

where m is the mass of an adsorbate and m_e the (effective) mass of a quasi particle. The thickness of the metal slab is denoted by W, and n_a and n_N are the concentration of adsorbates on the surface, and the number density of the normal fluid component in the bulk, respectively. The driving force F in the QCM measurements is the inertia force resulting from the QCM vibrations. The electric field E in the metal film results from the build up of charges at the two ends of the metal slab. Now, let us consider the steady state. In the normal state $(T > T_c)$ we must have $v = 0$ [which implies $E = -mn_a \eta \dot{u}/(n_N W e)$], and (8.162) reduces to

$$m\eta \dot{u} = F.$$

In the superconducting state $(T < T_c)$ we must have $E = 0$. From (8.163) it follows that

$$v = \frac{m\eta \dot{u}}{m\eta + m_e n_N W / n_a \tau}.$$

Substituting this in (8.162) gives

$$m\eta_\text{eff} \dot{u} = F,$$

where

$$\eta_{\text{eff}} = \frac{\eta}{1 + (n_a/n_N W)(m/m_e)\eta\tau} \ .$$

Thus, $\eta_{\text{eff}} < \eta$ so that the sliding friction is smaller in the superconducting state. If we put the Drude relaxation time $\tau = \infty$ we get $\eta_{\text{eff}} = 0$, in accordance with the qualitative arguments presented above. Let us estimate the change in the sliding friction expected for the system studied in [8.69]. In this case $n_a \approx 10^{19} \text{m}^{-2}$ and (just below T_c) $n_N \approx 10^{29} \text{m}^{-3}$ so that $n_a/n_N W \approx 10^{-3}$. The relaxation time τ can be estimated from the observed film resistivity at $T = 7\text{K}$. This gives $\tau \approx 10^{-13}\text{s}$ corresponding to a mean free path of order 1000Å. Since $\eta \approx 10^8 \text{s}^{-1}$ we get $(n_a/n_N W)(m/m_e)\eta\tau \approx 10^{-3}$. Thus, based on this calculation, one expects a negligible change in the electronic friction at $T = T_c$, in contrast to the factor of (at least) 2 observed in the experiments. In the discussion above we have implicitly neglected the temperature dependence of η. This is an excellent approximation in the normal state $(T > T_c)$, but is not valid in the superconducting state. Nevertheless, as we will now show, the temperature dependence of η is much weaker than that observed in the experiments by Krim and coworkers.

Temperature Dependence of η

Assume first that the sliding adsorbates have permanent dipole moments, p, so that there is a direct coulomb interaction between the sliding layer and the electrons in the metal film. In this case the friction coefficient is given by (8.161), and there will be a surface and a bulk contribution to η, depending on the momentum source for the excitation. Note that the ratio between the surface and bulk contributions to the friction force is typically of order $\sim ql$. Since $q \sim 1/d$ it follows that the *surface contribution* will *dominate* over the bulk contribution by a factor $\sim l/d \sim 100 - 1000$, where we have assumed a bulk mean free path $l \sim 1000\text{Å}$ and a dipole–surface separation d of a few Ångstroms. Thus, the bulk contribution can be neglected in most cases.

Let us study the temperature dependence of the bulk and surface contribution as the metal is cooled below T_c. Let us first focus on the bulk contribution. We again use the two-fluid model and write

$$\epsilon(\omega) = 1 - \frac{4\pi\sigma(\omega)}{i\omega} \ ,$$

where the conductivity σ consists of two additive terms, the first due to the superfluid component and the other due to the normal fluid component:

$$\sigma = \sigma_S + \sigma_N \ .$$

The normal component is given by the standard Drude formula

$$\sigma_N(\omega) = -\frac{n_N e^2}{m_e} \frac{1}{i\omega - 1/\tau} \ .$$

8.11 Friction and Superconductivity: A Puzzle*

The superfluid component experiences no frictional force due to disorder and in the long-wavelength limit (transverse field) behaves like

$$\sigma_S(\omega) = -\frac{n_S e^2}{m_e} \frac{1}{i\omega}.$$

Let us now write $n_N = nx$ and $n_S = n(1-x)$ so that $n_N + n_S = n$ is constant. We have $x = 1$ for $T > T_c$, while x decreases continuously from $x = 1$ at $T = T_c$ to $x = 0$ for $T = 0K$. An approximate formula for x is given by the Gorter–Casimir expression [8.73]

$$x(T) \approx (T/T_c)^4 \approx 1 - 4(T_c - T)/T_c,$$

for T close to T_c. Combining the equations above, and using [see (8.154) and (8.161)]:

$$\eta_{\text{bulk}} \sim \text{Im}\frac{\epsilon(\omega) - 1}{\epsilon(\omega) + 1},$$

gives for T close to (but below) T_c:

$$\eta_{\text{bulk}} \sim \frac{1}{1 + [(1-x)/\omega\tau]^2},$$

or,

$$\frac{\eta_{\text{bulk}}(T)}{\eta_{\text{bulk}}(T_c)} \approx \frac{(\Delta T)^2}{(T - T_c)^2 + (\Delta T)^2}, \qquad (8.164)$$

where

$$\frac{\Delta T}{T_c} = \frac{\omega\tau}{4}.$$

Since $\tau \approx 10^{-13}$s and $\omega \approx 10^7 \text{s}^{-1}$ we get $\Delta T/T_c \approx 10^{-6}$. Thus, (8.164) represents a smooth (but very rapid) drop in the friction when T is reduced below T_c.

Let us now study the temperature dependence of the surface contribution. The surface contribution is due to the coupling between the metal quasi-particles and the screened coulomb field induced by the external dipole within the last few Ångstroms at the surface (in fact, most of the coupling comes from the vacuum side of the last layer of ion cores). It has been shown that when calculating the screening of the external potential it is a very good approximation to neglect the time variation of the external potential [8.71]. Thus, since there is negligible influence of the superconducting condensate on the screening of a static electric field, we can take the exciting potential to be temperature independent [8.74]. This is similar to the calculation of the phonon frequencies in metals, where it is enough to use the statically screened crystal potential, which, to a good approximation, is temperature independent [8.70]. The derivation of the temperature dependence of the

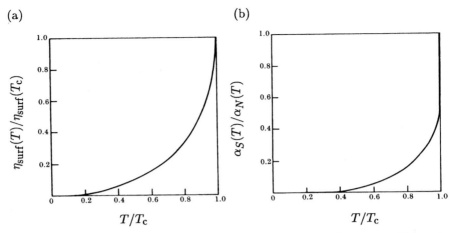

Fig. 8.55. (a) The surface contribution to the electronic friction coefficient in the superconducting state relative to that in the normal state. (b) The relative accoustic-attenuation coefficient for transverse waves in tin as measured by Bohm and Morse.

surface contribution to the electronic friction follows the analog derivation of acoustic attenuation rate of longitudinal acoustic phonons. The golden rule formula is used to calculate the rate of scattering of quasi-particles. For a BCS superconductor, this gives the temperature dependence [8.73].

$$\frac{\eta_{\text{surf}}(T)}{\eta_{\text{surf}}(T_c)} \approx \frac{2}{\exp[\Delta(T)/k_B T] + 1} , \qquad (8.165)$$

where $\Delta(T)$ is the temperature-dependent gap. Equation (8.165) describes a rapid drop in the friction with decreasing temperature (see Fig. 8.55a), but is still much too slow to explain the results of Krim et al.

The picture which emerges from the calculations above is very similar to the origin of the acoustic attenuation of transverse waves. A transverse phonon sets up a transverse electromagnetic field in addition to the crystal potential effect. While the screening of the crystal potential is essentially the same in the normal and superconducting states, the electromagnetic coupling is abruptly reduced at T_c. Thus, one expects an essentially discontinuous drop of the transverse acoustic attenuation rate upon entering the superconducting state at T_c (see Fig. 8.55b). This is similar to the bulk contribution discussed above, where transverse currents occur in the bulk when the metal is in its normal state, but which are abruptly screened out when the temperature is reduced below T_c. The remaining coupling, which in our case corresponds to the surface contribution to the friction, can be accurately treated by using the statically screened crystal potential. However, there is a fundamental difference between our case and the attenuation α of acoustic transverse waves: in the latter case the transverse currents give a factor of ~ 2 change in α at T_c, while the change in the sliding friction at T_c is typically less than 1%.

In the calculations above the friction is derived from the coulomb field of the permanent dipole moments on the adsorbates. However, this is, in general, not the most important coupling mechanism (see Sect. 8.2). For physisorbed inert molecules such as N_2 or C_2H_6, and for the light noble gas atoms Ne and Ar, the permanent dipole moments p are extremely small, and, as discussed in Sect. 8.2, the main contribution will instead come from the fluctuating (virtual) dipoles, which are also the origin of the (van der Waals) attraction between these molecules and the metal surface. For more reactive adsorbates, e.g., C_2H_4 and Xe, covalent effects (resulting from electrons temporarily trapped in orbitals on the adsorbates) will be very important (see Sect. 8.2). However, these processes will give the same temperature dependence of the friction as for the dipolar surface contribution discussed above. Thus, the abruptly jumping friction remains a puzzle, to which I wish to call attention, and for which I can offer no clear-cut explanation at present.

8.12 Layering Transition: Nucleation and Growth

For many applications involving lubrication, the stability under pressure of thin lubrications films is of major importance, since the complete squeezing out of the lubrication film from an interface may give rise to cold-welded junctions, resulting in high friction and catastrophically large wear. In Sects. 7.4 and 8.6.2 we have studied the nucleation and spreading of the $n \to n-1$ layering transition. In this section we present further experimental and theoretical results for this fundamental process. We first present the results of experiments in which the layering transition has been observed directly by imaging the lateral variation of the gap between the solid surfaces as a function of time. These results are for 2D-liquid-like layers, for which the theory developed in Sects. 7.4 and 8.6.2 is applicable. Next we present a detailed theoretical discussion of the dynamics of the boundary line separating the n and $n-1$ regions. Finally, we include results of computer simulations which illustrate the nature of the layering transition when the lubrication film is in a 2D-solid-like state.

Experimental Study of the Layering Transition

The dynamics of the layering transition has been studied with the Surface Forces Apparatus by imaging the gap width in two dimensions [8.75]. The experiment was performed with a chain alcohol ($C_{11}H_{23}OH$), where the unit of liquid that is expelled in a layering transition corresponds to a bilayer of molecules, with the OH-groups pointing towards each other, see Fig. 8.56. The mica surfaces are covered by monolayers of $C_{11}H_{23}OH$, chemically bound (via the OH-group) to the mica surfaces, leading effectively to a CH_3-terminated

Fig. 8.56. Surface bound $C_{11}H_{23}OH$ monolayers and an additional bilayer. From [8.75].

substrate for any additional material inside the gap. The coated surfaces are very inert (see Sect. 7.5), and the additional alcohol does not wet the surfaces. Shear experiments showed that the static frictional force vanishes when the passivated mica surfaces is separated by a lubricant bilayer. Thus, the bilayer film is likely to be in a 2D-liquid-like state, and the theory developed in Sects. 7.4 and 8.6.2 may be applicable to the present system. The data presented below show the expulsion of the last bilayer, i.e., the $n = 1 \to 0$ layering transition (see Fig. 8.56).

The experiment was performed at room temperature and variations in film thickness, to within one Ångstrom, were studied in real time by monitoring the transmitted light. The experiment was performed by starting at a separation of several micrometers, and then abruptly increasing the normal force. The surfaces came into contact and the contact area increased to its final radius $R = 40$ μm within a few seconds. Then, at an average pressure of $P_{\text{ext}} \approx 4$ MPa, the system remained in a metastable state for a few seconds, until the layering transition nucleated. The total time for the expulsion of the bilayer was ~ 1 s. The film thickness decreased from $37 \to 25$ Å at the transition.

Figure 8.57 shows the transmitted intensity averaged over the contact area as a function of time. Within the first 4 s (phase I), the intensity decreases as liquid drains out of the gap in a process governed by standard hydrodynamics (see Sect. 7.4). In phase II, the thickness of the liquid layer remains constant. After a time $t_{\text{II}} \approx 8$ s, the intensity drops by a discrete amount within a transition time $t_{\text{III}} < 1$ s. This abrupt transition marks the expulsion of the last unbound layer of alcohol molecules. In the final state (phase IV), the mica-bound $C_{11}H_{23}OH$ monolayers on each surface are in direct contact, and the thickness of this film remains constant indefinitely.

We first focus on the dynamics of the layering transition during phase III. Figure 8.58 shows a set of images recorded during a layering transition. Image A shows the starting configuration just prior to the nucleation. In image B the encircled darker area shows where the molecules are already expelled. This "island" expands with time and its boundary quickly moves across the contact area. During this process the local curvature of the boundary becomes negative in some areas. Some of these areas eventually detach from the boundary and leave behind pockets of trapped material in the final sate. In some cases, pockets that are located close to the rim of the contact area eventually disappear. These pockets move towards the edge as a whole. There they form little necks through which liquid is squeezed out, as seen at the bottom of

8.12 Layering Transition: Nucleation and Growth   299

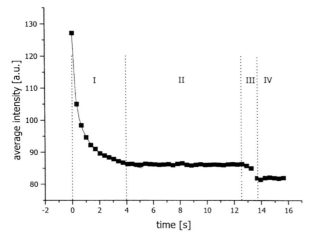

Fig. 8.57. Transmitted intensity, averaged over the contact area, as a function of time. From [8.75].

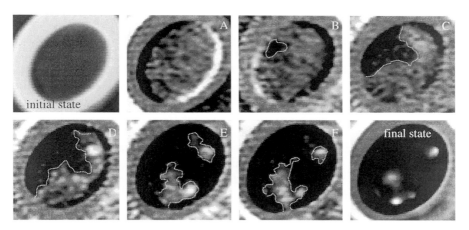

Fig. 8.58. Snapshots of the contact area during the layering transition. Darker color corresponds to smaller film thickness. Images A to F are recorded at 0.16 s time intervals. From [8.75].

image F. A series of approximately 100 subsequent expulsion events resulted in more than 20 non-overlapping locations of pockets, indicating that island formation is due not to pinning by defects but rather to intrinsic instabilities of the non-linear equations of motion of the boundary line (see below). The diameters range from 5 to 10 μm. Some pockets are just one bilayer thick while others are two layers thick.

Analysis of many squeeze-out events indicate that the nucleation of the squeeze-out is preceded by a well-defined "incubation time" τ. The physical origin of τ may be that the density of the layer is slightly enhanced due to the rapid compression, so that it will take some time τ before the 2D-pressure has decreased (by migration of molecules out from the contact area) to the point that a thermal fluctuation is able to nucleate the layering transition. Alternatively, small $n = 2$ islands may be initially trapped in the contact region, and only after these have been squeezed into a single layer (giving, initially, a compressed $n = 1$ layer), can the $n = 1 \to 0$ layering transition nucleate. (Note: if the radii of the $n = 2$ islands are below 1 μm they would not be observed in the present experiments.) By performing many experiments, and determining the probability distribution of times t_{II}, the incubation time τ can be determined, and from the spread in the observed t_{II} times, the average nucleation time for the layering transition was determined to be of order $\sim$ 1s. The nucleation seems to take place preferentially in the upper left section of the contact area. Most likely, this is due to inhomogeneity or contamination on one of the mica surfaces.

The dynamics of the layering transition observed in the experiments separates into two phases. In the first phase, the system is trapped in a metastable state at the initial film thickness, i.e. one bilayer of alcohol molecules between the substrate-bound monolayers. Thermal fluctuations of the two-dimensional density in the bilayer eventually lead to the formation of a hole with a radius that exceeds the critical radius R_c. Once the nucleus is formed the growth phase begins, and the rest of the bilayer is quickly expelled. Since the film seems to be in a liquid-like state (as manifested by the vanishing of the static friction force) we can analyze the data using the theory presented in Sect. 7.5. From the average nucleation rate we get $U(R_c) \approx 39 k_B T$. From the contact angle of $C_{11}H_{23}OH$ of freshly cleaved mica (52°) one can calculate the spreading pressure $p_0 \approx -0.65$ meV/Å^2. Using (7.31) and (7.38), we find the critical radius $R_c \approx 23$Å and the line tension $\Gamma \approx 15$meV. If Γ is estimated by multiplying the surface tension of liquid undecanol (27 mJ/m^2) with the thickness of the expelled bilayer (12 Å) we get $\Gamma \approx 20$ meV/Å^2.

Let us now consider the growth of the nucleus of the layering transition. In the experiments the bilayer was squeezed out over $\sim$ 1s, which is $\sim$ 100 times faster than for the $n = 2 \to 1$ transition for OMCTS (see Sect. 8.6.2). Thus, in this case (8.74) gives $\eta \approx 10^{13}$s^{-1}, which is two orders of magnitude smaller than for OMCTS. In contrast to the latter system, in the present case it is not necessary to slide the surfaces relative to one another, in order to induce the layering transition. This is consistent with the fact that the $C_{11}H_{23}OH$ bilayer is likely to be in a 2D-liquid-like state. We will show below that for solid-like, pinned layers the layering transition is very slow and sluggish, and in some cases it is only possible to squeeze out the lubricant layer if lateral sliding occurs, which tends to break up the pinning.

Dynamics of the Boundary Line

Let us discuss the evolution of the boundary of the hole-island during the layering transition $n = 1 \to 0$ when the nucleation of the layering transition occurs off-center. We assume, for simplicity, that the pressure in the contact area is constant (equal to P_0); the qualitative picture presented below does not change if the pressure varies with r as discussed in Sect. 8.6.2. Neglecting the non-linear and viscosity terms in (8.65), and assuming that the velocity field changes so slowly that the time derivative term can be neglected, gives

$$\nabla p + m n_a \bar{\eta} \boldsymbol{v} = 0 \ . \tag{8.166}$$

From this equation it follows that

$$\boldsymbol{v} = \nabla \phi \ , \tag{8.167}$$

and the continuity equation (8.64) gives

$$\nabla^2 \phi = 0 \ . \tag{8.168}$$

Substituting (8.167) in (8.166) gives

$$\phi = -p/m n_a \bar{\eta} \ . \tag{8.169}$$

Now, from (8.168), we see that the velocity potential can be interpreted as an electrostatic potential. Furthermore, since the pressure p is constant at both the (outer) boundary $r = r_0$ of the contact area, and at the (inner) boundary to the region $n = 0$, the problem of finding ϕ is mathematically equivalent to finding the electrostatic potential between two conducting cylinders at different potentials, $\phi_0 = -p_0/m n_a \bar{\eta}$ and $\phi_1 = -p_1/m n_a \bar{\eta}$. The outer cylinder has a circular shape (radius r_0), and the inner cylinder an unknown (time-dependent) shape to be determined.

Now, suppose that the initial nucleation of the $n = 1$ region occurs some distance away from the center of the contact region, as indicated by the small circle in Fig. 8.59a. The lines between the two circular regions in this figure indicate the velocity field of the two-dimensional fluid at this moment in time, constructed by analogy to the electrostatic field lines between two cylinders at different potentials. Now, a little later in time, the velocity field will result in a larger $n = 1$ region (dotted area) as indicated in Fig. 8.59b. Fig. 8.59c and d shows the further spreading of the $n = 1$ region as time increases, constructed on the basis of the analogy with electrostatics. These results for the evolution of the boundary line are in relatively good agreement with the experimental observations for $C_{11}H_{23}OH$, see Fig. 8.58. One difference, however, is that the boundary line in the experiment has "roughness" on a length scale beyond a few µm, while we have drawn smooth boundary lines in Fig. 8.59. However, we will show below that the model presented above indeed predicts rough boundary lines. That is, any small perturbation of the boundary line will be amplified, so that the boundary line is unstable with

302   8. Sliding of Adsorbate Layers

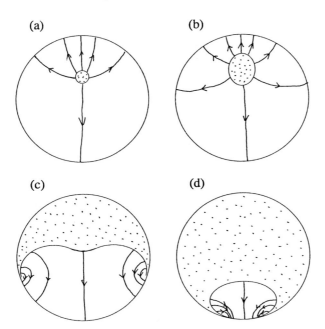

Fig. 8.59a–d. Snapshots (schematic) of the time evolution of the squeeze-out of a monolayer, $n \to n-1$, in the slow (Laplacian) limit of (8.64) and (8.65). The dotted area denotes the $n-1$ region. The lines denote the velocity field.

respect to arbitrarily small perturbations. In real systems such perturbations always exist, e.g., due to defects or thermal fluctuations.

In order to show that the boundary line is unstable with respect to small perturbations, let us first consider a perfectly smooth circular boundary line centered at the center of the contact area as indicated in Fig. 8.60a. Assume that due to a fluctuation a small "bump" is formed on the boundary line as indicated in Fig. 8.60b. By analogy to electrostatics, this will give rise to an enhanced "draining" velocity of the fluid at the bump, so that the boundary line at the bump will move faster towards the periphery than the other regions of the boundary line. This argument is valid for "bumps" of any size, and it follows that, within the model studied above, the boundary line will be *rough on all length scales*. However, we will demonstrate below that when the free energy per unit length, Γ (line tension), of the boundary line is taken into account (it was neglected in the study above), the boundary line will be smooth on all length scales below some critical length λ_c, while it will be rough on longer length scales. In a typical case we calculate λ_c to be of order a few μm, in good agreement with experiments (see Fig. 8.58).

We note that the instability described above is at the origin of many beautiful physical effects, such as the shape of snowflakes, electro-deposition

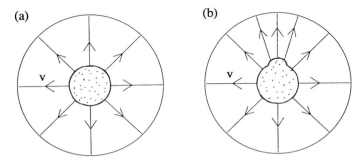

Fig. 8.60. Instability of the boundary line separating the n and $n-1$ (*dotted*) regions. (**a**) A circular boundary line. (**b**) A small bump on the boundary line results in locally enhanced squeeze-out, magnifying the pertubation.

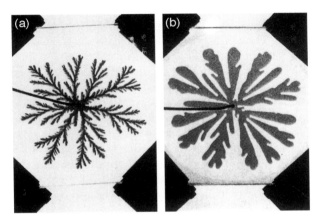

Fig. 8.61. Viscous fingering patterns in a thin gap between two glass plates. The less viscous fluid is injected into the center and displaces the more viscous fluid at high flow rate. (**a**) Water injected into glycerin. (**b**) Oil injected into glycerin. From [8.77].

or viscous fingering [8.76,77]. All these problems can be mapped onto the same (or slightly modified) problem as that studied above, involving a scalar field ϕ (e.g., the temperature field, the velocity potential or the electric potential) which satisfies the Laplace equation in the region between two boundaries, together with a set of boundary conditions, one at the outer (fixed) boundary, and another at an inner moving boundary. The problem of viscous fingering is particularly close to our subject, and is briefly described here.

Figure 8.61 shows the viscous fingering pattern in a thin gap between two glass plates. The less viscous fluid is injected into the center and displaces a more viscous fluid at a high flow rate. In (a) the injected liquid is water, and the displaced liquid is glycerine. In (b), water is replaced by an oil. The dendritic nature of the pattern is due to the instability discussed above. Note, however, that on a short length scale the boundary lines are smooth, which is particularly clear in case (b). The cut-off length λ_c below which the boundary lines are smooth is dependent on the line tension. We also note that the fingering-like growth of a cavity observed during the rapid separation of two solids in a high-viscosity fluid (see Fig. 7.18), may have the same origin as the instabilities described above.

Linear Stability Analysis

We now show why the boundary line separating the $n = 1$ region from the $n = 0$ region in Fig. 8.58 is rough on the length scale > 5 μm. A rough boundary line could, in principle, result from the influence of defects (pinning centers), or be a result of instabilities due to the non-linear nature of the equation of motion for the boundary line. We have shown above (see Fig. 8.60) that such instabilities are indeed expected. However, we will now show that instabilities only occur on long enough length-scale, typically larger than a few μm. Let us first show that the local 2D-pressure in the $n = 1$ region differs from $p_1 = p_0 + Pa$ by a term determined by the line tension Γ. The line-tension has a contribution from unsaturated bonds at the boundary line, and another much larger contribution from the energy stored in the elastic deformation field in the confining solids in the vicinity of the boundary line (see Fig. 8.62a). If a denotes the difference in the separation between the solid walls in the $n = 1$ and $n = 0$ regions (which is of order the thickness of a lubrication monolayer), then it follows from dimensional arguments that the elastic deformation energy per unit length of boundary line must be of order $\sim Ea^2$. In fact, an exact calculation (within the elastic continuum model) of

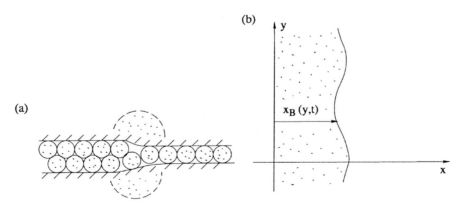

Fig. 8.62. (a) Elastic deformation energy is stored in the dotted region in the solids at the boundary line. (b) Boundary line $x_B(y, t)$ separated the $n = 0$ area (*dotted region*) from the $n = 1$ area.

8.12 Layering Transition: Nucleation and Growth

the change in the elastic energy when a $n=0$ hole of radius R is formed in a $n=1$ system is given by (for $R \gg a$)

$$U(R) = -\pi R^2 Pa + 2\pi R\Gamma,$$

where $\Gamma = Ea^2/2\pi(1-\nu^2)$. This gives rise to a 2D-pressure p_1' in the fluid layer at the boundary $r = R$, which can be obtained by calculating the adiabatic work needed to increase the radius from R to $R + \Delta R$. We find

$$U(R) - U(R+\Delta R) = 2\pi R(p_1' - p_0)\Delta R,$$

which gives

$$p_1' = p_0 + Pa - \Gamma/R = p_1 + \kappa\Gamma, \tag{8.170}$$

where $p_1 = p_0 + Pa$ is the pressure which would occur in the fluid at a straight boundary line, and $\kappa = -1/R$ the curvature, which is positive if the radius of curvature is located in the 2D-fluid region, and otherwise negative. Equation (8.170) is, of course, valid for an arbitrary curved boundary, i.e., not just a circular boundary. In the above study of the spreading of the $n=0$ hole we have neglected the line tension term. However, this term has a crucial influence on the stability of the boundary line: without it, as shown above, the boundary line is unstable on all length scales. Let us prove this for the simplest case when the unperturbed boundary is a straight line, rather than a circle. Let $x = x_B(y,t)$ be the equation for the boundary line, and assume that the $n=1$ fluid layer initially occupies the half plane $x > x_B(y,t)$, see Fig. 8.62b. The velocity potential ϕ satisfies [see (8.167)-(8.169)]:

$$\nabla^2 \phi = 0, \tag{8.171}$$

$$\hat{n} \cdot \nabla \phi = v_n \quad \text{for} \quad \boldsymbol{x} = \boldsymbol{x}_B, \tag{8.172}$$

$$\phi(\boldsymbol{x}_B) = -p_1'/m n_a \bar{\eta}, \tag{8.173}$$

where $\boldsymbol{x}_B = (x_B, y)$ is a point on the boundary line. The unperturbed boundary line is given by

$$x_B^0 = x_0 + v_0 t.$$

Let us now study the stability of this solution by adding a small perturbation:

$$x_B = x_B^0 + \xi(t)e^{iky}. \tag{8.174}$$

The most general solution to (8.171) which is consistent with this boundary line is

$$\phi = \phi_0(t) + v_0 x + \zeta(t)e^{iky - k[x - x_B(t)]}, \tag{8.175}$$

where we have used the fact that the velocity for large x must equal v_0. Substituting (8.174) and (8.175) into (8.172) gives to linear order in ξ and ζ:

$$\dot\xi = -k\zeta .\tag{8.176}$$

The curvature is

$$\kappa = \frac{d^2 x_B}{dy^2} = -k^2 \xi e^{iky},\tag{8.177}$$

so that, to linear order in ξ and ζ, (8.173) takes the form

$$\phi_0 + v_0 x_B^0 + v_0 \xi e^{iky} + \zeta e^{iky} = -\frac{p_1}{mn_a\bar\eta} + \frac{k^2 \Gamma}{mn_a\bar\eta}\xi e^{iky}.$$

Comparing the coefficients of the $\exp(iky)$ terms gives

$$\zeta = -\xi v_0 \left(1 - k^2 \Gamma/mn_a\bar\eta v_0\right),$$

so that, using (8.176),

$$\dot\xi = v_0 k \xi \left(1 - k^2 \Gamma/mn_a\bar\eta v_0\right).$$

Hence, for $k < k_c$, where

$$1/k_c = \left(\Gamma/mn_a\bar\eta v_0\right)^{1/2},$$

the boundary line is unstable, while it is stable for $k > k_c$. Thus, we expect that the boundary line is rough on the length scale $\lambda > \lambda_c = 2\pi/k_c$, but smooth on the length scale $\lambda < \lambda_c$. In the present case $\lambda_c \sim 10$ μm, which is about $1/10$ of the diameter of the contact area. The experimental boundary line for $C_{11}H_{23}OH$ is indeed rough on this length scale, while it is smooth on shorter length scale. Based on this result one may also argue that the linear size of trapped islands should be of order λ_c (or larger), which again agrees with observations.

Computer Simulations of the Layering Transition

The theory presented above and in Sects. 7.4 and 8.6.2, is based on the assumption that the lubricant film is in a 2D-fluid state. This seems to be the case in the experiments by Mugele et al. It may also be the case for solid lubricant films during sliding. In this section we present a computer simulation study for the more common case where the lubricant films are in a solid state. We consider three different cases, where the film forms (A) unpinned and (B) pinned incommensurate layers, and (C) a commensurate layer.

We investigate the late stages of the approach of two elastic solids limited by two curved surfaces, wetted by an atomic lubricant film of microscopic thickness [8.78]. Under these conditions, the behavior of the lubricant is mainly determined by its interaction with the solids, which induces a 2D order along the surfaces, and layering in the perpendicular direction. We focus on the atomic processes by which the thickness of the interface decreases by discontinuous steps, corresponding to the decrease in the number n of lubricant layers. For solid surfaces that approach one another without lateral sliding, separated by unpinned or weakly pinned (incommensurate) lubrication layers, fast and complete layering transitions occur. Commensurate or strongly pinned incommensurate layers lead to sluggish and incomplete transitions, often leaving islands trapped in the contact region. In fact, for commensurate layers we observe that it is virtually impossible to squeeze out the last few layers simply by increasing the perpendicular pressure. However, the squeeze-out rate is greatly enhanced by lateral sliding, since, in this case, the lubricant film can enter a fluidized or disordered state, facilitating the ejection of one layer.

We have performed simulations for three different cases. In all cases the lubricant is Xe, but we have varied the Xe–substrate interaction potential so that a monolayer film of lubrication atoms forms unpinned (case A) or pinned (case B) incommensurate layers, or a commensurate layer (case C). The block is 100 Å thick, with the same elastic modulus as for steel. The upper surface of the block is "glued" to a rigid surface profile with a cosine corrugation in the x-direction, and constant in the y-direction, and periodic boundary conditions are used in the xy-plane. The amplitude of the cosine corrugation is large enough that during all stages in the squeeze out, there is "empty" space in some regions between the block and the substrate, into which the lubrication fluid can be squeezed without building up a hydrostatic pressure. The substrate is only one atomic layer thick, and is "glued" onto a flat rigid surface, which is kept fixed in space. In all the computer simulations, the upper rigid surface profile moves with a constant velocity towards the substrate, and the distance displayed in Fig. 8.63 is the displacement of the upper rigid surface profile towards the substrate rigid surface, where $d = 0$ corresponds to an (arbitrary) starting point where the two elastic solids with lubrication layers are nearly in contact.

A) Incommensurate Layer (Unpinned)

Figure 8.63 (A) shows the average perpendicular stress acting on the substrate as a function of the displacement of the block towards the substrate, for two different temperatures, $T = 50$ K and 300 K, where (for clarity) the latter curve is displaced towards negative pressure by 0.2 GPa. The block and the substrate are initially separated by about four Xe-monolayers. The three "bumps" on the curves correspond to the layering transitions (with increasing pressure) $n = 4 \to 3$, $3 \to 2$ and $2 \to 1$. We observe that these transitions, in particular the $n = 2 \to 1$ transition, are rather abrupt, and are marked by a significant pressure drop. The latter implies that the squeeze-out occurs so rapidly that during the transition, the upper surface has moved only a small fraction of the diameter of the Xe-monolayer. Note that the layering transitions occur at higher pressures at low temperature, indicating that they are thermally activated.

308   8. Sliding of Adsorbate Layers

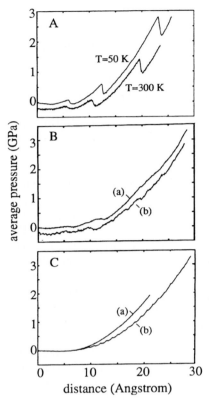

Fig. 8.63. The dependence of the average pressure on the distance the upper surface of the block has moved towards the bottom surface of the substrate. (**A**) Results for model A for two different temperatures, $T = 50$ K and $T = 300$ K, and with the squeeze velocity $v_z = 3$ m/s. (**B**) Results for model B at $T = 200$ K and for (a) squeezing with $v_z = 4.4$ m/s and (b) squeezing ($v_z = 4.4$ m/s) and sliding at $v_x = 17.7$ m/s. (**C**) Results for model C at $T = 80$ K and for (a) squeeze with $v_z = 4.6$ m/s and (b) squeezing ($v_z = 4.6$ m/s) and sliding at $v_x = 18.3$ m/s.

Figure 8.64 (top) shows a sequence of snapshots of the central interface region during squeeze at $T = 300$ K. The time (in natural units $\tau \approx 2.9$ ps) of each snapshot is indicated. The open circles denote the bottom layer of block atoms and top layer of substrate atoms.

Figure 8.64 (bottom) shows snapshots of the lubrication film during the nucleation of the squeeze-out $n = 2 \to 1$. Immediately before the nucleation of the layering transition the lubrication film in the central region has undergone a phase transformation and now exhibits fcc(100)-layers parallel to the solid surfaces. Since the fcc(100)-layers have a lower concentration of Xe atoms than the hexagonal layers (assuming the same nearest neighbor Xe–Xe distance), a fraction of the Xe–solid binding energy is lost during this trans-

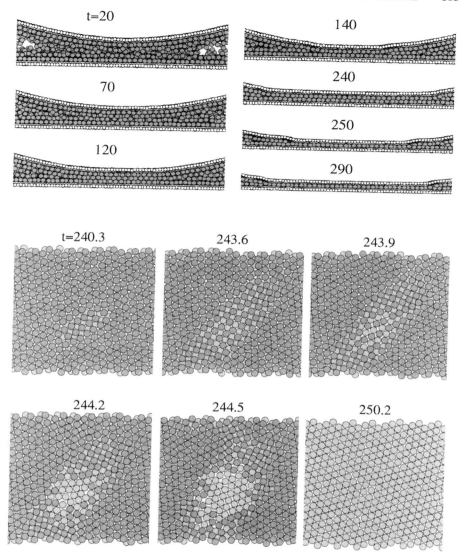

Fig. 8.64. *Top:* Snapshots during squeeze-out. The time of each snapshot is indicated. *Bottom:* Snapshots of the lubricant layer for times close to the point where the $n = 2 \to 1$ squeeze out transition occurs. The first and second monolayer atoms are indicated with different grey-scale. For model A at $T = 300$ K.

Fig. 8.65. Snapshots, at two different viewing angles, of the trapped island. For model B at $T = 200$ K.

formation. On the other hand the solid surfaces can now move closer to each other (since the distance between the fcc(100)-layers is smaller than that between the hexagonal layers) and in this way elastic energy is released. After the phase transformation, the layering transition $n = 2 \to 1$ can occur much more easily since density fluctuations (opening up of a "hole") require less energy in the more dilute fcc(100) layers than in the higher density hexagonal layers.

B) Incommensurate Layer (Pinned)

Figure 8.63 (B) shows the average perpendicular stress acting on the substrate (or block) as a function of the displacement of the block towards the substrate. The upper curve (a) is without lateral sliding ($v_x = 0$) while the lower curve (b) is for $v_x > 0$. For clarity the latter curve is displaced towards negative pressure by 0.2 GPa. Note that, in contrast to case (A), in the present case, where the lateral atomic corrugation experienced by the lubrication atoms is much higher, the squeeze out is more sluggish, and only very weak bumps corresponding to the $n = 4 \to 3$ and $n = 3 \to 2$ transitions can be detected in the upper curve.

In case (a) it is found that at the end of squeezing in Fig. 8.63 (B), a trapped $n = 2$ island occurs, surrounded by a single Xe-monolayer. Figure 8.65 shows snapshots of the trapped island. As discussed above, trapped islands have recently been observed experimentally for 2D-liquid-like lubrication films. In this latter case the island may be a result of dynamical instabilities of the line boundary line caused by the non-linearity of the equations of motion (see above). Similar instabilities may be the origin of the trapped island we observe in the present case. However, when the lubrication film is in a 2D-solid-like state, plastic deformation must occur during squeeze out in order to allow different parts of the lubrication film to move with different velocity relative to the solid surfaces. During squeezing *and* sliding the $n = 2 \to 1$ transition is complete, i.e., no $n = 2$ island remains trapped.

C) Commensurate Layer

Figure 8.63 (C) shows the average perpendicular stress acting on the substrate (or block) as a function of the displacement of the block towards the substrate.

The upper curve (a) is without lateral sliding ($v_x = 0$) while the lower curve (b) is for $v_x > 0$. Note that the commensurate adsorbate layers are strongly pinned, and even though the Xe–substrate binding energy in the present case is much smaller than for case A, it is difficult (if no lateral sliding occurs) to squeeze out the lubrication film. Thus, at the end of the squeeze-out process in Fig. 8.63 (C) (no sliding) the surfaces are still separated by four Xe-layers, just as at the beginning of squeeze-out. However, lateral sliding tends to break up the pinning (e.g., fluidization of the adsorbate layer may occur), and during sliding it is much easier to squeeze out the lubrication layer, and at the end of squeeze-out [curve (b)] only one Xe-layer remains between the surfaces in the high-pressure region.

9. Boundary Lubrication

It is very hard experimentally to directly probe the nature of the rapid processes which occur at a sliding interface. But some information can be inferred indirectly by performing sliding friction measurements on well-defined systems, and registering the macroscopic (e.g., center-of-mass) motion of the block as a function of time. For lubricated surfaces, such measurements have been performed during the past few years, using the Surface Forces Apparatus (Chap. 3) [9.1–5]. These studies usually use mica surfaces which can be produced atomically smooth (without a single step) over macroscopic areas. A spring is connected to the mica block and the "free" end of the spring is moved with some velocity v_s, which typically is kept constant but sometimes is allowed to change in time. The force in the spring is registered as a function of time and is the basic quantity measured in most of these friction studies. It is obvious that the time dependence of the spring force (and its dependence on v_s) contains information about the nature of the processes occurring at the sliding interface, but this information is very indirect.

The sliding friction probed in an experiment of the type discussed above can be considered as resulting from the process of eliminating (or "integrating out") physical processes which vary rapidly in space and time in order to obtain an effective equation of motion for the long-distance and long-time behavior of the system (see Chap. 1). For example, the first step may be to eliminate those processes in the lubricant film which vary rapidly in space and time, to obtain the effective surface stress that the lubricant layer exerts on the lower surface of the block. The next step is to study how this surface stress is transmitted to the upper surface of the block and, finally, how this affects the spring force studied in a typical sliding friction measurement.

In this section we discuss some simple aspects of boundary lubrication. We study a model which illustrates how the surface stress generated by the lubrication layer at the block–substrate interface is transmitted to the upper surface of the block. The study of this problem turns out to have important implications for the understanding of "starting" and "stopping". We discuss the origin of stick-slip motion and of the critical velocity v_c where steady motion switches to stick-slip motion. A more complete treatment of friction dynamics of boundary lubricated surfaces is presented in Chap. 12.

9.1 Relation Between Stress σ and Sliding Velocity v

We consider the model shown schematically in Fig. 9.1a. A spring with the force constant k_s is connected to an elastic block in the shape of a rectangular parallelepiped located on a flat substrate. The block has the thickness d and the side in contact with the substrate is a square with the area $A = D \times D$; the surfaces are assumed to be perfectly flat so that "contact" occurs over the whole area A. We assume that a molecularly thin layer of lubrication molecules separates the block from the substrate (boundary lubrication).

Let (x, y, z) be a coordinate system with the $z = 0$ plane in the top surface of the block and with the positive z-axis pointing towards the substrate, see Fig. 9.1a. The free end of the spring moves with the velocity v_s along the positive x-axis. We assume that the block is made from an isotropic elastic medium and that $D \gg d$, i.e., that the width and depth of the block are much larger than its height. Under these conditions, if the stress σ on the surface $z = d$ does not depend on x and y, the displacement field $\boldsymbol{u}(\boldsymbol{x}, t)$ in the block will, to a good approximation, depend only on z and t; close to the vertical sides of the block the displacement field will be more complicated but this region of space can be neglected if $D \gg d$. The field $\boldsymbol{u}(\boldsymbol{x}, t) = \hat{\boldsymbol{x}} u(z, t)$ satisfies the wave equation

$$\frac{\partial^2 u}{\partial t^2} - c^2 \frac{\partial^2 u}{\partial z^2} = 0 , \tag{9.1}$$

where c is the transverse sound velocity. The tangential stress exerted by the spring on the $z = 0$ surface of the block can be related to $u(z, t)$ via

$$\frac{1}{\kappa} \frac{\partial u}{\partial z}(0, t) = -\frac{F_s(t)}{A} = -\sigma_s(t) \tag{9.2}$$

where $1/\kappa = \rho c^2$ (ρ is the mass density of the block). If the stress exerted by the lubrication molecules on the surface $z = d$ of the block is denoted by $-\sigma(t)$, then

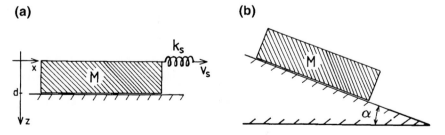

Fig. 9.1. (a) An elastic block on a substrate. The free end of the spring moves with the velocity v_s. (b) An elastic block on a tilted substrate.

$$\frac{1}{\kappa}\frac{\partial u}{\partial z}(d,t) = -\sigma(t) \ . \tag{9.3}$$

Note that $\sigma(t)$ is obtained from the microscopic stress $\sigma(\boldsymbol{x},t)$ by averaging over (or "integrating out") the rapid (in space and time) fluctuating part of the motion of the lubrication molecules. It is implicitly assumed that no spatial fluctuations of $\sigma(\boldsymbol{x},t)$ occur on the length scale d or longer; if such fluctuations did occur they would need to be taken directly into account, and the problem we study below would be much more complicated. If $-\sigma(t)$ is the (average) stress exerted by the lubrication molecules on the bottom surface of the block, then, according to Newton's law of action and reaction, the block must exert the (average) stress $\sigma(t)$ on the layer of lubrication molecules. This will, in general, lead to some drift motion of the lubrication layer. If one assumes that the time dependence of $\sigma(t)$ is slow compared with typical relaxation times associated with the motion of the lubricant molecules, then one can treat σ as a constant when determining the relation $v = f(\sigma)$ between σ and the speed v of the adsorbate layer. Note that the velocity v of the bottom surface of the sliding block is not identical to the drift velocity of the adsorbate layer. For example, if the block and the substrate are made from identical material, then the drift velocity of the adsorbate layer will be half of the velocity of the bottom surface of the block. This follows directly from symmetry: In a reference frame where the substrate moves with the velocity $-v/2$ and the block with the velocity $v/2$, the drift velocity of the adsorbate layer must, by symmetry, vanish.

As discussed in Sect. 8.9 and summarized below, the relation $v = f(\sigma)$ can have two qualitatively different forms. If the adsorbate layer is in a two-dimensional (2D) fluid state, which is always the case in some parts of the (θ,T) (θ is the adsorbate coverage) phase diagram, then the $v = f(\sigma)$ relation has the form indicated in Fig. 9.2a. In this case the drift velocity will be non-zero for arbitrarily small σ. This is, of course, exactly what one expects for a fluid: an arbitrarily weak external force can shear a fluid. Furthermore, no hysteresis is observed, i.e., the relation between σ and v does not depend on whether σ decreases from a high value or increases from zero. Hence, if the lubrication layer in a sliding friction experiment is in a 2D-fluid state, smooth sliding is expected (i.e., no stick-slip motion) for any spring velocity v_s. This is exactly what is observed experimentally. For example, *Yoshizawa* and *Israelachvili* [9.1] have studied a 12 Å thick hexadecane film between two smooth mica surfaces and found stick-slip motion when the temperature $T = 17°\mathrm{C}$ but smooth sliding for $T = 25°\mathrm{C}$, see Fig. 9.3. As will be shown below, stick-slip motion is observed (if v_s is small enough) when the adsorbate layer is in a pinned solid state at "stick". Hence, the melting temperature of the hexadecane film is somewhere between $17°$ and $25°\mathrm{C}$.

Assume now instead that the system is in a part of the (θ,T) phase diagram where the adsorbate layer is in a solid state which is commensurate (or at least pinned) by the substrate. In this case the $v = f(\sigma)$ relation has

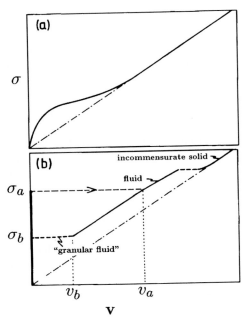

Fig. 9.2a,b. The drift velocity $\langle v \rangle$ of an adsorbate layer as a function of the external stress $\sigma = n_{\mathrm{a}} F$, where n_{a} is the adsorbate coverage and F the external force acting on each adsorbate. (a) The adsorbate layer is in a fluid state when $\sigma = 0$. (b) The adsorbate layer is in a pinned solid state when $\sigma = 0$.

the qualitative form shown in Fig. 9.2b. If the system is first thermalized with $\sigma = 0$ and then σ increased, the pinned solid structure will remain, and the drift velocity is zero ($v = 0$), until σ reaches some critical stress σ_{a}. At this point the adsorbate system fluidizes and the (steady state) drift velocity increases abruptly from $v = 0$ to v_{a}. If σ increases further the drift velocity continues to increase as indicated in the figure. If σ is reduced below σ_{a} the system does not return to the pinned solid state at $\sigma = \sigma_{\mathrm{a}}$ but continues to slide until σ reaches some lower critical stress σ_{b}.

As discussed in Sect. 8.9.1, the hysteresis shown in Fig. 9.2b can have two origins. The first follows from the fact that the temperature in the adsorbate system during sliding is higher than that of the substrate and might be so high that the fluid configuration rather than the solid pinned state is stable for $\sigma_{\mathrm{b}} < \sigma < \sigma_{\mathrm{a}}$. A more general explanation is the following. In Sect. 8.9.1 we showed that the return to the pinned solid state at $\sigma = \sigma_{\mathrm{b}}$ involves a nucleation process. But a drag force will act on a pinned island from the surrounding flowing 2D-fluid. Assuming a circular pinned island, and that the drag force is uniformly distributed over the adsorbates in the island, the drag force is so large that the island will fluidize if $\sigma > \sigma_{\mathrm{b}} \approx \sigma_{\mathrm{a}}/2$.

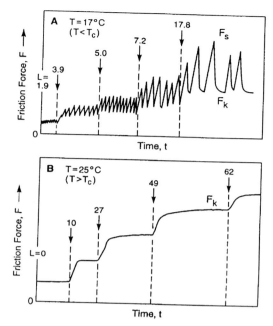

Fig. 9.3. Measured friction force as a function of time at increasing load L (in units of mN, shown by arrows) for two mica surfaces separated by a ~ 12 Å hexadecane film and sliding at a fixed velocity of $v_s = 0.4$ μm/s at (A) T=17°C, and (B) T=25°C. From [9.1].

The picture presented above has been obtained from numerical simulations on finite systems but can be understood on the basis of simple physical arguments and should therefore be very general. But these theoretical arguments also indicate that the transition to the pinned state at $\sigma = \sigma_b$ may be more complex than indicated by the simulations. In particular, we will now present both theoretical and experimental arguments that the return to the pinned state may occur as indicated by the thick dashed line in Fig. 9.2b, i.e., the $v = f(\sigma)$ relation has a wide region $0 < v < v_b$ where $\sigma \approx \sigma_b$.

The lubrication layer for $0 < v < v_b$ is likely to consist of a "granular fluid" with pinned solid islands and 2D-fluid. To see this, suppose we reduce the stress so that pinned islands start to occur. Now, if the islands are pinned by both of the sliding surfaces simultaneously, then, since it will take time for an island to grow and since the block and the substrate are in relative motion, during the growth of an island a force will build up on it due to the local (at the island) elastic deformations of the block and the substrate. If the force on the island becomes large enough, the island will fluidize. At high sliding velocity (but $v < v_b$) the force on the island increases so rapidly with time that this nucleation process can probably be neglected. On the other

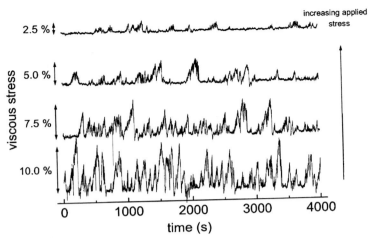

Fig. 9.4. Viscous stress fluctuations in time for four different applied stresses increasing from bottom to top. From [9.6].

hand, if an island is initially pinned by only one of the two sliding surfaces, with different islands being pinned by either of the two different surfaces, then "collisions" between pinned islands will occur during the sliding process which could result in fluidization of islands. This should result in a "granular" sliding state for the adsorbate layer, where pinned islands are continuously being formed and fluidized.

Experimental support for a granular sliding state at low sliding velocities comes from oscillatory shear experiments [9.6] on a thin film of squalane ($C_{30}H_{62}$, a branched hydrocarbon) confined between two atomically smooth surfaces of mica. In the slip regime close to the transition to stick, extensive viscous stress fluctuations, revealed as triangular spikes on a constant baseline, are observed. Figure 9.4 shows the kinetic friction force as a function of time. It is astonishing to observe that the kinetic friction force fluctuates by $\sim 10\%$; evidently some kind of coherent structures, extended over distances huge when compared with molecular dimensions, occur, otherwise they would have averaged out over the contact area whose diameter was 45 µm. One observes that the friction increases steadily at a well-defined rate and then collapses abruptly. This provides direct support for a "granular" state with pinned regions and fluid regions. Other oscillatory shear experiments have shown that close to the transition to stick significant elastic stress persisted, which decreased with increasing deflection amplitude [9.7]. The fact that elastic stresses occurred and that the elastic component decreased with increasing oscillation amplitude supports the existence of pinned islands; a larger oscillation amplitude would then imply stronger forces on the pinned islands and hence would tend to fluidize a larger fraction of them than would an oscillation of smaller amplitude.

9.2 Inertia and Elasticity: "Starting" and "Stopping"

Suppose that the free end of the spring in Fig. 9.1a is pulled very slowly. This will lead to elastic deformation of the block (and the spring) as indicated in Fig. 9.5a. As long as the tangential stress at $z = d$ is below σ_a the block will not move and the displacement field $u(z,t)$ is given by $u = -\kappa\sigma_s z$, so that the tangential stress in the block equals σ_s everywhere. When σ_s reaches the critical stress σ_a the lubricant layer will fluidize and the surface $z = d$ of the block will start to move with some velocity v_0 to be determined below. This change in the displacement field will propagate with the sound velocity c towards the upper surface ($z = 0$) of the block (Fig. 9.5b). To study this elastic wave propagation in detail [9.8], let $t = 0$ be the time at which the adsorbate layer fluidizes. The displacement field for times $0 < t < d/c$ can be written as (Fig. 9.5b,c)

$$u = -\kappa\sigma_a z, \quad 0 < z < d - ct, \tag{9.4}$$

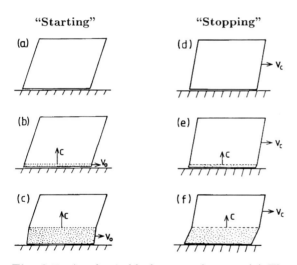

Fig. 9.5. An elastic block on a substrate. (a) The stress σ_s in the block induced by the external spring is just below the critical stress σ_a necessary to fluidize the adsorbate layer. (b) The spring stress has reached σ_a, the adsorbate layer has fluidized and the bottom surface of the block moves with the velocity v_0. (c) The region of "motion" propagates with the transverse sound velocity c towards the upper surface of the block. After the time interval d/c the elastic wave has reached the upper surface and the whole block moves with the velocity v_0. (d) The stress σ in the block is just above the critical stress σ_c necessary for the adsorbate layer to return to its pinned solid state. (e) The stress has reached σ_c, the adsorbate layer has returned to its pinned state and the bottom surface of the block has ceased moving. (f) The region of "stopped motion" propagates with the transverse sound velocity c towards the upper surface of the block. After the time interval d/c the elastic wave has reached the upper surface and the whole block has ceased moving.

320   9. Boundary Lubrication

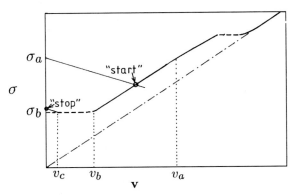

Fig. 9.6. "Starting": Graphical solution of equations (9.7) and (9.8). The "stopping" transition is determined by a similar graphical construction.

$$u = u_0 - \kappa\sigma_0 z + v_0 t, \quad d - ct < z < d, \tag{9.5}$$

where (u_0, σ_0, v_0) are constants. The displacement field u must be continuous for $z = d - ct$, which gives

$$u_0 - \kappa\sigma_0 z + v_0(d-z)/c = -\kappa\sigma_a z$$

or

$$u_0 + v_0 d/c = 0, \tag{9.6}$$

$$\kappa\sigma_0 + v_0/c = \kappa\sigma_a. \tag{9.7}$$

The displacement field given by (9.4–7) is continuous everywhere *and* satisfies the wave equation (9.1). In order to satisfy the correct boundary condition on the surface $z = d$, the stress σ_0 and the velocity v_0 occurring in (9.5) must satisfy

$$v_0 = f(\sigma_0). \tag{9.8}$$

In Fig. 9.6 we show the graphic solution to (9.7) and (9.8). Note that the system "jumps" [in the (v, σ) plane] from $v = 0$ to a finite velocity v_0. It is of crucial importance to note that this is possible because initially only a very thin (in the continuum model, infinitesimal) slab of solid at the bottom of the block (z=d) needs to be accelerated to the speed v_0. If the block were perfectly rigid then the whole block would have to change its velocity. This can only occur "slowly" in order not to generate enormous forces of inertia. But any real solid has a finite elasticity (and a finite sound velocity) and the transition indicated in Fig. 9.6 can occur practically instantaneously.

According to (9.4) and (9.5), at the moment ($t = c/d$) the elastic "starting" wave generated at the lower surface of the block reaches the upper surface, the whole block moves uniformly with the velocity $v_0 = (\sigma_a - \sigma_0)/\rho c$.

This is the same motion as exhibited by a rigid block and, in fact, the velocity of the elastic block at $t = d/c$ is exactly the same as that of a rigid block exposed to a surface stress σ_a on the upper surface and $-\sigma_0$ on the lower surface. This follows at once from Newton's equation of motion for the center of mass, $M\ddot{x} = A(\sigma_a - \sigma_0)$, so that $\dot{x} = A(\sigma_a - \sigma_0)t/M$. Substituting $t = d/c$ in this equation gives $\dot{x} = Ad(\sigma_a - \sigma_0)/Mc = (\sigma_a - \sigma_0)/\rho c$ where $\rho = M/Ad$ is the mass density. Hence, the motion of the block at time $t = d/c$ is identical to that of a rigid block, but for $t < d/c$ the results differ and, as shown above, this has important implications for the nature of "starting". Furthermore, in the rigid-block limit the stress σ_0 is not defined.

It is, in principle, possible to follow the motion of the block for all times by simply studying how elastic waves are generated and reflected from the surfaces $z = 0$ and $z = d$ of the block, accounting for the appropriate boundary conditions at these surfaces. But, except for the "starting" and the "stopping" (see below) time periods of duration d/c, such a treatment is in most cases equivalent to the motion of a rigid block.

After the time d/c, the elastic wave generated at the bottom surface ($z = d$) reaches the top surface ($z = 0$), and the whole block moves uniformly with velocity v_0. From here on we can accurately study the motion by assuming a rigid block. When the stress $\sigma(t)$ for $z = d$ finally decreases to σ_b, the lubricant layer will return to its pinned state. But just as in "starting", this "stopping" transition cannot be treated by using the rigid-block limit, as the force of inertia to retard a rigid macroscopic block from speed v_c to zero over a distance of order an atomic lattice spacing would be very high and would immediately fluidize the pinned structure. This causes no problem in the elastic continuum model, since we only need to "stop" the motion of the $z = d$ side of the block in order to return to the pinned state; the region of "stopped" motion then propagates as an elastic "stopping" wave with sound velocity c towards the upper surface ($z = 0$) of the block, and only after the time period d/c is the whole block stationary (Fig. 9.5d–f).

The discussion above is based on the assumption that the fluidization and freezing transitions of the lubrication film occur rapidly over the whole block–substrate interface; in reality this is unlikely to be the case if the transitions occur by nucleation and growth. But the analysis above can also be applied to solid islands, and shows that for even the most rapid fluidization and freezing of the lubrication layer, there are no problems related to the inertia of the sliding block.

9.3 Computer Simulations of Boundary Lubrication

Thompson and *Robbins* [9.9] have studied a very simple model of boundary lubrication. The model consists of two thin solid slabs separated by a thin film of fluid molecules interacting via Lennard–Jones pair potentials. The fluid was modeled as either spherical molecules or chain molecules. A Lennard–Jones potential (with modified parameters) was also used to model the interaction between the fluid molecules and the discrete wall atoms. For a thick fluid film, over a wide range of parameters, one or more layers of fluid molecules crystallized at the solid surfaces. This tendency to form a solid wetting phase increased with the degree of corrugation in the wall–fluid potential and with the degree to which the natural spacing of fluid molecules matched that of the wall. The walls were face-centered cubic solids with (111) surfaces. The walls were only a few atomic layers thick, and the wall atoms were coupled to the sites of a rigid lattice via stiff springs.

To mimic boundary lubrication experiments, simulations were performed in a constant load ensemble where the separation between the surfaces was allowed to vary. The top plate was coupled via a spring with constant k_s to a stage moving at a constant velocity v_s. As the stage moves forward, the spring stretches and the spring force F_s increases. If the film is in a liquid state, the top plate accelerates until the steady-state F_s balances the viscous friction force. If the film has a crystalline state when $F_s = 0$, stick-slip behavior is observed as illustrated in Fig. 9.7. Initially, the film responds elastically. The

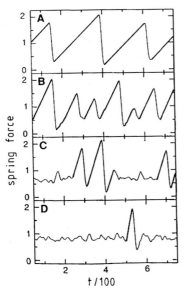

Fig. 9.7. The variation of the spring force with the velocity v_s of the drive. (A) $v_s = 0.1$, (B) $v_s = 0.2$, (C) $v_s = 0.3$, and (D) $v_s = 0.5$. From [9.9].

top plate remains stuck, and F_s increases linearly with time. Eventually F_s exceeds the yield stress of the film and the top plate begins to slide. Since the friction force is less in the sliding state, the plate accelerates to catch up with the stage and F_s decreases. At some lower value of the spring force, the top plate sticks once more, and the process repeats.

During sliding, the plate separation was found to increase by about 1 Å. Snapshots of the lubrication film show that during sliding it has a fluid-like configuration while crystalline order occurs during stick. As v_s increases, the maximum value of F_s remains fixed, but variations in the static friction force increase substantially and intermittent periods without stick-slip motion become more frequent. These changes occur because the film does not have enough time to fully order during each stick. Above a critical velocity v_c the film remains in a liquid state and slides smoothly.

Simulation performed with amorphous walls generated by rapid quenching of a fluid state leads to similar sliding dynamics as discussed above. What changes is the static and kinetic friction forces but stick-slip motion is always observed below a critical velocity v_c assuming that the lubrication film is not in a fluid state when $F_s = 0$. When the walls are amorphous, the order induced in the fluid at "stick" is no longer crystalline. Instead, a glassy structure that minimizes the wall–fluid interaction is observed. Glassy states may also be formed by long-chain molecules between ordered plates because the relaxation of these molecules is slow.

In the simulations the transition from smooth sliding to stick-slip motion occurred at some critical spring velocity v_c which was found to scale as $M^{-1/2}$ with the mass M of the moving wall. This fact can be explained as follows: In order for the block to stop, its kinetic energy must be converted into potential energy in the film. The maximum potential energy that can be stored in the film is on the order of aF_a where a is a lattice constant of the wall, and F_a the static friction force. Equating this to the kinetic energy at $v = v_c$ gives

$$v_c \sim (aF_a/M)^{1/2} . \tag{9.9}$$

Alternatively, this equation follows from the fact that the force of inertia generated when the block changes its momentum from Mv_s to zero during the time period $\tau \sim a/v_c$ is of order of $Mv_c/\tau = Mv_c^2/a$, and the condition that this force equals F_a gives (9.9). However, for a real physical sliding system this explanation for the occurrence of the critical speed v_c is *incorrect*, since the block as a whole will never stop abruptly but, as discussed in Sect. 9.2, initially only the bottom surface of the block stops and the inertia forces involved in this are negligible. The reason why $v_c \propto M^{-1/2}$ is observed in the simulations of *Thompson* and *Robbins* is the unphysical treatment of the elastic properties of the walls: By assuming that the wall atoms are connected to the sites of a *rigid* lattice, all the atoms in the block or wall must stop simultaneously at the return to the pinned state. Removing this unphysical assumption by using an elastic block gives a different mechanism for the origin of the critical velocity v_c as discussed in the next section and in Chap. 12.

9.4 Origin of Stick-Slip Motion and of the Critical Velocity v_C

In this section we discuss the origin of the transition from slip to stick. We consider first a block sliding on a tilted substrate (Fig. 9.1b) and then the more complex case of a block pulled by a spring (Fig. 9.1a).

Critical Velocity for a Block Sliding on a Tilted Substrate

Consider a solid block (mass M) sliding on a tilted lubricated substrate (Fig. 9.1b). When the tilt angle α is reduced to a critical value α_c the motion will stop. We show below that the motion of the block stops when α is reduced to the point where the steady-state sliding velocity reaches a critical value v_c, which is determined by the condition that the solid pinned state can be nucleated in the lubrication film. For the spring–block system (Fig. 9.1a), if the spring constant k_s is very weak, the transition from steady to stick-slip motion will occur at the same sliding velocity v_c.

Let us first return to equation (9.9) and point out that *most of the elastic energy is not stored in the lubrication layer*, as assumed in the derivation of (9.9), *but rather in the elastic deformation field of the walls of the confining solids*. This can be seen as follows. If a shear stress σ acts on the surface of an elastic solid within the circular area πR^2 then the center of the area will displace a distance u given by $F = k_{\text{eff}} u$, where $F = \pi R^2 \sigma$ is the total tangential force and $k_{\text{eff}} = \alpha E R \approx ER$ [where E is the elastic modulus and $\alpha = \pi/(2-\nu)(1+\nu)$ – see (9.12) – is of order unity], the effective spring constant. Using this formula, (9.9) must be replaced by

$$Mv_c^2/2 \approx F_s a/2 + 2k_{\text{eff}} u_0^2/2 \ .$$

If we use the fact that $k_{\text{eff}} u_0 = F_s$ we get

$$Mv_c^2/2 \approx F_s a/2 + F_s^2/k_{\text{eff}} \ .$$

Since $k_{\text{eff}} \approx ER$, and writing $F_s = \pi R^2 \sigma_s$, we obtain the condition

$$\frac{1}{2}Mv_c^2 \approx \frac{\pi}{2} R^2 \sigma_s a + \frac{\pi^2 R^3 \sigma_s^2}{E} \ ,$$

or

$$v_c \approx \left(\frac{\pi R^2 \sigma_s a}{M}\right)^{1/2} \left(1 + \frac{2\pi R \sigma_s}{aE}\right)^{1/2} \ . \tag{9.10}$$

If we write $\sigma_s = \mu_s \sigma_0$, where σ_0 is the perpendicular stress in the junction, then

$$\frac{2\pi R \sigma_s}{aE} = 2\pi \mu_s \frac{R}{a} \frac{\sigma_0}{E} \ .$$

9.4 Origin of Stick-Slip Motion and of the Critical Velocity v_c

In most cases $2\pi\mu_s \approx 1$ and $R/a \approx 10^4$, where we have assumed that R is a few μm and a a few Å. If we assume that σ_0 is close to the yield stress of the solid walls, then typically $\sigma_0/E \approx 0.01$, so that the ratio between the elastic energy stored in the solid walls and in the lubrication film becomes ~ 100. In some cases, e.g., for rubber on a rough substrate, $\sigma_0 \approx E$, and in this case the ratio becomes $\sim 10^4$.

As another application, let us consider a mica junction lubricated by one or two monolayers of hexadecane. In this case

$$\frac{2\pi R \sigma_s}{aE} \approx 400 >> 1,$$

where we have used the experimental contact radius $R = 30\,\mu\text{m}$, the static shear stress $\sigma_s \approx 10^7 \text{N/m}^2$, $a \approx 1\text{Å}$ and the elastic modulus $E \approx 10^{10} \text{N/m}^3$. Thus, the elastic energy stored in the lubrication film is completely negligible compared to the elastic energy stored in the solids. We now show that the critical velocity calculated from (9.10) is much larger then observed experimentally, and the mechanism considered above can therefore be ruled out. To prove this, note that for the parameter values quoted above (and with $M = 0.01$ kg) we have

$$\left(\frac{\pi R^2 \sigma_s a}{M}\right)^{1/2} \approx 17\,\mu\text{m},$$

so that according to (9.10), $v_c \approx 400\mu\text{m/s}$, which is a factor of ~ 1000 times higher than observed experimentally.

I now present what I believe is the correct explanation for the transition from slip to stick. First, both experimental and theoretical results have been presented suggesting that the transition from slip to stick involves nucleation of solid structures in the (fluidized) lubrication film. Let us assume that at time $t = 0$ a small circular solid region of radius R and area $\Delta A = \pi R^2$ has been formed due to a fluctuation. The solid island pins the two solid walls together. Now, even if the formation of the solid island were to occur instantaneously, there should be no "problems" related to the inertia of the sliding block since the maximum initial stress (which would result if the pinning were to occur perfectly abruptly) in the contact region is given by [see (9.7)] $\sigma = \rho v c_T$, where c_T is the transverse sound velocity and ρ the mass density of the solid. In a typical case this gives $\sigma \approx 10$ N/m^2, which is completely negligible compared to the yield stress of a solid domain, which is similar to the static frictional shear stress and typically of order $\sim 10^8$ N/m^2. Thus the initial increase in the shear stress at the solid island, associated with the elastic stopping waves in the confining solid walls, is negligible. For times $t > R/c_T \sim 10^{-12}$ s (where we have assumed an initial radius $R \sim 10$ Å) the shear stress increases monotonically with time according to $k_{\text{eff}} u$. Now if this increase in the shear force is faster than the increase in the pinning force which results from the growth of the solid island, then the solid island will shear melt and no transition to the stick state will occur. Let us analyze the problem in greater detail.

Assume that the radius $R(t) = v_1 t$ of an island increases linearly with time. In order to break the junction, the local surface stress at the island must reach the stress σ_s necessary for fluidization of the solid structure. But the force which acts on the junction equals $\pi R^2 \sigma_k + k_{\text{eff}} v_c t$ where σ_k is the kinetic frictional stress. Hence v_c is determined by the condition

$$\pi R^2 \sigma_s \approx \pi R^2 \sigma_k + k_{\text{eff}} v_c t ,$$

or, since $R = v_1 t$,

$$v_c \approx \frac{\pi(\sigma_s - \sigma_k)}{E} v_1 . \tag{9.11}$$

Equation (9.11) determines the velocity at which solid domains, which pin the two surfaces together, can form in the lubricant film. However, as shown below, this does not necessarily imply that the motion of the block will stop abruptly at $v = v_c$. Rather, if the block is "heavy", $M > M_c$, it will continue to slide a finite distance, while solid domains in the lubrication film form and fluidize. When the block reaches a lower critical velocity v_c' [given by (9.10)] it stops abruptly. If $M < M_c$ the motion of the block stops abruptly at $v = v_c$.

We now show that when the driving force (or the tilt angle α) has been reduced to the point where the sliding velocity of the block reaches v_c, the block stops abruptly if $M < M_c$, while it will continue to slide a finite distance if $M > M_c$, where the distance slid is roughly proportional to $M - M_c$. The basic idea is as follows: Assume that the driving force is reduced to the point where the sliding velocity becomes equal to v_c so that the solid, pinned phase can nucleate. Now, it will take some time τ for the whole contact area to get fully pinned (in the case studied above, this is the time it takes for the solid islands to grow and fill out the whole contact area). We may interpret τ as the time period over which the junction strengthens. If the solid film were to grow out from a single point (nucleus) as assumed above, $\tau \sim R/v_1$, but in a more complex case, where the solid film may nucleate at several different points at different times (within the time period τ), τ may be shorter. Now, it will take some time before the velocity of the center of mass of the block has decreased to zero. If this time is longer than τ, then, since the strengthening of the junction saturates for times larger than $\sim \tau$, while the shear stress in the contact area continues to increase (because the center of mass of the block still moves forwards) the elastic shear stress will finally (for $t > \tau$) reach the static shear stress, at which point the lubrication film shear melts, and local slip occurs. In this case the block will continue to slide for a while (but under retardation), while the contact area will perform stick-slip motion.

Let us study this problem in greater detail. Assume first that $M > M_c$ and $v = v_c$, and that the contact area with the substrate abruptly stops

9.4 Origin of Stick-Slip Motion and of the Critical Velocity v_c

moving at time $t = 0$. At short times we know that the shear force acting on the junction takes the form $F \approx k_{\text{eff}} vt + F_k$. However, this expression for F grows without limit as $t \to \infty$, and must break down for large t because of the decrease of the center of mass velocity of the block. Let us estimate the time dependence of the shear force F. If $x(t)$ denotes the center of mass coordinate of the block, where $t = 0$ correspond to the time when the contact area stops moving, we get for $t > 0$ with $x(0) = 0$:

$$M\ddot{x} \approx -k_{\text{eff}} x + F_k ,$$

where F_k is the kinetic frictional force for $v = v_c + 0^+$ (which is identical to the driving force). During stick we get

$$x(t) \approx A \sin(\omega t) + F_k/k_{\text{eff}} ,$$

where $\omega = (k_{\text{eff}}/M)^{1/2}$. Since $\dot{x}(0) = v = \omega A$ we get $A = v/\omega$. Thus the shear force acting in the contact area is

$$F \approx (k_{\text{eff}} v/\omega) \sin(\omega t) + F_k .$$

For short times ($\omega t \ll 1$) this formula reduces to $F \approx k_{\text{eff}} vt + F_k$, while the maximum $k_{\text{eff}} v/\omega + F_k$ occurs for $t = \pi/2\omega$. Assume now that the block-substrate pinning force, for the reasons discussed above, increases with the time of stationary contact up to the time τ, and then stays constant. It is clear that if $\tau > (M/k_{\text{eff}})^{1/2}$, the motion of the block will stop abruptly when v reaches v_c, while if $\tau < (M/k_{\text{eff}})^{1/2}$ the pinning will be broken after a time of order τ, after which the contact area will rapidly "jump" forward (because of the elongation of the "spring" k_{eff}), only to get pinned again. During this process energy will be dissipated at a rate (per unit length) which is higher than during steady sliding with $v = v_c + 0^+$. Thus, the effective kinetic friction force is larger than the driving force, and the block will slow down (retard), and finally the motion stops. The condition $\tau \approx (M_c/k_{\text{eff}})^{1/2}$ defines the critical mass M_c. Using $k_{\text{eff}} \approx RE$ this gives $M_c \approx ER\tau^2$.

To summarize, for $M < M_c$ the block stops abruptly when the driving force is reduced to the point where the steady state sliding velocity $v = v_c$. For $M > M_c$ the block will start to *decelerate* when the driving force is reduced to the point where the steady state sliding velocity $v = v_c$, but the motion does not stop until the velocity (at a constant driving force) has reached $v'_c < v_c$, where v'_c will be much smaller than v_c if the block is very heavy. Note that v'_c is determined by the arguments presented earlier, namely that the motion will stop when the kinetic energy of the block equals the elastic energy stored in the solid walls when the shear stress in the contact area equals the static shear stress. This gives $v'_c \approx \pi F_s/(MER)^{1/2}$. From the equations above it follows, for $M \geq M_c$, that $v'_c \leq v_c$, where the equality holds when $M = M_c$.

For the sliding configuration in Fig. 9.1a with $k_s > 0$, the transition from steady to stick-slip sliding will, in general, occur at a velocity below v_c. This follows from the observation that when the motion of the block has stopped, the spring force $k_s(v_s t - x)$ will increase, and if the rate of increase of the spring force is higher than the the rate of increase of the friction force due to the freezing of the lubrication layer, the motion of the block will not stop. This point was discussed in Sect. 3.1 but is reiterated below.

Critical Velocity for a Block–Spring System

Assume that after the return to the pinned state the static friction force depends only on the time of stationary contact t, $F_0 = F_0(t)$. We assume that $F_0(0) = F_b$, where F_b is the kinetic friction force at low sliding velocity just before stick, and that $F_0(t)$ increases monotonically with the time t of stationary contact, as shown in Fig. 9.8. The dashed lines in Fig. 9.8 show the increase in the *spring force*, $k_s v_s t$, as a function of the contact time for three different cases. In case (1) the spring force increases faster with t than the initial linear increase of the static friction force; hence the motion of the block will not stop and no stick-slip motion will occur. If the spring velocity v_s is lower than the critical velocity v_c [cases (2) and (3)] determined by the initial slope of the $F_0(t)$ curve [$k_s v_c = dF_0/dt(t=0)$], the spring force will be smaller than the static friction force $F_0(t)$ until t reaches the value t_2 [case (2)] or t_3 [case (3)], at which time slip starts. In these cases stick-slip motion will occur.

In the model above, the increase of $F_0(t)$ results from the growth of the solid phase at "stop". But there are also other mechanisms which could give rise to an increase in the static friction force with increasing time of stationary contact. One such mechanism is related to the increase in the contact area with the time of stationary contact (Chap. 5 and Sect. 13.2). This is caused by slow (thermally induced) plastic flow. In the measurements using the surface forces apparatus only elastic deformations occur, and the contact area is constant.

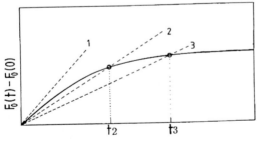

Fig. 9.8. Solid line are the variation of the static friction force with stopping time. Dashed lines are the variation of the spring force for three different (*1*)–(*3*) sliding velocities, $v_1 > v_c$ but $v_c > v_2 > v_3$.

9.5 Comparison with Experiments and Discussion

Using the surface force apparatus, *Berman* et al. [9.11] have studied the sliding of two mica crystals separated by a 12 Å thick tetradecane film (about 3 monolayers) at $T = 21°C$. Both overdamped and underdamped motion was observed. In this section we present a qualitative discussion of these results. In Chap. 12 we present a more complete analysis of lubricated friction dynamics based on a friction law constructed from the microscopic picture presented above.

Underdamped Motion

Underdamped sliding dynamics can occur under various conditions, e.g., when the spring constant k_s is large. In this case, when the block slides after having been in "stick", it will accelerate for a very short time period before the spring force equals the friction force, after which the block will decelerate. Hence, the block will never gain a high sliding velocity (assuming that v_s is "small") and the kinetic friction force can usually be treated as a velocity independent constant, i.e., $F_k(v) \approx F_b$. In this case, as discussed in Sect. 7.6, the spring force will oscillate between F_a and $2F_b - F_a$ and the slip time is approximately half the cycle time $T = 2\pi(M/k_s)^{1/2}$ of a free harmonic oscillator.

An example of underdamped motion is shown in Fig. 9.9. During slip, the spring force F_s varies with time according to $F_s = F_b + A\sin(2\pi t/T)$ (see dashed curve in Fig. 9.9), as expected if the the kinetic friction force can be approximated by a velocity independent constant [see (7.50)]. The slip period is close to $\pi(M/k_s)^{1/2}$.

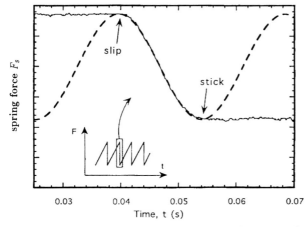

Fig. 9.9. Measured spring force F_s as a function of time for underdamped slip. The dashed line is a cosine fit to the data, which is the predicted motion of the slider using its inertial properties and assuming a constant intrinsic kinetic friction $F_k(v) \approx F_k(0)$. From [9.11].

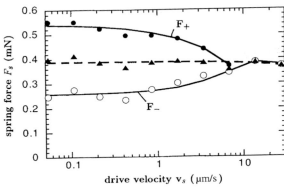

Fig. 9.10. The maximum F_+ and the minimum F_- of the spring force during stick-slip motion, as a function of the drive velocity for OMCTS confined between mica surfaces. F_+ decreases with increasing velocity because at higher drive speeds the sticking time is shorter, resulting in less complete freezing of the lubrication layer. The observed F_- increases as F_+ decreases, resulting in a nearly constant kinetic friction $F_k(v)$. The critical velocity is reached when F_+ decreases to the intrinsic friction F_k. From [9.11].

For the phase transition model the static friction force increases with the time of stationary contact because of the growth of the solid state. Thus, if the stick time is very short, the system will not reach the maximally pinned state. Hence, higher sliding velocity, which gives rise to shorter stick times, should result in a lower static friction force and thus a smaller amplitude of the stick-slip oscillations, see Sect. 12.1 (Fig. 12.4, right). That this is indeed the case is shown in Fig. 9.10, where the maximum F_+ and the minimum F_- spring force (during a stick-slip cycle) are plotted as a function of the drive velocity v_s. The maximum spring force F_+ equals the static friction force F_a while the minimum spring force F_- should equal $2F_b - F_a$ if the model with a velocity-independent kinetic friction force is correct. Note that F_+ decreases and F_- increases with increasing drive velocity v_s until the two are equal at the critical velocity v_c, beyond which only smooth sliding occurs. Using $F_- = 2F_b - F_a$ we can calculate F_b (dashed line in Fig. 9.10), which turns out to be nearly constant over the velocity range in Fig. 9.10.

Figure 9.11 shows the critical velocity v_c as a function of the spring constant k_s for OMCTS between mica surfaces. As expected, the critical velocity decreases with increasing k_s; the solid line is derived in Sect. 12.1 [see (12.10)] and is in relatively good agreement with the experimental data.

Overdamped Motion

Let us now consider the case of overdamped motion. This may occur, e.g., if the spring k_s is very soft. In this case the spring is strongly elongated at the beginning of slip and the block will accelerate for a relatively long

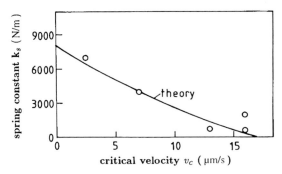

Fig. 9.11. Critical velocity as a function of the spring constant k_s for OMCTS between mica surfaces. The load between the surfaces is held constant at 24.5 mN. From [9.11].

time before the spring force equals the kinetic friction force. In this case the block will acquire a high sliding velocity and the velocity dependence of the kinetic friction force cannot be neglected. This is easy to see in the limit of an extremely weak spring where the spring force will be (nearly) equal the static friction force (which it had at the beginning of slip) for such a long time that the block reaches such a high velocity that the kinetic friction force becomes nearly equal to F_a, before the block starts to be retarded. Furthermore, the time period of retardation will be much longer than the time of acceleration.

Figure 9.12 shows an example of overdamped motion. In this case the total slip time is of order 10 s, in contrast to the case of underdamped motion where the slip time was in the millisecond range. In this regime, slip begins with a rapid acceleration followed by a lengthy deceleration before resticking occurs. Note also that the stick-slip amplitude is independent of the driving velocity v_s almost all the way up to the critical velocity v_c^+ (Fig. 9.12, inset). The transition from stick-slip to smooth sliding with increasing v_s goes from regular, periodic stick-slip at $v_s < v_c^+$ through a non-periodic stick-slip regime as v_s approaches v_c^+, to smooth sliding at $v_s > v_c^+$. Note also that in the present case $F_- \approx F_b$. The underdamped and overdamped sliding scenarios considered above represent extreme cases; in general the sliding dynamics will be somewhere between these limiting cases.

Motion of the Area δA of Real Contact

The discussion above has focused on the spring force which reflects the motion of the center of mass of the block. It is important to note, however, that in the surface forces apparatus, the diameter $D = 2R$ of the circular contact area ($\delta A = \pi R^2$), is very small compared with the linear size L of the sliding block (typically $D \sim 0.01$ cm while $L \sim 1$ cm) and the motion of the surface area δA, *where the friction force is generated*, is not the same as that of the center of mass of the block.

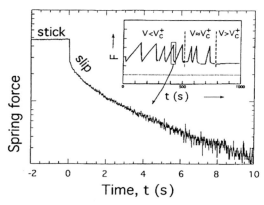

Fig. 9.12. Overdamped stick-slip motion. During slip, the friction drops rapidly at first and then more slowly. *Inset*: The measured F_a and F_b are nearly independent of the drive velocity below the critical velocity. At drive velocity $v_s > v_c^+$ the smooth sliding friction is equal to the same F_b as deduced from the stick-slip regime. This indicates that in stick-slip, the friction does drop to the steady state kinetic friction before sticking again in the cycle. From [9.11].

Assume that an isotropic elastic medium occupies the half-space $z > 0$. On the surface $z = 0$ we assume that a tangential stress σ acts within a small circular region with radius R centered at a material point P. This will give rise to an elastic deformation of the solid, and P will displace parallel to the surface by the amount [9.12]

$$u = \frac{1+\nu}{2\pi E} \sigma \int_{r<R} d^2x \frac{1}{r} \left(2(1-\nu) + 2\nu \frac{x^2}{r^2} \right) ,$$

where $r^2 = x^2 + y^2$ and where ν is the Poisson's ratio and E the Young's modulus. The integral is trivial to perform, giving

$$u = (2 + \nu - \nu^2)\sigma R/E = (1 - \nu/2)(\sigma R/\rho c^2) ,$$

where c is the transverse sound velocity. If we define a force constant by $k_{\text{eff}} u = F$, where $F = \pi R^2 \sigma$ is the total force, then

$$k_{\text{eff}} = \pi \rho c^2 R/(1 - \nu/2) = 2\rho c^2 D^* , \tag{9.12}$$

where $D^* = (\pi/2)R/(1 - \nu/2)$. For metals, $\nu \approx 0.3$, so that $D^* \approx 2R = D$ and $k_{\text{eff}} \approx 2\rho c^2 D$.

For semi-quantitative purposes, we may replace the real physical system in Fig. 9.13a, with the fictive system shown in Fig. 9.13b, where the spring connecting the miniblock to the big block (mass M) has the bending force constant $k_{\text{eff}} \approx \rho c^2 D$ and the mass of the miniblock equals $M^* \approx \rho D^3$. The

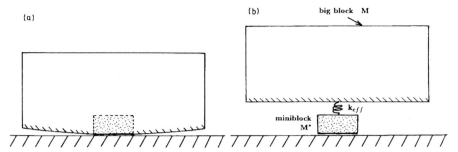

Fig. 9.13. (a) A mica block on a mica surface. The mica block has a slightly curved surface with a single contact area with the substrate. The dashed volume element defines the miniblock. (b) A small block of mass M^* is connected to a big block with mass M by a spring with the bending force constant k_{eff}. The free end of the spring k_s moves with the velocity v_s.

big block has the (lower) surface area A and the miniblock the surface area δA, where $\delta A \ll A$. As before, a spring with force constant k_s is connected to the big block and the free end of the spring is moved with the speed v_s, which we assume is below v_c. Acting on the big block are the forces from the springs k_s and k_{eff}. We assume that $k_{\text{eff}} \gg k_s$ and $M \gg M^*$.

The motion of the miniblock and of the surface area δA is assumed to be as follows. The spring k_{eff} first deforms elastically until the critical surface stress σ_a has been reached (point A in Fig. 9.14), at which point the lubrication film fluidizes. Within a very short time period the miniblock reaches its maximum velocity v_0 (point B) after which its velocity first decreases (overdamped motion) while the velocity of the big block increases. After the short time period τ, the average velocity $v(t)$ of the surface $z = d$ of the big block (which corresponds to the spring velocity for the miniblock) and the velocity of the miniblock will (nearly) coincide; we denote $v(\tau) = v_\tau$. For $t > \tau$ the motion of the miniblock will follow that of the big block. The time period τ is so short that the spring force $k_s[v_s\tau - x(\tau)]$ is nearly equal to the value F_a it had at the start of slip. Hence, since the kinetic friction force acting on the big block (via the spring k_{eff}) is approximately $F_k(v_\tau) \approx F_b$, the big block will accelerate until the spring force equals the friction force, after which retardation occurs; the "turning point" is denoted by D in Fig. 9.14. If the spring k_s is very stiff (resulting in underdamped motion), the turning point D occurs close to $v = 0$ where the spring force is $\approx F_b$. If the the spring k_s is very soft (overdamped motion), the turning point D occurs when $v \approx v_a$ where $F_k(v) \approx F_a$. During a full stick-slip cycle, the miniblock goes through the sequence $F - A \to B - C - D - C - E \to F$.

We note that in most practical cases the velocity v_τ is negligibly small compared to the highest velocity attained by the big block at later times.

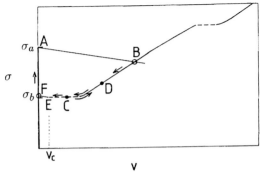

Fig. 9.14. The relation between the stress σ at the bottom surface of the miniblock and the velocity v of the same surface. After the fluidization of the adsorbate layer (transition $A \to B$) a rapid relaxation of the velocity of the miniblock occurs ($B - C$). During this time period the big block accelerates and at point C the big block and the miniblock have the same velocity. From here on the big block and the miniblock move together with almost the same velocity, both accelerating to reach the maximum velocity at point D. The system then decelerates and when the velocity reaches the critical velocity v_c (point E) the bottom surface of the miniblock returns to the pinned state (transition $E \to F$) and the whole cycle $F - A \to B - C - D - C - E \to F$ repeats itself. For a stiff spring k_s the big block accelerates for a very short time and point D occurs at such a low velocity that the velocity dependence of the kinetic friction force can be neglected during the motion $C \to D \to E$ (underdamped motion). If the spring k_s is very weak, the big block accelerates for a very long time and acquires a very high velocity (point D occurs close to $v = v_a$) and the velocity dependence of the kinetic friction force cannot be neglected (overdamped motion).

That is, the motion of the big block will, after the initial rapid relaxation of the mini block, start from $v = v_\tau \approx 0$ and only gradually increase and then decrease again. Hence, the motion of the big block can be studied by solving Newton's equation for a rigid block, with the initial condition $v \approx 0$ at the start of the slip.

10. Elastic Interactions and Instability Transitions

The study in Chap. 9 assumed that the lateral corrugation of the adsorbate–substrate interaction potential was so weak that the adsorbate layer could fluidize as a result of the shear force stemming from the external force acting on the block. For this case we have argued that the kinetic frictional force at low sliding velocities is likely to involve the formation and fluidization of solid structures. But in many cases the corrugation of the adsorbate–substrate potential energy surface is so strong that no fluidization of the adsorbate layer can occur. This often seems to be the case when fatty acids are used as boundary lubricants: the polar heads of fatty acid molecules bind strongly to specific sites on many metal oxides and the sliding occurs now between the inert hydrocarbon tails as indicated schematically in Fig. 2.5. In Sect. 10.1 we discuss some simple models which illustrate the fundamental origin of the friction force when no fluidization can occur. As will be shown below, if the corrugated substrate potential is large enough compared with the local elasticity, an elastic instability will occur, which will result in a kinetic friction force which remains finite as the sliding velocity $v \to 0$ (we assume zero temperature, so that no thermally activated creep motion occurs). In Sect. 10.2 the elastic coherence length ξ is introduced and calculated for a semi-infinite elastic solid exposed to a random surface stress. In Sect. 10.3 we consider the sliding or "pushing" of islands and big molecules on surfaces where the non-zero elasticity again is of fundamental importance. Finally, Sect. 10.4 presents an exact solution for the force necessary to cause the sliding a 1D elastic chain over a defect.

10.1 Elastic Instability Transition

Consider the sliding system shown in Fig. 10.1a. A particle with mass m is connected via a spring (bending force constant k) to a block. The particle experiences the corrugated "substrate" potential $U(q)$. The equation of motion for the particle is

$$m\ddot{q} = -U'(q) - m\gamma\dot{q} + k(x-q), \tag{10.1}$$

where the displacement coordinates x and q are defined in the figure. The friction force $-m\gamma\dot{q}$ in (10.1) originates from the coupling of the adsorbate

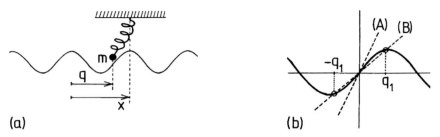

Fig. 10.1. (a) A particle (mass m) connected by a spring (bending force constant k) to a block sliding on a corrugated substrate (interaction potential U). (b) Graphical solution of (10.5). *Solid line*: the force $-U'(q)$ from the substrate. *Dashed lines*: the spring force kq for (A) a "stiff" and (B) a "soft" spring, k.

to the substrate excitations as discussed in Sect. 8.2. That is, $-U'(q)$ is the conservative, and $-m\gamma\dot{q}$ the dissipative part of the adsorbate–substrate interaction. [In principle, a friction term of the form $-m\gamma'(\dot{q} - \dot{x})$, caused by the interaction with the upper solid, will also exist but the exact nature of the friction force is irrelevant for what follows.] Let us define the effective potential

$$U_{\text{eff}}(x, q) = U(q) + k(x - q)^2/2 \, . \tag{10.2}$$

We first study the equilibrium position of the particle as a function of x. Assume for simplicity that

$$U(q) = U_0 \cos(2\pi q/a) \, . \tag{10.3}$$

At equilibrium $\dot{q} = 0$ and (10.1) gives

$$U_0(2\pi/a)\sin(2\pi q/a) + k(x - q) = 0 \, . \tag{10.4}$$

Suppose that the spring is centered over a local maxima in the potential $U(q)$, see Fig. 10.1a. In this case (10.4) reduces to

$$U_0(2\pi/a)\sin(2\pi q/a) - kq = 0 \, . \tag{10.5}$$

The graphical solution to this equation is shown in Fig. 10.1b for two different cases, namely, for a "stiff" spring (A) and for a "soft" spring (B). In case (A) only $q = 0$ is a solution to (10.5) while in case (B) three solutions exit, namely, $q = 0$ and $q = \pm q_1$. More than one solution exists only if the initial slope of the function $U_0(2\pi/a)\sin(2\pi q/a)$ is larger than the slope of the function kq, i.e., if $\kappa = U_0/\mathcal{E} > 1$, where the *elastic energy* $\mathcal{E} = ka^2/4\pi^2$. If $\kappa < 1$, $q = 0$ and the spring is straight as in Fig. 10.2a. If $\kappa > 1$ at least three solutions exist, but since $U''_{\text{eff}}(0) = -U_0(2\pi/a)^2 + k < 0$, if $\kappa > 1$ the $q = 0$ solution is unstable and the particle will occupy a displaced position ($q = q_1$ or $-q_1$) as indicated in Fig. 10.2b. The discontinuous flipping from one stable position

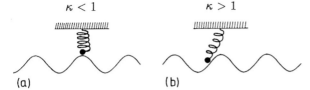

Fig. 10.2. Stable equilibrium positions for the particle when (a) $\kappa < 1$ and (b) $\kappa > 1$.

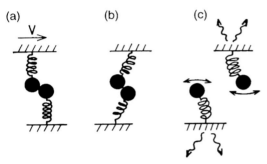

Fig. 10.3. As two surfaces slide relative to each other, atomic groups will interlock (a), deform elastically (b), and finally undergo a rapid slip process (c). The rapid local motion is damped by the emission of sound waves.

to another, which occurs during sliding at low drive velocities if $\kappa > 1$, is called an *elastic instability transition* [10.1].

Suppose now that the upper surface moves parallel to the substrate with the velocity v, i.e., $x = vt$. The resulting sliding dynamics differs in a profound manner according to whether $\kappa > 1$ or $\kappa < 1$. If $\kappa < 1$ the particle will perform a slow (if v is small) oscillatory motion with frequency $\omega = kv$ superimposed on a uniform translational motion vt and the maximum particle velocity will decrease monotonically towards zero as $v \to 0$. The force acting on the block from the substrate will oscillate in time; if we define the kinetic friction force F_k as the time average of this fluctuating force then $F_k \to 0$ as $v \to 0$. On the other hand, for $\kappa > 1$, "fast" motion of the particle will always occur for some time intervals, independent of how small the driving velocity v is. This is indicated in Fig. 10.3 where first "slow" (if v is small) elastic deformation occurs until q reaches q_c (the displacement at the onset of elastic instability), at which point the particle flips rapidly to another potential well, where it will perform a damped oscillatory motion (if γ is large the motion will be overdamped). Thus, if $\kappa > 1$, after averaging over time, a finite kinetic friction force results even when $v \to 0$.

The model studied above considers only one particle (or several non-interacting particles). However, in practice very many interacting atoms or

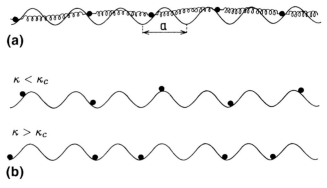

Fig. 10.4. (a) A one-dimensional string of point particles connected via harmonic springs and located on a corrugated substrate (potential U). (b) The distribution of particles when $\kappa \ll \kappa_c$ and $\kappa \gg \kappa_c$ (schematic).

molecules will be present at the contact areas between the block and the substrate. A model which contains some aspects of this situation is the Frenkel–Kontorova model [10.2]. This model consists of a one-dimensional string of point particles connected via harmonic springs and located on a corrugated substrate (Fig. 10.4a). The particles obey the equation of motion

$$m\ddot{q}_i = -U'(q_i) - m\gamma \dot{q}_i + k(q_{i+1} + q_{i-1} - 2q_i) + F , \qquad (10.6)$$

where F is the driving force. We assume (as is usually the case in practice) an incommensurate system, i.e., that the ratio b/a between the "lattice constant" a of the substrate and the natural distance b between two nearby particles on the string is an irrational number. Consider first the ground state. The system described by (10.6) undergoes an elastic instability transition if $\kappa = U_0/\mathcal{E}$ (where $\mathcal{E} = ka^2/4\pi^2$) is large enough. The critical κ depends on the ratio b/a. For example, if $b/a = (1+\sqrt{5})/2 = 1.6180\ldots$ then the transition occurs at $\kappa_c \approx 0.88$. Figure 10.4b shows schematically the distribution of atoms for $\kappa > \kappa_c$ and $\kappa < \kappa_c$. For $\kappa \gg \kappa_c$ the amplitude U_0 of the corrugated substrate potential is so large compared with the elastic energy that the particles tend to accommodate close to the bottom of the potential wells, while for $\kappa \ll \kappa_c$ the particles are more or less uniformly distributed along the corrugated substrate, with some particles close to the local maxima of $U(q)$.

Let us now consider the dynamical properties of the system in Fig. 10.4a. For $\kappa < \kappa_c$ the static friction force vanishes, i.e., an arbitrarily weak external force will allow the adsorbate layer to slide along the corrugated surface. For $\kappa > \kappa_c$ the static friction force is non-vanishing. This behavior is plausible from Fig. 10.4b: For $\kappa < \kappa_c$, as some particles move downhill during sliding other particles will move uphill in such a manner that there is no net energy barrier towards sliding and an arbitrarily weak external force will induce sliding. In contrast, if $\kappa \gg \kappa_c$ all particles are initially close to the bottom of

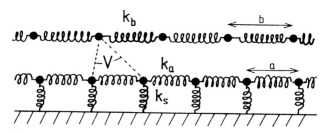

Fig. 10.5. Sliding system studied by *Matsukawa* and *Fukuyama* [10.4].

the potential wells and moving particles from one potential well to another will involve surmounting the intervening barrier, thus producing a non-zero static friction force.

Elastic instability transitions occur also in more complex and realistic model systems than the ones studied above [10.3]. For example, Fig. 10.5 shows the sliding system studied by *Matsukawa* and *Fukuyama* [10.4] This model differs from the Frenkel–Kontorova model mainly in the fact that the rigid substrate potential $U(q)$ has been replaced by a string of atoms connected to a rigid substrate via elastic springs (with bending force constants k_s). The particles in the upper string interact with the substrate particles via a pair potential $V(x) = V_0 \exp[-4(x/a)^2]$ where x is the separation between a particle on the string and a particle on the substrate. With this $V(x)$, when the lower substrate atoms are fixed (i.e., $k_s \to \infty$), the force acting on a string atom is almost a sinusoidal function so in this case the present model reduces essentially to the Frenkel–Kontorova model. In the results below, $b/a = 1.619$.

Figure 10.6 shows the velocity v of the center of mass of the string as a function of the external force F (which equals the friction force per particle). In case (a) the interaction V_0 is so small that no elastic instability transition has occurred and the static friction force vanishes and the kinetic friction force goes continuously to zero as $v \to 0$. In cases (b) and (c) V_0 is larger and an elastic instability transition has occurred. This implies that the static friction force F_s is non-zero and that the kinetic friction force F_k does not vanish as $v \to 0$, but approaches the static friction force in this limit. The latter result differs from the case of sliding of adsorbate layers where the kinetic friction force $F_k < F_s$ as $v \to 0$ and where a velocity gap occurs, i.e., a velocity region where no steady sliding motion is possible. This difference is related to the fact that, during the transition from the sliding state to the pinned state, there is a discontinuous change in the particle system in the case of sliding of adsorbate layers, while in the present case the transition is continuous. Snapshots of the particle system show that when V_0 is large atoms jump (rapidly) from one metastable position to another as time passes, while for small V_0 the motion occurs slowly if v is small.

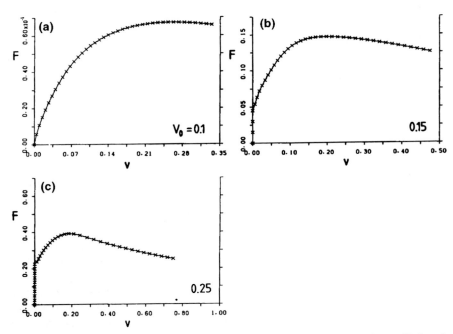

Fig. 10.6. The sliding velocity v as a function of the driving force F for the system in Fig. 10.5 for V_0 equal to (a) 0.1, (b) 0.15 and (c) 0.25. Spring constants $k_b = k_s = 1$, $k_a = 0$. From [10.4].

Caroli and *Noziéres* [10.5] have developed a model of dry friction based on a purely elastic response of interacting asperities. The asperities are treated as point defects that are swept along as the slider moves. When two asperities collide they will deform elastically; one can define an elastic spring constant k so that the elastic displacement u of an asperity is related to the acting force via $F = ku$. For elastically "hard" solids, where k is large, no elastic instability occurs and the kinetic friction force vanishes continuously as the driving velocity $v \to 0$. However, for elastically soft materials an elastic instability will occur and (at zero temperature) the kinetic friction force continuously approaches to the static friction force $F_s > 0$ as $v \to 0$. In the latter case, independent of the driving speed v, rapid processes will occur during some fraction of the sliding process.

10.2 Elastic Coherence Length

In a typical sliding friction experiment a block is pulled over a substrate. The block is usually treated as a mass point subject both to a friction force and to the pulling force. In this section we address the question under what conditions this picture is accurate and what modifications to the sliding dynamics occur when it no longer holds [10.6]. We show that if the block (say a cube of side L) is smaller than a characteristic length ξ, the elastic coherence length, the block can be treated as a point mass, while if $L > \xi$ it is necessary to divide the block into cells of linear size ξ which are elastically coupled to each other but pinned individually at the block–substrate interface.

The concept of an elastic coherence length ξ was first developed in the context of flux line lattices [10.7] and charge density waves [10.8]. Both these systems can, under some conditions, be treated as (anisotropic) elastic solids pinned by imperfections, e.g., by grain boundaries or dislocations. When an external driving force F acts on it, the elastic solid will not move until F reaches a critical value F_p. However, during motion (at low sliding velocity, i.e., F close to F_p) the solid will not translate as a single unit but only the volume elements of size $\sim \xi$ will displace coherently. In Sect. 14.4.1 we will present a detailed discussion of the concept of elastic coherence length.

Consider an elastic block on a substrate. The contact between the two bodies will occur at (almost) randomly distributed contact areas (junctions). During sliding at low velocities the contacting asperities will first deform elastically and then (when the local stress has reached the static shear stress) they will slip rapidly (we neglect creep motion). Since the asperities have different sizes, if the sliding motion is stopped and the external force reduced to zero the contacting asperities will, in general, be in an elastically deformed state as indicated in Fig. 10.7a, i.e., there is a residual stress. The lower surface on the block will therefore be acted upon by an (almost) random tangential surface stress $(\sigma_{xz}, \sigma_{yz})$ with $\langle \sigma_{xz} \rangle = \langle \sigma_{yz} \rangle = 0$ [note: (x, y, z) is a coordinate system with the (x, y)-plane in the block–substrate interface]. We now show that the surface stress σ_{i3} (where $i = 1, 2, 3$ denote x, y, z) will induce an instability transition in the block (and the substrate) which will "split" the block at the interface into "correlated volumes" which behave elastically independently and are pinned individually (Fig. 10.7b).

The stress $\sigma_{i3}(\boldsymbol{x})$, which acts on the lower surface of the block, will give rise to the displacement field $u_i(\boldsymbol{x})$ in the block. Consider the correlation function

$$I(\boldsymbol{x}) = \langle [\boldsymbol{u}(\boldsymbol{x}) - \boldsymbol{u}(\boldsymbol{0})]^2 \rangle = 2[\langle \boldsymbol{u}^2(\boldsymbol{0}) \rangle - \langle \boldsymbol{u}(\boldsymbol{0}) \cdot \boldsymbol{u}(\boldsymbol{x}) \rangle] \,, \tag{10.7}$$

where $\boldsymbol{x}$ is a point on the surface $z = 0$, and where $\langle ... \rangle$ stands for ensemble averaging over different realizations of the (random) distribution of contact junctions. If N denotes the number of contact "points" and A the surface area, then the concentration of contact points is $n_\mathrm{a} = N/A$ and the fraction

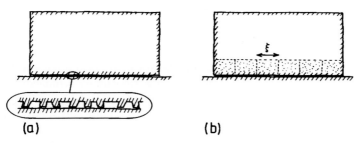

Fig. 10.7. (a) Elastic block on a substrate. The asperities of the block which contact the substrate will, in general, be in an elastically deformed state and a random tangential surface stress will act on the lower surface of the block and on the substrate. For simplicity, the substrate is assumed to be perfectly smooth. (b) The surface stress at the block substrate interface will "divide" the block (and the substrate) at the interface into "correlated volumes" which behave elastically independently and are pinned individually.

of the surface area where real contacts occur is $P_\mathrm{a} = N\delta A/A$, where δA is the (average) contact area of a junction. In order to evaluate (10.7), we need the correlation function $\langle \sigma_{i3}(\boldsymbol{x})\sigma_{j3}(\boldsymbol{x'}) \rangle$ on length scales large compared with the average distance between two nearby contact points. In this case we can approximate

$$\langle \sigma_{i3}(\boldsymbol{x})\sigma_{j3}(\boldsymbol{x'}) \rangle = K_{(i)}\, \delta_{ij}\, \delta(\boldsymbol{x} - \boldsymbol{x'}) \;, \tag{10.8}$$

where

$$K_1 = n_\mathrm{a}\langle \sigma_{xz}^2 \rangle \delta A^2, \quad K_2 = n_\mathrm{a}\langle \sigma_{yz}^2 \rangle \delta A^2, \quad K_3 = n_\mathrm{a}\sigma_0^2 \delta A^2 \;. \tag{10.9}$$

Here we have assumed that the perpendicular stress $\sigma_{zz} = \sigma_0$ at each contact area (in most cases σ_0 is the plastic or brittle fracture yield stress) while the tangential stress $(\sigma_{xz}, \sigma_{yz})$ fluctuates randomly from one contact area to another with $\langle \sigma_{xz} \rangle = \langle \sigma_{yz} \rangle = 0$. We define $\sigma_\parallel^2 = \sigma_{xz}^2 + \sigma_{yz}^2$. The surface stress σ_{i3} gives rise to the displacement u_i given by

$$u_i = \int d^2x'\, G_{ij}(\boldsymbol{x} - \boldsymbol{x'})\sigma_{j3}(\boldsymbol{x'}) \;. \tag{10.10}$$

In what follows, we only need u_x and u_y on the surface plane $z = 0$. In this case [10.9] ($i = 1, 2;\ j = 1, 2, 3$)

$$G_{ij}(\boldsymbol{x}) = \frac{1+\nu}{2\pi E}\frac{1}{|\boldsymbol{x}|}\left[(2\nu - 1)\frac{x_i}{|\boldsymbol{x}|}\delta_{j3} + 2(1 - \nu)\delta_{ij} + 2\nu\frac{x_i x_j}{|\boldsymbol{x}|^2}\right]\;, \tag{10.11}$$

where ν and E are Poisson's number and Young's modulus, respectively. Using (10.7,8,10,11) gives

$$I(\boldsymbol{x}) = 2\sum_{ij} K_{(j)} \int d^2x'\left[G_{ij}^2(-\boldsymbol{x'}) - G_{ij}(\boldsymbol{x} - \boldsymbol{x'})G_{ij}(-\boldsymbol{x'})\right] \;. \tag{10.12}$$

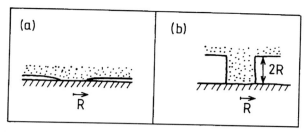

Fig. 10.8. Two models (a) and (b) of a contact area between the block and the substrate.

Now, note that

$$\int d^2x' \frac{1}{|x'|} \left(\frac{1}{|x'|} - \frac{1}{|x-x'|} \right)$$

$$= \int_0^{2\pi} d\varphi \int_a^\infty d\rho \left(\frac{1}{\rho} - \frac{1}{(\rho^2 + x^2 - 2\rho|x|\cos\varphi)^{1/2}} \right)$$

$$= \int_0^{2\pi} d\varphi \int_{a/|x|}^\infty d\eta \left(\frac{1}{\eta} - \frac{1}{(\eta^2 + 1 - 2\eta\cos\varphi)^{1/2}} \right)$$

$$\to 2\pi \ln(|x|/a) \quad \text{for} \quad |x|/a \gg 1, \tag{10.13}$$

where, in order for the ρ-integral to converge, we have introduced the short-distance cut-off a, which is on the order of the length scale over which the shear stress is highly correlated, which in turn we take to be on the order of the asperity junction contact diameter, typically a few micrometers. Substituting (10.11) into (10.12) and using (10.13) gives for $|x|/a \gg 1$

$$I(x) = \frac{(1+\nu)^2}{\pi E^2} \left[K_3(1-2\nu)^2 + 2(K_1+K_2)(2-2\nu+\nu^2) \right] \ln\left(\frac{|x|}{a}\right). \tag{10.14}$$

The random surface stress σ_{i3} splits the lattice into "correlated volumes" $V_c \sim \xi^3$ that behave elastically independently and are pinned individually. We obtain ξ from the requirement [10.10] $I(\xi) \approx u^2$ where u is the magnitude of the lateral displacement necessary for the local stress to reach the static friction stress, and which depends on the shape of a junction (Sect. 14.4.1 gives a more detailed discussion of the concept of an elastic coherence length). We consider the two cases shown in Fig. 10.8. If the tangential stress $\sigma_1 \hat{x}$ acts within the area $\delta A = \pi R^2$, then the center of δA will be displaced by an amount u where, from elastic continuum theory,

$$u = C \sigma_1 R/E. \tag{10.15}$$

In case (a) we have $C = 2 + \nu - \nu^2 \approx 2.2$, while in case (b) we take C to correspond to uniform shear within a cylinder (radius R and height $2R$)

added to case (a), $C \approx 4 + 3\nu - \nu^2 \approx 4.8$, where for the Poisson number we have used $\nu = 0.3$. Using (10.9,14,15) gives

$$\xi = a\, e^{Q/P_a}, \tag{10.16}$$

where

$$Q = \frac{2C^2}{(1+\nu)^2}\left[\frac{(1-2\nu)^2}{\mu^2} + 2(2 - 2\nu + \nu^2)\frac{\langle\sigma_\parallel^2\rangle}{\sigma_1^2}\right]^{-1}, \tag{10.17}$$

with $\mu = \sigma_1/\sigma_0$ the static friction coefficient and σ_1 the tangential stress necessary to break a junction. If we assume $\langle\sigma_\parallel^2\rangle/\sigma_1^2 = 1/3$ and $\mu = 0.5$ we get with $\nu = 0.3$, $Q = 3.5$ in case (a) and $Q = 7.3$ in case (b).

Let us apply (10.16) to a few examples. First, consider a steel cube with 1 dm side on a steel table. The plastic yield stress of steel is $\sigma_c \sim 10^9\,\text{N/m}^2$ so that $P_a = Mg/\sigma_c A \sim 10^{-5}$ and $\xi \sim 10^{100000}$ m, i.e., the coherence length is much larger than the side of the block, and the block will slide as a single unit. Next, consider applications to earthquakes. It has been suggested [10.11] that the smallest earthquakes involve blocks of order $(100\,\text{m})^3$ so that we must have $\xi \approx 100\,\text{m}$ (see also [10.12,13] for other estimates of the smallest earthquakes). Using (10.16) with $a \sim 1$ μm gives $Q/P_a \approx 18$. Small earthquakes occur typically at $h \sim (10 \pm 5)$ km depth where the pressure $\sim \rho g h \sim (3 \pm 1.5) \times 10^8\,\text{N/m}^2$. The yield pressure of granite is $\sim (1.5 \pm 0.5) \times 10^9\,\text{N/m}^2$ so that $P_a \sim 0.2 \pm 0.1$. Hence, $Q = 4 \pm 2$ which is very reasonable. It is interesting to note that small earthquakes typically do not occur close to the Earth's surface [10.14] (the upper ~ 2 km of crust forms a "dead" layer); this is expected from the model above since the "low" pressures in this region make P_a much smaller and hence ξ much larger than at larger depth. But the surface region can participate in large earthquakes where a large fraction of the fault moves. Note, however, that some materials, e.g., limestone, are much softer than granite and if they are present close to the surface, P_a may still be relatively large and relatively small earthquakes can then also occur close to the Earth's surface.

When studying the sliding behavior of a block that is larger than the coherence length ξ the block cannot be treated as a point mass, but must be divided into cells of size ξ coupled elastically to each other.

To summarize, we have calculated the elastic coherence length ξ at the interface between an elastic block and a substrate, where the random block–substrate interaction potential generates a pinning potential. We found $\xi = a\exp(Q/P_a)$ where a is of order the length scale over which the shear stress is highly correlated, and Q a number which depends on the nature of the contact areas but which typically is somewhere in the range 3–8, and P_a is the fraction of the apparent block–substrate contact area where real contact occurs. In most practical applications ξ is much larger than the sliding block (linear size L) and the block can be treated as a point mass. However, for very soft materials under heavy load, P_a may be large and $\xi < L$, and in these cases it is necessary to discretize the block into cells of linear size ξ and study the motion of each cell. For example, the Brinnell hardness of lead is $400 \, \text{kg/cm}^2$ and a mass load of order $130 \, \text{kg}$ on a cubic lead crystal with $1 \, \text{cm}$ sides, would therefore result in $P_a \approx 0.3$ and $\xi \sim 1 \, \text{cm}$. Hence, for loads above $\sim 100 \, \text{kg}$ the lead cube would not slide as a point mass but local slip processes would occur at the block–substrate interface. To test the theory presented above, we suggest recording the elastic waves emitted from the sliding junction as a function of the load. The emitted waves will have a high-frequency cut-off at $\omega_c \sim c/\xi$ (for $\xi < L$) and increasing the load will decrease ξ and increase ω_c. A particularly interesting and important application of the theory above is in earthquake dynamics. This topic will be discussed in Sect. 14.1.

10.3 Sliding of Islands, Big Molecules, and Atoms

The forces necessary to slide islands and large molecules on surfaces have been measured using the surface force microscope. In these measurements a sharp tip is brought into contact with one edge of an island or a large molecule and, by applying a lateral force F, the island (or molecule) can be displaced parallel to the surface.

Sliding of Islands

Figure 10.9 shows the motion of a C_{60} island (diameter $\sim 2000 \, \text{Å}$) adsorbed on a NaCl(001) surface at room temperature [10.15]. In the surface force microscopy images (a–d) the cluster mainly rotates, whereas in (e–g) it translates. The forces needed to move two different islands are shown in Fig. 10.10 as a function of the lateral displacement. The force needed to move an island is proportional to the area and depends on the location on the substrate. On flat terraces the islands can be easily moved, whereas at step edges the lateral force is higher and the motion may even be inhibited.

During lateral sliding, when the probing tip (oxidized Si) comes in contact with the perimeter of the island, the tip first deforms elastically until the lateral force reaches F_s at which point the island starts to move. During motion the force decreases to $F_k \sim F_s/2$. When the island approaches an imperfection on the substrate, it stops sliding and the lateral force increases again until the tip can rise over the island. Since the area A of the cluster is known, it is possible to calculate the shear stress $\sigma = F/A$. For island *1* in Fig. 10.9a, $A = 7.5 \times 10^6 \, \text{Å}^2$. Thus the shear stress to initiate sliding $\sigma \approx 1 \times 10^5 \, \text{N/m}^2$ while during steady sliding $\sigma \approx 0.5 \times 10^5 \, \text{N/m}^2$. These shear stresses are typically of a factor 100 lower than those observed with the surface forces apparatus for mica surfaces covered by grafted chain molecules. In another study it was found that islands of C_{60} on graphite could not be

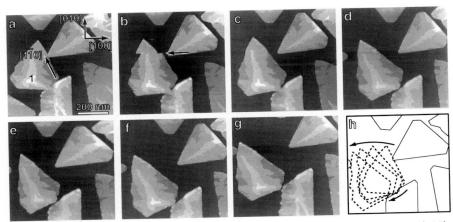

Fig. 10.9. A sequence of surface force microscopy images of C_{60} on NaCl(100). The dark areas correspond to NaCl(100) substrate and the bright areas to C_{60} islands. A C_{60} island [labeled as **1** in **a**] is moved in steps (**a–g**) by the action of the probing tip. The image (**b**) represents a typical shear event when an island is moved during the scan process [indicated by an arrow in **b**]. (**h**) shows a summary of the movement of the C_{60} island. From Ref. [10.15].

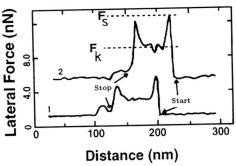

Fig. 10.10. Force as a function of sliding distance when two different C_{60} clusters, *1* and *2*, are slid on a NaCl(100) substrate. The sliding occurs from right to left. Note that the maximum lateral force F_s (the static friction force) is about a factor of 2 larger than the kinetic friction force F_k observed during steady sliding. At "stop" the motion of the C_{60} island stops and the tip slides over the island. The small "bump" in the friction force after "stop" corresponds to the friction force on the C_{60} island, which is higher than on the clean NaCl surface. From Ref. [10.15].

slid as a whole but were pinned so strongly by the substrate that they instead disrupted when the lateral force reached some critical value.

In order to determine the origin of the friction forces it is important to know how they depend on the temperature T, on the size of the island, and the sliding velocity v. No study of the temperature dependence has been performed yet. However, the friction force was found to be the same at $v = 0.1$ µm/s and 1 µm/s, indicating that it is independent of the sliding velocity at low sliding velocities. Based on this observation we propose that the kinetic friction force has the following origin. The C_{60} lattice is likely to be incommensurate with the substrate lattice. In this case, for an infinite system, if the ratio $U_0/\mathcal{E}$ between the lateral corrugation amplitude U_0 of the adsorbate–substrate interaction potential and an elastic energy $\mathcal{E}$ associated with the C_{60} island (Sect. 10.1) is below a critical value, the static friction force vanishes and the kinetic friction force is proportional to the sliding velocity (at low sliding velocity); see Sect. 10.1. However, if the ratio $U_0/\mathcal{E}$ is large enough an *elastic instability* occurs resulting in a non-vanishing static friction force, and a kinetic friction force which is non-zero and (at zero temperature) equals the static friction force as $v \to 0$. In the present case we postulate that this second limit is obeyed. The fact that the static friction is larger than the kinetic friction is most likely due to substrate imperfections: As pointed out above, the actual force needed to move an island depends on its location on the substrate. Thus, it is likely that $F_s > F_k$ because the island is initially pinned at, e.g., a substrate step. (It would have been interesting to stop sliding when the friction force had its low value F_k – the force to start sliding should then be equal to F_k, i.e., not the factor of 2 larger F_s found in the experiments. No such experiment was performed.) For C_{60} islands on graphite the static friction force is larger than on NaCl, and no sliding is possible before the applied force becomes so large that the island disrupts.

The discussion above is for an infinite adsorbate layer. For a finite island one would always expect some pinning due to relaxation of the positions of the atoms or molecules at the periphery of the island. However, the fact that the friction force is found to be proportional to the area A of the island rather than the length of its periphery, indicates that this boundary relaxation has a negligible influence on the friction force.

It has been suggested that C_{60} islands could be used as transport devices in the fabrication of nanometer-sized machines. Large molecules (e.g., biomolecules) could be deposited on such a "nanosled" and then transported to the desired location.

Sliding of Big Molecules

Not only islands but also big molecules can be "pushed" on a surface by a sharp tip. This was recently demonstrated by the positioning of porphyrin-based molecules in a stable predefined pattern without disrupting the internal molecular bonds [10.16]. The molecule studied is shown in Fig. 10.11 and has four "legs" that are oriented out of the plane of the porphyrin ring because of

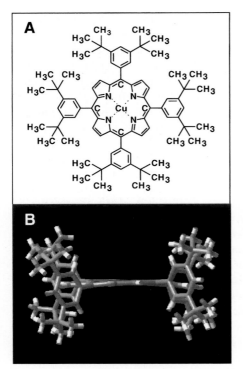

Fig. 10.11. (A) Structure and (B) conformation of Cu-TBP-porphyrin. From Ref. [10.16].

steric repulsion effects. Consequently, the interaction between the molecule and the substrate is mainly via the legs. In the experimental study a Cu(100) surface was used, for which the molecule–metal interaction is mainly of the van der Waals type.

Figure 10.12 shows the construction of a hexametric ring by pushing the molecules along the surface. Computer simulations have confirmed the crucial role of the legs in the motion of the molecule. During sliding the legs show (nearly) uncorrelated oscillatory lateral and torsional motion; the translational motion resembles a stick-slip type action of the individual legs with the surface. This effectively lowers the barrier for lateral displacement as compared with that of a rigid molecule – a crucial aspect of the sliding for this molecule.

Sliding of Atoms

The sliding of molecules (and islands) discussed above was performed by moving the tip in contact with one side of the molecule and "pushing" it along the surface. This is, of course, only possible with big molecules. However, the

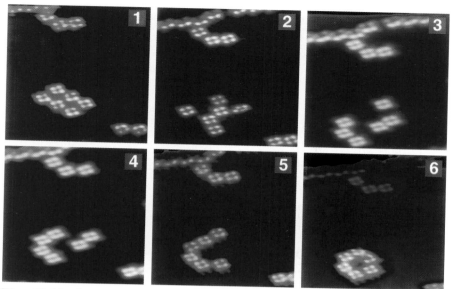

Fig. 10.12. Stages in the construction of a hexametric ring. Each molecule is individually positioned. From Ref. [10.16].

tip–substrate forces, even the weak van der Waals force, can be controlled and used to slide atoms and molecules laterally on a surface. The energy barrier to lateral motion depends on the nature of the surface and of the adsorption bond. Metal surfaces are relatively smooth, and the adsorption bonds are not very directional. As a result, the diffusional barriers are low, typically $\sim 10\%$ of the energy of the adsorption bond. At semiconductor surfaces, on the other hand, adsorption bonds are strong and directional, so that for semiconductor surfaces with open structures, the barriers for diffusion are of comparable magnitude to the binding energy. Indeed, experiments on Si(111) have shown that the tip can induce desorption and lateral displacement with comparable probabilities. Metals, however, are much better substrates for performing atom sliding. This capability was demonstrated in an elegant experiment by *Eigler* and *Schweizer* [10.17], who were able to arrange Xe atoms in a variety of patterns on a Ni(110) surface; see Fig. 10.13a. Xe is weakly adsorbed on Ni(110), with a binding energy ~ 250 meV, while the energy barrier for diffusion along the close-packed direction is ~ 20 meV. So in order to maintain a given arrangement of Xe atoms on the surface, the experiment was performed at 4K.

The mechanism by which the Xe atoms can be moved is illustrated by the potential energy curves in Fig. 10.13b [10.18]. The figure shows the potential energy as a function of the lateral displacement of Xe in the absence and presence of the tip. When the tip is far away, the Xe atom experiences a weakly

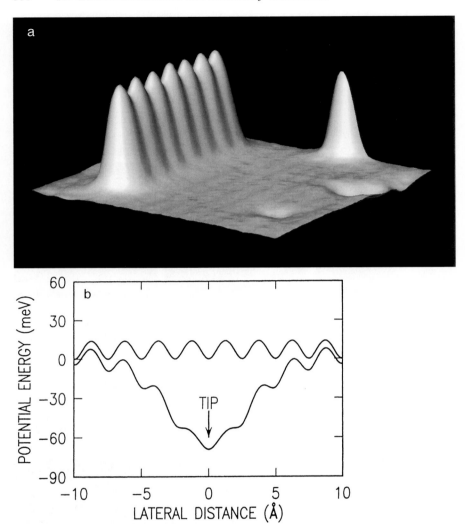

Fig. 10.13. (a) A linear chain of Xe atoms on Ni(110) obtained by sliding individual Xe atoms. The Xe atoms appear as 1.6 Å-high protrusions. From [10.17]. (b) The potential energy as a function of the lateral displacement without and with a tip. From [10.18].

corrugated potential along the surface. When the tip is lowered (tip–sample distance ∼ 4 Å), a deep van der Waals minimum develops which confines the Xe atom to the region under the tip. If the tip is moved parallel to the surface without the tip–sample distance being changed, the Xe atom will follow. When the desired location on the surface is reached, the tip is withdrawn, leaving the Xe atom in that position. We can call this arrangement for atom sliding a "van der Waals atom trap". No measurement of the force needed to slide Xe atoms has been presented, but it is clear that if elastic instabilities occur during the sliding process, the friction force (at low temperature) would be independent of the sliding velocity. Finally, we note that the same scheme as presented above, but with smaller tip–sample distances, can be used to move more strongly adsorbed species.

Rolling and Sliding of Carbon Nanotubes

For macroscopic cylindrical and spherical bodies, rolling on a flat substrate usually requires forces which are typically 100 to 1000 times *smaller* than for sliding of the same bodies. The reason for this is well understood and will be discussed in Sect. 14.8. For micro- or nanosized objects, where adhesional interactions are very important, the prevailing information is that the rolling friction may be as large, or even larger, than the sliding friction. We will illustrate this with experimental data for carbon nanotubes sliding and rolling on a graphite surface.

The experiments were performed using a sharp silicon tip to apply lateral forces at different locations along the tube. A nanotube can be pushed from the end or from the side. End-on pushes produce only sliding and stick-slip features. Side-on pushes produce either sliding or rolling. When sliding occurs the static and kinetic frictional forces are equal (see inset in Fig. 10.14b), i.e., no stick-slip spikes are observed.

Let us consider the side-on case in detail. We describe the sliding case first. One class of outcomes features a relatively smooth lateral force jump (inset in Fig. 10.14b). In this case the nanotube performs an in-plane rotation about a point which depends on the location of the tip during the push. This behavior can be explained by assuming a uniform (over the length of the

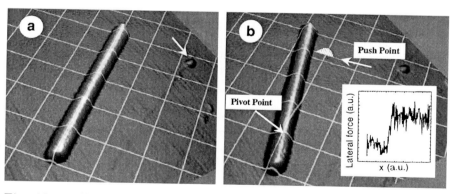

Fig. 10.14. Sliding carbon nanotube on graphite. (a) The tube in its original position. (b) The tube's position after manipulation. The pivot point and push point are indicated by the bottom and top arrows, respectively. The inset shows the lateral force trace during a sliding manipulation. From [10.20].

tube), velocity-independent shear stress acting on the nanotube during sliding. This result is very surprising, since the interaction between the carbon tube and the graphite surface is very weak (mainly of the van der Waals nature), while the internal bonds in the graphite planes (and the carbon tube) are extremely strong, suggesting that elastic instabilities should be absent. However, in the absence of elastic instabilities one expects the frictional force to be proportional to the sliding velocity and, in particular, the static friction should vanish. The reason for the finite friction may be due to the fact that the carbon tube has a finite length ("end" point effects), but in such a case it would not be correct to assume a frictional shear stress which is constant along the nanotube. Alternatively, there may be contamination on the surfaces, which could result in pinning. However, because of its inert nature, most molecules from the atmosphere do not adsorb on a graphite surface.

Let us now consider rolling, which is observed in the side-on manipulation of large nanotubes. In this case, the frictional force is found to exhibit stick-roll-stick, with a periodicity corresponding to the nanotube circumference (see Fig. 10.15g). In the experiment, the nanotube was $\sim$ 5000 Å long and had a radius of about 100 Å. The result in Fig. 10.15g can be understood as follows: After contact with the tube, the tip first deforms elastically until a critical lateral force has been reached, after which rolling occurs. During this rapid roll event the bending of the tip (and the force on the tip from the nanotube) decreases abruptly. If the tip is pushed against a tube, but is turned around before a sudden decrease in the tip-force has occurred (no rolling), the subsequently acquired topographical data indicated no resulting movement of the carbon tube. Thus, in this case the tip–tube interaction is purely elastic and no energy is "dissipated".

If the tip is stopped just after a stick-roll event, the subsequently acquired topographical image shows that the carbon tube has rotated a fixed amount, see Fig. 10.15a–f. The change in the tube-end shape from image to image indicates a rolling reorientation. Note that in Fig. 10.15b,e and c,f, tube-end shapes are similar, indicating a similar rolling angular orientation. In addition to the tube-end shape, the topographical data indicate a linear translation between images, which is consistent with rolling without slip.

What is the origin of the stick-roll-stick motion? The direct correspondence between lateral force features and the nanotube angular orientation ϕ implies that the dependence of the tube–substrate interaction energy, $E(\phi)$, on the rotation angle ϕ may be the dominant mechanism responsible for the stick-roll dynamics. For example, the amplitude of the individual stick-roll peaks in Fig. 10.15g may be related to the varying contact area between the structured tube and the substrate during the course of its roll. The magnitude of this effect can be related to the van der Waals adhesion of the nanotube with the substrate. One can estimate the substrate–tube van der Waals force to be $\sim$ 1000 nN. This force, which completely separates the tube from the substrate, should represent an upper bound for the force required for rolling.

10.3 Sliding of Islands, Big Molecules, and Atoms   353

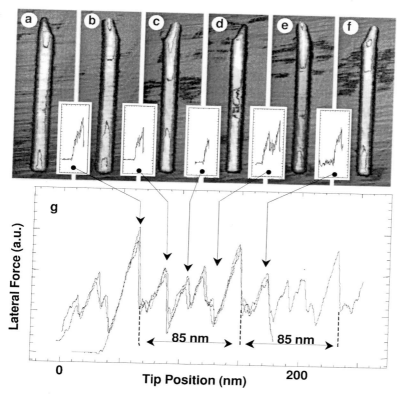

Fig. 10.15. Rolling carbon nanotube on graphite. (**a**)–(**f**) The tupe as it is manipulated from left to right. The insets shows the lateral force during each manipulation. (**g**) Three overlapping lateral force signals from separate rolling trials, as the tube is pushed through several revolutions of stick-roll motion. From [10.20].

In fact, the stick-roll peaks in Fig. 10.15g (in the range 20 − 50 nN) are only a few percent of this force. If the area of real contact is estimated using the JKR theory, the frictional stress which must be overcome to start rolling can be calculated to be of the order of a few MPa.

Rolling behavior has been found to be accompanied by a preferential, threefold, in-plane orientation that indicates nanotube/graphite lattice registry. In these cases there may be a barrier to sliding which is larger than that for rolling and may preclude sliding.

10.4 Sliding Friction: Contribution from Defects*

In Sect. 8.10.2 we studied the influence of substrate defects on the sliding friction [10.19]. For 1D and 2D elastic solids the treatment was only valid for high sliding velocities. In this section we present an exact solution, valid for arbitrary temperature and defect strength, for the sliding dynamics of a 1D solid (elastic chain) interacting with a point defect, see Fig. 10.16. The starting point is the equation of motion (8.127):

$$\rho\left(\frac{\partial^2 u}{\partial t^2} + \eta\frac{\partial u}{\partial t} - c^2\frac{\partial^2 u}{\partial x^2}\right) = f(x,t) + n_\mathrm{a} F - \rho\eta v - \lambda\sin[k(vt+u)]\delta(x) \,, \tag{10.18}$$

where

$$\lambda = 2U_\mathrm{d} k^2 f(k) \,. \tag{10.19}$$

The fluctuating force $f(x,t)$ satisfies (8.125)

$$\langle f(x,t)f(x',t')\rangle = 2\rho k_\mathrm{B} T\eta\delta(x-x')\delta(t-t') \,. \tag{10.20}$$

Let us write

$$f(x,t) = \frac{1}{(2\pi)^2}\int dq\,d\omega\, f(q,\omega)\mathrm{e}^{\mathrm{i}(qx-\omega t)} \,. \tag{10.21}$$

Substituting (10.21) into (10.20) gives

$$\langle f(q,\omega)f(q',\omega')\rangle = 2\rho k_\mathrm{B} T\eta(2\pi)^2\delta(q+q')\delta(\omega+\omega') \,. \tag{10.22}$$

A single impurity in an infinite system cannot affect the sliding friction *per particle* so that $F = m\bar\eta v = m\eta v$. Thus $v = F/m\eta$ and the term $n_\mathrm{a} F - \rho\eta v$ which occurs in (10.18) vanishes. The solution to (10.18) for $x = 0$ can be written as

$$u(0,t) = u_\mathrm{T}(0,t) + \int_{-\infty}^{t} dt'\, G(t-t')\lambda\sin k[vt' + u(0,t')] \,, \tag{10.23}$$

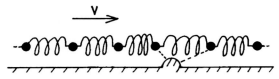

Fig. 10.16. An elastic string sliding on a flat "substrate" with a point imperfection.

where

$$G(t) = \frac{1}{2\pi\rho c} \int_0^1 ds \frac{e^{-\eta t s}}{(s-s^2)^{1/2}}, \tag{10.24}$$

and where u_T is the contribution to u from the fluctuating force $f(x,t)$:

$$u_T(0,t) = \frac{1}{(2\pi)^2} \int dq d\omega \frac{f(q,\omega)e^{-i\omega t}}{\rho(c^2 q^2 - \omega^2 - i\omega\eta)}. \tag{10.25}$$

Let us take the time derivative of (10.23). We get with $u(0,t) = u(t)$,

$$\frac{du}{dt} = \frac{du_T}{dt} + \lambda G(0) \sin k[vt + u(t)]$$

$$- \eta \int_{-\infty}^t dt' H(t-t') \lambda \sin k[vt' + u(t')], \tag{10.26}$$

where

$$H(t) = \frac{1}{2\pi\rho c} \int_0^1 ds \frac{s e^{-\eta t s}}{(s-s^2)^{1/2}}. \tag{10.27}$$

It is easy to calculate

$$G(0) = 1/2\rho c, \quad H(0) = 1/4\rho c. \tag{10.28}$$

Note that

$$\langle \dot{u}_T(t)\dot{u}_T(0) \rangle = \frac{1}{(2\pi)^4} \int dq dq' d\omega d\omega' (-\omega\omega') e^{-i\omega t}$$

$$\times \frac{2\rho\eta k_B T (2\pi)^2 \delta(\omega+\omega')\delta(q+q')}{\rho^2(c^2 q^2 - \omega^2 - i\omega\eta)(c^2 q'^2 - \omega'^2 - i\omega'\eta)}$$

$$= \frac{k_B T}{\rho c} \delta(t). \tag{10.29}$$

We assume that η is "small" and calculate the contribution from the defect to the friction force to first-order in η.

To first-order in η we can put $\eta = 0$ in the integral $H(t)$. Using $H(0) = 1/4\rho c$ this gives

$$\frac{du}{dt} = \frac{du_T}{dt} + (\lambda/2\rho c) \sin k[vt + u(t)] - \eta(\lambda/4\rho c)$$

$$\times \int_{-\infty}^t dt' \sin k[vt' + u(t')]. \tag{10.30}$$

Now, note that to zero-order in η (10.23) gives

$$(\lambda/2\rho c) \int_{-\infty}^t dt' \sin k[vt' + u(t')] = u - u_T. \tag{10.31}$$

Fig. 10.17. The dynamics when a 1D elastic solid slides over a defect is equivalent to a particle in a periodic potential pulled by a harmonic spring.

Substituting this into (10.30) gives to first-order in η

$$\frac{du}{dt} = \frac{du_T}{dt} + (\lambda/2\rho c)\sin k[vt + u(t)] - (\eta/2)(u - u_T) . \qquad (10.32)$$

We introduce $X = k(vt + u)$, $F^* = kv$, $f^* = k\dot{u}_T$ and $U_d^* = k\lambda/2\rho c$ so that

$$\frac{dX}{dt} = F^* + f^* + k\eta u_T + (\eta/2)(kvt - X) + U_d^* \sin X , \qquad (10.33)$$

where

$$\langle f^*(t) f^*(0) \rangle = 2T^* \delta(t) ,$$

$$T^* = \frac{k_B T}{\rho c} k^2 .$$

In (10.33) $k\eta u_T$ gives a small contribution to the fluctuating force f^* which has no important physical effects and we neglect this term. If we interpret $\eta/2 = k^*$ as a spring constant, then (10.33) takes the form

$$\frac{dX}{dt} = f^* + k^*(v^* t + X_0 - X) + U_d^* \sin X , \qquad (10.34)$$

where

$$X_0 = F^*/k^*, \quad k^* = \eta/2, \quad v^* = kv . \qquad (10.35)$$

Equation (10.34) describes a particle connected to a spring (spring constant k^*) where the free end of the spring moves with the velocity v^*. The particle moves in the potential $U^* = U_d^* \cos X$ and under the influence of the fluctuating force f^* (Fig. 10.17). Since the force from the spring in (10.34) increases linearly with time, after a long enough time, the spring force will be large enough to overcome the pinning barrier. Note that the contribution $k^*(v^* t + X_0)$ to the spring force in the original quantities takes the form

$$(F/2)(2ct/a) + 2\rho cv . \qquad (10.36)$$

To understand this expression, consider an elastic string which slides (velocity v) on a surface (along the x-axis) with one defect at $x = 0$. Assume that at time $t = 0$, the string atoms at the defects get pinned. This will generate a *stopping wave* which propagates to the right and to the left with

10.4 Sliding Friction: Contribution from Defects*

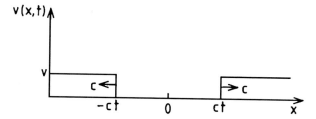

Fig. 10.18. In the limit $\eta \to 0$ and for $v < U_d k^2 f(k)/\rho c$ the solid is pinned at the defect and the "friction" is due to the "stopping wave" emitted from the defect, propagating to the right and to the left with the sound velocity c.

the sound velocity c (Fig. 10.18). To zero-order in η the displacement field $u(x,t)$ is easy to calculate. For example, for $x > 0$,

$$u_0 = vt \quad \text{for} \quad x > ct,$$

$$u_0 = vx/c \quad \text{for} \quad x < ct,$$

or

$$u_0 = vt + (v/c)(x - ct)\theta(ct - x). \tag{10.37}$$

Thus the force on the defect from the stopping wave occurring to the right of the defect is $\rho c^2 \partial u/\partial x(0,t) = \rho c v$. Similarly the force from the stopping wave from the region to the left of the defect equals $\rho c v$. Thus the total force equals $2\rho c v$ which gives the zero-order (in η) term in (10.36). Note that, if the force from the stopping waves is bigger than the force $\lambda = 2U_d k^2 f(k)$ necessary to overcome the pinning barrier, no pinning will occur. Thus to zero-order in η the de-pinning threshold is determined by $2\rho c v = 2U_d k^2 f(k)$ or $v = U_d k^2 f(k)/\rho c$.

The first term in (10.36) is of first-order in η and has the following physical origin: On the moving atoms for $x > ct$ (and $x < -ct$) act the driving force F and the friction force $m\eta v$, and since the atoms do not accelerate these forces are equal. Now, on the "stopped" atoms for $-ct < x < ct$ the latter force vanishes but the external force F still acts and must in equilibrium be balanced by the spring forces between the particles. Since the number of "stopped" atoms at time t equals $2ct/a$, the total force (derived from the external force F) acting on the defect equals $(F/2)(2ct/a)$. The presence of the factor $F/2$ rather than F results from the fact that when the stopping wave arrives at $x = ct$ it takes a time $t = x/c$ for that "message" to propagate back to the defect. Thus, effectively, only the external force acting on half of the atoms contained in the "stopped" region will contribute to the force acting on the defect.

Let us discuss the origin of the first term in (10.36) in detail. We calculate the displacement field u to first-order in η. The zero-order solution (in η) u

is given by (10.37). Now, let us write $u = u_0 + u_1$ and require that, to linear order in η, u satisfies the equation of motion

$$\rho \left(\frac{\partial^2 u}{\partial t^2} + \eta \frac{\partial u}{\partial t} - c^2 \frac{\partial^2 u}{\partial x^2} \right) = n_a F , \qquad (10.38)$$

and the boundary conditions $u(0,t) = 0$ and $u(ct,t) = vt$. This gives

$$u_1 = (\eta v / 2c^2) x (ct - x) \theta(ct - x) . \qquad (10.39)$$

The force acting on the defect from the elastic solid to the right of it is given by

$$\rho c^2 \partial u / \partial x (0, t) = \rho c v + (F/2)(ct/a) , \qquad (10.40)$$

where $F = m\eta v$. The force from the elastic solid to the left of the defect is also given by (10.40). Thus the total force is the same as obtained above [see (10.36)].

Let us discuss the nature of the solutions of (10.34). For high sliding velocity we can write

$$X = v^* t + \xi , \qquad (10.41)$$

where ξ is small. Substituting this in to (10.34) gives to leading order in $1/v^*$

$$\xi = -(U_d^*/v^*) \cos v^* t . \qquad (10.42)$$

The force on the defect is

$$F_0 = -\lambda \langle \sin X \rangle = -\lambda \langle \sin(v^* t + \xi) \rangle \approx -\lambda \langle \sin v^* t + \xi \cos v^* t \rangle . \qquad (10.43)$$

Substituting (10.42) into (10.43) and performing the time average gives the kinetic friction force

$$F_0 = \lambda U_d^* / 2v^* . \qquad (10.44)$$

Substituting (10.19) in this expression gives the same high-velocity result as derived earlier [see (8.137) with $G = \pm 2\pi/a$ included in the sum]. It is clear that for very large v the velocity of the solid at the defect has only a small modulation $\propto \cos(kvt)$ around the steady drift velocity v occurring far away from the defect. Associated with these periodic oscillations will be a damping caused by the emission of sound waves.

Let us now consider very low sliding velocities. Assume first zero temperature. We must consider two different cases. If the spring k^* is weak or the amplitude U_d^* of the defect potential large, an elastic instability occurs and the motion of the particle will always be rapid during some time periods of the sliding, independent of how low the driving velocity v^* is. On the other hand, if the spring is stiff or U_d^* small, no elastic instability will occur and the

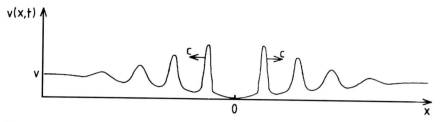

Fig. 10.19. At low driving force F the solid at the defect will perform stick-slip motion. This results in a series of wave pulses emitted from the defect, propagating both to the right and to the left of the defect with the sound velocity c. Because of the finite friction η the wave pulses are damped (the damping and the width of the wave pulses are exaggerated).

velocity $\dot{X}$ of the particle will always be of order v^*. For a cosinus potential, $U^* = U_d^* \cos X$, the elastic instability occurs if $U_d^* > k^*$ or

$$k\lambda > \rho c\eta \,. \tag{14.45}$$

In most cases of interest this inequality is satisfied. In such cases the elastic solid at the defect will perform stick-slip motion, where (for small v^*) "long" stick periods are interrupted by rapid slip. During stick the spring force increases continuously with increasing time, while X is nearly constant, until the critical force necessary to overcome the pinning barrier is reached at which point rapid slip starts. This will result in a series of wave pulses emitted from the defects. In each pulse the solid moves with a high velocity, but because of the damping η the "height" of the pulses decreases so that far away from the defect the displacement field asymptotically approaches $u = vt$ (Fig. 10.19). Since the motion is overdamped the "particle" relaxes down into the next potential well in the periodic potential $U_d^* \cos X$, i.e., the *elastic solid in each stick-slip cycle displaces a single lattice constant*. The physical origin of the overdamped motion is the emission of sound waves during the rapid local slip. If we define the kinetic friction force as the force on the defect averaged over time, it is enough to include the stick time in the average as the slip occurs very fast on the time scale of a complete stick-slip period. During a stick period the spring force increases (approximately) linearly with time. If the sliding velocity is v the time of a stick-slip period is a/v. The spring force at the onset of slip is approximately λ. During slip v^*t is nearly constant while X increases by 2π corresponding to a displacement by one lattice constant. Thus, during slip, the spring force decreases to $\lambda(1 - 2\pi k^*/U_d^*) = \lambda - mc\eta$. The kinetic friction force is therefore

$$F_0 \sim \lambda(1 - \pi k^*/U_d^*) = \lambda - mc\eta/2 \,, \tag{14.46}$$

which, for small η, is nearly equal to λ, i.e., the kinetic friction is nearly equal to the static friction, and is independent of the sliding velocity. The latter is a consequence of the fact that, independent of how small the driving velocity v is, rapid processes always occur during sliding.

Let us analyze the condition found above for an elastic instability to occur: $k\lambda > \rho c\eta$. When an elastic instability occurs the lattice at the defect performs local stick-slip motion. During "stop" an elastic stopping wave is emitted from the defect. The lattice atoms in the "stopped" area exert a force on the defect which increases in proportion to the stopping time t [see the second term in (10.40)], which can be written (using $F = m\eta v$) as:

$$F_1(t) = \rho \eta c v t.$$

The force necessary for the lattice to surmount the barrier at the defect equals λ. Thus the stopping time t_0 is determined by $F_1(t_0) = \lambda$ or $\rho\eta c v t_0 = \lambda$. Comparing this equation with (10.45) gives $kvt_0 > 1$ as the condition for local stick-slip instability to occur. That is, no local stick-slip motion occurs if the force on the defect from the lattice reaches the pinning force λ during a time period which is shorter than the time it takes for the lattice to move a distance $1/k = a/2\pi$. This condition is physically very appealing and should also be valid for 2D and 3D-systems.

The picture above is modified at non-zero temperature. For $T > 0\,\mathrm{K}$, if v is low enough, the "particle" will jump over the barrier by thermal excitation. If the velocity is very small the only role of the external spring force is to slightly "tilt" the potential energy surface so that the "particle" jumps slightly more often in the direction of the driving force than in the opposite direction. This results in "creep motion", where the friction force depends linearly on the sliding velocity as $v \to 0$.

We close this section by presenting some numerical results which illustrate the analytical results above. It is convenient to introduce a new time variable $\bar{t} = k^* t$ and define $\bar{U}_\mathrm{d} = U_\mathrm{d}^*/k^*$, $\bar{v} = v^*/k^*$ and $\bar{f} = f^*/k^*$. In these variables (10.34) takes the form

$$\frac{dX}{d\bar{t}} = \bar{f} + (\bar{v}\bar{t} + X_0 - X) + \bar{U}_\mathrm{d} \sin X$$

where

$$\langle \bar{f}(\bar{t}) \bar{f}(0) \rangle = 2\bar{T}\delta(\bar{t})$$

with $\bar{T} = T^*/k^*$. In the original variables $\bar{U}_\mathrm{d} = 2U_\mathrm{d} k^3 f(k)/\rho c\eta$ and $\bar{T} = 2k_\mathrm{B} T k^2/\rho c\eta$. Figure 10.20 shows the dependence of the friction force $\bar{F}_0$ on the sliding velocity $\bar{v}$ for different barrier heights. For $\bar{U}_\mathrm{d} > 1$ (i.e., $U_\mathrm{d}^* > k^*$) an elastic instability occurs and the friction force $\bar{F}_0(0)$ is greater than zero, while it vanishes for $\bar{U}_\mathrm{d} < 1$. The dashed line indicates the high-velocity expansion (10.44) (which takes the form $\sim \bar{U}^2/2\bar{v}$ in the present units) for the case $\bar{U}_\mathrm{d} = 3.5$. Figure 10.21 shows the friction force at low sliding velocity $[\bar{F}_0(0) = \lim_{v \to 0} \bar{F}_0(\bar{v})]$ as a function of the barrier height. The solid line is the exact result while the dashed line is the high-$\bar{U}$ expansion (10.46), which

10.4 Sliding Friction: Contribution from Defects*

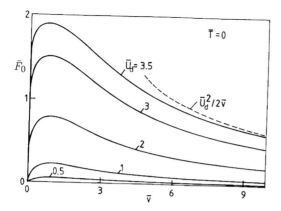

Fig. 10.20. The dependence of the friction force, $\bar{F}_0 = -\langle \bar{U}_d \sin X \rangle$, on the sliding velocity v for different barrier heights. The dashed line indicates the high-velocity expansion (10.44) (which takes the form $\sim \bar{U}_d^2/2\bar{v}$ in the present units) for the case $\bar{U}_d = 3.5$.

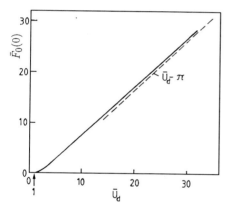

Fig. 10.21. The friction force at low sliding velocity $[\bar{F}_0(0) = \lim_{\bar{v} \to 0} \bar{F}(\bar{v})]$, as a function of the barrier height. The solid line is the exact result while the dashed line is the high-$\bar{U}_d$ expansion (10.46), which, in the present units, takes the form $\bar{F}_0 = \bar{U}_d - \pi$.

in the present units takes the form $\bar{F}_0 = \bar{U}_d - \pi$. Finally, Fig. 10.22 shows the relation between $\bar{F}_0$ and $\bar{v}$ for three different temperatures and for (a) $\bar{U}_d = 3$ and (b) $\bar{U}_d = 1$. In the former case, $\bar{F}(0)$ is non-zero when $\bar{T} = 0$, while it increases linearly with $\bar{v}$ (for small $\bar{v}$) for $\bar{T} > 0$.

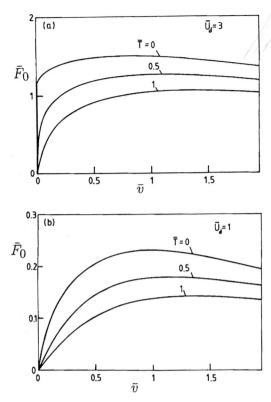

Fig. 10.22. The temperature dependence of the relation between friction force $\bar{F}_0$ and the sliding velocity $\bar{v}$ for (**a**) $\bar{U}_d = 3$ and (**b**) $\bar{U}_d = 1$.

11. Stress Domains, Relaxation, and Creep

In this section we consider creep and other slow (thermally induced) relaxation processes which occur at low sliding velocity [11.1]. We first argue that the lubrication film at low sliding velocities has a granular structure, with pinned adsorbate domains accompanied by elastic stress domains in the block and substrate. At zero temperature, the stress domains form a "critical" state, with a continuous distribution $P(\sigma)$ of local surface stresses σ extending to the critical stress σ_a, necessary for fluidization of the pinned adsorbate structure. During sliding adsorbate domains will fluidize and refreeze. During the time for which an adsorbate domain remains in a fluidized state, the local elastic stresses built up in the elastic bodies during "stick" will be released, partly by emission of elastic wave pulses (sound waves) and partly by shearing the lubrication fluid. The role of temperature-activated processes (relaxation and creep) will be studied and correlated with experimental observations. In particular, the model explains in a natural manner the logarithmic time dependence observed for various relaxation processes; this time dependence follows from the existence of a sharp step-like cut-off at $\sigma = \sigma_\mathrm{a}$ in the distribution $P(\sigma)$ of surface stresses. A simple experiment is suggested to test directly the theoretical predictions: By registering the elastic wave pulses emitted from the sliding junction, e.g., using a piezoelectric transducer attached to the elastic block, it should be possible to prove whether, during uniform sliding at low velocities, rapid fluidization and refreezing of adsorbate domains occur at the interface. Finally, we present a detailed discussion of plastic creep and relaxation in solids, which forms the basis for calculating how the area of real contact between two solids changes with the time of stationary contact. This topic is, of course, also of central importance in many engineering applications as is illustrated with a few examples.

11.1 The Model

We consider the sliding of a block on a lubricated substrate. Assume that the temperature and coverage are such that the lubrication layer at "stick" is in a solid state which is commensurate or at least pinned by the substrate. As discussed in Chap. 9, at very low sliding velocity, the lubrication layer is likely to consist of *solid domains* which pin the surfaces together and which, during sliding, fluidize and freeze in an incoherent or stochastic manner. Since the block and the substrate are in relative motion, during the growth of an island there will be a force on the island building up due to the local (at the island) elastic deformations of the block and the substrate. If the force on the island becomes large enough, the island will fluidize. At very low sliding velocity (creep) the lubrication layer consists almost entirely of pinned adsorbate domains and the associated stress domains in the block and the substrate. We note that direct observation of the contact areas during sliding of transparent materials (glass and plastic) has shown that the surface asperities in the contact areas perform a slow creep motion, i.e., no rapid slip involving the whole asperities occurs during steady sliding at very low (typically below 10 μm/s) sliding velocities [11.2].

Associated with each adsorbate domain is a local stress domain in the elastic block and in the substrate (Fig. 11.1a). During sliding, the surface stress at a stress domain increases continuously until it reaches the critical value σ_a. At this point the adsorbate layer locally fluidizes, followed by a rapid local slip, during which the local stress in the block and the substrate drops to a value close to zero. The elastic energy originally stored in the stress domain is partly radiated as elastic wave pulses into the block and substrate, and partly "consumed" in shearing the lubrication "fluid". After some short time period the lubrication film refreezes and the whole process repeats itself.

We now present a simple model which captures the essential physics of the scenario outlined above. Assume that the width of a typical adsorbate domain (and accompanying stress domain) is D (we assume that D is independent of the sliding velocity v). The extent of the stress domain into the solid is of order D. We replace the original system (Fig. 11.1a) with a model system (Fig. 11.1c) where each stress domain is replaced by a block ("stress" block) of mass $m = \rho D^3$, where ρ is the mass density. A stress block is connected to the main (or "big") block by a spring k_1 and to the other nearby stress blocks with springs of magnitude k_2. The magnitudes of k_1 and k_2 can be estimated as follows. Let us first consider the block without the substrate. If a tangential surface stress of magnitude σ acts within the surface area $\delta A = D \times D$ it will induce a tangential displacement of δA by the amount $u \sim F/(\rho c^2 D)$ where $F = \sigma D^2$. This relation follows directly from elastic continuum theory, see Sect. 9.5. In the elastic continuum theory the elastic displacement field at distance D away from the center P of δA is about one third of the displacement of P. In the model system (Fig. 11.1c), if a force F

11.1 The Model

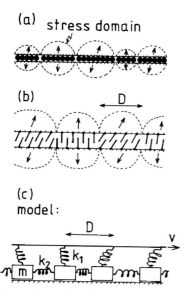

Fig. 11.1. (a) During sliding at very low velocity the adsorbate layer forms solid pinned domains. Associated with each domain is a stress field in the substrate and block. The arrows indicate the directions of thin cylindrical volume elements of the solids, which would be perpendicular to the sliding surfaces when the stress fields vanishes. (b) The same as a but for long-chain molecules (e.g., fatty acid molecules) bound to the solid surfaces at one end. (c) A mechanical model used to describe the sliding properties of the interfaces in (a) and (b).

acts on block 0, the resulting displacements q_i of the blocks can be determined by force equilibrium

$$k_1 q_i + k_2 (2q_i - q_{i-1} - q_{i+1}) = F\delta_{i0} .$$

It is easy to calculate the displacements q_0 and $q_{\pm 1}$:

$$q_0 = \frac{F}{k_1 + 2k_2(1-\xi)} ,$$

$$q_{\pm 1}/q_0 = \xi ,$$

where

$$\xi = 1 + (k_1/k_2) - \left[2(k_1/k_2) + (k_1/k_2)^2\right]^{1/2} .$$

We now require that the displacement q_0 and the ratio $q_0/q_{\pm 1}$ agree with the elastic continuum model where these quantities are $\sim F/(\rho c^2 D)$ and ~ 3, respectively. This gives $k_1 \sim k_2 \sim \rho c^2 D$. Now, note the following: suppose first that the surface area δA is displaced a distance u by an applied force F. At time $t = 0$ the force F is abruptly removed. This result in an elastic wave pulse emitted into the elastic medium, while u decays towards zero.

We can include this damping mechanism in the model system (Fig. 11.1c) by introducing a damping force $-m\gamma(\dot{q}_i - \dot{x})$ in the equation of motion of the stress-blocks ($\dot{x}$ is the velocity of the surface of the big block connecting the stress blocks to the big block). The damping γ can be estimated from (8.12), $\gamma \sim k_1^2/(m\rho c^3)$. Using $k_1 \sim \rho c^2 D$ and $m \sim \rho D^3$, this gives $\gamma \sim c/D$. Note that this damping is of similar magnitude to the resonance frequency $\omega_0 = (k_1/m)^{1/2} \sim c/D$. The strong "damping" is, of course, expected – any elastic displacement field set up at time $t = 0$ within some volume $\sim D^3$ somewhere in the surface region of the big block and then let free, will rapidly spread into the bulk of the big block: It will have left the volume D^3 after the time period τ taken for an elastic wave to propagate the distance D, i.e., $\tau \sim D/c$ so that the damping $\gamma \sim 1/\tau \sim c/D$.

We have performed computer simulations and analytical calculations based on the model shown in Fig. 11.1. We assume that the adsorbate domain related to stress block q_i is in a pinned solid state, $\dot{q}_i = 0$, until the elastic force

$$F = k_1(x - q_i) + k_2(q_{i+1} + q_{i-1} - 2q_i)$$

from the springs connected to block q_i reaches the critical force $F_a = \delta A \sigma_a$ necessary for fluidization of the pinned adsorbate structure. After fluidization the motion of q_i satisfies Newton's equation

$$\begin{aligned} m\ddot{q}_i = &-m\gamma(\dot{q}_i - \dot{x}) - m\gamma'\dot{q}_i + k_1(x - q_i) \\ &+ k_2(q_{i+1} + q_{i-1} - 2q_i) \, . \end{aligned} \quad (11.1)$$

In the calculations below we assume that the adsorbate layer returns to the pinned state when $\dot{q}_i = 0$ at the end of the rapid local slip process.

In the application below we focus on very small $\dot{x}$, and we can neglect the $\dot{x}$-term in (11.1). Hence the damping term in (11.1) is $m(\gamma + \gamma')\dot{q}_i = m\bar{\gamma}\dot{q}_i$, which defines the effective damping coefficient $\bar{\gamma}$. From now on, we measure time in units of $(m/k_1)^{1/2}$, $\bar{\gamma}$ in units of $(k_1/m)^{1/2}$, distance in units of F_a/k_1, velocity in units of $F_a(mk_1)^{-1/2}$, force in units of F_a, and spring constants in units of k_1. In these units (11.1) becomes

$$\ddot{q}_i = -\bar{\gamma}\dot{q}_i + (x - q_i) + k_2(q_{i+1} + q_{i-1} - 2q_i) \, . \quad (11.2)$$

Note that from the estimates presented above $k_2 \sim 1$ and $\bar{\gamma} \sim 1$. We have solved the equation of motion (11.2) numerically. In all calculations presented below we have chosen $\bar{\gamma} = 1$ and $k_2 = 1$. We have also performed calculations with other parameter values but the qualitative picture remains the same as presented below.

11.2 Critical Sliding State at Zero Temperature

Figure 11.2a shows the friction force (per stress block) as a function of time when the sliding velocity $v_s = 0.005$. In the calculation we have used $n = 10,000$ stress blocks which initially ($t = 0$) were distributed very regularly on the surface (the initial coordinates were taken as $q_i = 0.01 \times r_i$, where r_i is a random number between zero and one) and many stick-slip periods occur before the system reaches a steady state. The function $h(t)$ shown in Fig. 11.2b is the fraction of the blocks which are moving at time t. Note that during the first slip period all stress blocks are moving (i.e., $h = 1$) while

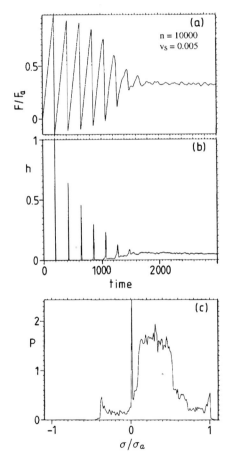

Fig. 11.2. (a) The friction force as a function of time. Initially the stress blocks are distributed very regularly on the surface and many stick-slip periods occur before the system reaches a steady state. (b) The function $h(t)$ is the fraction of the blocks which are moving at time t. (c) The distribution of surface stresses in the steady state. In the calculation $k_1 = k_2 = \bar{\gamma} = 1$, $v_s = 0.005$ and $n = 10,000$ stress blocks.

during the second slip period only $\sim 70\%$ of the blocks move. The first few slip periods are separated by time intervals where no blocks are moving. But for $t > 1000$ some blocks are always moving and, in particular, when the steady state has been reached ($t > 1800$) about 10% of the blocks move at any time. This does not mean, of course, that the same blocks are moving the whole time, but each individual block performs a stick-slip type of motion (Sect. 11.5) but in such a manner that about 10% of the blocks move at any given time. Finally, Fig. 11.2c shows the distribution of surface stresses in the steady state. Note that the distribution $P(\sigma)$ is *critical*, i.e., a continuous distribution of local surface stresses σ exists, which *extend to the critical stress σ_a*, necessary for fluidization of the pinned adsorbate structure. This implies that thermal processes will be very important at low sliding velocities (see next section). Note that the surface stress σ is negative for about 5% of the blocks; this effect results from inertia and corresponds to stress blocks which, during sliding, pick up enough kinetic energy to "overshoot" so that when returning to the stick-state the sum of the spring forces acting on such a block is negative. The sharp peak close to $\sigma \approx 0$ in Fig. 11.2c corresponds to a small fraction of the blocks which are sliding with a velocity close to the driving velocity $v_s = 0.005$ so that a kinetic friction force of magnitude $\bar{\gamma}\dot{q} \sim 0.005$ acts on these blocks. In contrast, most of the blocks perform a very rapid motion during slip, where velocities of order 1 occur. At sliding velocities lower than 0.005, the sharp peak in Fig. 11.2c disappears and the distribution $P(\sigma)$ converges towards a velocity-independent distribution which looks virtually identical to the one in Fig. 11.2c, i.e., $v_s = 0.005$ is small enough to be practically in the asymptotically small v_s regime. This velocity region will be our main concern below.

Figure 11.3 shows the same quantities as Fig. 11.2 but for a higher sliding velocity, $v_s = 0.03$. Note the sharp peak in Fig. 11.3c which carries about 50% of the blocks, i.e., about half of the blocks now perform a slow drift motion $\dot{q} \sim v_s$ where the surface stress is $\sigma \sim \bar{\gamma}v_s = 0.03$. But some of the blocks still perform rapid slip motion giving rise to the broad distribution in $P(\sigma)$, extending from $\sigma = -0.4\sigma_a$ to σ_a. For sliding velocities above $v_s = 0.05$ all stress blocks slide with the speed v_s and the stress distribution $P(\sigma)$ is a Dirac delta function centered at $\sigma_s = \bar{\gamma}v_s$.

The dashed lines in Fig. 11.4a and b show, in the steady state sliding regime, the dependence of the friction force and the fraction of sliding blocks on the natural logarithm of the sliding velocity. These curves have been constructed from many computer simulations of the type shown in Figs. 11.2 and 11.3. Note the following: (a) For $\ln v_s < -5$, i.e., $v_s < 0.006$, the friction force is velocity independent. As discussed above, in this velocity region the stress distribution $P(\sigma)$ is velocity independent and the number of stress blocks moving at any given time is proportional to the sliding velocity (not shown). Hence, a velocity independent sliding friction force is indeed expected. For $\ln v_s > -3$ or $v_s > 0.05$ all stress blocks move with the velocity v_s and the

11.2 Critical Sliding State at Zero Temperature   369

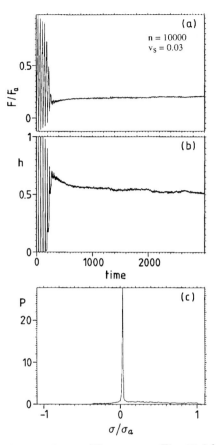

Fig. 11.3a–c. The same as Fig. 11.2 but with $v_\mathrm{s} = 0.03$.

friction force is given by $\bar\gamma v_\mathrm{s}$. As a function of $\ln v_\mathrm{s}$ this gives the rapidly increasing friction force shown by the dashed line in Fig. 11.4a for $\ln v_s > -3$. As pointed out above, in this case $P(\sigma)$ is a Dirac delta function centered at $\bar\gamma v_\mathrm{s}$. Finally, in the "transition region" $-5 < \ln v_\mathrm{s} < -3$, the friction force varies smoothly from its small-v_s asymptotic value to the high velocity behavior where all blocks slide with the velocity v_s. In this transition region $P(\omega)$ has both a broad continuum extending from $\sim -0.4\sigma_\mathrm{a}$ to σ_a as well as a Dirac delta function contribution centered at a stress close to the friction stress experienced by a block which slides at the speed v_s.

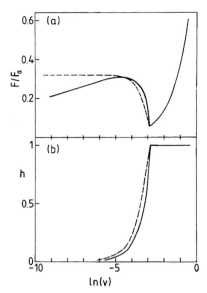

Fig. 11.4. (a) The friction force in the steady state as a function of the natural logarithm of the sliding velocity. (b) The fraction of sliding blocks as a function of the natural logarithm of the sliding velocity. The solid lines include thermally activated processes ($\nu = 0.1$ and $\beta = 40$) while the dashed lines are for zero temperature.

11.3 Relaxation and Creep at Finite Temperature

The calculations presented in Sect. 11.2 were for zero temperature where no thermal excitations can occur. Now, we have seen that during sliding at low velocity v the distribution $P(\sigma)$ is critical, i.e., the distribution $P(\sigma)$ of local surface stresses has a sharp step-like cut off at $\sigma = \sigma_a$ (Fig. 11.2c). But this implies that thermally activated processes will be very important at low sliding velocities, since those stress blocks which have surface stresses in the vicinity of $\sigma = \sigma_a$ may be thermally exited over the small elastic energy barrier which separates them from the sliding state. In fact, if v is small enough this means of "surmounting the barrier" will occur before a stress block is driven over the barrier by the motion $x = x(t)$. Similarly, thermal processes are crucial during "stop": If no thermal excitations were to occur (i.e., zero temperature), the distribution $P(\sigma)$ at "stop" would be time independent and of the form shown in Fig. 11.2c with some stress blocks with σ just below σ_a. At non-zero temperature, an arbitrarily small thermal fluctuation can "kick" these blocks over the barrier to the sliding state. In this section we introduce thermal excitations in a realistic manner and study their influence on the sliding dynamics.

We assume that during sliding a stress block satisfies the same equation of motion (11.2) as in the last section. This is certainly a good approximation

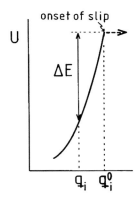

Fig. 11.5. The elastic potential energy as a function of the coordinate q_i of block i ($q_{i\pm 1}$ are fixed). At the critical displacement, $q_i = q_i^0$, slip starts.

since during local sliding large velocities occur and small thermal fluctuations should be of no importance. Consider now a pinned stress block. In the last section it was assumed that the block remains pinned until the force on it from the springs connected to it reaches the critical value F_a. At this point the adsorbate layer fluidizes and local slip motion occurs. We now generalize this assumption in order to allow the stress blocks to be thermally excited over the barrier to motion. Let us consider stress block i with the coordinate q_i. The potential energy stored in the springs connected to this block is

$$U = \frac{1}{2}k_1(x - q_i)^2 + \frac{1}{2}k_2(q_{i+1} - q_i)^2 + \frac{1}{2}k_2(q_{i-1} - q_i)^2 , \tag{11.3}$$

and the force on this block i is

$$F = k_1(x - q_i) + k_2(q_{i+1} + q_{i-1} - 2q_i) ,$$

which must be below F_a in order for the block to be pinned. Now, consider $U = U(q_i)$ as a function of q_i (i.e., $q_{i\pm 1}$ fixed). $U(q_i)$ has the form indicated in Fig. 11.5. If we define the "critical" displacement q_i^0 by

$$F_a = k_1\left(x - q_i^0\right) + k_2\left(q_{i+1} + q_{i-1} - 2q_i^0\right) ,$$

i.e.

$$q_i^0 = \frac{k_1 x + k_2(q_{i+1} + q_{i-1}) - F_a}{k_1 + 2k_2} , \tag{11.4}$$

then the elastic barrier $\Delta E = U(q_i^0) - U(q_i)$ must be overcome by thermal excitation in order to initiate sliding, see Fig. 11.5. According to statistical mechanics, the probability rate w for thermal excitation over a barrier of height ΔE can be written

$$w = \nu e^{-\beta \Delta E},$$

where $\beta = 1/k_\mathrm{B} T$. Let us estimate the magnitude of the prefactor ν. We can visualize the thermal excitation over the barrier ΔE as resulting from a bulk phonon wave packet arriving at the stress block and locally increasing the stress to the level necessary for the block to start to slide. Since the motion of a stress block is overdamped, we use Kramers formula in the overdamped limit to obtain

$$\nu \approx \omega_0^2/(2\pi\gamma),$$

where (Sect. 11.1) $\omega_0^2 = k_1/m \sim c^2/D^2$ and $\gamma \sim c/D$. Hence

$$\nu \sim c/(2\pi D).$$

Assuming $c \sim 2000\,\mathrm{m/s}$ and $D \sim 300\,\text{Å}$ gives $\nu \sim 10^{10}\,\mathrm{s}^{-1}$. In the natural units defined above $\nu \sim 1/2\pi$.

When studying the time evolution of the system, time is discretized in steps of Δt. The probability that the stress block jumps over the barrier during the time Δt is $w\Delta t$. Since the actual excitation over the barrier occurs in a stochastic manner as a function of time we use random numbers to determine when the jump occurs. That is, if r is a random number between 0 and 1 (generated by a random number generator on a computer) then if $r < w\Delta t$ the jump is assumed to have occurred during the time period Δt, whereas the stress block remains in the pinned state if $r > w\Delta t$. In the simulations Δt is so short that $\nu \Delta t < 1$ which implies $w\Delta t < 1$. That is, for any stress block, even those which have a stress just below σ_a (i.e., with $\beta \Delta E \approx 0$), there is a probability less than 1 that it will be thermally excited over the barrier during the time period Δt. In all calculations below we have used $\nu = 0.1$ and $\beta = 40$. We have also performed calculations for other parameter values but the same qualitative picture emerges.

Figure 11.6 is obtained with the same model parameters as Fig. 11.2 except that thermal excitation is included as discussed above. Note that when thermal excitation is included, the onset of the first slip is at a lower static friction force; F_max is reduced from $0.98\,F_\mathrm{a}$ in Fig. 11.2a to $0.70\,F_\mathrm{a}$ in Fig. 11.6a. Furthermore, the number of stick-slip oscillations before the steady-state sliding regime is reached is reduced. Even more dramatic is the change in the fraction of blocks which move at time t; compare Figs. 11.2b and 11.6b. Without thermal excitation all the stress blocks move during the first slip period, while only $\sim 30\,\%$ move when thermal excitation is included. The stress distribution $P(\sigma)$ in the steady-state region (Fig. 11.6c) is qualitatively similar to that without relaxation (Fig. 11.2c).

The solid line in Fig. 11.4a shows the friction force in the steady-state sliding regime, as a function of the natural logarithm of the sliding velocity, as obtained from many computer simulations of the type shown in Fig. 11.6. This result differs most importantly from the zero-temperature results (dashed

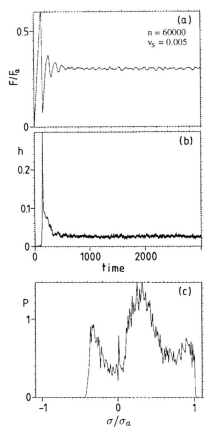

Fig. 11.6a–c. The same as Fig. 11.2 except that thermally activated processes are included as discussed in the text. $\nu = 0.1$ and $\beta = 40$.

line, from Fig. 11.4a) in the *linear increase of the friction force with increasing* $\ln v$; $F/F_a = 0.44 + 0.026 \ln v$ for $\ln v < -5.5$. Note that, in this velocity region, when thermal excitation is taken into account, a smaller external force generates the same sliding velocity as a larger force in the absence of thermal processes. This effect results from thermal excitation of stress blocks from close to $\sigma = \sigma_a$, over the barrier to the sliding state. Hence, a given block does not need to be pulled over the barrier by the motion v, but can jump over the barrier by thermal excitation. The solid line in Fig. 11.4b shows the fraction of sliding blocks as a function of the natural logarithm of the sliding velocity.

Figure 11.7 illustrates the slow relaxation which occurs if the sliding velocity is abruptly reduced from a non-zero value to zero. The time variation of the friction force is shown in Fig. 11.7a and that of the fraction of mov-

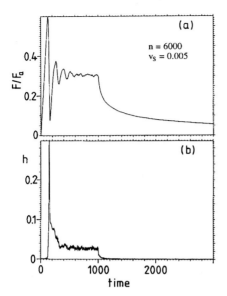

Fig. 11.7. The time variation of (a) the friction force and (b) the fraction of moving blocks. For $t < 1000$ the sliding velocity $v_s = 0.005$. At $t = 1000$ the sliding velocity is put equal to zero ($v_s = 0$). The time variation of F and h for $t > 1000$ is due to thermally excited processes (relaxation). In the calculation $\nu = 0.1$ and $\beta = 40$.

ing blocks in Fig. 11.7b. For $t < 1000$ the sliding velocity $v_s = 0.005$. At $t = 1000$ the sliding velocity is set equal to zero ($v_s = 0$). The decay of F and h with increasing time for $t > 1000$ is due to thermal excitation (relaxation). Figure 11.8a shows the friction force (from Fig. 11.7a) during the relaxation time, as a function of the natural logarithm of the stopping time $\tau = t - 1000$. Note that except for very short and very long times, F decays proportionally to the logarithm of the stopping time τ; the dashed line is given by $F(\tau)/F_a = 0.38 - 0.044 \ln \tau$. Finally, Fig. 11.8b shows the stress distribution at the end of the relaxation process in Fig. 11.8a, i.e., for $t = 3000$. Note that the distribution is no longer critical, i.e., all the stress blocks which were close to $\sigma = \sigma_a$ have been removed by thermal excitation. As $\tau \to \infty$ the distribution $P(\sigma) \to \delta(\sigma)$ (assuming low temperature) and $F \to 0$, but it is not possible to obtain this fully relaxed state in a computer simulation because of the long stopping time τ necessary.

Figure 11.9 shows another manifestation of relaxation in a computer simulation of "stop-and-start". For $t < 600$ the sliding velocity $v_s = 0.01$. For $600 < t < 2600$ the sliding velocity vanishes ($v_s = 0$). For $t > 2600$ the sliding velocity $v_s = 0.01$. In (a) the friction force is shown as a function of time. Note the striction spike of height ΔF. The fraction of blocks moving at time t is shown in (b). The stress distribution at the end of the sliding process ($t = 3000$) is shown in (c). The circles in Fig. 11.10 show the height ΔF of

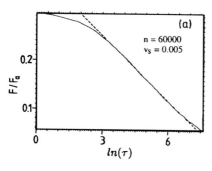
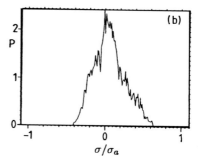

Fig. 11.8. (a) The friction force (from Fig. 11.7a) during the relaxation period, as a function of the natural logarithm of the stopping time $\tau = t - 1000$. The dashed line is given by $F(\tau)/F_a = 0.38 - 0.044 \ln \tau$. (b) The stress distribution at the end of the relaxation process in Fig. 11.7a, i.e., for $t = 3000$. $\nu = 0.1$ and $\beta = 40$.

the striction spikes as a function of the stopping time τ from several computer experiments of the type shown in Fig. 11.10. The solid line is determined by $\Delta F/F_a = -0.19 + 0.05 \ln \tau$.

It is possible, with a simple model calculation, to understand why the relaxation processes studied above have the asymptotic time-dependency $\ln \tau$ and why the creep motion depends linearly on $\ln v$. Let us first consider how the force acting on the big block decreases with time after the sliding motion has abruptly been stopped at time $t = 0$ (Fig. 11.7a). We will use a mean-field type of approximation to estimate the elastic barrier ΔE which a pinned stress block must overcome by thermal excitation before local sliding can occur. The exact elastic barrier for block i in Fig. 11.1c was calculated above and depends on the positions of blocks $i \pm 1$

$$\Delta E = U(q_i^0) - U(q_i) \,.$$

In the expression for U we replace $q_{i\pm1}$ by their average values q. Using (11.3) and (11.4) this gives

$$\Delta E = \frac{F_a^2 - F^2}{2(k_1 + 2k_2)} \,,$$

where $F = k_1(x - q_i) + 2k_2(q - q_i)$ is the mean-field force acting on block i. In terms of the surface stress $\sigma = F/\delta A$ we have

$$\Delta E = \varepsilon \left[1 - (\sigma/\sigma_a)^2\right] \,, \tag{11.5}$$

where

$$\varepsilon = \frac{(\delta A \sigma_a)^2}{2(k_1 + 2k_2)} = \frac{F_a^2}{2(k_1 + 2k_2)} \,. \tag{11.6}$$

Assume that no slip has occurred at time $t = 0$. The probability that the block has not jumped over the barrier ΔE at time $t > 0$ is

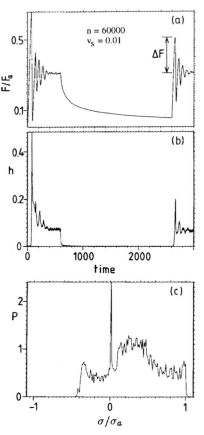

Fig. 11.9. A computer simulation of "stop-and-start". For $t < 600$ the sliding velocity $v_s = 0.01$. For $600 < t < 2600$ the sliding velocity vanishes ($v_s = 0$). For $t > 2600$ the sliding velocity $v_s = 0.01$. (a) The friction force as a function of time. Note the striction spike of height ΔF. (b) The fraction of blocks moving at time t. (c) The stress distribution at the end of the sliding process ($t = 3000$).

$$P(t) = e^{-wt}, \tag{11.7}$$

where the rate coefficient w has the form

$$w = \nu e^{-\beta \Delta E}. \tag{11.8}$$

Now, assume that at the interface between the big-block and the substrate there are $N_0 \gg 1$ stress blocks. The total number of stress blocks which, at time $t > 0$, remain in their original ($t = 0$) positions is given by

$$N(t) = \sum_i P_i(t) = \sum_i e^{-w_i t}. \tag{11.9}$$

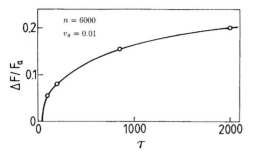

Fig. 11.10. *Circles*: The height ΔF of the striction spikes as a function of the stopping time τ from computer experiments of the type shown in Fig. 11.9. The solid line is determined by $\Delta F = -0.19 + 0.05 \ln \tau$.

Let $P(\sigma)$ be the distribution of surface stresses at time $t = 0$. Let us first assume that $P(\sigma)$ is uniform, i.e., that $P(\sigma) = 1/\sigma_a$ for $0 < \sigma < \sigma_a$ and zero otherwise. The actual distribution is not uniform but the only relevant result for the present study is that there is a sharp step-like cut-off in the distribution $P(\sigma)$ at $\sigma = \sigma_a$. In this case, using (11.5,7–9), we get

$$N(t) = N_0 \int_0^{\sigma_a} d\sigma P(\sigma) e^{-\nu t \exp\{-\beta\varepsilon[1-(\sigma/\sigma_a)^2]\}} , \qquad (11.10)$$

where $N_0 = N(0)$ is the total number of stress blocks. If we introduce $\sigma = \sigma_a(1-\xi)$ and approximate

$$\sigma_a^2 - \sigma^2 \approx 2\sigma_a^2 \xi ,$$

and assume $P(\sigma) = 1/\sigma_a$ for $0 < \sigma < \sigma_a$ and zero otherwise, we get from (11.10)

$$N(t) \sim N_0 \int_0^1 d\xi e^{-\nu t \exp(-\beta 2\varepsilon \xi)} .$$

If we put

$$\eta = e^{-\beta 2\varepsilon \xi}$$

then

$$N(t) \sim \frac{N_0}{\beta 2\varepsilon} \int_{e^{-\beta 2\varepsilon}}^1 \frac{d\eta}{\eta} e^{-\nu t \eta} . \qquad (11.11)$$

Now, if $\beta\varepsilon \gg 1$ and $\nu t \gg 1$ we can accurately approximate (11.11) by

$$N(t) \sim \frac{N_0}{\beta 2\varepsilon} \int_{e^{-\beta 2\varepsilon}}^{1/\nu t} \frac{d\eta}{\eta} = N_0 \left(1 - \frac{\ln \nu t}{\beta 2\varepsilon}\right) . \qquad (11.12)$$

The asymptotic result $N_0 - N(t) = \Delta N(t) \sim \ln \nu t$ is valid whenever the probability distribution $P(\sigma)$ has a step-like cut-off. On the other hand, if $P(\sigma)$ goes continuously to zero at $\sigma = \sigma_a$ then this asymptotic time dependence no longer holds. For example, if

$$P(\sigma) \propto (\sigma_a - \sigma)^n$$

as $\sigma \to \sigma_a$ from below, while $P = 0$ for $\sigma > \sigma_a$, then it is easy to show that

$$\Delta N \propto [\ln \nu t]^n .$$

The formula (11.12) explains the $\ln t$-dependence found in the simulations above and is in good quantitative agreement with the simulations. For example, consider the results in Fig. 11.7a for $F(t)/F_a$. To apply (11.12) to this case let us note that when a stress block "jumps" over the elastic barrier, the force on the big block changes by roughly $\sigma_a \delta A$. Hence $F(0) - F(t) \approx \delta A \sigma_a \Delta N(t)$ and since $F_a = N_0 \delta A \sigma_a$ we get $(F(0) - F(t))/F_a \approx \Delta N/N_0$ or, using (11.12),

$$\frac{F(t)}{F_a} \approx \frac{F_k}{F_a} - \frac{1}{\beta 2\varepsilon} \ln \nu t , \qquad (11.13)$$

where $F(0) = F_k = F_a/2$ is the kinetic friction force in the present model. Let us compare the mean-field result (11.13) with the results in Fig. 11.7a. The dashed line in Fig. 11.8a is given by

$$F/F_a = 0.38 - 0.044 \ln t ,$$

which we can write as $F(t)/F_a = F_k/F_a - B \ln bt$ where $F_k/F_a = 0.3$ (from Fig. 11.7a), $B = 0.044$ and $b = 0.15$. In the simulations $\beta = 40$ and, since $\varepsilon = 1/6$, the values for B and b in the mean field model equals $1/2\beta\varepsilon = 0.075$ and $b = \nu = 0.1$, in remarkably good agreement with the results of the simulation.

The analysis presented above assumed that the motion of the block is abruptly stopped at time $t = 0$, i.e., $x(t) = $ const. for $t > 0$. More complex sliding and relaxation problems can be studied on the basis of an equation of motion for the distribution function $P(\sigma, t)$. It can be shown that the distribution $P(\sigma, t)$ of surface stresses at time t obeys the equation [11.1]

$$\frac{\partial P}{\partial t} = -\frac{k_1 v}{\delta A} \frac{\partial P}{\partial \sigma} - \nu P e^{-\beta \Delta E(\sigma)} + \delta(\sigma)$$
$$\times \left(\int d\sigma' \nu P(\sigma', t) e^{-\beta \Delta E(\sigma')} + \frac{k_1 v}{\delta A} P(\sigma_a, t) \right) . \qquad (11.14)$$

The first term on the rhs of this equation describes the rate of change of $P(\sigma, t)$ due to the motion of the big block [velocity $v = \dot{x}(t)$]. The second term describes the probability rate for a stress block to be thermally excited over the barrier $\Delta E(\sigma) = U(\sigma_a) - U(\sigma)$. Finally, the last term, which is non-zero only if $\sigma = 0$, describes the increase in the number of stress blocks

11.3 Relaxation and Creep at Finite Temperature

with zero surface stress which results from all the stress blocks which have been thermally excited over the barrier $\Delta E(\sigma) = U(\sigma_a) - U(\sigma)$ or driven over the barrier by the motion $x(t)$; we assume that after fluidization of the adsorbate layer a block rapidly relaxes down into the state where the surface stress equals zero, after which the adsorbate layer refreezes and the block returns to the pinned state.

Note that (11.14) conserves probability: integrating (11.14) over σ from $-\sigma_a$ to σ_a and using $P(-\sigma_a, t) = 0$ (we assume sliding along the positive x-direction) gives

$$\frac{d}{dt} \int_{-\sigma_a}^{\sigma_a} d\sigma P(\sigma, t) = 0 .$$

Hence, if the normalization condition

$$\int_{-\sigma_a}^{\sigma_a} d\sigma P(\sigma, 0) = 1 \tag{11.15}$$

is satisfied at time $t = 0$, it will be automatically satisfied for all other times.

Let us study steady sliding. In this case $\partial P / \partial t = 0$ and (11.14) reduces to an ordinary differential equation for $P(\sigma, t) \equiv P(\sigma)$ with the solution

$$P(\sigma) = Ce^{-\frac{\delta A \nu}{k_1 v} \int_0^\sigma d\sigma' \exp[-\beta \Delta E(\sigma')]} \tag{11.16}$$

if $0 \leq \sigma \leq \sigma_a$, and zero otherwise. The number C is determined by the normalization condition (11.15). The force F which gives rise to the sliding velocity v is determined by the equation

$$F = \int_{-\sigma_a}^{\sigma_a} d\sigma A\sigma P(\sigma) . \tag{11.17}$$

Let us first consider the large-v limit. Assume that $\alpha \ll 1$ where $\alpha = F_a \nu / (2k_1 v \beta \varepsilon)$. In this case, (11.16,17) give, to leading order in $\alpha \sim 1/v$,

$$\frac{F}{F_a} \sim \frac{1}{2} - \frac{\alpha}{4\beta\varepsilon} = \frac{1}{2} - \frac{F_a \nu}{2k_1 v} \left(\frac{k_B T}{2\varepsilon} \right)^2 .$$

Next, let us consider sliding at small velocities, which we will refer to as creep motion. Assume that $\alpha \gg 1$ but $\alpha \exp(-2\beta\varepsilon) \ll 1$. In this case (11.16,17) give

$$\frac{F}{F_a} \sim \frac{1}{2} + \frac{1}{4\beta\varepsilon} \ln\left(\frac{2\beta\varepsilon k_1 v}{F_a \nu} \right) = \frac{F_k}{F_a} + \frac{1}{4\beta\varepsilon} \ln\left(\frac{2\beta\varepsilon k_1 v}{F_a \nu} \right) . \tag{11.18}$$

Let us compare the prediction of this formula with Fig. 11.4. In the creep region, $\ln v < -5$ in Fig. 11.4, F varies linearly with $\ln v$ in accordance with (11.18). The creep region in Fig. 11.4 is well described by

$$F/F_a = 0.44 + 0.026 \ln v$$

which we can write as $F/F_a = F_k/F_a + B \ln bv$ where $F_k/F_a = 0.32$, $B = 0.026$, and $b = 118$. The corresponding values for B and b in our mean-field model (11.18) are 0.037 and 133, in remarkably good agreement with the simulations.

Finally, in the extremely low v limit, where $\alpha \exp(-2\beta\varepsilon) \gg 1$, (11.16,18) give

$$F \sim (F_a/2\alpha\beta\varepsilon)e^{\beta\varepsilon} = N_0(k_1v/\nu)e^{\beta\varepsilon} \ . \tag{11.19}$$

Note that $F \ll F_a$ and that the friction force is proportional to the sliding velocity v. Formula (11.19) is of very limited practical use, however, since extremely low sliding velocities v are necessary for it to be valid.

Slip-Size Distribution

What type of local slip events occur during sliding at low velocities? The fact that, in the steady state, the function $h(t)$ is very smooth (i.e., very weak noise) in Figs. 11.2 and 11.6, implies that the largest slip events (avalanches) must be very small compared with the total number of blocks used in the simulations (10,000 and 6000, respectively). To study the distribution of sizes of slip events, we consider a system consisting of only $n = 100$ blocks and a very low sliding velocity $v_s = 0.0002$. In this case the individual slip events can be resolved in time.

Figure 11.11a,b shows the time variation of (a) the friction force and (b) the fraction of moving blocks at zero temperature. Note that, at a given time, at most two blocks are sliding. However, most of the time no block or only one block is sliding. In fact, when two blocks slide these are not necessarily two nearby blocks but the blocks could be separated by other non-sliding blocks. Hence, we conclude that only very localized slip events occur in the present model. When thermal processes are included, the same qualitative picture emerges (Fig. 11.11c,d where $\nu = 0.1$ and $\beta = 40$).

These results differ from those of *Carlson* and *Langer* [11.3] who found a wide distribution of the sizes of slip events in their model calculations. However, Carlson and Langer only considered systems where the ratio $k_2/k_1 \gg 1$. In fact, they argued that in the context of earthquakes the limit $k_2/k_1 \to \infty$ is relevant. In Fig. 11.11e,f I show the time variation of (e) the friction force and (f) the fraction of moving blocks at zero temperature when $k_1 = 1$ and $k_2 = 50$ (all earlier calculations were with $k_2 = 1$). Note that in this case the sliding force F exhibits stick-slip oscillations, and the fraction of moving blocks $h(t)$ shows a very wide distribution of sizes of the slip events. The largest slip event involves about 50% of the blocks while the smallest involves a single block. These results are similar to those of Carlson and Langer. It is easy to understand why large slip events can occur when $k_2 \gg k_1$. It is clear that if $k_2 = \infty$ all blocks must slide together while in the opposite limit, $k_2 = 0$, the blocks are completely independent of each other and only slip events involving a single block can occur.

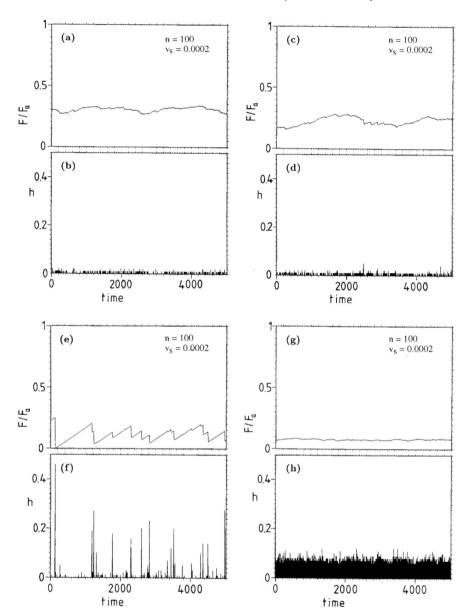

Fig. 11.11. (a,c,e,g) The time variation of the friction force and (b,d,f,h) the fraction of moving blocks. (a,b) are for zero temperature and (c,d) for a finite temperature ($\nu = 0.1$ and $\beta = 40$) and for $k_1 = k_2 = \bar{\gamma} = 1$. (e,f) and (g,h) the same as (a,b) and (c,d), respectively, but now for $k_2 = 50$. In the calculations the number of blocks is $n = 100$ and the sliding velocity $v_s = 0.0002$.

Experimental Implications

In this section we discuss how the results obtained above may be tested experimentally. There are in principle two different ways to probe the nature of the lubrication layer during sliding. One way is to use X-ray scattering to obtain information about the orientation of the lubrication molecules at the interface. Using this method it may be possible to probe whether the lubrication molecules, during steady sliding at low velocity, occur in a 2D-fluid state or if a granular state occurs with solid domains. Another interesting method would be to study the sound waves generated at the sliding interface. If the lubrication layer were to have a granular structure, consisting of pinned solid islands, the sound wave emitted from the interface would consist of a sequence of wave pulses. The rate of emission of wave pulses from the interface equals the rate of fluidization of pinned islands, which in turn depends on the sliding velocity, assuming smooth sliding. If the individual wave pulses could be studied experimentally, information about the detailed fluidization process of a single island could be gained. Let us estimate how many islands are fluidized per unit time. Assume that the contact area has a radius $R = 10$ μm. If an average pinned island has the diameter $D = 2000$ Å, then $N \sim 10^4$ islands would exist at the interface. Now, an island will exist for a time τ before the surface stress reaches the critical value σ_a necessary for fluidization. At the time of fluidization the local elastic displacement of the solids at the island equals $\Delta l \approx D^2 \sigma_a / k_1 \approx 1$ Å. Hence we must have $v\tau = \Delta l$ or $\tau \approx 10^{-2}$ s if $v = 0.01$ μm/s. Hence, the rate of emission of wave pulses from the sliding interface will be on the order of 10^6 per second. The duration of each wave pulse is on the order of $D/c \sim 10^{-11}$ s. These estimates are very rough but indicate that it may be possible to resolve the individual wave pulses.

11.4 Time-Dependent Plastic Deformation in Solids

When a block is placed on a substrate, plastic deformation occurs at the contact points, and a short time after contact the local stresses equal the plastic yield stress. Now, plastic deformation can be considered as resulting from shear "melting" and "refreezing" of small volume elements (stress blocks); a stress block melts when the elastic stresses it is exposed to from the surrounding stress blocks satisfies some yield criteria (e.g., the von Mises yield condition) and when the stress block refreezes the elastic (shear) stresses in it are reduced. This model of plastic deformation is very similar to the model studied above and one therefore expects that immediately after the rapid plastic deformations which occur when the block is put on the substrate, the stress blocks in the vicinity of the contact area will be in a "critical" state with a distribution of local stresses such that some stress blocks are almost ready to undergo plastic deformation. Just as in the discussion in Sect. 11.3

11.4 Time-Dependent Plastic Deformation in Solids

this implies that thermal processes will be of crucial importance; slow relaxation (creep) will occur and the contact area will increase slowly with time. In this section we develop a general (mean-field) theory of creep in solids which can be used not only to study how the contact area grows with the time of stationary contact, but which may also be used for other applications involving creep and relaxation of stress distributions in solids [11.4].

It has been found experimentally that plastic deformation and creep only depend on the shear stresses and that the volume of the body does not change during plastic deformation or creep. Thus, if we introduce the hydrostatic pressure $p = -\sigma_{kk}/3$, then plastic deformation does not depend on p but only on the stress deviator

$$s_{ij} = \sigma_{ij} + p\delta_{ij} .$$

Note that $s_{ii} = 0$. Similarly, if ε_{ij} denotes the strain tensor, the strain deviator is defined by

$$e_{ij} = \varepsilon_{ij} - \delta_{ij}\varepsilon_{kk}/3 .$$

When only elastic deformation occurs

$$e_{ij} = s_{ij}(1+\nu)/E ,$$

where E is the elastic modulus and ν the Poisson number. Since the onset of plastic yield does not depend on p, the yield condition is a function only of s_{ij}. Since $s_{ii} = 0$, the simplest non-zero invariant (under rotations) which can be formed from s_{ij} is $s_{ij}s_{ij}$. In the theory of von Mises, which has been found to be in very good agreement with the experiments, the plastic yield criterion is taken to be $s_{ij}s_{ij} = s_0^2$, where s_0 is a constant which can be related to the yield stress σ_a during uniaxial tension, where the stress tensor and the stress deviator have the form

$$\sigma_{ij} = \begin{pmatrix} \sigma & 0 & 0 \\ 0 & 0 & 0 \\ 0 & 0 & 0 \end{pmatrix} ; \quad s_{ij} = \begin{pmatrix} 2\sigma/3 & 0 & 0 \\ 0 & -\sigma/3 & 0 \\ 0 & 0 & -\sigma/3 \end{pmatrix} .$$

Thus $s_{ij}s_{ij} = 2\sigma^2/3$ so that $2\sigma_\mathrm{a}^2/3 = s_0^2$. The von Mises yield criterion can be expressed as follows: A body flows plastically if the elastic energy per unit volume due to shear deformations reaches a critical value. This result is trivial to prove if one notes that the elastic energy per unit volume is given by

$$\frac{dU}{dV} = \frac{1}{2}\sigma_{ij}\varepsilon_{ij} = \frac{1}{2}(s_{ij} - p\delta_{ij})(e_{ij} + \delta_{ij}\varepsilon_{kk}/3) = \frac{1}{2}s_{ij}e_{ij} - \frac{1}{2}p\varepsilon_{kk} .$$

Since ε_{kk} is the increase in the volume per unit volume due to the elastic deformations, the last term in the expression above is the elastic energy due to the volume change, while the first term is the elastic energy due to the shear deformations only, which can be written as

$$\frac{dU_1}{dV} = \frac{1+\nu}{2E} s_{ij} s_{ij} \ .$$

Thus plastic flow occurs when $dU_1/dV = (1+\nu)\sigma_a^2/3E$.

Let us now introduce the stress (probability) distribution function $P(s_{ij}, t)$ which satisfies

$$\frac{dP}{dt} = \frac{\partial P}{\partial t} + \frac{\partial P}{\partial s_{ij}} \dot{s}_{ij} = -\nu_0 e^{-\beta \Delta E(s_{ij})} P + \dot{N} \delta(s_{ij}) \ , \tag{11.20}$$

where

$$\delta(s_{ij}) = \delta(s_{11})\delta(s_{12})\ldots\delta(s_{33})$$

and where $\dot{N}$ is determined so that probability is conserved, i.e.,

$$\int ds P(s_{ij}, t) = 1 \ , \tag{11.21}$$

where $ds = ds_{11} ds_{12} \ldots ds_{33}$ is a volume element in the 9-dimensional s-space. In (11.20) we have assumed that after shear melting the shear stress in a stress block is reduced to zero. Integrating (11.20) over s-space gives

$$\frac{\partial}{\partial t} \int ds P = -\dot{s}_{ij} \int ds \frac{\partial P}{\partial s_{ij}} - \nu_0 \int ds e^{-\beta \Delta E(s_{ij})} P + \dot{N} \ .$$

Using (11.21) the left-hand side of this equation vanishes. Using Gauss' theorem we get

$$\int ds \frac{\partial P}{\partial s_{ij}} = \int_S dS P n_{ij} \ ,$$

where dS is a surface element of the 8-dimensional sphere S determined by $s_{ij} s_{ij} = s_0^2$, and where the unit "vector" (in s-space) $n_{ij} = s_{ij}/s_0$ with $s_{ij} s_{ij} = s_0^2$, is everywhere perpendicular to the surface S. Thus we obtain

$$\dot{N} = \dot{s}_{ij} \int_S dS P n_{ij} + \nu_0 \int ds e^{-\beta \Delta E(s_{ij})} P \ . \tag{11.22}$$

We can interpret the two terms in this equation as the rate at which a stress block shear-melt (a) by being driven over the barrier towards shear melting by the increase in the elastic stress (first term), or (b) as a result of thermal excitation over the barrier (second term). We can visualize process (b) as occurring when a (thermally exited) elastic wave packet reaches the stress block and locally increases the shear stress $s_{ij} \to s_{ij}^0$ to the yield condition $s_{ij}^0 s_{ij}^0 = s_0^2$. Using the elastic continuum model one can estimate the energy barrier $\Delta E(s_{ij})$ as follows. Assume that a stress block has the form of a sphere with the radius R which may be of order ~ 100 Å. By doing an external work ΔE on the surface $r = R$ of the stress block it is possible to change the shear stress in the stress block from s_{ij} to s_{ij}^0. Using the theory of elasticity one can calculate [11.4]

$$\Delta E = A(s_{ij}^0 - s_{ij})s_{ij}^0 , \qquad (11.23)$$

where

$$A = \frac{5\pi R^3(1-\nu^2)}{E(4-5\nu)} . \qquad (11.24)$$

The energy barrier $\Delta E(s_{ij})$ in (11.20) is obtained by minimizing (11.23) with respect to s_{ij}^0 with the constraint that $s_{ij}^0 s_{ij}^0 = s_0^2$ and $s_{kk}^0 = 0$. To solve this problem let us form

$$F = A(s_{ij}^0 - s_{ij})s_{ij}^0 + \alpha(s_{ij}^0 s_{ij}^0 - s_0^2) + \beta s_{kk}^0 ,$$

where α and β are Lagrange multipliers. Minimizing F with respect to s_{ij}^0 gives

$$s_{ij}^0 = \frac{As_{ij}}{2A + 2\alpha} , \qquad (11.25)$$

and from the condition $s_{ij}^0 s_{ij}^0 = s_0^2$ we get

$$\frac{A}{2A+2\alpha} = \frac{s_0}{(s_{ij}s_{ij})^{1/2}} . \qquad (11.26)$$

Substituting (11.25) and (11.26) into (11.23) gives

$$\Delta E = As_0^2 \left[1 - (s_{ij}s_{ij})^{1/2}/s_0\right] \equiv \varepsilon \left[1 - (s_{ij}s_{ij})^{1/2}/s_0\right] . \qquad (11.27)$$

Finally, we must specify $\dot{s}_{ij}$ occurring in (11.20), which is the rate of increase of the shear stress s_{ij} due to local *elastic deformations*. We will show below that

$$\dot{s}_{ij} = \frac{E}{1+\nu}(\dot{e}_{\text{el}}^{\text{loc}})_{ij} = \frac{E}{1+\nu}[(\dot{e}_{\text{el}})_{ij} + (\dot{e}_{\text{pl}})_{ij}] \equiv \frac{E}{1+\nu}\dot{e}_{ij} . \qquad (11.28)$$

If we assume that the elastic shear strain $(1+\nu)s_{ij}/E$ in a stress block is completely converted into plastic strain during the shear melting process, then the rate of plastic deformation is

$$(\dot{e}_{\text{pl}})_{ij} = \frac{1+\nu}{E}\left(\int_S dS s_{ij} P n_{kl} \dot{s}_{kl} + \nu_0 \int ds e^{-\beta \Delta E(s_{kl})} s_{ij} P\right) . \qquad (11.29)$$

The two terms in this equation are the product of the rate at which a stress block shear-melts and the plastic strain gained in such a process [which equals $(1+\nu)s_{ij}/E$]. The first term in (11.29) results from the stress block being driven over the barrier towards shear melting by the increase in the elastic stress, while the second term results from thermal excitation over the barrier

$\Delta E(s_{ij})$. Equations (11.20–22) and (11.27–29) constitute a complete set of equations from which creep and thermally induced relaxation processes can be studied. From the theory the stress distribution function $P(s_{ij}, t)$ can be deduced, and from this one can obtain the average stress deviator

$$\langle s_{ij} \rangle = \int ds\, s_{ij} P(s_{kl}, t)$$

and

$$(\dot{e}_{el})_{ij} = \frac{1+\nu}{E} \frac{\partial}{\partial t} \langle s_{ij} \rangle . \tag{11.30}$$

Let us multiply (11.20) with s_{ij} and integrate over s-space:

$$\frac{\partial}{\partial t} \int ds\, s_{ij} P = -\int ds\, s_{ij} \frac{\partial P}{\partial s_{kl}} \dot{s}_{kl} - \nu_0 \int ds\, e^{-\beta \Delta E(s_{kl})} s_{ij} P . \tag{11.31}$$

But

$$\int ds\, s_{ij} \frac{\partial P}{\partial s_{kl}} = \int ds\, \frac{\partial}{\partial s_{kl}} (s_{ij} P) - \delta_{ik}\delta_{jl} \int ds\, P$$

$$= \int_S dS\, s_{ij} P n_{kl} - \delta_{ik}\delta_{jl} , \tag{11.32}$$

where we have used Gauss' theorem and the normalization condition (11.21). Substituting (11.32) into (11.31) gives

$$\frac{\partial}{\partial t} \int ds\, s_{ij} P = \dot{s}_{ij} - \int_S dS\, s_{ij} P n_{kl} \dot{s}_{kl} - \nu_0 \int ds\, e^{-\beta \Delta E(s_{kl})} s_{ij} P . \tag{11.33}$$

Comparing (11.33) to (11.29) and using (11.30) gives (11.28). Thus, (11.20) takes the form

$$\frac{\partial P}{\partial t} + \frac{E}{1+\nu} \frac{\partial P}{\partial s_{ij}} \dot{e}_{ij} + \nu_0 e^{-\beta \Delta E(s_{ij})} P - \dot{N}\delta(s_{ij}) = 0 . \tag{11.34}$$

This equation can be solved using the method of characteristic curves. Here we consider the simplest (but very important) case of one-dimensional creep.

One-Dimensional Creep

Assume that a rectangular solid block is exposed to a constant surface stress σ_0 on two opposite surfaces as indicated in Fig. 11.12. If $l(t)$ denotes the length of the block at time t,

$$\dot{\varepsilon} = \dot{l}/l(t) .$$

After an initial relaxation period τ, during which the stress distribution $P(s_{ij}, t)$ changes from the initial $t = 0$ form (which is determined by the

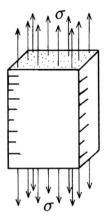

Fig. 11.12. A solid block under uniaxial tension.

previous history of deformations which the block has been exposed to) to its steady-state form, we expect $\dot{\varepsilon}$ to depend only on σ_0. Thus

$$\dot{\varepsilon} = f(\sigma_0) ,$$

so that

$$l(t) = l(0)e^{f(\sigma_0)t} .$$

Note that we have assumed that σ_0 is constant, and since the area $A(t)$ depends on time in such a way that the volume $V = A(t)l(t)$ is constant (plastic deformations usually occur without any change in the volume of the body), it follows that the total force which acts on the surface A decreases with time.

We now determine the function $f(\sigma_0)$. For uniaxial tension as in Fig. 11.12 the strain rate $\dot{\varepsilon}_{ij}$ has the form

$$\dot{\varepsilon}_{ij} = \begin{pmatrix} \dot{\varepsilon} & 0 & 0 \\ 0 & -\dot{\varepsilon}/2 & 0 \\ 0 & 0 & -\dot{\varepsilon}/2 \end{pmatrix} , \qquad (11.35)$$

and it is then easy to show that (11.34) is satisfied with

$$P(s_{ij}) = (3/2)P(\sigma)\delta(s_{22} + \sigma/3)\delta(s_{33} + \sigma/3)\prod_{i \neq j}\delta(s_{ij}) , \qquad (11.36)$$

where $\sigma = 3s_{11}/2$. Thus s_{ij} and the stress tensor have the form

$$s_{ij} = \begin{pmatrix} 2\sigma/3 & 0 & 0 \\ 0 & -\sigma/3 & 0 \\ 0 & 0 & -\sigma/3 \end{pmatrix} ; \qquad \sigma_{ij} = \begin{pmatrix} \sigma & 0 & 0 \\ 0 & 0 & 0 \\ 0 & 0 & 0 \end{pmatrix} .$$

Substituting (11.36) into (11.34) gives the following equation for $P(\sigma)$

$$\frac{3E\dot{\varepsilon}}{2(1+\nu)}\frac{\partial P}{\partial \sigma} + \nu_0 e^{-\beta \Delta E(\sigma)} P - \dot{N}\delta(\sigma) = 0 , \tag{11.37}$$

where

$$\Delta E(\sigma) = \varepsilon\left[1 - \sigma/\sigma_a\right]$$

for $\sigma \geq 0$. The general solution of (11.37) is

$$P = Ce^{-[2\nu_0(1+\nu)/3E\dot{\varepsilon}]\int_0^\sigma d\sigma' \exp[-\beta\varepsilon(1-\sigma'/\sigma_a)]} \tag{11.38}$$

for $0 \leq \sigma < \sigma_a$, and zero otherwise. Let us introduce $\alpha = 2\nu_0(1+\nu)\sigma_a/3E\dot{\varepsilon}\beta\varepsilon$ and $x = \sigma'/\sigma_a$ so that

$$P = Ce^{-\beta\varepsilon\alpha \int_0^{\sigma/\sigma_a} dx \exp[-\beta\varepsilon(1-x)]} . \tag{11.39}$$

Thus the average stress

$$\langle\sigma\rangle = \sigma_0 = \int d\sigma \sigma P(\sigma) , \tag{11.40}$$

in units of σ_a, depends only on the two parameters α and $\beta\varepsilon$. Let us first consider the high-strain-rate limit where $\alpha \ll 1$. In this case (11.39) can be expanded to leading order in α and substituting the result into (11.40) gives to leading order in $1/\dot{\varepsilon}$:

$$\langle\sigma\rangle = \frac{\sigma_a}{2}\left(1 - \frac{\alpha}{\beta\varepsilon}\right) = \frac{\sigma_a}{2}\left(1 - \frac{2\nu_0(1+\nu)\sigma_a}{3E\dot{\varepsilon}(\beta\varepsilon)^2}\right) . \tag{11.41}$$

Next, let us consider the small $\dot{\varepsilon}$ limit where $\alpha \gg 1$. In this case we get, if $\alpha\exp(-\beta\varepsilon) \ll 1$,

$$\langle\sigma\rangle = \frac{\sigma_a}{2}\left(1 - \frac{1}{\beta\varepsilon}\ln\alpha\right) = \frac{\sigma_a}{2}\left[1 + \frac{1}{\beta\varepsilon}\ln\left(\frac{3E\dot{\varepsilon}\beta\varepsilon}{2\nu_0(1+\nu)\sigma_a}\right)\right] . \tag{11.42}$$

In the extremely small $\dot{\varepsilon}$ limit where $\alpha \gg 1$ and $\alpha\exp(-\beta\varepsilon) \gg 1$ we find

$$\langle\sigma\rangle = \frac{\sigma_a}{\alpha\beta\varepsilon}e^{\beta\varepsilon} = \frac{3E\dot{\varepsilon}}{2\nu_0(1+\nu)}e^{\beta\varepsilon} . \tag{11.43}$$

Comments

In the theory above we have made a few assumptions and approximations, which we now comment on. We have treated the media as isotropic, which (for a polycrystalline material) is usually a good approximation with respect to the long-wavelength deformations of interest in most practical applications. However, in the present case this approximation is less accurate since the volume elements (stress blocks) which deform and yield are typically much smaller than the grain sizes. We note, however, that the von Mises type of yield criterion applied to the stress blocks should still be accurate. First, it is clear that even for anisotropic materials the yield criterion depends only on the stress deviator s_{ij} and not on the hydrostatic pressure p. Thus, for example, a highly anisotropic body such as a shrimp does not "yield" even when it lives 1000 m below the water surface, where the hydrostatic pressure $p \sim 10^7 \, \text{N/m}^2$. However, common experience shows that during uniaxial tension a shrimp yields at a stress of only about $\sim 10^5 \, \text{N/m}^2$. Secondly, molecular dynamics calculations [11.5] of the yielding of nanoscale junctions between a Ni tip and a gold substrate (single crystals) have shown that yield occurs when the quantity $(3s_{ij}s_{ij}/2)^{1/2}$ reaches a critical value $\sigma_a \approx 5 \times 10^9 \, \text{N/m}^2$ (which, for reasons discussed in Sect. 5.2, is much higher than the macroscopic yield stress of gold), i.e., the von Mises yield criterion may be at least approximately valid also for single crystals.

Another assumption made above is that when a stress block shear-melts, the local shear stress drops to zero. This is unlikely to be true in general. Thus, during the nanoscale yielding processes discussed above, the quantity $(3s_{ij}s_{ij}/2)^{1/2}$ changed (abruptly) from 5GPa to 3GPa in the region where the yield occurred. However, it is easy to modify the theory to take into account only a partial release of shear stress during local shear melting.

It is interesting to note that although plastic deformation does not depend on the hydrostatic pressure, the *ductility* of a solid, i.e., the amount of plastic deformation possible before fracture, often increases very strongly with increasing hydrostatic pressure. The reason for this is as follows. The fracture of a solid usually occurs by the growth of cracks and voids, which may have been generated during the plastic deformation. A superimposed hydrostatic stress will close the cracks and increase the (friction) stress necessary for sliding over the crack faces [11.6].

We have assumed a well-defined pinning barrier ε. In reality one may expect a distribution of pinning barriers, and a corresponding distribution of local yield stresses.

Finally, we note that the theory above is of the mean-field type. However, using the finite-element method, with the basic volume elements chosen as stress blocks, it should be possible to go beyond the mean field approximation as in the study in Sect. 11.3. However, because of the small size of the stress blocks it is not possible to study creep and relaxation in large macroscopic bodies using this method.

We now present experimental data which illustrate the theoretical results presented above.

Thermally Activated Creep in Metals

The pinning energy for dislocations in metals is typically $\varepsilon \sim 1\,\text{eV}$. In order to reduce the creep rate as much as possible, the pinning energy ε should

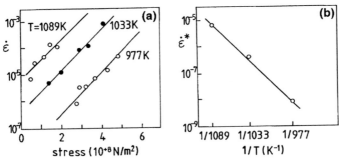

Fig. 11.13. (a) Steady state creep rate $\dot{\varepsilon}$ (in units of h^{-1}) for the superalloy Waspaloy as a function of stress. (b) Strain-rate $\dot{\varepsilon}^*$ (in units of h^{-1}) as a function of the inverse of the temperature, $1/T$ ($\dot{\varepsilon}^*$ is the strain rate extrapolated to zero stress). From [11.8].

be as large as possible. Special "superalloys" have been developed, which exhibit only small creep even at relatively high temperatures, and which are used, e.g., for jet engine turbine blades. Figure 11.13a shows the relation between the strain rate $\dot{\varepsilon}$ and the stress σ for three different temperatures for a nickel-based superalloy (Waspaloy). Note that σ depends linearly on $\ln \dot{\varepsilon}$ in accordance with (11.42). In order to determine the activation energy ε we write (11.42) in the form

$$\sigma = \sigma_c \left(1 + \frac{k_B T}{\varepsilon} \ln \left(\frac{\dot{\varepsilon}}{\dot{\varepsilon}_0} \right) \right) , \tag{11.44}$$

where $\sigma_c = \sigma_a/2$ is the macroscopic yield stress and

$$\dot{\varepsilon}_0 = \frac{8\nu_0(1+\nu)\sigma_c}{3E\beta\varepsilon} . \tag{11.45}$$

Thus, if the curves in Fig. 11.13a are extrapolated to $\sigma = 0$, the resulting creep rate $\dot{\varepsilon} = \dot{\varepsilon}^*$ should satisfy

$$1 + \frac{k_B T}{\varepsilon} \ln \left(\frac{\dot{\varepsilon}^*}{\dot{\varepsilon}_0} \right) = 0$$

or

$$\ln(\dot{\varepsilon}^*/\dot{\varepsilon}_1) = \ln(\dot{\varepsilon}_0/\dot{\varepsilon}_1) - \varepsilon/k_B T , \tag{11.46}$$

where $\dot{\varepsilon}_1$ is an arbitrarily chosen reference strain rate (e.g., $\dot{\varepsilon}_1 = 1\,\text{s}^{-1}$). In a typical case $\dot{\varepsilon}_0 \sim 10^8\,\text{s}^{-1}$ so that, if $\dot{\varepsilon}_1 \sim 1\,\text{s}^{-1}$, even if $\dot{\varepsilon}_0$ is proportional to T, the temperature dependence of $\ln(\dot{\varepsilon}_0/\dot{\varepsilon}_1)$ is completely negligible. Thus, from (11.46) we expect a linear relation between $\ln(\dot{\varepsilon}^*/\dot{\varepsilon}_1)$ and $1/T$, as is indeed is observed, see Fig. 11.13b. From the slope of the curve in Fig. 11.13b we deduce $\varepsilon \approx 5.5\,\text{eV}$, which is higher than for "normal" steel. Using $\varepsilon = As_0^2$

and (11.24) one can calculate the diameter D of a stress block, which turns out to be $D \sim 30\,\text{Å}$. This gives $\nu_0 \sim c/2\pi D \sim 10^{11}\,\text{s}^{-1}$ and $\dot{\varepsilon}_0 \sim 10^8\,\text{s}^{-1}$.

Creep Enhancement of the Area of Real Contact

We have pointed out several times that immediately after the rapid plastic deformations which occur when the block is put on the substrate, the stress blocks in the vicinity of the contact area will be in a "critical" state with a distribution of local stresses such that some stress blocks are almost ready to undergo plastic deformation. This implies that thermal processes will be of crucial importance; slow relaxation (creep) will occur and the contact area will increase slowly with time. The theory developed above can be used to study how the area of real contact depends on time t, and in Sect. 13.2 we show that $A = A_0[1 + B\ln(1 + t/\tau_0)]$ to leading order in $B = k_\text{B}T/\varepsilon$. This logarithmic increase in the contact area with the time of stationary contact has been observed for many systems (see Sect. 13.2) and is of crucial importance for dry friction dynamics.

Thermally Activated Creep of Clayey Samples

The theory developed above and in Sect. 11.3 also seems to be valid for granular multi-component systems like natural soils [11.7]. Creep and relaxation have been studied for natural or reconstituted samples of water-saturated clay or similar densified mud. These systems consist of grains of size ranging from about 10^{-7} m to 10^{-5} m; the mineral is mainly silicate. Experiments are performed on samples confined within cylindrical or plane walls; see Fig. 11.14.

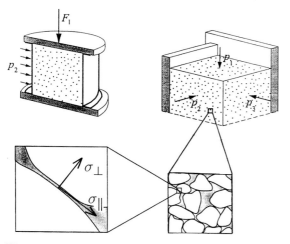

Fig. 11.14. Devices for homogeneous deformation of grain skeletons. The inset shows a few grains in contact with the perpendicular and parallel shear stresses indicated in one contact area.

Stationary Creep

Stationary creep experiments are performed with constant volume and skeleton pressure components. The cylindrical sample is shortened at a constant rate $\dot{\varepsilon} = \dot{l}/l$ while the system expands in the other two spatial directions so that the volume is constant. After a transition time period the system reaches a stationary state where the stress does not depend on the initial state of the system, and where the difference between the skeleton pressures p_1 and p_2, along the cylinder axis and the radial direction, respectively, satisfies

$$p_1 - p_2 = C\left(1 + \frac{k_B T}{\varepsilon} \ln\left(\frac{\dot{\varepsilon}}{\dot{\varepsilon}_0}\right)\right),$$

where C and ε are temperature independent, while the reference strain rate $\dot{\varepsilon}_0$ is proportional to T.

Volumetric Creep

Volumetric creep is observed by keeping the external pressure constant. The void ratio $e = (1-\alpha)/\alpha$, where α is the solid volume fraction, is found to have the time dependence

$$e(0) - e(t) = C' \frac{k_B T}{\varepsilon} \ln(1 + t/\tau),$$

where C' and ε are independent of the temperature T.

Relaxation

Relaxation is observed by keeping cuboidal samples fixed within plates after different deformation histories. The skeleton pressure components p_i return to equilibrium according to characteristic relaxation curves when plotting p_i as a function of $\ln(t/\tau)$, where the relaxation time τ is defined by extrapolating the straight section of this plot to the equilibrium value. It is found that τ is independent of the deformation history.

These experimental results can be understood on the basis of the theoretical results presented in this section and in Sect. 11.3. Thus, the particle system deforms slowly (creep and relaxation) by (a) thermally activated slip between particles in response to the tangential surface stresses in the contact areas (inset in Fig. 11.14), as described by the theory presented in Sect. 11.3, and by (b) time-dependent plastic deformation of the solid particles (mainly due to the perpendicular contact stresses), as described by the theory above (see also Sect. 13.2). Both processes give relaxation $\propto \ln(1 + t/\tau)$ and creep rates $\propto \ln(\dot{\varepsilon}/\dot{\varepsilon}_0)$, with the same temperature dependence as observed experimentally. From the experimental data $k_B T/\varepsilon = 0.02$ and 0.05 for lean and fat clay, respectively, corresponding to $\varepsilon \approx 1.3\,\mathrm{eV}$ and $0.5\,\mathrm{eV}$, respectively. The value of $k_B T/\varepsilon$ for lean clay (which consists mainly of quartz or calcite powder) is similar to that which can be deduced from indentation experiments for quartz (Fig. 13.4), which gives ~ 0.02.

The stationary creep results presented above are not valid if the driving stress and resulting strain rate are extremely low. In the latter case the theory [see (11.19) and (11.43)] predicts that the strain rate should be directly proportional to the stress. This limiting case cannot be probed in laboratory experiments because of the low strain rates involved, but may show up in nature on geological time scales since clay deposits can exist for millions of years.

Other Manifestations of Creep

We mention here some other manifestations of creep. First, consider a block of glass containing a gas bubble. Glass is usually considered as a brittle material but the response to a small stress is not brittle fracture but slow creep motion. The air bubble therefore moves slowly upwards and on the time scale of ~ 1000 years glass behaves as a typical viscous fluid. Similarly, old glass windows are thicker at the lower edges than at the upper edges because of a slow flow of the glass due to the gravitational force. In both these cases the driving forces (or stresses) are so small that the stress σ is likely to be linearly related to the strain rate $\dot{\varepsilon}$ [(11.19) and (11.43)], as expected for a fluid. As another example, consider the Earth's crust, which is continuously exposed to fluctuating stresses. This gives rise to a slow creep motion, but in most cases the stress and strain rates are so high that the relation $\dot{\varepsilon} = f(\sigma)$ is non-linear [(11.18) and (11.42)]. One has discovered mine tunnels in Yugoslavia from Roman times, where the tunnel height has been so much reduced by creep that it is now impossible for humans to move in them. Another proof that stone creeps under load is the Abraham Spring close to Jerusalem, which is now strongly deformed plastically. Other manifestations of creep are the formation of mountains (over millions of years) and the creep motion at fault lines (Sect. 14.1).

12. Lubricated Friction Dynamics

Consider the sliding of a block on a substrate under boundary lubrication conditions. To the block is connected a spring (force constant k_s) and the free end of the spring is moving with the velocity v_s. Depending on the parameter values (k_s, v_s) the motion of the block may be either steady or of stick-slip nature. In this chapter we discuss the nature of the (k_s, v_s) phase diagram, i.e., the regions in the (k_s, v_s)-plane where the motion is steady and oscillatory are determined and the nature of the transition between the two types of sliding behavior is discussed.

We consider two different cases, namely when the lateral corrugation of the interaction potential between the lubrication molecules and the substrate is "small" and "large". The first class of systems includes saturated hydrocarbons on metal or mica surfaces, where, at the onset of sliding, we believe that the lubrication film shear-melts or fluidizes. The second class of systems includes grafted monolayer films, e.g., fatty acid monolayers on many metal oxides or n-hexadecanethiol ($C_{16}H_{33}SH$) on gold. In these cases the corrugation of the adsorbate–substrate potential energy surface is so large that no fluidization can occur and the sliding takes place at the interface between the monolayer films adsorbed on the two surfaces.

Based on a microscopic picture of the physically most relevant processes occurring in the lubrication film during sliding, we present constitutive equations or friction "laws". These equations depend not only on the center of mass coordinate $x(t)$ of the block, which is treated as a rigid object, but also on a state variable, which we denote by θ or ϕ, which describes the state of the lubrication film. Thus, in addition to Newton's equation of motion for $x(t)$ [which depends on the state variable $\theta(t)$ or $\phi(t)$ via the friction force], the theory also involves an equation of motion for the state variable, which describes the average or mean field response of the lubrication film to the motion of the block. The solution of the two coupled equations describes the time evolution of $x(t)$ and of the state variable of the lubrication film.

12.1 Small Corrugation: Shear-Melting and Freezing

In Chap. 9 we presented a qualitative discussion of boundary lubrication for the case when the adsorbate–substrate interaction potential was so weak that the adsorbate layer could fluidize at the onset of sliding. We discussed the origin of the critical velocity v_c for a block sliding on a tilted substrate and presented a qualitative discussion of the mechanism behind the transition from stick-slip to steady sliding for the spring–block system.

In this section we consider the model of *Carlson* and *Batista* [12.1] which is a first attempt to present a constitutive relation for friction between lubricated surfaces. We present a linear instability analysis of the equations of motion and discuss the physical origin of stick-slip motion.

Model

The theory is based on the picture presented in Sect. 9.1 and introduces a new dynamical variable or order parameter θ which can take values between zero and one, where $\theta = 1$ corresponds to the fully solid lubrication layer at stick and $\theta = 0$ to the uniform fluid state at "high" sliding velocity. In general, the state variable θ represents the degree of melting of the lubrication layer.

The friction force F_0 is the surface stress acting on the block from the lubrication layer, integrated over the contact area, and it is determined by the microscopic forces which the lubrication molecules exert on the bottom surface of the block. The basic assumption behind the phenomenological approach is that the system is large enough that the microscopic forces are sufficient self-averaging to produce a macroscopic force law which is deterministic and depends only on the macroscopic variables associated with the motion. In this equation the state variable θ is associated with a history-dependent mean-field value, characteristic of some average property of the interface.

The equation of motion for the block is

$$M\ddot{x} = k_s(v_s t - x) - F_0 \ . \tag{12.1}$$

The friction force F_0 is assumed to depend on the velocity $\dot{x}$ and on the state variable θ of the lubrication layer,

$$F_0 = F_b + \Delta F_0 \theta + M\gamma\dot{x} \ , \tag{12.2}$$

where $\Delta F_0 = F_a - F_b$. Hence, for a stationary block with a frozen adsorbate layer, $\dot{x} = 0$ and $\theta = 1$ giving $F = F_a$, the static friction force. On the other hand, at high sliding velocity the lubrication layer is in a smooth fluid state and $\theta = 0$ and hence $F_0 = F_b + M\gamma\dot{x}$, where the linear velocity contribution to F_0 arises from the processes discussed in Sect. 8.2. The state variable θ is assumed to satisfy

$$\dot{\theta} = (1/\tau)\theta(1 - \theta) - \theta\dot{x}/D \ . \tag{12.3}$$

12.1 Small Corrugation: Shear-Melting and Freezing

This equation can be interpreted as follows. The first term on the right-hand side describes the behavior when the block is at rest ($\dot{x} = 0$). The stationary solutions to (12.3) when $\dot{x} = 0$ are $\theta = 0$ or $\theta = 1$. These solutions correspond to the fluid and the frozen states, respectively. According to (12.3), from any initial state with $\theta > 0$ the film will tend to freeze at a characteristic rate proportional to $1/\tau$. The first term in (12.3) is the simplest one possible which correctly describes this property. The second term on the right-hand side of (12.3) is the simplest term which can be used to describe the tendency of the film to melt as a consequence of relative shear. The quantity D is a characteristic length over which the melting transition takes place.

It is, in principle, possible to integrate (12.3) to obtain $\theta(t)$ as a function of $x(t')$, $t' \leq t$. Substituting the result for $\theta(t)$ into (12.2) gives rise to a *memory term* in the equation of motion for the center of mass coordinate $x(t)$, i.e., a term which depends on the coordinate $x(t)$ not just at time t but for all *earlier* times, i.e., on $x(t')$ with $t' \leq t$. However, from both a practical and a physical point of view it is better to use the systems of equations (12.2) and (12.3) directly, rather than (12.2) with a complicated memory term.

Assume that the block suddenly stops moving at time $t = 0$. In this case it is easy to integrate (12.3) to yield

$$\theta = \frac{\theta_0}{\theta_0 + (1 - \theta_0)\mathrm{e}^{-t/\tau}},$$

where $\theta_0 = \theta(0)$. Thus,

$$F_\mathrm{s}(t) = F_\mathrm{b} + \frac{\Delta F_0 \theta_0}{\theta_0 + (1 - \theta_0)\mathrm{e}^{-t/\tau}}.$$

This function is shown in Fig. 12.3 below when $\theta_0 = 0.05$. Note that if θ_0 is small, F_s increases rather abruptly for $t \sim \tau$, as is indeed observed in stop-start experiments (Sect. 3.1).

Linear Instability Analysis

In this section we determine the region in the $(k_\mathrm{s}, v_\mathrm{s})$ plane in which stick-slip motion occurs when starting with steady sliding and reducing v_s. The line separating steady sliding from stick-slip sliding is determined by a linear instability analysis [12.2].

We consider first the steady state $\dot{x} = v_\mathrm{s}$ and $\dot{\theta} = 0$. Using (12.2) and (12.3) we get, when $v_\mathrm{s} < D/\tau$,

$$\theta = 1 - \tau v_\mathrm{s}/D, \tag{12.4}$$

$$F_0 = F_\mathrm{a} - (\tau \Delta F_0/D - M\gamma)v_\mathrm{s}. \tag{12.5}$$

If $v_\mathrm{s} > D/\tau$ then $\theta = 0$ and $F_0 = F_\mathrm{b} + M\gamma v_\mathrm{s}$. We now investigate when the steady sliding becomes unstable. Let us write

$$x = x_0 + v_\mathrm{s}t + \xi,$$

$$\theta = \theta_0 + \eta,$$

where $\theta_0 = 1 - \tau v_s/D$. Substituting these quantities into (12.1–3) gives, to linear order in ξ and η,

$$M\ddot{\xi} = -k_s\xi - \Delta F_0\eta - M\gamma\dot{\xi}, \tag{12.6}$$

$$\dot{\eta} = (1/\tau)(1 - 2\theta_0)\eta - (\theta_0/D)\dot{\xi} - (v_s/D)\eta. \tag{12.7}$$

Assuming that

$$\xi = \xi_0 e^{\kappa t}, \qquad \eta = \eta_0 e^{\kappa t}$$

and substituting into (12.6,7) gives

$$\left[M\kappa^2 + M\gamma\kappa + k_s\right]\xi_0 + [\Delta F_0]\eta_0 = 0,$$

$$[\theta_0\kappa/D]\xi_0 + [\kappa + (2\theta_0 - 1)/\tau + v_s/D]\eta_0 = 0.$$

This system of linear equations has a nontrivial solution only if the coefficient determinant vanishes, i.e.,

$$\left[M\kappa^2 + M\gamma\kappa + k_s\right][\kappa + (2\theta_0 - 1)/\tau + v_s/D] - [\Delta F_0][\theta_0\kappa/D] = 0,$$

or, using $\theta_0 = 1 - \tau v_s/D$,

$$\kappa^3 + A\kappa^2 + B\kappa + C = 0 \tag{12.8}$$

where

$$A = \gamma + \frac{1}{\tau} - \frac{v_s}{D}, \qquad C = \frac{k_s}{M}\left(\frac{1}{\tau} - \frac{v_s}{D}\right)$$

$$B = \frac{k_s}{M} + \left(\frac{1}{\tau} - \frac{v_s}{D}\right)\left[\gamma - \frac{\tau\Delta F_0}{MD}\right].$$

Equation (12.8) has three roots. The qualitative nature of the roots as a function of the coefficient B of the linear κ-term in (12.8) is shown in Fig. 12.1. For $B > B_0$ the real parts of all roots are negative and the perturbations ξ and η to the steady motion will decay with increasing time, i.e., the steady sliding state is stable with respect to *small* perturbations. On the other hand, for $B < B_0$ two roots will have a positive real part, and the steady motion is unstable. Hence, $B = B_0$ determines the line $v_s = v_c(k_s)$ in the (k_s, v_s) plane that separates steady sliding motion from stick-slip motion, when starting with steady sliding and reducing v_s. It is easy to determine this phase boundary since for $B = B_0$ the three roots are of the form κ_1 and $\pm i\omega_c$ where κ_1 and ω_c are real. Hence for $B = B_0$ the left-hand side of (12.8) must have the form

$$(\kappa - \kappa_1)(\kappa - i\omega_c)(\kappa + i\omega_c) = \kappa^3 - \kappa_1\kappa^2 + \omega_c^2\kappa - \kappa_1\omega_c^2.$$

Comparing this with (12.8) gives

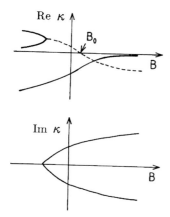

Fig. 12.1. The variation of the real and imaginary parts of the solutions of (12.8) with the coefficient B of the linear term in the polynomial (12.8). The solid and dashed lines in the upper figure show $\operatorname{Re}\kappa$ for purely real and for complex roots, respectively. In the latter case the imaginary parts of the solutions are given in the lower figure.

$$A = -\kappa_1, \quad B = \omega_c^2, \quad C = -\kappa_1 \omega_c^2 \ . \tag{12.9}$$

Thus $AB = C$ and with the expressions for A, B, and C given above,

$$\frac{k_s}{M} = \frac{1}{D^2}\left(\frac{D}{\tau} - v_c\right)\left(\frac{D}{\tau} - v_c + \gamma D\right)\left(\frac{\tau \Delta F_0}{M\gamma D} - 1\right) \ . \tag{12.10}$$

If we introduce

$$v_0 = D/\tau$$

and

$$k_0 = \frac{M}{\tau^2}\left(\frac{\tau \Delta F_0}{MD\gamma} - 1\right) \ ,$$

then (12.10) takes the form

$$\frac{k_s}{k_0} = \left(1 - \frac{v_c}{v_0}\right)\left(\gamma \tau + 1 - \frac{v_c}{v_0}\right) \ . \tag{12.11}$$

Note that $v_c \to v_0$ as $k_s \to 0$ and in Sect. 9.4 we derived an expression for v_0 based on a microscopic picture of the nucleation process.

Figure 12.2 shows the (k_s, v_s) phase diagram when $\gamma\tau = 10$. A particularly interesting limiting case is when the motion is overdamped and when inertia is negligible, i.e., when $\gamma\tau \gg 1$ and $\tau\Delta F_0/MD\gamma \gg 1$. In this case (12.10) reduces to

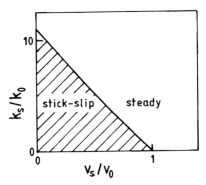

Fig. 12.2. Dynamical phase diagram for $\gamma\tau = 10$.

$$\frac{k_s}{M} = \frac{\tau \Delta F_0}{MD^2}\left(\frac{D}{\tau} - v_c\right). \tag{12.12}$$

The physical origin of this equation can be understood as follows: During steady sliding, where the friction force equals F_k, the state variable is $\theta_0 = 1 - \tau v_s/D$. Assume that the system makes a transition to the pinned state. The static friction force $F_s(t)$ increases monotonically with the pinning time t with $F_s(0) = F_k$. The increase of $F_s(t)$ with the time of stationary contact is determined by the term $\Delta F_0 \theta(t)$ in (12.2). Using (12.3) with $\dot{x} = 0$ and $\theta(0) = \theta_0$ we get

$$\theta = \frac{\theta_0}{\theta_0 + (1 - \theta_0)e^{-t/\tau}}. \tag{12.13}$$

For small times, θ increases linearly with time, $\theta \approx \theta_0 + \theta_0(1-\theta_0)t/\tau$. For large times θ approaches unity. In Fig. 12.3 we show $F_s(t) - F_k = \Delta F_0[\theta(t) - \theta_0]$ as a function of t. The dashed line in the same figure shows the increase in the *spring force*, $k_s v_s t$, as a function of the contact time for three different cases (1)–(3). In case (1) the spring force increases faster with the time t of stationary contact than the initial linear increase of the static friction force; hence the motion of the block will not stop and no stick-slip motion will occur. If the spring velocity v_s is lower than the critical velocity v_c [case (3)] determined by the initial slope of the $F_s(t)$ curve $[k_s v_c = dF_s/dt(t = 0)]$, the spring force will be smaller than the static friction force $F_s(t)$ until t reaches the value t_3, at which time slip starts. In this case stick-slip motion will occur. In the model studied above $dF_s/dt(t = 0) = \Delta F_0 \theta_0(1 - \theta_0)/\tau$ and the criteria $k_s v_c = dF_s/dt(0)$ gives (12.12). Figure 12.2 suggests that the transition is discontinuous, in agreement with experiments for overdamped motion.

The linear instability analysis presented above determines the phase boundary separating stick-slip motion from steady sliding, when starting from steady motion. If the transition is continuous, the same result is obtained when one starts in the stick-slip region. However, if the transition is

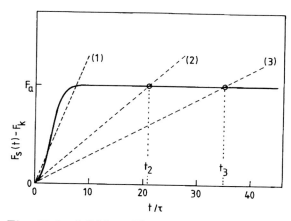

Fig. 12.3. *Solid line*: The increase in the static friction force with the time t of stationary contact when $\theta_0 = 0.05$. *Dashed lines*: The variation of the spring force for three different sliding velocities, (*1*): $v_1 > v_c$, (*2*): $v_2 = v_c$, and (*3*): $v_3 < v_c$.

discontinuous, as in the overdamped case, this is not necessarily the case, and hysteresis may occur, e.g., as a function of the drive velocity v_s. However, for the present system this does not seem to be the case. Thus the analytical results presented above are in perfect agreement with the numerical results of *Carlson* and *Batista* [12.1] (as obtained by increasing v_s from the stick-slip region), even though the transition is discontinuous. The solid line in Fig. 9.11 is calculated from (12.10) and is in relatively good agreement with the experimental data.

Numerical Results

The linear stability analysis presented above determines the boundary line in the (k_s, v_s)-plane at which uniform sliding becomes unstable with respect to *small* perturbations. However, the analysis says nothing about the form of the resulting stick-slip motion. In this section we present numerical results which illustrate the nature of the transition between stick-slip and steady motion.

Let us introduce dimensionless variables. We measure time in units of $(M/k_s)^{1/2}$ and length in units of $\Delta F_0/k_s$. In these units (12.1–3) take the form

$$\ddot{x} = (\bar{v}_s t - x) - A - \theta - \bar{\gamma}\dot{x},$$

$$\dot{\theta} = \theta(1-\theta)/\bar{\tau} - \theta\dot{x}/\bar{D},$$

where

$$\bar{D} = Dk_s/\Delta F_0, \quad \bar{\tau} = \tau(k_s/M)^{1/2},$$

$$\bar{\gamma} = \gamma(M/k_s)^{1/2}, \quad \bar{v}_s = (v_s/\Delta F_0)(Mk_s)^{1/2}, \quad A = F_b/\Delta F_0.$$

Note that $\bar{\gamma} \to \infty$ as $k_s \to 0$, i.e., for low enough spring constant the motion will always be overdamped. The physical origin of this was discussed in Sect. 9.5: for a very weak spring the spring force will change very slowly with time and the block will accelerate until the friction force nearly reaches the value it had at the onset of sliding (the static friction force), and then only very slowly decays towards the steady kinetic friction value.

In the simulations presented below we have used $M = 20\,\text{g}$ and $k_s = 440\,\text{N/m}$ which are typical values for experiments with the surface forces apparatus. We have used $F_b = \Delta F_0$ but the sliding dynamics do not depend on F_b. Figure 12.4 shows the sliding dynamics for overdamped motion (left, with $\gamma = 400\,\text{s}^{-1}$ and $D = 1 \times 10^{-7}\,\text{m}$) and underdamped motion (right, with $\gamma = 50\,\text{s}^{-1}$ and $D = 1 \times 10^{-6}\,\text{m}$). The relaxation time τ was chosen so that the critical velocity $v_c = 0.4\,\mu\text{m/s}$ as observed experimentally for overdamped motion (see Fig. 3.6a). This gives $\tau = 0.24\,\text{s}$ and $\tau = 1.4\,\text{s}$ for overdamped and underdamped motion, respectively. Figure 12.4 shows the time dependence of (a) the spring force, (b) the state

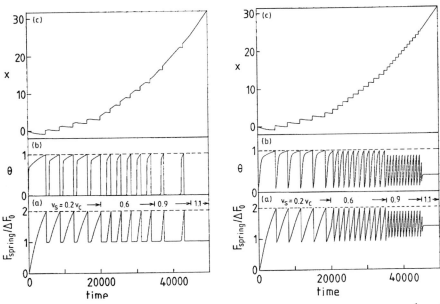

Fig. 12.4. Sliding dynamics for overdamped motion (*left*, with $\gamma = 400\,\text{s}^{-1}$ and $D = 1 \times 10^{-7}\,\text{m}$) and underdamped motion (*right*, with $\gamma = 50\,\text{s}^{-1}$ and $D = 1 \times 10^{-6}\,\text{m}$). The relaxation time τ was chosen so that the critical velocity $v_c = 0.4\,\mu\text{m/s}$ as observed experimentally for overdamped motion. This gives $\tau = 0.24\,\text{s}$ and $\tau = 1.4\,\text{s}$ for overdamped and underdamped motion, respectively. The time dependence of (**a**) the spring force (**b**), the state variable θ, and (**c**) the coordinate x of the center of mass **c**. As indicated in the figure, the spring velocity has been increased in steps from below v_c to above v_c. With $M = 20\,\text{g}$ and $k_s \approx 440\,\text{N/m}$ which are typical for surface forces apparatus measurements. All quantities are in the natural units defined in the text.

variable θ, and (c) the coordinate x of the center of mass. As indicated in the figure, the spring velocity has been increased in steps from below v_c to above v_c. For overdamped motion, in both the model (Fig. 12.4, left) and the experiment (Fig. 3.6a) the transition from stick-slip to steady sliding is discontinuous with the amplitude of the stick-slip spikes being nearly independent of the spring velocity. However, the detailed form of the stick-slip spikes are not in very good agreement with experiment. *Carlson* and *Batista* have presented results for much higher damping γ (with $\gamma \approx 4000\,\mathrm{s}^{-1}$) where the stick-slip spikes are in better agreement with experiment. However, in the latter case the damping γ is so large that the steady sliding friction in the studied spring velocity range becomes strongly velocity dependent, in contrast to what is observed experimentally. It is clear that the system of equations (12.2) and (12.3) need modifications in order to be able to accurately reproduce the experimental data.

A New Friction Law

We now propose a modification of (12.3) which is built directly on the picture of formation and fluidization of solid islands. When the surfaces are stationary the solid phase forms and grows by a constant interface process so that the radius R of an island satisfies

$$dR/dt = v_1 \ .$$

If an island nucleates at time $t = \tau$ then its radius at time $t > \tau$ is $R = v_1(t-\tau)$ and the area $A(t;\tau) = \pi v_1^2(t-\tau)^2$ (we have assumed that the initial radius R_0 of the nucleated island is negligibly small). As individual islands grow, they impinge on one another; one way to account for impingement is to consider the "extended" fractional coverage θ^* of the solid phase, which is a fictitious area fraction to which the islands would grow if it were not for impingement (Fig. 12.5). Here θ^* can be larger than 1. *Avrami* [12.3] has shown that the actual fraction θ of the total area occupied by the solid phase is

$$\theta = 1 - e^{-\theta^*} \ . \tag{12.14}$$

If there is a nucleation rate I per unit area and time,

$$\theta^*(t) = \int_0^t d\tau I A(t;\tau) = \int_0^t d\tau I \pi (t-\tau)^2 = \pi v_1^2 I t^3 / 3 \ . \tag{12.15}$$

Substituting (12.15) into (12.14) gives

$$\theta = 1 - e^{-(t/3\tau)^3} \ , \tag{12.16}$$

where $\tau = \left(9\pi I v_1^2\right)^{-1/3}$. Now, let us assume that for a general case the variable θ is determined by the equation

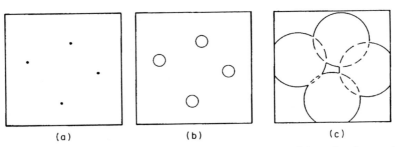

Fig. 12.5. Schematic representation of phase growth: (**a**) nucleation at short times at a fixed number of sites, (**b**) growth with no additional nucleation, and (**c**) extended area (*dashed lines*) and actual area (*solid lines*) after impingement of growing islands.

$$\dot\theta = F(\theta) - \theta\dot x/D ,\qquad(12.17)$$

where the last term has the same physical meaning as before. For stationary surfaces $\dot x = 0$ and we can write $d\theta/F(\theta) = dt$. Using (12.16) we get

$$t = 3\tau\left[-\ln(1-\theta)\right]^{1/3}$$

and thus

$$dt = \frac{d\theta\,\tau}{(1-\theta)\left[-\ln(1-\theta)\right]^{2/3}} .$$

Comparing this with $dt = d\theta/F(\theta)$ gives

$$F(\theta) = (1/\tau)(1-\theta)\left[-\ln(1-\theta)\right]^{2/3} .\qquad(12.18)$$

Substituting (12.18) into (12.17) gives

$$\dot\theta = (1/\tau)(1-\theta)\left[-\ln(1-\theta)\right]^{2/3} - \theta\dot x/D .\qquad(12.19)$$

Note that when $\theta \ll 1$ (high sliding velocity) then (12.19) reduces to

$$\dot\theta = (1/\tau)(1-\theta)\theta^{2/3} - \theta\dot x/D ,$$

which differs from (12.3) in that the term $(1-\theta)\theta$ in (12.3) is replaced by $(1-\theta)\theta^{2/3}$. Studies of the friction dynamics which result from the friction law (12.19) are in progress [12.4].

12.2 Large Corrugation: Interdiffusion and Slip at Interface

In this section we study sliding dynamics for the case where the corrugation of the adsorbate–substrate interaction potential is so large that no shear melting or fluidization of the adsorbate layers occurs during sliding [12.5]. This is the case, e.g., for many fatty acid–metal oxide systems, where the sliding is likely to occur between the tails of the grafted hydrocarbon chains. For these systems the static friction force $F_s(t)$ increases with the time t of stationary contact because of interdiffusion and other relaxation processes as sketched in Fig. 12.6.

Model

We study sliding dynamics for systems where the static friction force $F_s(t)$ increases with the time t of stationary contact because of interdiffusion (Fig. 12.6). In the simplest possible case, one assumes that the increase in the static friction force $F_s(t)$ with the time t of stationary contact is characterized by a single relaxation time τ and that

$$F_s(t) = F_b + \Delta F_0 \left(1 - e^{-t/\tau}\right) , \qquad (12.20)$$

where $\Delta F_0 = F_a - F_b$. Note that $F_s \to F_a$, the maximum static friction force, as $t \to \infty$. The equation of motion for the center of mass of the block is

$$M\ddot{x} = k_s(v_s t - x) - F_0 \qquad (12.21)$$

where the friction force F_0 is taken to be

$$F_0 = F_b + \Delta F_0 \left(1 - e^{-\phi/\tau}\right) + M\gamma\dot{x} . \qquad (12.22)$$

The "contact age" variable $\phi(t)$ is defined by [12.6–9]

$$\phi(t) = \int_0^t dt' \exp\left(-\frac{x(t) - x(t')}{D}\right) , \qquad (12.23)$$

(a) 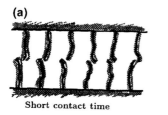 (b)

Short contact time · · · · · · · · · Long contact time

Fig. 12.6. Two surfaces covered by monolayers of grafted chain molecules. (a) Short time after contact. (b) After a long contact time.

where $t = 0$ is the time at which the two solids first came into contact. In (12.23) D is the (average) displacement of the center of mass of the block (treated as a rigid object) necessary to break a molecular-sized junction, i.e., D is of order a few Ångstroms. Note that for stationary contact, $\dot{x} = 0$, (12.23) gives $\phi = t$, i.e., ϕ equals the time of stationary contact, in which case (12.22) reduces to (12.20). On the other hand, for uniform sliding $x = vt$ and (12.23) gives $\phi = D/v$ which is the (average) time a junction survives before being broken by the sliding motion. Note that (12.23) is equivalent to

$$\dot{\phi} = 1 - \dot{x}\phi/D . \tag{12.24}$$

Equations (12.21,22,24) are a "minimal" model which is consistent with experiment, but a more detailed quantitative comparison with experimental data may require certain higher order terms to be added to the equations of motion. For example, we have assumed that the relaxation processes at the interface are characterized by a single relaxation time τ; in general a wide distribution of relaxation processes, characterized by different relaxation times, may occur.

Linear Instability Analysis

We now analyze when the uniform sliding state becomes unstable with respect to *small* perturbations. We write

$$x = x_0 + v_s t + \xi , \tag{12.25}$$

$$\phi = D/v_s + \eta , \tag{12.26}$$

where ξ and η are "small". Instead of considering the special friction "law" (12.22), let us assume that $F_0 = f(\phi, \dot{x})$. Substituting (12.25) and (12.26) into the equations of motion (12.21) and (12.24) and expanding to linear order in ξ and η gives

$$\ddot{\xi} = -\omega_0^2 \xi - a\eta - b\dot{\xi} , \tag{12.27}$$

$$\dot{\eta} = -\dot{\xi}/v_s - v_s \eta/D , \tag{12.28}$$

with

$$a = \frac{1}{M}\frac{\partial f}{\partial \phi}, \quad b = \frac{1}{M}\frac{\partial f}{\partial \dot{x}} , \tag{12.29}$$

where the partial derivatives are evaluated for $\phi = D/v_s$ and $\dot{x} = v_s$. In (12.27), $\omega_0^2 = k_s/M$. Note that with the form for $f(\phi, \dot{x})$ given by (12.22), one has

$$a = \frac{\Delta F_0}{M\tau}e^{-D/v_s\tau}, \quad b = \gamma . \tag{12.30}$$

12.2 Large Corrugation: Interdiffusion and Slip at Interface

Assume now that $\xi = \xi_0 \exp(\kappa t)$ and $\eta = \eta_0 \exp(\kappa t)$. Substituting this into (12.27) and (12.28) gives

$$\left(\kappa^2 + \omega_0^2 + b\kappa\right)\xi_0 + a\eta_0 = 0 \, ,$$

$$\kappa\xi_0/v_s + (\kappa + v_s/D)\,\eta_0 = 0 \, .$$

This system of equations has a non-trivial solution only if the determinant of the coefficient matrix vanishes, which gives

$$\kappa^3 + A\kappa^2 + B\kappa + C = 0 \, , \tag{12.31}$$

where

$$A = b + v_s/D, \qquad C = \omega_0^2 v_s/D \, ,$$
$$B = \omega_0^2 + bv_s/D - a/v_s \, .$$

This equation is of exactly the same general form as the one studied in Sect. 12.1. Thus we conclude that the curve $v_s = v_c(k_s)$ in the (k_s, v_s)-plane where steady sliding is replaced by stick-slip sliding is determined by $AB = C$, which gives

$$\frac{k_s}{M} = \left(1 + \frac{v_c}{bD}\right)\left(\frac{a}{v_c} - \frac{bv_c}{D}\right) \, . \tag{12.32}$$

If we assume that a and b are given by (12.30) then from (12.32)

$$\frac{k_s}{M} = \left(1 + \frac{v_c}{\gamma D}\right)\left(\frac{\Delta F_0}{M\tau v_c}e^{-D/v_c\tau} - \frac{\gamma v_c}{D}\right) \, . \tag{12.33}$$

If we introduce $v_0 = D/\tau$ and $k_0 = \Delta F_0/D$ then (12.33) can be written as

$$\frac{k_s}{k_0} = \left(1 + \frac{v_c}{v_0}\frac{1}{\gamma\tau}\right)\left(\frac{v_0}{v_c}e^{-v_0/v_c} - \frac{v_c}{v_0}\kappa\right) \, , \tag{12.34}$$

where

$$\kappa = \gamma\tau(Mv_0^2/k_0 D^2) \, . \tag{12.35}$$

The solid line in Fig. 12.7 shows the boundary line $v_s = v_c(k_s)$ separating stick-slip motion (dashed area) from steady sliding for the case $\gamma\tau = 950$ and $\kappa = 1.52 \times 10^{-7}$, as would result if, e.g., $\tau = 5\,\mathrm{s}$, $\gamma = 190\,\mathrm{s}^{-1}$, $M = 20\,\mathrm{g}$, $\Delta F_0 = 1 \times 10^{-3}\,\mathrm{N}$ and $D = 2\,\mathrm{Å}$. In what follows we denote the highest spring velocity for which stick-slip motion occurs by v_c^0, which corresponds to the case where $k_s = 0$; see Fig. 12.7. With the parameter values above, $v_c^0 \approx v_0/\kappa^{1/2} \approx 1.026 \times 10^{-7}\,\mathrm{m/s}$.

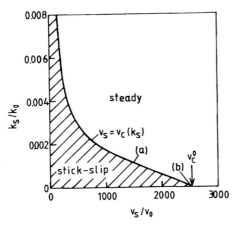

Fig. 12.7. Dynamical phase diagram for $\gamma\tau = 950$ and $\kappa = 1.52 \times 10^{-7}$.

A particularly interesting limiting case is when $v_0\gamma\tau \gg v_c$ and $v_c^2\kappa \ll v_0^2$. In this case (12.33) reduces to

$$k_s = \frac{\Delta F_0}{v_c\tau} e^{-D/v_c\tau} . \tag{12.36}$$

This equation can be understood as follows. First note that during steady sliding, where the friction force equals F_k, the contact age variable $\phi_0 = D/v_s$. Assume that the system makes a transition to the pinned state. The static friction force $F_s(t)$ increases monotonically with the time t of stationary contact, with $F_s(0) = F_k$. In the model studied above, $\phi = D/v_s + t$ and

$$F_s - F_k = \Delta F_0 \left(1 - e^{-t/\tau}\right) e^{-D/v_s\tau} . \tag{12.37}$$

This function of t is shown in Fig. 12.8. The dashed lines in Fig. 12.8 show the increase in the *spring force*, $k_s v_s t$, as a function of the contact time for three different cases. In case (1) the spring force increases faster with t than the initial linear increase of the static friction force; hence the motion of the block will not stop and no stick-slip motion will occur. If the spring velocity v_s is lower than the critical velocity v_c [case (3)] determined by the initial slope of the $F_s(t)$ curve [$k_s v_c = dF_s/dt(0)$], the spring force will be smaller than the static friction force $F_s(t)$ until t reaches the value t_3, at which time slip starts. In this case stick-slip motion will occur. In case (2), $v_s = v_c$. Using (12.37) gives $dF_s/dt(0) = (\Delta F_0/\tau)\exp(-D/v_s\tau)$ and the criteria $k_s v_c = dF_s/dt(0)$ gives (12.36).

We note that (12.33) gives the phase boundary separating stick-slip from steady sliding when starting with steady sliding and reducing v_s. If the amplitude of the stick-slip spikes vanishes continuously at $v_s = v_c(k_s)$, then the transition between stick-slip and steady sliding does not depend on whether

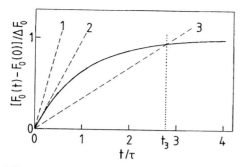

Fig. 12.8. *Solid line*: The increase in the static friction force with the time t of stationary contact. *Dashed lines*: The variation of the spring force for three different sliding velocities, (1): $v_1 > v_c$, (2): $v_2 = v_c$ and (3): $v_3 < v_c$.

v_s is *increased* from the stick-slip region or *decreased* from the steady sliding regime, and the phase boundary $v_s = v_c(k_s)$ is, in both cases, given by the linear instability analysis. However, if the transition is discontinuous, i.e., if the amplitude $\Delta F(v_c)$ of the stick-slip spikes is finite at $v_s = v_c$, then hysteresis may occur, where the transition from stick-slip to steady sliding (with increasing v_s) occurs at a spring velocity $v_c^+ > v_c$. The linear instability analysis presented above can, of course, not determine whether the transition is continuous or discontinuous. To study the nature of the transition one must perform numerical calculations (see below) or a non-linear analysis. Such an analysis shows that in the present case the transition is discontinuous for some values of the spring constant and continuous for other values. For example, if $k_s = 5651\,\text{N/m}$ [corresponding to $k_s/k_0 = 1.13 \times 10^{-3}$ and $v_c/v_0 = 1500$, point (a) in Fig. 12.7], then the transition is continuous as shown in Fig. 12.9a. However, if $k_s = 360\,\text{N/m}$ [corresponding to $k_s/k_0 = 7.21 \times 10^{-5}\,\text{N/m}$ and $v_c/v_0 = 2500$, point (b) in Fig. 12.7] then the transition is discontinuous with a hysteresis loop ($v_c^+ \approx 5.4 v_c$) as illustrated in Fig. 12.9b.

Numerical Results

The linear stability analysis presented above determines the boundary line in the (k_s, v_s)-plane at which the uniform sliding state becomes unstable with respect to *small* perturbations. However, the analysis says nothing about the form of the resulting stick-slip motion. In this section we present numerical results which illustrate the nature of the transition between stick-slip and steady motion.

Let us introduce dimensionless variables. Let us measure time in units of $(M/k_s)^{1/2}$ and length in units of $\Delta F_0/k_s$. In these units (12.21) and (12.24) take the form

$$\ddot{x} = (\bar{v}_s t - x) - A - \left(1 - e^{-\phi/\bar{\tau}}\right) - \bar{\gamma}\dot{x},$$

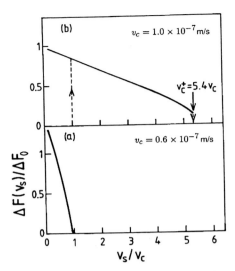

Fig. 12.9. The height $\Delta F(v_\mathrm{s})$ of the stick-slip spikes as a function of $v_\mathrm{s}/v_\mathrm{c}$. (a) $k_\mathrm{s} = 5651\,\mathrm{N/m}$ ($k_\mathrm{s}/k_0 = 1.13 \times 10^{-3}$) corresponding to $v_\mathrm{c}/v_0 = 1500$. (b) $k_\mathrm{s} = 360\,\mathrm{N/m}$ ($k_\mathrm{s}/k_0 = 7.21 \times 10^{-5}$) corresponding to $v_\mathrm{c}/v_0 = 2500$. In all calculations $M = 20\,\mathrm{g}$, $\gamma = 190\,\mathrm{s}^{-1}$, $\tau = 5\,\mathrm{s}$, $\Delta F_0 = 0.001\,\mathrm{N}$ and $F_\mathrm{b} = 0.0005\,\mathrm{N}$.

$$\dot\phi = 1 - \dot x \phi/\bar D\;,$$

where

$$\bar D = k_\mathrm{s} D/\Delta F_0\;,\quad \bar\tau = \tau(k_\mathrm{s}/M)^{1/2}\;,$$
$$\bar\gamma = \gamma(M/k_\mathrm{s})^{1/2}\;,\quad \bar v_\mathrm{s} = (v_\mathrm{s}/\Delta F_0)(Mk_\mathrm{s})^{1/2}\;,\quad A = F_\mathrm{b}/\Delta F_0\;.$$

Note that the motion is overdamped ($\bar\gamma > 2$) when γ is "large" or k_s "small". In the simulations presented below we have used $M = 20\,\mathrm{g}$, $\tau = 5\,\mathrm{s}$, $\Delta F_0 = 1 \times 10^{-3}\,\mathrm{N}$ and $\gamma = 190\,\mathrm{s}^{-1}$ which are typical for measurements with the surface forces apparatus. We have also taken $D = 2\,\mathrm{Å}$. The value of F_b does not affect the sliding dynamics; we have taken $F_\mathrm{b} = \Delta F_0/2$ so that $A = 0.5$. Figure 12.10 shows the spring force as a function of time for 5 different spring constants ($k_\mathrm{s} = 186, 360, 2385, 5651$, and $14650\,\mathrm{N/m}$) as the spring velocity is gradually increased from below v_c to above v_c, where $v_\mathrm{c}(k_\mathrm{s})$ is the boundary line in the ($k_\mathrm{s}, v_\mathrm{s}$)-plane separating stick-slip from steady sliding, as obtained from the linear instability analysis (Fig. 12.7).

Figure 12.11a shows the height $\Delta F(v_\mathrm{c})$ of the stick-slip spikes for $v_\mathrm{s} = v_\mathrm{c}$, as a function of $v_\mathrm{c}/v_\mathrm{c}^0$, where $v_\mathrm{c}^0 \approx v_0/\kappa^{1/2} \approx 1.026 \times 10^{-7}\,\mathrm{m/s}$ is the highest spring velocity for which stick-slip motion is observed when v_s is reduced from steady sliding (Fig. 12.7). Note that $\Delta F(v_\mathrm{c})$ equals zero for $0.3 < v_\mathrm{c}/v_\mathrm{c}^0 < 0.7$; thus for this range of spring constants the transition is continuous and no hysteresis occurs as a function of v_s. Figure 12.11b shows

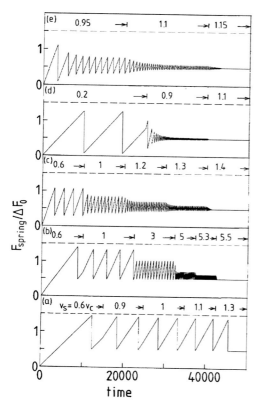

Fig. 12.10. Spring force as a function of time for 5 different spring constants. In each case the spring velocity is increased from a value below v_c to a value above v_c in a sequence of steps as indicated in the figure. In all calculations $M = 20\,\mathrm{g}$, $\gamma = 190\,\mathrm{s}^{-1}$, $\tau = 5\,\mathrm{s}$, $\Delta F_0 = 0.001\,\mathrm{N}$ and $F_\mathrm{b} = 0.0005\,\mathrm{N}$. The spring constant k_s equals (a) $186\,\mathrm{N/m}$ (giving $v_c = 1.013 \times 10^{-7}\,\mathrm{m/s}$), (b) $360\,\mathrm{N/m}$ ($v_c = 1 \times 10^{-7}\,\mathrm{m/s}$), (c) $2385\,\mathrm{N/m}$ ($v_c = 8.5 \times 10^{-8}\,\mathrm{m/s}$), (d) $5651\,\mathrm{N/m}$ ($v_c = 6 \times 10^{-8}\,\mathrm{m/s}$) and (e) $14650\,\mathrm{N/m}$ ($v_c = 2 \times 10^{-8}\,\mathrm{m/s}$).

v_c^+, which is the highest sliding velocity for which stick-slip motion occurs when v_s is increased from below v_c. As expected, $v_c^+ = v_c$ when $\Delta F(v_c) = 0$. Note the sharp drop in the hysteresis as v_c approaches v_c^0. The drop is related to the transition to overdamped sliding dynamics; in the present case $\bar{\gamma} = 2$ when $k_\mathrm{s} \approx 180.5\,\mathrm{N/m}$, corresponding to $v_c/v_c^0 \approx 0.987$.

Let us compare theory with experiment. *Yoshizawa* et al. [12.10] have studied the sliding dynamics for mica surfaces with two close-packed DMPE (a grafted chain molecule) monolayer (Fig. 3.6b). In the experiments the mass and the force constant are similar to those in Fig. 12.10b. Comparing Fig. 3.6b with Fig. 12.10b we conclude that the theory is in qualitative agreement with experiment. Note, in particular, that the sliding exhibits hysteresis, i.e.,

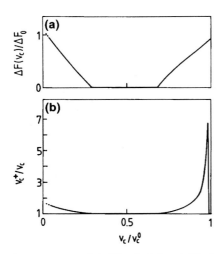

Fig. 12.11. (a) The height $\Delta F(v_c)$ of the stick-slip spikes when $v_s = v_c$, as a function of v_c/v_c^0, where $v_c^0 \approx 1.026 \times 10^{-7}$ m/s is the highest possible spring velocity for which stick-slip motion is observed (see Fig. 12.7). Note that $\Delta F(v_c)$ is equal to zero for $0.3 < v_c/v_c^0 < 0.7$; thus for this range of spring constants the transition is continuous and no hysteresis occurs as a function of v_s. (b) v_c^+ (the highest sliding velocity for which stick-slip motion occurs when v_s is increased from below v_c) as a function of v_c/v_c^0. Note that $v_c^+ = v_c$ when $\Delta F(v_c) = 0$. In all calculations $M = 20$ g, $\gamma = 190$ s^{-1}, $\tau = 5$ s, $\Delta F_0 = 0.001$ N and $F_b = 0.0005$ N.

$v_c^+ > v_c$. This observation agrees with experiments, although no detailed study of the magnitude of the hysteresis loop has been presented. It would be interesting to extend the experimental study by varying the spring constant k_s and compare the observed sliding dynamics with the theoretical results presented in Figs. 12.7, 10, and 11.

Discussion

In the model studied above stick-slip disappears if τ is below a critical value which we denote by τ_0. To prove this, note that the right-hand side of (12.34) can be written as the product of a function

$$f(x) = x^2 e^{-x} - \kappa$$

where $x = v_0/v_s$, and another function of x which is positive definite. Now, if κ is large enough, the function $f(x)$ will be negative for *all* $x > 0$. Since the spring constant k_s is positive this implies that only steady sliding is possible. Let us determine the smallest possible κ for which only steady sliding is possible. This is obviously determined by the conditions $f'(x) = 0$ and $f(x) = 0$. Since $f'(x) = (2x - x^2)\exp(-x)$ we get $x = 2$. Substituting this into $f(x) = 0$ gives $\kappa = 4e^{-2}$ or, using (12.35), $\tau = \tau_0$ where

12.2 Large Corrugation: Interdiffusion and Slip at Interface

$$\tau_0 = M\gamma D\mathrm{e}^2/4\Delta F_0 \approx 1.847 M\gamma D/\Delta F_0 \ .$$

For $\tau > \tau_0$ stick-slip will occur in some regions of the (k_s, v_s)-plane while only steady sliding is possible for $\tau < \tau_0$. However, in a typical case τ_0 is very short, e.g., using the same parameter values as in Figs. 12.10 and 12.11, $\tau_0 = 1.4 \times 10^{-6}$ s, and it is unlikely that the present formalism is valid in this case. Instead, as discussed in Sect. 3.2, the grafted monolayer film is likely to be in an *liquid-like* state, and the sliding dynamics will be similar to that of a non-Newtonian liquid [12.10]. Thus, the mechanism for the disappearance of stick-slip observed experimentally for boundary lubrication films when τ is very short is likely to differ from that of the model calculation above.

It is easy to understand the physical origin of the critical time τ_0. Note first that $x = 2$ corresponds to $v_s = v_0/2$. The steady friction force is

$$F_k = F_b + \Delta F_0 \left(1 - \mathrm{e}^{-D/v_s\tau}\right) + M\gamma v_s \ ,$$

which must equal the spring force. Now, at stick the maximum friction force equals $F_b + \Delta F_0$. The steady sliding state is stable if the spring force is higher than this maximum static friction force, i.e., if

$$\Delta F_0 \left(1 - \mathrm{e}^{-D/v_s\tau}\right) + M\gamma v_s > \Delta F_0 \ .$$

Substituting $v_s = v_0/2 = D/2\tau$ gives

$$M\gamma D/2\tau > \Delta F_0 \mathrm{e}^{-2} \ ,$$

which, except for a factor of 2, is exactly the same condition as derived above.

13. Dry Friction Dynamics

In Chap. 12 we studied sliding dynamics for lubricated surfaces. We have emphasized several times that most "real" surfaces are lubricated, e.g., by a layer of grease, even if no lubrication fluid has been intentionally added to the system. Nevertheless, in some cases the increase of the static friction force with the time of stationary contact, due to the processes discussed in Chap. 12, may be negligible; we will refer to these systems as exhibiting "dry" friction. Dry friction may prevail for many surfaces lubricated by fatty acids; because of the high local pressures which occur in the contact areas between, e.g., two steel surfaces, the relaxation time for interdiffusion and other rearrangement processes may be very long (Sects. 7.2,6) so that a negligible increase in the static friction force may occur during a typical stick-time period, in which case the static friction nearly equals the kinetic friction, as indeed observed in many cases (Table 7.5). But even in the case of "dry" friction, the static friction force may increase with time due to an increase in the area of real contact which occurs in many real systems for the reasons discussed in Chap. 5 and below. For hydrophilic materials, the formation of capillary bridges will also result in an increase of the static friction force with the time of stationary contact; see Sect. 7.5.

The results presented below are likely to be very general but are illustrated with the detailed measurements of *Baumberger* et al. [13.1–4] where the surfaces in contact were made of Bristol board. This unusual sliding system was selected because of the weakness of wear effects and the high reproducibility of the observed effects. Similar effects as described below have, however, been observed for other sliding systems involving stone-on-stone and metal-on-metal, and are likely to be robust characteristics of the dynamics of dry friction between macroscopic solid surfaces.

One fundamental difference between sliding friction studies using the surface forces apparatus and most practical sliding systems is that in the former case the normal pressure in the contact area is well below that necessary for plastic or brittle deformation of the block and substrate; the deformations are purely elastic. But in most practical sliding systems the normal pressure at a junction is determined by the yield strength of the material. As discussed in Chap. 5 and below, when the normal stress is close to the yield stress, thermal processes will be of crucial importance; slow (perpendicular) relaxation (creep) will occur and the contact area will increase slowly with time.

13.1 A Case Study: Creep and Inertia Motion

Baumberger et al. [13.1–4] have studied the sliding of a block on a substrate where both the block and the substrate were made from aluminum and the two surfaces in contact were covered by a 2-mm-thick paper (Bristol board), glued to the aluminum surfaces with cyanoacrylate. The Bristol board was made from cellulose fibres (diameter ∼ 10 μm) strongly glued together and showed no sign of anisotropy. This unusual system displayed very regular sliding behavior, reproducible at constant hygrometry for several hundreds of runs.

Figure 13.1 shows the (k_s, v_s) phase diagram. The dotted area denotes the regime where oscillatory sliding motion was observed while steady sliding occurred in the other part of the (k_s, v_s)-plane. The dashed line in Fig. 13.1 separates two different regions, the "creep" regime (A) at "low" speed, and the "inertia" regime (B) at "high" speed. The transition between oscillatory and steady sliding occurs in a continuous manner in the creep region (A), while the transition is discontinuous in the inertia region (B).

Consider first the creep region. Figure 13.2b shows the spring force in the stick-slip region on approaching the bifurcation line (the line separating stick-slip motion from steady sliding) [arrow (b) in Fig. 13.1]. Consider first the bottom curve and note that just before the rapid slip the spring force shows a marked rounding off. This indicates that there is no true stick phase and the elastic stress partially relaxes through a creep process over a distance $D \sim 1$ μm. Figure 13.2b shows that on approaching the bifurcation line the amplitude of stick-slip oscillations tends *continuously* to zero.

Next, consider the inertia regime, where the transition from stick-slip to steady sliding occurs at a finite stick-slip amplitude (Fig. 13.2a). Even near the transition, the slip is abrupt, and occurs over the time $T_\text{inertia} \sim \pi(M/k_s)^{1/2}$ (Sect. 7.6). Close to the bifurcation, the system is very sensitive

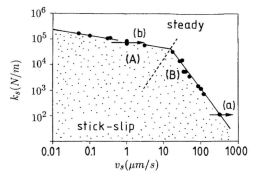

Fig. 13.1. Dynamical phase diagram. The dotted area shows the region in the (k_s, v_s)-plane where stick-slip motion occurs. In the experiment $M = 1.2$ kg. From [13.2].

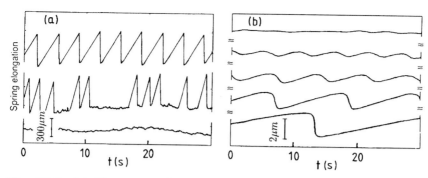

Fig. 13.2. (a) The spring elongation (proportional to the spring force) in the inertial regime [arrow a in Fig. 13.1] for three different spring velocities, $v_s = 890\,\mu\text{m/s}$ (*bottom*), $525\,\mu\text{m/s}$ (*middle*), and $275\,\mu\text{m/s}$ (*top*). In the measurements $k = 100\,\text{N/m}$ and $m = 1.2\,\text{kg}$. (b) The same as (a) but now in the creep region [arrow b in Fig. 13.1] with, from lower to upper curve, $v/v_c = 0.25, 0.42, 0.59, 0.75$, and 1.0. From [13.2].

to noise, either intrinsic or extrinsic (external vibrations). Noise may lead to fluctuations between stick-slip and steady sliding. If the noise level is reduced, a hysteresis loop between stick-slip and steady sliding is observed. Within the loop the system can be brought either into the stick-slip or the steady sliding regime by perturbations.

The dashed line in Fig. 13.1 separates the creep region from the inertial region and is determined roughly by the following condition: in (A) the time of creep over the length D is $T_{\text{creep}} \sim D/v_s$. In (B) T_{inertia} is the characteristic time. The dashed line is determined by $T_{\text{creep}} = T_{\text{inertia}}$.

The change of regime between creep and inertial motion is accompanied by changes in the velocity dependence of the (steady) friction force (Fig. 13.3). In the creep region F_k decreases approximately linearly with $\ln v_s$ while it increases linearly with v_s for large enough velocities. The $\ln v_s$-dependence in the creep region has its origin in two (compensating) effects. First, as discussed in Chap. 11, for low velocities creep motion occurs parallel to the interface and the friction force *increases* linearly with $\ln v_s$. As shown in the next section, creep motion also occurs in the direction perpendicular to a sliding junction resulting in an increase in the area of real contact with decreasing sliding velocity, giving rise to an approximate $\ln v_s$ contribution to the friction force, but with *opposite sign* to that originating from the parallel creep motion. For the present system (and many other systems as well) the latter contribution dominates so that the net friction force *decreases* linearly with $\ln v_s$ with increasing sliding velocity.

At first, the results for the friction force presented in Figs. 13.1,2 may seem paradoxical: Why is uniform sliding possible in the creep region, as observed if k_s is large enough, when the friction force is a *decreasing* function of the velocity v_s? As discussed in Sect. 7.6, if F_k decreases with increasing $\dot{x}$, the uniform sliding state is unstable and stick-slip motion is expected

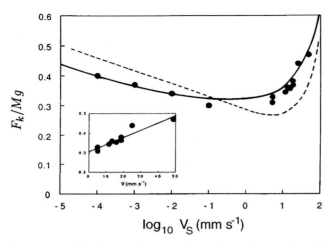

Fig. 13.3. The kinetic friction force F_k in the steady sliding regime. A linear fit to the data in the inertial region is shown in the inset. The dashed and solid lines are the steady-state friction force which follows from two different friction laws (see text). From [13.2].

(Fig. 7.40). The resolution of this problem has already been indicated in Sects. 3.2, 7.6 and 9.4. The static friction force F_s depends on the time t of stationary contact. If the spring force $k_s v_s t$ increases faster with time than the initial linear increase of $F_s(t)$ with time (which is *always* the case if k_s is large enough), then the motion of the block will not stop and only steady sliding is possible. Stated differently, the friction force in Fig. 13.3 is measured during steady sliding, i.e., for $\dot{x} = \text{const}$. But during stick-slip motion the velocity $\dot{x}$ varies with time and it is not correct to use the friction force measured under constant sliding velocity conditions in the equation of motion for the block. This is in particular true in the present case where the perpendicular relaxation introduces a *memory term* in the equation of motion of the block.

13.2 Memory Effects: Time Dependence of Contact Area

In Chap. 12 we discussed how sliding dynamics depends on relaxation processes occurring in the *lubrication film*. For "dry" friction systems these relaxation processes are assumed to have a negligible influence on the sliding dynamics, i.e., a negligible increase in the static friction force occurs as a result of, e.g., interdiffusion. But in cases where the local pressure in the contact areas is close to the plastic yield stress of the solids, the area of real

13.2 Memory Effects: Time Dependence of Contact Area

contact will increase with time. This will give an increase of the static friction force with the stick-time which has a very important influence on the sliding dynamics at low spring velocities. In this section we estimate the time dependence of the area of real contact.

When a block is placed on a substrate, plastic deformation occurs at the contact points, and a short time after contact the local stresses equal the plastic yield stress. Now, as discussed in Sect. 11.4, plastic deformation can be considered as resulting from shear "melting" and "refreezing" of small volume elements (stress blocks); a stress block melts when the elastic stresses it is exposed to by the surrounding stress blocks satisfies a yield criterion (e.g., the von Mises yield condition) and when the stress block refreezes the elastic (shear) stresses in the stress block are reduced. One therefore expects that immediately after the rapid plastic deformations which occur when the block is put on the substrate, the stress blocks in the vicinity of the contact area will be in a "critical" state with a distribution of local stresses such that some stress blocks are almost ready to undergo plastic deformation. This implies that thermal processes will be of crucial importance; slow relaxation (creep) will occur and the contact area will increase slowly with time.

Following the discussion in Sect. 11.4 we now present arguments to show that the contact area between the two solids increases with the logarithm of the contact time. Let us treat a surface asperity as a cylindrical rod which is pushed with the force F on a rigid substrate, see Fig. 13.4a. Let $l(t)$ be the length and $A(t)$ the cross sectional area of the cylinder, with $l(0)$ and $A(0)$ the length and cross sectional area a short time after contact has occurred. Since the volume is constant during the creep motion we have

$$V = l(t)A(t) = \text{const}.$$

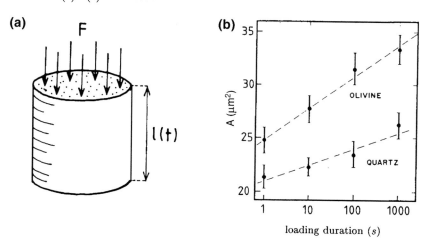

Fig. 13.4. (a) The creep-deformation of a cylindrical rod exposed to a constant external load $F = \sigma(t)A(t)$. (b) Real contact area versus loading time measured in micro-indentation experiments. From [13.5].

so that the strain rate is

$$\dot{\varepsilon} = \dot{l}/l = -\dot{A}/A \ . \tag{13.1}$$

If $A_0 = A(0)$ we write

$$A(t) = A_0 + a(t) \ .$$

Substituting this result into (13.1) gives

$$\dot{\varepsilon} = -\frac{\dot{a}/A_0}{1 + a/A_0} \ .$$

It is convenient to introduce $\xi = a(t)/A_0$ so that

$$\dot{\varepsilon} = -\dot{\xi}/(1+\xi) \ . \tag{13.2a}$$

Since the load F is constant, the absolute magnitude of the stress $\sigma = -F/A(t)$ decreases with time. We have

$$\sigma = -\frac{F}{A} = -\frac{F/A_0}{1 + a/A_0}$$

But since $F/A_0 = \sigma_c$ is the yield stress, we get

$$\frac{\sigma}{\sigma_c} = -\frac{1}{1+\xi} \ . \tag{13.2b}$$

Assume now that the stress σ is related to the strain rate $\dot{\varepsilon}$ according to the creep law (11.44)

$$\sigma/\sigma_c = -1 - B\ln(-b\dot{\varepsilon}) \ , \tag{13.3}$$

where $b = 1/\dot{\varepsilon}_0$ and $B = k_B T/\epsilon$. Substituting (13.2a,b) into (13.3) gives

$$B\ln\frac{b\dot{\xi}}{1+\xi} = -\frac{\xi}{1+\xi} \ . \tag{13.4}$$

The solution ξ to this can be written as a power series in the small parameter B. To first-order in B, (13.4) reduces to

$$B\ln(b\dot{\xi}) = -\xi$$

or

$$b\dot{\xi} = e^{-\xi/B}$$

with the solution

$$\xi = B\ln(1 + t/\tau_0) \tag{13.5a}$$

with $\tau_0 = bB$. One can show that to second-order in B:

$$\xi = (B - B^2)\ln(1 + t/\tau_0) + B^2 \ln^2(1 + t/\tau_0) + B^2 t/(\tau_0 + t). \quad (13.5b)$$

For metals, the pinning energy of dislocations is typically $\epsilon \sim 1\,\text{eV}$ giving $B \sim 0.03$ at room temperature. Similarly, for metals $\rho \sim 10^4\,\text{kg/m}^3$, $c \sim 10^3\,\text{m/s}$ and $\sigma_c \sim 10^9\,\text{N/m}^2$ giving $b = 1/\dot{\varepsilon}_0 \sim 10^{-8}\,\text{s}$. Thus $\tau_0 = bB$ is an extremely short time and $t/\tau_0 \gg 1$ for all relevant times, and we can therefore replace $1 + t/\tau$ by t/τ in (13.5).

The calculation above is based on the creep law (13.3) which only holds if the stress $\sigma(t)$ changes very slowly with time. This is a good approximation if the time of stationary contact is long enough, but will break down at high sliding velocities. Thus while many polymers (and other materials as well) deform plastically when exposed to large forces that vary slowly with time, they deform elastically when exposed to rapidly fluctuating forces as is the case at high sliding velocities, where junctions are rapidly formed and broken. An extreme example which illustrates this effect is provided by the remarkable toy-plastics which behave as a perfect elastic solid when dropped on the floor but which slowly flow out to a pancake shaped object when put on a table for a few minutes. However, an analysis of the time-dependent deformations of junctions, based on the full equation of motion (11.20) for the stress distribution function $P(\sigma, t)$, should describe the transition from elastic deformations for short junction lifetimes, to plastic-creep behavior for long junction lifetimes.

As discussed in Sect. 5.1, the slow increase in the contact area with increasing time of stationary contact has been studied using optical techniques for rough surfaces of glass and acrylic plastic (Fig. 5.5), where it was found that the area of real contact increases as $\ln(t/\tau_0)$. As another illustration, Fig. 13.4b shows two examples of how the area of real contact varies with the loading time, measured in microindentation experiments [13.5]. In both cases $A(t) - A(0) \propto \ln(t/\tau_0)$ is a good approximation. Similarly, indentation experiments for ice show that the contact area increases linearly with $\ln(t/\tau_0)$ (Sect. 14.2).

13.3 Theory

Model

As discussed above, when two surfaces are brought into contact, the area of real contact increases with time. During uniform sliding (velocity v) a junction survives on the average for a time D/v, where D is the (average) displacement of the center of mass of the block necessary in order to break a junction. We define the "contact age" variable $\phi(t)$ via (12.24):

$$\dot{\phi} = 1 - \dot{x}\phi/D. \quad (13.6)$$

For stationary contact [$x = \text{const.}$] this gives $\phi = t$, i.e., ϕ equals the contact time. On the other hand, for uniform sliding $x = vt$ and (13.6) gives $\phi = D/v$.

Assume that the area of real contact can be written as (with $\tau_0 = \phi_0$)

$$A(t) = A_0 \left[1 + B \ln(\phi/\phi_0)\right], \tag{13.7}$$

which reduces to (13.5a) for stationary surfaces. Consider now sliding at very low sliding velocities where thermal activation occurs. The friction force F_0 is the product of the real contact area $A(t)$ and the shear stress $\sigma(t)$ which we assume depends on the sliding velocity according to (11.18). Thus in the creep region $\sigma = \sigma_k[1 + B_\parallel \ln(\dot{x}/v_1)]$, and the friction force is

$$\begin{aligned} F_0 = A(t)\sigma(t) &= A_0 \left[1 + B \ln(\phi/\phi_0)\right] \sigma_k \left[1 + B_\parallel \ln(\dot{x}/v_1)\right] \\ &\approx A_0 \sigma_k \left[1 + B_\perp \ln(\phi/\phi_0) + B_\parallel \ln(\dot{x}/v_1)\right] \\ &= \text{const.} + A_0 \sigma_k (B_\perp \ln \phi + B_\parallel \ln \dot{x}), \end{aligned} \tag{13.8}$$

where

$$B_\parallel = \sigma_a/(\sigma_k 4\beta\epsilon), \qquad B_\perp = B,$$
$$v_1 = F_a \nu/(2\beta\epsilon k_1).$$

At high sliding velocities (because of the processes discussed in Sect. 8.2), F_0 increases linearly with $\dot{x}$; we can take this into account by adding the term $M\gamma\dot{x}$ to the rhs of (13.8) where $\gamma \approx 33\,\text{s}^{-1}$ from Fig. 13.3. This gives

$$F_0 \approx \text{const.} + A_0 \sigma_k (B_\perp \ln \phi + B_\parallel \ln \dot{x}) + M\gamma\dot{x}. \tag{13.9}$$

The friction force F_k during steady sliding is obtained by substituting $\phi = D/v$ in (13.9),

$$F_k = \text{const.} + A_0 \sigma_k \left[1 - B_\perp \ln(v\tau_0/D) + B_\parallel \ln(v/v_1)\right] + M\gamma v.$$

This function is indicated by the dashed line in Fig. 13.3 for the case where $B_\parallel = 0.03$, $B_\perp = 0.08$, and $\mu_k^0 = A_0 \sigma_k/Mg = 0.4$. The equation of motion of the block is given by

$$M\ddot{x} = k_s(v_s t - x) - F_0. \tag{13.10}$$

We note that adding the friction term $M\gamma\dot{x}$ to (13.8) results in an expression for F_k which has approximately the correct high and low velocity behavior. But the detailed nature of the transition from a logarithmic dependence at low sliding velocities to a linear dependence at high sliding velocities is likely to be more complex.

To illustrate the universal nature of the constitutive friction law presented above, Fig. 13.5 shows the effect of steps in the spring velocities on the coefficient of friction μ in various materials [13.6]. The top curve gives the response predicted by (13.6,9,10). The memory length D varies with the surface roughness and with the characteristics of layers of comminuted material separating the surfaces (gouge). It is important to note, however, that while (13.6,9) correctly describe the qualitative sliding dynamics for a

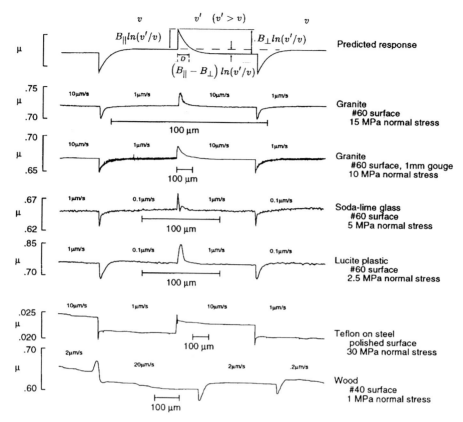

Fig. 13.5. Effects of steps in spring velocity on the coefficient $\mu = F(t)/L$, where $F(t)$ is the spring force and L the load, for various materials. The top curve gives the response predicted by (13.6,9,10). The length D depends on the surface roughness and on characteristics of layers of comminuted material separating the surfaces (gouge). From [13.6].

large class of dry friction systems, a detailed comparison with experimental data may require certain higher order terms to be added to the equations of motion. This will be illustrated below by the paper on paper system.

Linear Instability Analysis

The linear instability analysis presented in Sect. 12.2 is also valid in the present case. Thus, the boundary line in the (k_s, v_s)-plane, separating stick-slip motion from steady sliding, is determined by

$$\frac{k_s}{M} = \left(1 + \frac{v_c}{bD}\right)\left(\frac{a}{v_c} - \frac{bv_c}{D}\right), \tag{13.11}$$

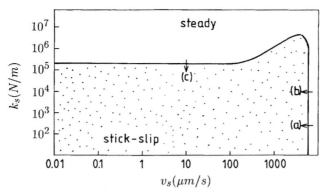

Fig. 13.6. Theoretical dynamical phase diagram calculated with $B_\perp = 0.08$, $B_\parallel = 0.03$, $\gamma = 33\,\text{s}^{-1}$, $D = 1 \times 10^{-6}\,\text{m}$, $\mu_k^0 = 0.4$, and $M = 1\,\text{kg}$.

where in the present case

$$a = \frac{1}{M}\frac{\partial F_0}{\partial \phi} = \frac{g\mu_k^0 B_\perp}{D} v_c, \qquad b = \frac{1}{M}\frac{\partial F_0}{\partial \dot{x}} = \frac{g\mu_k^0 B_\parallel}{v_c} + \gamma. \qquad (13.12)$$

Substituting these expressions for a and b into (13.11) gives

$$\frac{k_s}{M} = \frac{g\mu_k^0}{D}\left(B_\perp - B_\parallel - \frac{\gamma v_c}{g\mu_k^0}\right)\left(1 + \frac{v_c^2}{D(g\mu_k^0 B_\parallel + \gamma v_c)}\right). \qquad (13.13)$$

Figure 13.6 shows the phase boundary (13.13) for the case $\mu_k^0 = 0.4$, $M = 1.0\,\text{kg}$, $D = 1 \times 10^{-6}\,\text{m}$, $\gamma = 33\,\text{s}^{-1}$, $B_\parallel = 0.03$ and $B_\perp = 0.08$.

Numerical Results

The linear stability analysis presented above determines the boundary line in the (k_s, v_s)-plane where the uniform sliding state becomes unstable with respect to *small* perturbations. However, the analysis says nothing about the form of the resulting stick-slip motion. In this section we present numerical results which show the nature of the stick-slip motion close to the phase boundary in Fig. 13.6. We consider the three cases indicated by arrows in Fig. 13.6.

Let us write (13.10) in dimensionless units. If we measure distance in units of F_k^0/k_s (where $F_k^0 = Mg\mu_k^0$) and time in units of $(M/k_s)^{1/2}$, (13.9), (13.10) and (13.6) take the form

$$\ddot{x} = (\bar{v}_s t - x) - (A + B_\perp \ln \phi + B_\parallel \ln \dot{x}) - \bar{\gamma}\dot{x}, \qquad (13.14)$$

$$\dot{\phi} = 1 - \dot{x}\phi/\bar{D}, \qquad (13.15)$$

where

$$\bar{D} = k_s D/F_k^0, \qquad \bar{v}_s = v_s(Mk_s)^{1/2}/F_k^0, \qquad \bar{\gamma} = \gamma(M/k_s)^{1/2}. \qquad (13.16)$$

In the calculations below we have used $A = 2$ but the A-term in (13.16) will only give rise to a time-independent contribution to the spring force and has no influence on the sliding dynamics.

Figure 13.7a shows the time dependence of the spring force as the spring velocity v_s is decreased from above v_c to below v_c when $k_s = 291\,\text{N/m}$. Figure 13.7b shows the same transition for a stiffer spring, $k_s = 10^4\,\text{N/m}$. Finally, Fig. 13.7c shows how the spring force varies as the spring constant k_s is reduced from above k_c to below k_c when $v_s = 10\,\mu\text{m/s}$. The first two transitions exhibit hysteresis as a function of sliding velocity; when the sliding velocity increases from the stick-slip region the return to steady sliding does not occur until $v_s = v_c^+ = 1.45 v_c$ in case (a) and $1.7 v_c$ in case (b). This is shown in Fig. 13.8 where the amplitude ΔF of the stick-slip oscillations is plotted as a function of v_s/v_c both as the sliding velocity decreases from above v_c and as it increases from below v_c. The sliding friction in Fig. 13.7(c) exhibits no hysteresis as a function of k_s.

Figure 13.9 (left) shows the detailed form of the spring force during stick-slip motion in Fig. 13.7(a) and (b) for $v_s = 0.95 v_c$ and in Fig. 13.7(c) for $k_s = 0.98 k_c$. Figure 13.9 (right) shows results for larger thermal activation (with $B_\parallel = 0.05$ instead of 0.03). We have also performed a few calculations without thermal activation (i.e., with $B_\parallel = 0$). In this case the transition (c) is continuous with periodic oscillations $\sim \sin(\omega_c t)$ when $k_s/k_c < 1$ (but close to unity). When k_s/k_c is decreased to below about 0.9 period doubling occurs and then, finally, for $k_s/k_c \sim 0.8$ chaotic motion.

Discussion

We now show how the phase boundary in Fig. 13.6 and the qualitative form of the stick-slip curves in Fig. 13.9 can be understood on the basis of simple physical arguments.

Consider first the phase boundary in Fig. 13.6 which has a roughly rectangular form. The nearly vertical section of the phase boundary at "high" spring velocities v_s is caused by a transition from a velocity weakening to a velocity strengthening kinetic friction force. To prove this, note that, according to (13.13), $k_c \to 0$ as $v_s \to v_{\max} = g\mu_k^0(B_\perp - B_\parallel)/\gamma$ and for spring velocities above $v_{\max}$ only steady sliding is possible independent of the magnitude of the spring constant k_s. But $v_{\max}$ is just the velocity where the friction force F_k of steady sliding changes from velocity weakening to velocity strengthening, i.e., where $dF_k/dv = 0$. This follows from $F_k = \text{const} + Mg\mu_k^0(B_\parallel - B_\perp)\ln v + M\gamma v$ which gives $dF_k/dv = Mg\mu_k^0(B_\parallel - B_\perp)/v + M\gamma = 0$, i.e., $v = v_{\max}$.

Next, consider the nearly horizontal part of the phase boundary which occurs at "low" spring velocities. Assume for simplicity that no parallel creep motion occurs so that $B_\parallel = 0$. In this case (13.13) reduces (as $v_s \to 0$) to:

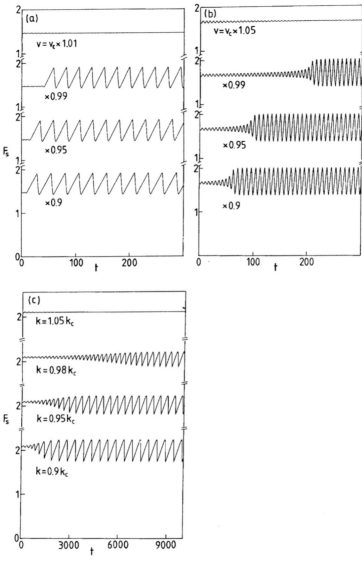

Fig. 13.7. The time dependence of the spring force on crossing the bifurcation line along arrows (**a**), (**b**), and (**c**) in Fig. 13.6. All quantities in natural units (see text).

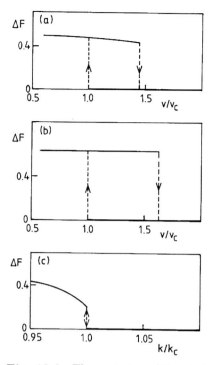

Fig. 13.8. The variation of the stick-slip amplitude ΔF of the spring force when crossing the bifurcation line (in both directions) in Fig. 13.6 along arrows (**a**), (**b**), and (**c**). In cases (**a**) and (**b**) hysteresis is observed, while no hysteresis occurs in case (**c**).

$$\frac{k_\mathrm{s}}{M} = \frac{g\mu_\mathrm{k}^0 B_\perp}{D}. \tag{13.17}$$

Now, during steady sliding, where the friction equals F_k, the contact-age variable $\phi_0 = D/v_\mathrm{s}$. Assume that the system makes a transition to the pinned state. The static friction $F_\mathrm{s}(t)$ increases monotonically with the pinning time t with $F_\mathrm{s}(0) = F_\mathrm{k}$. In the case studied above, $\phi = D/v_\mathrm{s} + t$ and $F_\mathrm{s} - F_\mathrm{k} = Mg\mu_\mathrm{k}^0 B_\perp \ln(\phi/\phi_0)$; this function of t is shown in Fig. 13.10. The dashed lines in Fig. 13.10 show the increase in the *spring force*, $k_\mathrm{s}v_\mathrm{s}t$, as a function of the contact time for two different cases. In case (1) the spring force increases faster with t than the initial linear increase of the static friction force; hence the motion of the block will not stop and no stick and slip motion will occur. If the spring velocity v_s is lower than the critical velocity v_c [case (2)] determined by the initial slope of the $F_\mathrm{s}(t)$ curve [$k_\mathrm{s}v_\mathrm{c} = dF_\mathrm{s}/dt(0)$], the spring force will be smaller than the static friction force $F_\mathrm{s}(t)$ until t reaches the value t_2, at which time slip starts. In this case stick-slip motion will oc-

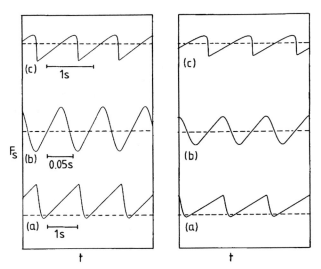

Fig. 13.9. The spring force as a function of time in cases (**a**),(**b**), and (**c**) in Fig. 13.6 for $v_s/v_c = 0.95$ [cases (**a,b**)] and $k_s/k_c = 0.98$ [case (**c**)]. The dashed lines denote the friction force during steady motion at the same spring velocities v_s. Left: $B_\parallel = 0.03$; right $B_\parallel = 0.05$.

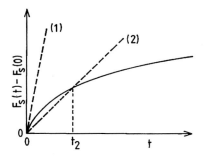

Fig. 13.10. *Solid line*: The dependence of the static friction force F_s on the contact time t. *Dashed lines*: The dependence of the spring force on contact time in two different cases (*1*) and (*2*). In case (*2*) stick-slip motion occurs while steady motion occurs in case (*1*).

cur. In the model studied above $dF_s/dt = Mg\mu_k^0 B_\perp v_s/D$ and the criteria $k_s v_c = dF_s/dt(0)$ gives (13.17).

We now discuss the qualitative form of the stick-slip spikes in Fig. 13.9. Consider first crossing the bifurcation line as indicated by arrow (b) in Fig. 13.6. The stick-slip spikes in Fig. 13.9b are very similar to those in Fig. 7.40d which result from the assumption of a velocity independent kinetic friction force F_k and a static friction force $F_s > F_k$; in this case the amplitude of the stick-slip spikes equals $\Delta F = 2(F_s - F_k)$ and during sliding it fluctuates by $\pm \Delta F/2$ around the friction force associated with steady sliding (dashed line in Fig. 13.9). In the present case F_k corresponds roughly to the friction force close to the minimum in Fig. 13.3.

Consider now the stick-slip spikes in Fig. 13.9a, which differ from those in case (b) in two important ways: (1) the motion is nearly overdamped, i.e., the minimum spring force during stick-slip motion is just below that of uniform sliding at the velocity v_s (dashed line in Fig. 13.9) and (2) there is a drastic difference between "starting" and "stopping" in a stick-slip cycle in that the starting occurs rather abruptly while the stopping takes a much longer time [compare with case (b) where a nearly perfect symmetry exists between starting and stopping].

Both of these effects result from the small spring constant k_s used in case (a). A small k_s implies that after sliding has started during a stick-slip cycle, the spring force will (nearly) equal F_s for a long time period and the block will rapidly accelerate to a high velocity where the kinetic friction force nearly equals F_s; see Fig. 13.11a. The block will then very slowly decelerate until it finally returns to the pinned state. This implies that "starting" occurs much more rapidly than "stopping" and also that the motion is nearly overdamped. The latter is formally proved as follows: During rapid slip (13.14) reduces to a good approximation to

$$\ddot{x} = x_0 - x - \bar{\gamma}\dot{x} ,$$

where x_0 can be treated as a constant. Assuming $x = x_0 + C\exp(\kappa t)$ gives

$$\kappa^2 + \bar{\gamma}\kappa + 1 = 0 ,$$

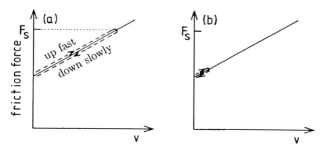

Fig. 13.11. The kinetic friction force during stick-slip in the inertia region. (a) Weak spring: After the beginning of slip the spring force will (nearly) equal F_s for a long time and the block will rapidly accelerate to such a high speed that the kinetic friction force (nearly) equals F_s. This is followed by a slow decrease of the the friction force (and the spring force which now nearly equals friction force) towards the kinetic friction force corresponding to steady sliding at the spring velocity v_s. (b) Stiff spring: The block will accelerate for a very short time. The block never acquires a high sliding velocity and during the whole slip period the friction force stays close to that for steady sliding at the spring velocity v_s.

or

$$\kappa = -\bar{\gamma}/2 \pm \left[(\bar{\gamma})^2/4 - 1\right]^{1/2}.$$

Now, since $\bar{\gamma} = \gamma(M/k_s)^{1/2}$, for a weak spring k_s or a large mass M, $\bar{\gamma} > 2$ and the motion will be overdamped. In fact, in Fig. 13.2a, $M = 1.2\,\text{kg}$ and $k_s = 100\,\text{N/m}$ and since $\gamma = 33\,\text{s}^{-1}$ this gives $\bar{\gamma} = 3.6 > 2$ and the motion during slip is overdamped. In the theoretical calculation (Fig. 13.9a) we have used $M = 1\,\text{kg}$ and $k_s = 291\,\text{N/m}$ giving $\bar{\gamma} = 1.93$, i.e., the damping is slightly below that necessary for overdamped motion; this explains why the spring force at the end of a slip is slightly below that of steady siding (dashed line in Fig. 13.9a).

13.4 Non-linear Analysis and Comparison with Experiments

The model studied above is the simplest possible which is likely to contain the basic physics relevant for many real dry friction systems. However, a detailed quantitative comparison with experimental data may require certain higher order terms to be added to the equations. Thus for the paper on paper system, (13.9) needs modifications in order for the (k_s, v_s)-diagram to agree with the experimental data of *Baumberger* et al. [13.1–4]. First, in the creep region the bifurcation line $v_s = v_c(k_s)$ in the (k_s, v_s)-plane is not a horizontal line as in the model studied above (Fig. 13.6) but a decreasing function of v_s. As shown by *Caroli* et al. [13.1,7], the bifurcation line in the creep region is accurately described if the term $B_\perp \ln \phi$ in (13.9) is replaced with

$$B_\perp^{(1)} \ln(\phi/\phi_0) + B_\perp^{(2)} [\ln(\phi/\phi_0)]^2. \tag{13.18}$$

In Fig. 13.12 we show the dynamical phase diagram obtained using this modification (with $B_\perp^{(1)} = 0.06$, $B_\perp^{(2)} = 0.002$ and $\phi_0 = 1\,\text{s}$) but keeping all the other parameters the same as before. The new phase boundary is in much better agreement with experiments and, more important, the transition (c) is no longer discontinuous as in Fig. 13.8 but continuous (Fig. 13.13c) in agreement with the experiment. The solid line in Fig. 13.3 shows the prediction for the kinetic friction force, which is also in good agreement with the experimental data. We note that the quadratic term in (13.18) is in accordance with (13.5b); even the magnitude of the prefactor $B_\perp^{(2)}$ is, within a factor of 2, the same as predicted by the theory in Sect. 13.2.

Caroli et al. [13.1,7] have studied the nature of the transition (c) in Fig. 13.12 analytically by performing a non-linear analysis close to the bifurcation line. According to the linear bifurcation analysis

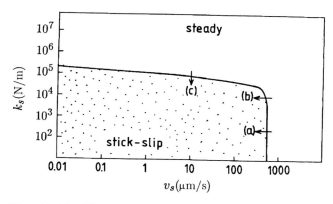

Fig. 13.12. Theoretical dynamical phase diagram calculated with $B_\perp^{(1)} = 0.06$, $B_\perp^{(2)} = 0.002$, $B_\parallel = 0.03$, $\gamma = 33\,\mathrm{s}^{-1}$, $D = 1 \times 10^{-6}\,\mathrm{m}$, $\mu_k^0 = 0.4$, and $M = 1\,\mathrm{kg}$.

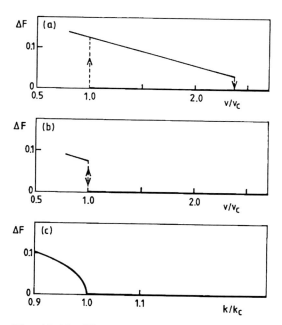

Fig. 13.13. The variation of the stick-slip amplitude ΔF of the spring force when crossing the bifurcation line (in both directions) in Fig. 13.12 along arrows (a), (b), and (c).

$$\delta x(t) \propto \cos \omega_\mathrm{c} t \qquad (13.19)$$

where for "small" v_s [using (12.9)]

$$\omega_\mathrm{c} = \left(\frac{k_\mathrm{s} v_\mathrm{c}^2}{Mg\mu_\mathrm{k}^0 B_\parallel D}\right)^{1/2}. \qquad (13.20)$$

Note that $\omega_\mathrm{c}/v_\mathrm{c}$ is proportional to $(k_\mathrm{s}/M)^{1/2}$, in agreement with the experiments [13.1,3]. The non-linear corrections to (13.19) can be obtained by expanding the center of mass coordinate $x(t)$ and the contact-age variable $\phi(t)$ in powers of the distance from the bifurcation line [13.3]:

$$x = x_0 + v_\mathrm{s} t + \delta x(t), \qquad \phi = D/v_\mathrm{s} + \delta\phi,$$

where

$$\delta x = \sum_n \epsilon^n x_n, \qquad \delta\phi = \sum_n \epsilon^n \phi_n.$$

The expansion parameter is

$$\epsilon = [1 - k_\mathrm{s}/(k_\mathrm{s})_\mathrm{c}]^{1/2}.$$

To first-order in ϵ this expansion is equivalent to the linear instability analysis presented above. But the linear instability analysis cannot determine the amplitude A of the stick-slip oscillations. To determine A (and the frequency shift $\delta\omega$) to leading order in ϵ it is necessary to perform an expansion to third-order in ϵ which gives

$$\delta x = A \cos[(\omega_\mathrm{c} + \delta\omega)t + \psi],$$

where

$$A^2 = 8D^2 k_\mathrm{s} \frac{\omega_\mathrm{c}^2 + v_\mathrm{c}^2/D^2}{\omega_\mathrm{c}^2(\mu_2 + \mu_3)} \epsilon^2 \qquad (13.21)$$

and

$$\delta\omega = -\frac{1}{2}\omega_\mathrm{c}\epsilon^2 \left(1 + \frac{2}{3} Dk_\mathrm{s} \frac{D^2 \omega_\mathrm{c}^2/v_\mathrm{c}^2}{(\mu_2 + \mu_3)} \left[1 + \frac{v_\mathrm{c}^2}{\omega_\mathrm{c}^2 D^2} + \frac{\mu_2}{\mu_1}\left(\frac{\mu_2}{\mu_1}\right)^2\right]\right), \qquad (13.22)$$

where $\mu_n = d^n F_\mathrm{k}/d(\ln v)^n$. Since the parameters which enter in A and $\delta\omega$ can be determined by comparing the result of the linear instability analysis with the experimental data, the results of the non-linear analysis for A and $\delta\omega$ represent a rigorous test of the theory in the creep region. In [13.3] it has been shown that the predictions of the non-linear bifurcation analysis is in excellent agreement with experiment.

Since A^2 must be positive, the analysis resulting in (13.21,22) is only valid if $\mu_2 + \mu_3 > 0$. If $\mu_2 + \mu_3 < 0$ the ϵ-expansion breaks down, which implies that the bifurcation transition is discontinuous, i.e., the amplitude

13.4 Non-linear Analysis and Comparison with Experiments

of the stick-slip oscillations changes abruptly when crossing the bifurcation line. Note that $\mu_2 + \mu_3 = -\phi_0^2[2F_s''(\phi_0) + \phi_0 F_s'''(\phi_0)]$.

As pointed out above, if $B_\parallel$ is small enough, the sliding motion becomes chaotic in some regions of the (k_s, v_s)-phase diagram. This may be the reason for the difference between the measurements of *Baumberger* et al. with paper and the measurements by *Johansen* et al. [13.8] with steel, brass and aluminum, where *irregular* stick-slip was observed. Chaotic stick-slip motion has also been observed in several 1D-models of sliding friction, see ref. [13.9].

14. Novel Sliding Systems

In this section we discuss in detail a few sliding systems, some of which are usually not included in books on tribology. The aim is to illustrate the general theoretical concepts and results obtained in the earlier part of this book, and to show how different physical phenomena are governed by similar physical principles. In addition, each of the topics treated below is itself of great intrinsic interest.

14.1 Dynamics of Earthquakes

The Earth's crust consists of several tectonic plates in relative motion, driven by large-scale convective flow in the mantle. The "collisions" between the plates result in the intricate pattern of cracks that we know as earthquake faults. Earthquakes results from frictional stick-slip instabilities at the faults, and on any given fault a wide distribution of earthquake sizes occurs, but the very largest earthquakes account for almost all the relative motion. Since large earthquakes typically involve slip length on the order of 1 m and since the drift velocities of the tectonic plates are on the order of 1 cm/year (or 1 Å/s), large earthquakes occur at roughly 100-year intervals, as illustrated in Fig. 14.1a with observations from a fault in Japan. Stick-slip motion resulting in earthquakes is, however, not observed for all faults; an example of an aseismic fault segment is shown in Fig. 14.1b (recorded at Cienega Winery, central California) where slow creep motion (11 mm/year) occurs [14.1].

Laboratory experiments [14.2,3] have shown that the friction relations governing the sliding of stone on stone are of the same general form as those discussed in Chap. 13. The linear instability analysis in Sect. 13.3 shows that at low driving velocity stick-slip motion can occur if $B_\perp - B_\parallel > 0$ while only creep motion is possible if $B_\perp - B_\parallel < 0$. Obviously, for the fault segment in Fig. 14.1a, $B_\perp - B_\parallel > 0$, at least in some regions of the fault area, while $B_\perp - B_\parallel < 0$ for the fault segment in Fig. 14.1b. These observations can be rationalized by assuming that the sign of $B_\perp - B_\parallel$ depends on the type of stone, the temperature and the pressure, and on pore fluid. For wet granite, $B_\perp - B_\parallel$ has been measured in laboratory experiments (Fig. 14.2) and for the temperatures and pressures which occur in a surface layer of thickness

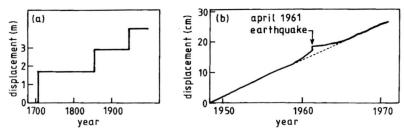

Fig. 14.1. (a) History of uplift at Muroto point in response to Nankaido earthquakes, located close to the ancient Japanese capital of Kyoto. (b) Creep record for Cienega Winery (central California). Fault creep occurs at a steady rate of 11 mm/year. In 1961 a modest earthquake occurred nearby and produced a surface offset at Cienega Winery. Note that creep ceased following the earthquake and did not resume until the strain had reaccumulated sufficiently to restore the stress to its previous level, as shown by the merging of the creep record into its previous rate. Note also the preseismic acceleration of creep. From [14.1]

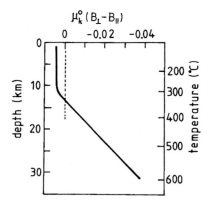

Fig. 14.2. Laboratory data for $\mu_k^0(B_\perp - B_\parallel)$ for wet granite as a function of temperature (*right scale*) and as a function of the depth into the earth (*left scale*). From [14.2]

~ 10 km, $B_\perp > B_\parallel$. For the high temperatures which occur below ~ 15 km, $B_\perp < B_\parallel$, and only creep motion is possible. This is consistent with the observation that earthquakes usually do not nucleate below ~ 15 km.

The distribution of energy released during earthquakes has been found to obey the famous *Gutenberg–Richter* [14.4] law which states that the number of earthquakes having the magnitude somewhere in the interval M to $M + \Delta M$ is proportional to $M^{-b} \Delta M$, where b is a positive number of order unity. It has recently been found that this law results from very simple spring-block models [14.5–7] and from even simpler cellular automaton stick-slip type models [14.8], and is a natural result of what is now called self-organized critical-

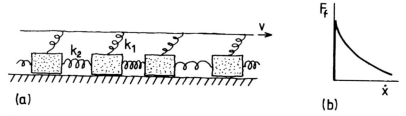

Fig. 14.3. (a) The Burridge–Knopoff model of earthquake dynamics. (b) A friction law used in many studies of model (a).

ity [14.9]. For example, the *Burridge–Knopoff* model [14.5] studied by *Carlson* and *Langer* [14.7], consists of a one-dimensional string of blocks connected to each other and to a rigid drive by harmonic springs (Fig. 14.3a). Using a very simple friction law (Fig. 14.3b), which does not allow for creep motion, the predictions of this model, as well as those of more realistic two- and three-dimensional versions of it, are consistent with the Gutenberg–Richter law. (Figure 11.11e,f shows a sequence of stick-slip oscillations which has been obtained with a different friction law to that used in [14.7] but the resulting distribution of slip events is very similar to that found by Carlson and Langer.) However, recently *Rice* [14.10] has performed calculations of fault dynamics using a three-dimensional Burridge–Knopoff model but with a realistic friction law which allows for creep motion. This introduces a new length scale, the creep or nucleation length l_c. Now, in order to correctly describe the continuum limit, Rice argued that the size l of the cells (or blocks) of the discretized elastic solid, must be much smaller than any relevant physical length scale of the problem and, in particular, $l \ll l_c$. However, simulations performed when $l \ll l_c$ exhibit only periodically repeated large earthquakes, in conflict with experiments. However, if "oversized" cells are used, $l \gg l_c$, a wide spectrum of event sizes occurs, similar to what is found with simpler friction laws. In this section we argue that the length l_c (if positive) is in fact irrelevant for earthquake dynamics, and that the cell size to be used in a discretized version of the sliding problem is determined by the elastic coherence length ξ (which is much larger than l_c) introduced in Sect. 10.2.

Let us first calculate the nucleation or creep length [14.10] l_c, which has the following physical meaning: l_c determines the minimum diameter of a patch of fault area which must slip simultaneously in order for a stick-slip instability to develop. To derive an expression for l_c, consider a cubic volume element at the fault with the fault area $l \times l$ and mass $M = \rho l^3$. The volume element is connected to the surrounding solid by an effective elastic spring (Sect. 9.5) $k_{\text{eff}} \sim \rho c^2 l$, where c is the sound velocity and ρ the mass density. At low driving velocity (i.e., when the solid region to which the volume element is connected performs a slow steady motion) the volume element will perform steady sliding if $k_{\text{eff}} > k_c$ while stick-slip motion occurs if $k_{\text{eff}} < k_c$, where the critical force constant k_c is determined by (13.13)

$$k_c = l^2 \sigma^0 (B_\perp - B_\parallel)/D, \qquad (14.1)$$

where $\sigma^0 = \rho g h \mu_k^0$ is a kinetic friction stress ($\mu_k^0 \approx 0.5$) and $D \approx 10^{-5}$ m the average distance the fault must move in order to break a junction. The condition $k_{\text{eff}} = k_c$ determines the critical cell size l_c:

$$l_c = \frac{D\rho c^2}{\sigma^0 (B_\perp - B_\parallel)}. \qquad (14.2)$$

A patch of fault with a diameter $l < l_c$ can only slip together with the surrounding solid, while a patch with $l > l_c$ will perform individual stick-slip motion. That is, cells larger than l_c can "fail" independent of one another, while those smaller than l_c cannot slip unstably alone and can do so only as part of a cooperating group of cells. For wet granite at the pressures and temperatures which occur ~ 10 km below the Earth's surface, $\mu_k^0(B_\perp - B_\parallel) \approx 0.003$ (Fig. 14.2) and hence $l_c \sim 1$ m.

Now, let us for a moment neglect the existence of the elastic coherence length ξ. In a discretization of the elastic solid at a fault, in order to obtain the correct continuum limit, it would then be necessary to choose the cells (say cubes of volume l^3) much smaller than any relevant length scale in the problem. This would require $l \ll l_c$. As stated above, Rice has performed numerical simulations and found that *if $l \ll l_c$ only periodically repeated large earthquakes occur*. However, if a grid with oversized cells ($l > l_c$) is used, that is, with the cell size too large to validly represent the underlying continuous system of equations, richly complex slip occurs, with a wide spectrum of event sizes, as observed experimentally.

The results of Rice can be understood if one notes that the cell size to be used in the simulations is not $l \ll l_c$ but rather the elastic coherence length ξ defined in Sect. 10.2. Since $\xi \gg l_c$, the effective spring constant $k_{\text{eff}} \sim \rho c^2 \xi \ll k_c = \xi^2 \sigma^0 (B_\perp - B_\parallel)/D$ and *the blocks can perform individual stick-slip motion* as in the Burridge–Knopoff model, which results in the Gutenberg–Richter law. It is important to note here that studies of different models show that the Gutenberg–Richter law is obtained under very general conditions, i.e., it is a very robust property.

Note that in an "exact" theory of the sliding dynamics one would discretize the solids into cells which are smaller than any relevant length in the problem (e.g., ξ and l_c) but much larger than the lattice constant so that continuum elastic theory is valid. But in this case it would not be correct to use a block-substrate friction law which is laterally uniform (as in the model study of Rice); instead, one would need to explicitly take into account the actual (nearly) random distribution of contact areas (pinning centers) at the fault surface. Such a calculation would, of course, be nearly impossible to perform owing to the large number of degrees of freedom. The approach outlined above, where the block (and the substrate) are discretized in cells of size ξ, may be the only practical scheme, and exhibits the relevant physical concepts more transparently that a full calculation would do.

Finally, note that the concept of elastic coherence length discussed in Sect. 10.2 also explains why there seems to be a minimum size of earthquakes, [14.11] involving blocks with the size of order $10-100$ m: in Sect. 10.2 we proposed that this cut-off is determined by the elastic coherence length which at a depth of ~ 10 km is about 100 m. It also explains why earthquakes usually do not nucleate within the first ~ 2 km into the Earth: close to the Earth's surface, because of the low pressure, P_a is small and, according to (10.16), ξ is extremely large. Thus the surface region of a fault will only participate in very large earthquakes where essentially the whole fault area moves.

14.2 Sliding on Ice and Snow

Ice is an interesting substance from the friction point of view since its kinetic friction can be lower by an order of magnitude than that of most other crystalline solids. It was long believed that the low friction observed during sliding on ice and snow has its origin in pressure melting leading to a thin layer of water which would act as a lubricant. This idea was proposed by *Reynolds*. However *Bowden* has shown that this explanation is incorrect [14.12,13]. Instead of pressure melting, the liquid water layer is usually produced by *friction heating*.

Experiments have shown that during sliding on ice and snow the friction force is proportional to the load [14.12], indicating that the real area of contact is also proportional to the load. Let us consider the sliding of ice on a body with a hard hydrophilic surface such as aluminum (covered by aluminum oxide) or granite. Since ice has a much smaller yield stress than these materials, the area of real contact should be determined by the yield stress of ice. Furthermore, since the interactions between the H_2O molecules and the polar aluminum oxide or granite surfaces are very strong and since the potential energy surface has a strong lateral corrugation, one would not expect any slip between the first layer of water molecules at the solid–ice interface. That is, the yield strength of a solid–ice junction should be determined essentially by the yield stress of ice. Under these conditions one would, according to the theory of Bowden and Tabor, expect the static friction coefficient to be $\mu_s \sim \tau_c/\sigma_c \sim 1$ as is indeed observed experimentally. This, in itself, is a proof that the area of real contact cannot be very much larger than that determined from the assumption that the normal stress in each contact area is given by the yield stress of ice.

The friction of ice is strongly dependent on the sliding velocity v. For example, for an aluminum block sliding on ice at $-10°C$, *Bowden* [14.12] found the static friction coefficient to be $\mu_s = 0.38$ while $\mu_k = 0.34$ at $v = 3$ cm/s and $\mu_k = 0.04$ at $v = 5\,\text{m/s}$. If a layer of water were to exist between the surfaces due to pressure melting, the static friction would be much smaller and no strong velocity dependence would be expected. The velocity dependence of the friction force results instead from the generation of a layer of melt-water due to the friction heating. Let us estimate what sliding velocity is necessary for the aluminum/ice system in order for the temperature at the interface to increase by $10°C$, the amount necessary in order to produce melt-water. The temperature increase is given by (8.98)

$$\Delta T = \frac{2J}{\lambda}\left(\frac{\kappa t}{\pi}\right)^{1/2}.$$

Because of the high thermal conductivity of aluminum, most of the heat produced at the interface will be conducted into the aluminum block and we can accurately estimate ΔT at the interface by assuming that the total heat current $J = \sigma_k v$ passes into the aluminum block. The initial frictional shear stress at an ice–aluminum junction, before a layer of melt-water has been formed, is $\sigma_k = \sigma_0 \mu_k$, where $\mu_k = 0.38$. The yield stress of ice is $\sigma_0 \approx 4 \times 10^7\,\text{N/m}^2$ giving $J = \sigma_k v \approx 1.5 \times 10^7\,\text{W/m}^2$ at the sliding velocity $v = 1\,\text{m/s}$. For aluminum $\rho = 2700\,\text{kg/m}^3$, $C_v = 900\,\text{J/Km}$, and $\lambda = 220\,\text{W/Km}$. Finally, assuming an asperity diameter $D = 100\,\mu\text{m}$, this gives $\Delta T \approx 7v^{1/2}$ where ΔT and v are measured in $°C$ and m/s, respectively. Hence we expect a sharp drop in μ_k to occur when v reaches $\sim 1.5\,\text{m/s}$. The thin ($\sim 30\,\text{Å}$) aluminum oxide layer which occurs on the aluminum surface has a negligible effect on this estimate.

The exact thickness of the water layer is not easy to calculate, since when the water layer starts to form the sliding friction drops rapidly, resulting in a similar drop in the friction heat production. Furthermore, for melting to occur a latent heat is required which must be taken into account in the energy balance equation. However, because of the relatively low viscosity of water ($\mu = 1.8 \times 10^{-3}\,\text{Ns/m}^2$ at $T = 0°C$), a $d = 20\,\text{Å}$ thick water layer between two *smooth* surfaces (with "stick" boundary condition) would already result in the kinetic friction coefficient $f_k \approx \mu v/d\sigma_0 \approx 0.02$ at $v = 1\,\text{m/s}$ sliding velocity.

Assume that a thin water layer has been formed at a junction. Because of the perpendicular stress σ_0 at the junction, the water layer will tend to be squeezed out from the contact area. We now show that this "squeeze-out" effect is negligible during the short time that a junction exists during sliding. If $h(t)$ is the thickness and D the diameter of the water layer, then according to (7.27)

$$\frac{1}{h^2(t)} - \frac{1}{h^2(0)} = \frac{16t\sigma_0}{3\mu D^2}.$$

If the initial water film were $20\,\text{Å}$ thick then after $t = 1 \times 10^{-4}\,\text{s}$ the film would have the thickness $19.95\,\text{Å}$, i.e., squeeze-out is completely negligible.

The temperature increase at a junction depends on the thermal conductivity and the heat capacitance of the block being slid. For a metal block with a high thermal conductivity, a relatively high sliding velocity is necessary in order to form a layer of melt water. But for materials with a low thermal conductivity, a relatively low speed is enough if the temperature is not too low. For example, for ice sliding on ice $\mu_k \approx 0.3$ at low sliding speeds which results in a temperature increase of 10°C already for $v \approx 0.1$ m/s (see Table 8.7 with $\Delta T \propto v^{1/2}$). Other materials, such as glass or ebonite, have lower thermal conductivity than ice, and in these cases most of the friction heat will flow into the ice and a temperature increase of 10°C occurs already when $v \approx 5$ cm/s. This conclusion is confirmed by many experiments [14.12–14]. For example, Fig. 14.7a below shows that for ice sliding on glass the decrease in the kinetic friction starts when v is only a few cm/s (where $\mu_k \approx 0.4$) and is reduced to ~ 0.02 at $v \approx 50$ cm/s. In other experiments performed with ice sliding on brass no change in the kinetic friction is observed between 0.1 cm/s and 1 m/s (the highest sliding velocity studied), as indeed expected because of the high thermal conductivity of brass [14.15].

We now discuss in greater detail some topics related to sliding friction on ice.

Area of Real Contact

When a block of a hard material is put on an ice surface, the ice asperities in contact with the block first deform "instantaneously" by pressure melting, and then deform more slowly by creep motion. The latter is illustrated in Fig. 14.4b which shows the time dependence of the contact area between a diamond crystal with a 1 kg load and an ice surface [14.12]. Results are shown for two different temperatures, $T = -5$°C and -20°C, and in both cases

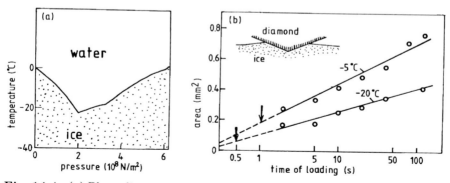

Fig. 14.4. (a) Phase diagram of the water–ice system. (b) Variation of the contact area with the contact time in a Vickers diamond hardness test (1 kg load). Results are shown for two different temperatures. The vertical arrows indicates the contact areas expected if the perpendicular stress in the contact area were equal to that necessary for pressure melting of ice. From [14.12].

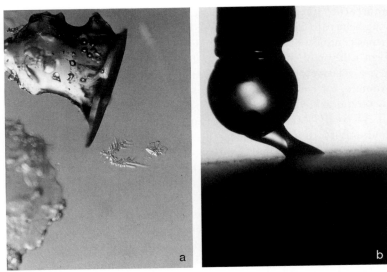

Fig. 14.5. (a) A polished ice grain. (b) A capillary attachment. The water in this photograph forms bridges between the base of a ski and a glass bead. The hysteresis in the contact angle is visible due to the motion of the water over the dry ski. From [14.17].

the contact area increases approximately linearly with the logarithm of the loading time as also found for other systems. In the figure the vertical arrows indicate the sizes of the contact areas expected if only pressure melting were to occur. In this case the contact area ΔA would be determined approximately by $P\Delta A = L$, where L is the load and P the pressure necessary to melt ice at the particular temperature in question, which can be read off from the ice–water phase diagram in Fig. 14.4a. Because of the creep motion, the area of real contact, and hence the static friction coefficient, will increase with the time of stationary contact. When repeated sliding occurs over the same area of snow, e.g., in a ski track, the ice grains will be "polished" by the repeated formation of a thin melt-water layer. Under such circumstances the area of real contact will be larger than that determined by the yield stress of ice. Figure 14.5a shows an ice grain taken from a ski track which clearly exhibits a large flat area generated by freezing of melt-water.

Static Friction

Figure 14.6 shows the variation with temperature of the static friction coefficient for skis on snow [14.16]. Results are shown for skis covered by polytetrafluoroethylene (teflon), and ski lacquer, which is a pigmented nitrocellulose compound. Teflon is an extremely inert and hydrophobic (contact angle 126°) material. The low static friction for the teflon ski must be related to slip at

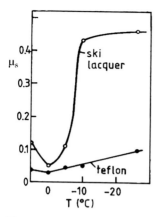

Fig. 14.6. The dependence of temperature on the static friction coefficient for a ski on snow. Results are shown for skis covered by ski lacquer and by teflon. From [14.12].

the ice–ski interface, i.e., the H$_2$O molecules at the ice–teflon interface experience a very weak adhesion and pinning potential. In contrast, ski lacquer is strongly hydrophilic (complete wetting, contact angle 0°) and in this case the yield stress of a junction is likely to be the yield stress of ice. Note that the static friction has a minimum at $T = 0°$C and that for wet snow (slush) the friction for the lacquer ski is much higher than for dry snow at $T = 0°$C. The reason for this is the formation of water bridges (Sect. 7.5) as illustrated in Fig. 14.5b [14.17]. Because of the strong hydrophobic nature of teflon, no water bridges occur for the teflon ski and the sliding friction for wet snow is similar to that of dry snow at $T = 0°$C.

Kinetic Friction and Creep

As pointed out before, the concept of a static friction force is somewhat ill-defined since even a small external force will give rise to some motion (creep). This is in particular true for sliding on ice because of the relatively high creep rate of ice even for a small applied stress (see Fig. 14.4b for perpendicular creep motion). To illustrate this, Fig. 14.7 shows the friction coefficient for the sliding of granite and glass on ice over a large velocity interval 10^{-8} m/s $< v < 10^{-1}$ m/s [14.14].

Figure 14.7a shows how the friction coefficient depends on the sliding speed at $-11.75°$C. Note that at low sliding velocities the friction force varies as $\ln v$ with $B_\parallel - B_\perp > 0$ so that no stick-slip motion is expected for very low sliding velocities. The high friction for granite on ice at low temperature is due to strong adhesion which gives an interfacial strength greater than the shear strength of ice so that shearing occurs in the ice itself. Thus, when ice is slid on granite surface at low temperatures a thin film of ice is formed on

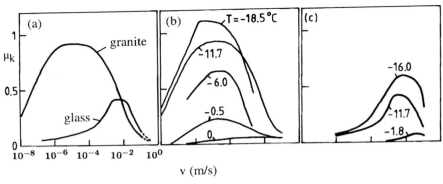

Fig. 14.7. (a) Friction coefficient of polycrystalline ice sliding on granite and glass at $-11.7°$C. The granite surface has a surface roughness of ~ 1.5 μm and stick-slip motion is observed for $v > 10^{-6}$ m/s. The glass surface has a surface roughness of ~ 0.1 μm and no stick-slip motion is observed. Parts (b) and (c) show the velocity dependence of the friction coefficient for granite and glass, respectively, for several different temperatures. From [14.14].

the rock, implying that shearing occurs within the ice itself and the stress required is a measure of its bulk properties. For $T < -10°$C the ice film is about 100 μm thick, but no ice film was observed at temperatures above $-10°$C suggesting that interfacial adhesion was then less effective.

The most important of the results in Fig. 14.7a is the velocity dependence of the sliding friction. As a function of v the friction coefficient first increases until it reaches a maximum, after which it decreases monotonically towards a value ~ 0.02 for $v \sim 1$ m/s. The initial increase is due to thermal activation (creep), while the decrease at "high" velocities is caused by melt-water generated by the friction heat production. The creep motion of ice is important in explaining the motion of glaciers [14.18].

Let us note that in skiing on snow at very low temperatures, say below $-35°$C, the frictional heating is too small to raise the temperature at the sliding interface to the melting point of ice. In this case the sliding friction coefficient is always very high (~ 0.4) and, in fact, nearly the same as for sliding on sand [14.12]. This fact is well-known to skiers in the arctic regions. For example, D.B. Macmillan (1925) writes: "At 36° below zero, the sledges (with steel runners) dragged hard over young ice covered with an inch of granular snow. Sand could hardly have been worse".

It is a remarkable observation that even a monolayer of water can strongly influence the friction properties of surfaces. We have already discussed the influence of water on the friction and adhesive properties of monolayer films of grafted chain molecules (Sect. 7.6). As another example [14.19] we note that with only one or two layers of water between mica surfaces the friction coefficient is ~ 0.02–0.03, a value that is as low as that for sliding on ice. Similarly, when small amounts of water (less than 0.01% by weight) are present in organic liquids between mica surfaces, the kinetic friction sometimes falls dramatically to 0.01–0.02, i.e., to values even lower than for sliding on ice [14.19].

Finally, let us comment on premelting. It is widely accepted that at the ice–vapor interface there is a disordered surface layer at temperatures well below the freezing point [14.20]. However, it not likely that such a layer also exists when ice is in contact with a solid. If a layer of premelted ice were to exist at an interface between a solid and ice, and if the viscosity of the "fluid" were similar to that of bulk water, then the static friction force would vanish, and the kinetic friction coefficient would be very small at low sliding velocities, contrary to experimental observations.

14.3 Lubrication of Human and Animal Joints

One of the most rapidly growing applications of tribology is in the general field of bio-systems, in particular the lubricating mechanism prevalent in living and artificial human and animal joints. Via the process of natural selection, nature appears to have outclassed the engineer with her natural joints. Thus normal healthy human joints have friction coefficients in the range 0.001–0.03, which is remarkably low even for hydrodynamically lubricated journal bearings. Nevertheless, the lubrication of joints cannot be explained by hydrodynamic lubrication alone since the relative sliding velocity between the bone surfaces is never greater than a few centimeters per second and often much less than this.

Figure 14.8 shows a schematic drawing of a human joint. The load-transmitting structural members are the bones, the ends of which are sufficiently blunt or globular to provide some form of bearing area. Within the joint, the bone surfaces are covered with a layer of relatively soft and porous bearing material. The bearing material on the upper and lower bones is separated by a lubrication fluid which is enclosed by a membrane.

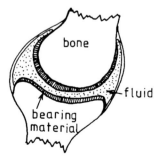

Fig. 14.8. Schematic diagram of a human joint. From [14.21].

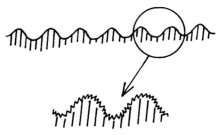

Fig. 14.9. Typical surface roughness of the bearing material in a joint of a young healthy human.

The bearing material consists of cells distributed throughout a three-dimensional network of collagen fibrils embedded in a ground substance called chondroitin sulphate. The bearing material is porous, with the average pore size of about 60 Å. The pores contain a liquid which plays an active part in the mechanism of weeping lubrication at the bearing surface. One remarkable observation is that the bearing surfaces in a human or animal joint have a surface roughness which is much larger than in engineering bearings (Chap. 4). Figure 14.9 shows schematically the surface of a bearing from a young healthy human. The surfaces have "ripples" (amplitude ~ 3 μm, wavelength ~ 40 μm) with superimposed micro-roughness (amplitude ~ 0.3 μm, wavelength ~ 0.5 μm).

The lubrication fluid is a clear, yellowish, and tacky substance, in which mucin forms the walls of a honeycomb-like network with a watery component in between. Thus, in a healthy state, the fluid appears to form a sponge-like structure. Its chemical composition indicates that it is a dialysate of blood plasma with the addition of a long-chain polymer (hyaluronic acid) with high molecular weight. The lubrication fluid is a non-Newtonian liquid having the property of shear-thinning; its viscosity decreases almost linearly with the shear rate. As a boundary lubricant, the polymer molecules can be expected to influence the friction characteristics when the opposing bearing surfaces are separated by its chain length ~ 1 μm.

It appears likely that under pressure, in the contact areas between the bearing surfaces, the volume fraction of the watery component in the lubrication fluid is reduced by seepage through the pores of the bearing material (Fig. 14.10), leading to the formation of a thin-film "fluid" having a much higher viscosity than the bulk fluid. Support for this conclusion comes from squeeze experiments. When two bearing surfaces are squeezed together, the separation between the surfaces decreases much more slowly than is calculated from the bulk viscosity of the lubrication fluid using (7.27). This can be interpreted to imply a strongly enhanced film viscosity (viscosity $\mu \sim 2\,\mathrm{Ns/m^2}$) for a thin ($\sim 1$ μm) lubrication film as compared with the bulk fluid ($\mu \sim 0.001\,\mathrm{Ns/m^2}$). The resulting gel-like fluid becomes trapped in the depressions on the surface bearing material, forming reservoirs to maintain

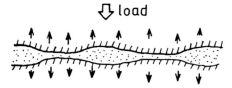

Fig. 14.10. Schematic representation of the lubrication of the bearing in a joint.

boundary lubrication action when necessary. During walking, the duration of loading is usually so short that very little diminution of the squeeze film occurs before replenishment through hydrodynamic action during the phase of swing under light load. During periods of long standing, the squeeze-film effect generates a thickened gel substance to provide a boundary lubricating effect, so that a reasonably low coefficient of friction (about 0.15) is still preserved.

14.4 Sliding of Flux-Line Systems and Charge Density Waves

In this section we discuss the sliding of flux-line systems and charge density waves. These topics have been studied intensively for several years [14.22,23], but a full theoretical understanding has not yet been reached, and below we focus mainly on a few important but well-established concepts and ideas related to sliding friction.

14.4.1 Flux-Line Systems

If a so called type-II superconductor is placed in an external magnetic field $\boldsymbol{H}$ there is a field range $H_{c1} < H < H_{c2}$ where there is partial penetration of flux and the material is in a mixed state with both normal and superconducting regions. In this state the field partly penetrates the sample in the form of thin filaments of flux [14.24]. Within each "flux line" the field is high, and the material is not superconducting. Outside the core of the filaments, the material remains superconducting. The flux lines repel each other, and for a perfect system they usually form a hexagonal lattice in order to minimize the free energy. For small displacements of the flux lines away from the minimum free energy configuration, the system behaves as an elastic solid. Here we focus on very thin samples (sheets), with $\boldsymbol{H}$ perpendicular to the sheets, which behave as 2D solids from the point of view of flux dynamics. With respect to long wavelength fluctuations in the positions of the flux lines we may use a continuum description of the motion of the flux lines [14.22,25]. If an external force density (force per unit area) $\boldsymbol{F}$ acts on the lattice one has for the displacement field $\boldsymbol{u}(\boldsymbol{x},t)$

$$nm^* \left(\frac{\partial^2 \boldsymbol{u}}{\partial t^2} + \eta \frac{\partial \boldsymbol{u}}{\partial t} \right) = K \nabla^2 \boldsymbol{u} + K' \nabla \nabla \cdot \boldsymbol{u} + \boldsymbol{F} + \boldsymbol{g} , \qquad (14.3)$$

where n is the number of flux lines per unit area. The force field $\boldsymbol{g}$ arises from the interaction with defects. The "force constants" K and K' depend on the concentration of flux lines which can be varied by changing the strength of the external magnetic field $\boldsymbol{H}$. In what follows we first put $K' = 0$ for simplicity, since this assumption does not influence the basic physics. The "inertia term" (the first term) in (14.3) can be neglected in most cases (the effective mass m^* is very small). The friction coefficient η can be obtained from an analysis of the dissipation processes inside and around the vortex cores [14.26].

When a current $\boldsymbol{J}$ flows perpendicular to the flux lines, the Lorentz force $\boldsymbol{F} = \boldsymbol{J} \times \boldsymbol{B}$ (force per unit volume) acts on the flux lines. If the flux lines were able to move freely, energy would be "dissipated" (via the friction η) and a voltage drop would occur. However, real crystals have defects which tend to pin the flux line lattice so that for low current density J it will not move and no energy "dissipation" will occur (except for a small contribution from creep motion). But when J reaches a critical value J_c the Lorentz force is large enough to break the pinning and the flux line lattice will move. The physical mechanisms which determine the pinning force (i.e., the static friction force) F_p and the nature of the sliding state for $F > F_p$ have been studied for many years [14.22].

Pinning Force F_p

Let us first estimate the pinning force in the limiting case where the interaction between the flux line lattice and the defects is very weak [14.22,27]. The basic idea, due to *Larkin* and *Ovchinnikov*, has already been presented in Sect. 8.4 but is repeated here. Because of its conceptual importance, we present two different derivations of the main results. Assume that the system is so thin as to behave as a 2D-system. In a certain area $A_c = \xi^2$, where ξ is the elastic coherence length (also called the Larkin length), there is short-range order and the flux lines form an almost perfect hexagonal lattice (Fig. 14.11). Since inside the area A_c the lattice is almost regular and the pinning centers randomly distributed, the pinning forces acting on the lattice nearly compensate each other. From random walk arguments, the pinning force acting on the area A_c is of order $g_0 N_d^{1/2}$ where g_0 is the root-mean-square force of interaction of an individual center with the lattice, and N_d is the number of pinning centers in the area A_c, i.e., $N_d = n_d A_c$, where n_d is the concentration of defect centers. The pinning force *per unit area* is therefore

$$F_p = g_0 N_d^{1/2} / A_c = g_0 (n_d / A_c)^{1/2} . \qquad (14.4)$$

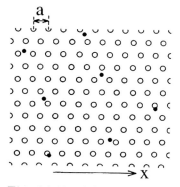

Fig. 14.11. A hexagonal lattice of flux lines (*circles*) with a random distribution of pinning centers (*black dots*).

Note that $F_p \to 0$ as $A_c \to \infty$, i.e., the *pinning force per particle vanishes for a rigid lattice* (for which $\xi = \infty$).

The result (14.4) can also be derived as follows: Let us consider the variation of the potential energy as the flux line lattice is displaced along the x-axis in Fig. 14.11. The interaction potential between one defect and the flux line lattice must be a periodic function of x with the period a equal to the lattice constant of the flux line lattice, and for simplicity we assume it to be of the form $U_d \cos(kx + \phi)$ where $k = 2\pi/a$. Consider now a surface area A with $N_d = n_d A$ randomly distributed pinning centers. The total pinning potential is $U_T = \sum_i U_i \cos(kx + \phi_i)$ where the phases ϕ_i and the amplitudes U_i are assumed to be (independent) random variables with ϕ_i uniformly distributed in the interval $0 < \phi < 2\pi$. Let $\langle \ldots \rangle$ denote ensemble averaging. For a *rigid lattice* the ensemble averaged pinning potential vanishes since

$$\langle U_T(x) \rangle = \langle \sum_i U_i \cos(kx + \phi_i) \rangle$$

$$= N_d \langle U \rangle \frac{1}{2\pi} \int_0^{2\pi} d\phi \cos(kx + \phi) = 0 , \qquad (14.5)$$

where $\langle U \rangle = \langle U_i \rangle$ is independent of i. However,

$$\langle U_T(x) U_T(0) \rangle = \langle \sum_{ij} U_i U_j \cos(kx + \phi_i) \cos(\phi_j) \rangle$$

$$= \sum_i \langle U_i^2 \cos(kx + \phi_i) \cos(\phi_i) \rangle$$

$$= N_d \langle U^2 \rangle \frac{1}{2\pi} \int_0^{2\pi} d\phi \cos(kx + \phi) \cos(\phi)$$

$$= N_d \langle U^2 \rangle \cos kx , \qquad (14.6)$$

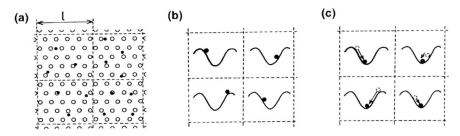

Fig. 14.12. (a) A flux-line system divided into sub-systems of linear size l. (b) Each sub-system experiences a pinning potential with amplitude $(n_d l^2 \langle U^2 \rangle)^{1/2}$. For a rigid lattice (or $l \ll \xi$) the sub-systems occupy random positions in the local pinning potential wells. (c) For a lattice with finite elasticity and for $l \sim \xi$, where ξ is the elastic coherence length, the sub-systems relax towards the bottom of the local pinning potential wells.

where $\langle U^2 \rangle = \langle U_i^2 \rangle$ is independent of i. Equations (14.5,6) tell us that even if the ensemble averaged pinning potential vanishes, in a given realization the amplitude of the pinning potential is non-vanishing and of order $(N_d \langle U^2 \rangle)^{1/2}$ and fluctuates over the typical length scale a. Now, let us "divide" the physical system into subsystems or blocks of linear size l, as illustrated in Fig. 14.12a. For a rigid lattice, each subsystem of area l^2 experiences a pinning potential with an amplitude of order $(n_d l^2 \langle U^2 \rangle)^{1/2}$. In order for the ensemble averaged pinning potential to vanish, $\langle U_T \rangle = 0$, the subsystems must occupy random positions in the local pinning potential wells; see Fig. 14.12b.

Let us now remove the assumption of a rigid lattice and consider the influence of elastic displacements. Since the periodicity of the pinning potential of the subsystems (Fig. 14.12b) is a, it is clear that if the elastic energy needed to displace a subsystem a distance of order a is less than the heights of the pinning potentials experienced by the subsystems, then the subsystems will relax towards the bottom of the local pinning potential wells; see Fig. 14.12c. The condition for such local relaxation to occur can also be stated as follows: According to (14.6), if the lattice were rigid, the pinning centers would act on a subsystem with a force of order $k(n_d l^2 \langle U^2 \rangle)^{1/2}$. If the flux lattice has a finite elasticity, effective local relaxation will occur if this force is strong enough to displace the subsystem a distance of order a.

Assume that the (constant) force density $\boldsymbol{F}$ acts within a subsystem S of size $l \times l$. This will give rise to an elastic displacement of S in the direction of $\boldsymbol{F}$ by an amount which can be easily calculated from the equations of motion for the 2D elastic medium. Assume that the displacement field $\boldsymbol{u}(\boldsymbol{x})$ satisfies [from (14.3) with $K' = 0$]

$$K\nabla^2 \boldsymbol{u} = -\boldsymbol{F}$$

where $\boldsymbol{F} = \text{const.}$ for $\boldsymbol{x} \in S$ and zero otherwise. This equation is easily integrated to give

14.4 Sliding of Flux-Line Systems and Charge Density Waves

$$\boldsymbol{u} = \frac{\boldsymbol{F}}{\pi^2 K} \int d^2 q \frac{\sin(q_x l/2) \sin(q_y l/2)}{q^2 q_x q_y} e^{i\boldsymbol{q}\cdot\boldsymbol{x}}. \tag{14.7}$$

This integral has an infrared divergence ($q \to 0$) but for a finite system of size L we must limit the q-integral to $q > 1/L$. The main contribution to (14.7) comes from $q < 1/l$ and we can therefore introduce a large-q cut off at $q = 1/l$ and expand $\sin(q_x l/2) \approx q_x l/2$ and $\sin(q_y l/2) \approx q_y l/2$. Substituting these results into (14.7) gives the displacement $\boldsymbol{u}(\boldsymbol{0})$ of S:

$$\boldsymbol{u}(\boldsymbol{0}) \approx \frac{\boldsymbol{F} l^2}{2\pi K} \ln(L/l) \ .$$

With $F = k(n_\mathrm{d} l^2 \langle U^2 \rangle)^{1/2}/l^2$ the condition $u(\boldsymbol{0}) \sim a$ determines the elastic coherence length $l = \xi$:

$$\xi^2 = \frac{2\pi^2 K^2 a^2}{g_0^2 n_\mathrm{d} [\ln(L/\xi)]^2}, \tag{14.8}$$

where $g_0^2 = \langle U^2 \rangle k^2/2$. For $l \ll \xi$ negligible elastic relaxation occurs and the subsystems occupy random positions in the local pinning potential wells as in Fig. 14.12b. For $l \sim \xi$ the subsystems can relax towards the minimum in the local pinning potentials (Fig. 14.12c). In this case, the external force needed for a sub-system to surmount its local pinning barrier [of magnitude $\Delta E \approx (n_\mathrm{d} \xi^2 \langle U^2 \rangle)^{1/2}$] is $\sim k(n_\mathrm{d} \xi^2 \langle U^2 \rangle)^{1/2}$. Hence the pinning force per unit area equals

$$F_\mathrm{p} \approx k(n_\mathrm{d} \xi^2 \langle U^2 \rangle)^{1/2}/A_\mathrm{c} = \frac{n_\mathrm{d} g_0^2}{\pi a K} \ln(L/\xi) \ . \tag{14.9}$$

Let us now present a second derivation of the coherence length ξ. Let us focus on (14.3):

$$n m^* \left(\frac{\partial^2 \boldsymbol{u}}{\partial t^2} + \eta \frac{\partial \boldsymbol{u}}{\partial t} \right) = K \nabla^2 \boldsymbol{u} + \boldsymbol{g} \ . \tag{14.10}$$

In most cases $\boldsymbol{g}$ is nearly random so that

$$\langle g_i(\boldsymbol{x}) g_j(\boldsymbol{x}') \rangle = n_\mathrm{d} g_0^2 \delta_{ij} \delta(\boldsymbol{x} - \boldsymbol{x}') \ . \tag{14.11}$$

Let us expand

$$\boldsymbol{u} = \sum_q \boldsymbol{u}_q e^{i\boldsymbol{q}\cdot\boldsymbol{x}}, \qquad \boldsymbol{g} = \sum_q \boldsymbol{g}_q e^{i\boldsymbol{q}\cdot\boldsymbol{x}} \ . \tag{14.12a, b}$$

Substituting (14.12b) in (14.11) and using

$$\delta(\boldsymbol{x} - \boldsymbol{x}') = (2\pi)^{-2} \int d^2 q \, e^{i\boldsymbol{q}\cdot(\boldsymbol{x}-\boldsymbol{x}')} \ ,$$

$$\sum_q \to A(2\pi)^{-2} \int d^2q$$

gives for the components g_q^i of the vector $\boldsymbol{g}_q$,

$$\langle g_q^i g_{q'}^j \rangle = \frac{(2\pi)^2 n_d g_0^2}{A^2} \delta_{ij} \delta(\boldsymbol{q}+\boldsymbol{q}') \,. \tag{14.13}$$

From (14.10) one gets

$$\boldsymbol{u}_q = \boldsymbol{g}_q / K q^2 \,. \tag{14.14}$$

Using (14.12a,13,14) gives

$$\langle [\boldsymbol{u}(\boldsymbol{x}) - \boldsymbol{u}(0)]^2 \rangle = \frac{2 n_d}{K^2} \frac{g_0^2}{(2\pi)^2} \int d^2 q \frac{1}{q^4} \mid e^{i\boldsymbol{q}\cdot\boldsymbol{x}} - 1 \mid^2 \,.$$

This integral has an infrared (i.e., $q \to 0$) divergence. However, since the physical system has a finite extent $\sim L$, we must cut off the q-integration for $q \sim 1/L$. Furthermore, since the main contribution to the integral comes from $q \mid \boldsymbol{x} \mid \ll 1$ we can cut off the large q integration for $q \sim 1/|\boldsymbol{x}|$ and expand $\exp(i\boldsymbol{q}\cdot\boldsymbol{x}) \approx 1 + i\boldsymbol{q}\cdot\boldsymbol{x}$ to get

$$\langle [\boldsymbol{u}(\boldsymbol{x}) - \boldsymbol{u}(0)]^2 \rangle = \frac{2 n_d}{K^2} \frac{g_0^2}{(2\pi)^2} \int_{1/L}^{1/|\boldsymbol{x}|} d^2 q \frac{1}{q^4} (\boldsymbol{q}\cdot\boldsymbol{x})^2$$

$$= \frac{n_d}{K^2} \frac{g_0^2}{2\pi} \boldsymbol{x}^2 \ln(L/\mid \boldsymbol{x} \mid) = I(\mid \boldsymbol{x} \mid) \,.$$

We can define the elastic coherence length ξ such that $I(\xi) \sim a^2$ where a is the lattice constant. This gives

$$\xi^2 = \frac{\pi a^2 K^2}{n_d g_0^2 \ln(L/\xi)} \,. \tag{14.15}$$

This result agrees with (14.8) except for a factor of $2\pi/\ln(L/\xi)$ which, in most cases, is of order unity. This latter method of estimating ξ was used in Sect. 10.2 when we calculated the elastic coherence length for a semi-infinite elastic solid under the influence of a random surface stress.

Let us compare the theoretical prediction (14.9) for the critical force F_p with experimental data for thin films (thickness ~ 1 µm) of amorphous Mo_3Si (similar results have been obtained for amorphous Nd_3Ge and Nb_3Si films). Experiments have shown that these films are so thin that the flux lines are nearly straight and the flux line lattice behaves as a 2D system [14.28]. The various symbols in Fig. 14.13 show the variation of the pinning force F_p with the reduced magnetic field $b = B/B_{c2}$ for several different temperatures $[T/T_c = 0.75$ (circles), 0.50 (triangles), 0.25 (squares)]. In order to compare these results with theory, we need to know the dependence of K, a and g_0 on the external magnetic field. A detailed study shows that [14.29] $K \sim$

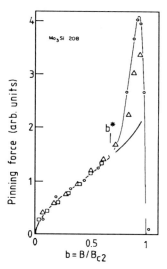

Fig. 14.13. The pinning force (per unit area) as a function of $b = B/B_{c2}$. The results are for a 0.46 μm thick amorphous Mo$_3$Si sample and for $T/T_c = 0.75$ (*circles*), 0.50 (*triangles*) and 0.25 (*squares*). From [14.28].

$b(1 - 0.29b)(1 - b)^2$. The lattice constant of the flux line lattice is $a \sim b^{-1/2}$ and [14.30] $g_0^2 \sim b(1-b)^2$. Substituting these results in (14.9) gives the (thick) solid line in Fig. 14.13 which is in remarkably good agreement with experiments for $b < b^*$. The deviation between theory and experiments for $b^* < b < 1$ can be understood as follows: The strength of the pinning potential vanishes as $g_0 \sim (1-b)$ as $b \to 1$ while the elastic constant $K \sim (1-b)^2$. Therefore as the external magnetic field is increased, for b close to 1 the pinning potential dominates over the vortex–vortex interaction potential. Hence, as the external field is increased towards B_{c2}, instead of purely elastic deformations, plastic flow will occur. As the vortex lattice is highly defective in the plastic flow regime, one would not expect the theory developed above to be valid. The crossover from elastic instabilities to plastic flow should therefore be characterized by a departure in the measured pinning force from the prediction (14.9).

Direct experimental support for this picture of plastic deformations for $b > b^*$ comes from history or memory effects observed in studies of the relation between the driving force F and the drift velocity v [14.28]. Curve (a) in Fig. 14.14 depicts the relation between F and v when the magnetic field is below b^*. Now, if b is increased above b^* the flux line structure "remembers" the structure at the starting field below b^*. The number of topological defects is small, as is the critical force. This state apparently is not stable in a moving flux line lattice since on first increasing and then decreasing F a hysteresis loop is observed as shown in Fig. 14.14b. The new critical force F'_p is higher

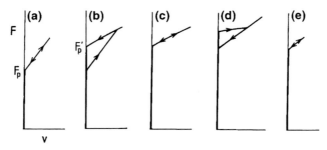

Fig. 14.14. History or memory effects of five measured $I-V$ curves at fixed fields b. See text for more details. From [14.28].

because of the greater disorder. A second force sweep (Fig. 14.14c) shows that the new state with more defects is stable.

The opposite process can be observed if the field is decreased from a value close to the peak maximum in Fig. 14.13. In the initial state the flux line lattice is strongly disordered, and the pinning force large. If b is reduced, a large initial pinning force is measured (Fig. 14.14d) but during flux flow the flux line lattice relaxes to a more stable state. This time the relaxation leads to a removal of flux-line defects and to a lower F_p value. The new vortex structure is stable as indicated in Fig. 14.14e.

This qualitative picture is supported by computer simulation results of *Brass* et al. [14.31] They studied the deformation of two-dimensional flux line systems by a random pinning potential, as a function of the strength of the pinning potential. Figure 14.15 shows the variation of the pinning force (per unit area) as a function of the strength of the pinning potential. The simulations are for a finite system (size L) and one can identify three different regions:

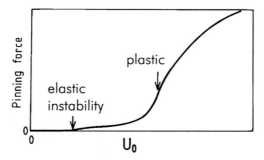

Fig. 14.15. The pinning force as a function of the strength U_0 of the pinning potential. Above the arrow "plastic", plastic deformation of the flux-line lattice occurs. From [14.31].

14.4 Sliding of Flux-Line Systems and Charge Density Waves

a) For very weak pinning, $L < \xi$ and only reversible elastic deformations of the lattice occur. In this case the pinning force (averaged over lateral displacements) vanishes. However, the size of this weak-pinning region goes to zero with increasing size of the system and this region has no relevance for macroscopic systems.
b) An intermediate regime where the pinning centers are strong enough that $\xi < L$ but weak enough that no plastic deformation occurs. In this region the formulas derived above are expected to apply.
c) For strong pinning plastic deformation and flow take place. The critical force F_p is now much higher than for weak pinning centers.

The discussion above has focused on the critical force F_p necessary to start sliding. Let us now discuss the nature of the relation between the sliding velocity and the driving force.

Sliding Dynamics: the $v = f(F)$ Relation

When the driving force F is small (and assuming zero temperature, i.e., no creep motion), the flux line lattice is stuck in one of a host of locally stable configurations. Increasing the force results in a series of avalanches which get larger and larger, until the critical force F_p is reached, above which the flux line system slides with a non-zero steady-state velocity. Before discussing the general form of the $v = f(F)$ relation, let us first consider the relation between v and F (for a 2D system) at high sliding velocity. If $V = \sum_{ij} v(\boldsymbol{r}_i - \boldsymbol{r}_j)$ denotes the flux line–flux line interaction potential and $U = U_d \sum_{in} f(\boldsymbol{r}_i - \boldsymbol{R}_n)$ the interaction between the flux lines and the pinning centers (positions $\boldsymbol{R}_n$, $n = 1, 2, \ldots, N_d$), then the equation of motion of the flux line $\boldsymbol{r}_i(t)$ is

$$m^* \ddot{\boldsymbol{r}}_i + m^* \eta \dot{\boldsymbol{r}}_i = -\frac{\partial U}{\partial \boldsymbol{r}_i} - \frac{\partial V}{\partial \boldsymbol{r}_i} + \boldsymbol{f}_i + \boldsymbol{F} , \qquad (14.16)$$

where $\boldsymbol{F}$ is the external driving force and $\boldsymbol{f}_i$ a stochastically fluctuating force which describes the influence on particle i from the irregular thermal motion of the surroundings. Now, when F is very large the flux line system will form a nearly perfect hexagonal lattice. This follows from the fact that, for large F, the sliding velocity of the flux line system is very high, and the flux lines have no time to adjust to the rapidly fluctuating forces from the pinning centers – hence the "particle" trajectories will be nearly straight lines. Due to the flux line–flux line interactions the flux lines will therefore form a nearly perfect hexagonal structure as expected in the absence of pinning centers (note: in the present case we assume that the temperature and the external magnetic field are such that the hexagonal structure is stable in the absence of pinning centers). Now, as F decreases the flux line velocities will decrease and the flux lines will, in response to the forces from the pinning centers, oscillate with increasing amplitude around the perfect lattice sites. In Sect. 8.10.2 we derived the high-velocity limiting behavior for $\bar{\eta}$ and $\boldsymbol{u}_i$, for a 2D model identical to the one described above:

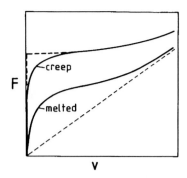

Fig. 14.16. The force–sliding velocity relation (schematic) of a flux-line system for "weak" pinning. The thick dashed line indicates the $v = f(F)$ relation at zero temperature. At non-zero temperature, creep motion occurs at low sliding velocity. If the flux-line system is in a fluid or melted state when $F = 0$, then the sliding velocity at very small F increases linearly with F. The thin dashed line denotes the relation between F and v for a flux-line system without pinning centers.

$$\bar{\eta} = \eta + \frac{(2\pi)^2 U_d^2 n_d}{m^2} \sum_s \int d^2q \frac{\eta q_x^2 (\boldsymbol{q} \cdot \boldsymbol{e}_s)^2 \mid f(q) \mid^2}{\omega_s^4(\boldsymbol{q}) + \eta^2 q_x^2 v^2} \tag{14.17}$$

$$\boldsymbol{u}_i = (\boldsymbol{u}_T)_i + \frac{iU_d}{m} \sum_s \int d^2q \frac{G(\boldsymbol{q}) f(q) e^{i\boldsymbol{q} \cdot (\boldsymbol{x}_i + \boldsymbol{v}t)}}{\omega_s^2(\boldsymbol{q}) - i\eta q_x v} \boldsymbol{q} \cdot \boldsymbol{e}_s \boldsymbol{e}_s , \tag{14.18}$$

where we have neglected the inertia terms, which is a good approximation for flux-line systems.

Based on (14.17,18) one can now give the following qualitative discussion of the $v = f(F)$ relation:

a) If the defect potential is "weak" (i.e., U_d is small) compared with the interaction between the flux lines, one expects (neglecting creep) the $v = f(F)$ relation to have the qualitative form shown by the thick dashed line in Fig. 14.16. In this case the transition to the pinned state is *continuous* and, for all F, the flux line system behaves as an elastic solid. The sliding dynamics for $F \approx F_p$ have been studied analytically using the renormalization group approach and it has been shown [14.32–34] that $v \propto (F - F_p)^\theta$ with $\theta \approx 0.64 \pm 0.03$ for a 2D-system and $\theta \approx 0.81$ for 3D-systems. In many real flux line systems, if b is below 0.5, the pinning potential is weak compared with the flux-line lattice and the $v = f(F)$ relation at low temperature is observed to have the form shown by the thick dashed line in Fig. 14.16.

b) If U_d is "large" compared with the interaction between the flux lines, the relation between v and F is likely to involve discontinuous dynamical phase transitions, as for the adsorbate systems studied in Sect. 8.9.1. If F is very large, the drift velocity is very high and the flux lines have no time to adjust to the rapidly fluctuating force from the pinning centers. This results in a solid flux line system with hexagonal structure. As F is reduced, the denominator in (14.18) decreases and the magnitude of $\langle(\Delta\boldsymbol{u})^2\rangle = \langle(\boldsymbol{u}_i - \boldsymbol{u}_j)^2\rangle$ (where i and j are nearest neighboring sites) increases. When this quantity becomes large enough (which will first happen in the vicinity of the defects) the solid will deform plastically or melt. This is essentially the Lindemann melting criteria. When the driving force is reduced further, a second dynamical phase transition may occur as the melted flux line system refreezes into a pinned solid structure. This sliding scenario has been confirmed by many experiments and calculations. Thus *Bhattacharya* and *Higgins* [14.35] (see also [14.36]) have studied the current–voltage characteristics of the layered superconductor $2H-NbSe_2$ and deduced the dynamical phase diagram shown in Fig. 14.17a. In the (b, F) plane this superconductor has a pinned solid state when F is below $F = F_p(b)$; note that this function exhibits a peak around $b \approx 0.85$, just as for the amorphous Mo_3Si system considered before. When $b < 0.7$, upon increasing F, the flux line system makes a direct transition to the flowing elastic solid as assumed in the Larkin–Ovchinnikov theory. However, for larger b, where the flux line lattice is soft (i.e., the pinning potential is strong compared with the interaction between the flux lines) the system goes from the pinned solid state for $F < F_p(b)$ to a fluidized or melted state for $F > F_p$. If the driving force is increased further the system finally makes a transition to the solid flowing state. It is interesting to note that the flowing solid state is more "perfect" than the pinned solid state; the latter is deformed by the pinning centers and may be better classified as a glassy solid phase rather than a crystalline solid. On the other hand, the flowing solid state may be a nearly perfect crystalline solid if the driving force is high enough: if F is large the flux lines have no time to "adjust" to the rapidly fluctuating forces from the pinning centers and a nearly perfect hexagonal crystalline state results, where the flux line interaction energy is minimized.

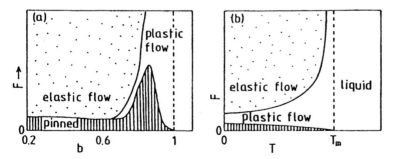

Fig. 14.17. (a) A non-equilibrium phase diagram of the flux-line dynamics for $2H - NbSe_2$. The driving force F and the reduced magnetic field $b = H/H_{c2}$. From [14.35]. (b) Numerically obtained non-equilibrium phase diagram for a strong pinning system. The driving force F and the temperature T. The melting temperature of the flux-line system (when $F = 0$) is denoted by T_m. From [14.39].

When the pinning is weak [case (a) above], close to the onset of vortex motion the drift motion of the flux line system occurs via the motion of bundles of vortices, with a size determined by the elastic coherence length. However, when the pinning potential is strong, the onset of vortex motion takes place through "channels" of plastic vortex flow [14.31,37,38].

The discussion above is for "low" temperature. If the temperature is above a critical value T_m, the flux line system is in a melted state. Figure 14.17b shows the expected (T, F) dynamical phase diagram. It is clear that when the temperature T approaches the melting temperature T_m, $F_\mathrm{p} \to 0$ while the force necessary for reaching the moving crystalline state diverges (Fig. 14.17b). The latter follows from the Lindemann melting criterion: when T is close to T_m the thermal contribution $\langle(\Delta u)^2\rangle_T$ to $\langle(\Delta u)^2\rangle$ is very close to the critical value $(\Delta u)_L^2$, for which melting occurs. Thus, the second contribution in (14.18), derived from the interactions with the pinning centers, will make $\langle(\Delta u)^2\rangle > (\Delta u)_L^2$ unless the drift velocity v is very large which corresponds to a very large driving force F [14.39].

The discussion above assumes that the flux line system is in a solid state when $F = 0$. This is usually the case but for the new high temperature superconductors, e.g., YBCO or BiSCCO, the flux line systems are in a fluid or melted state in a large fraction of the (T, b) phase diagrams [14.22]. If a flux line system is in a fluid state, the sliding dynamics will be exactly as for fluid adsorbate layers. That is, an arbitrarily small driving force gives rise to a finite sliding velocity and the $v = f(F)$ relation exhibits no hysteresis, see Fig. 14.16. For both "small" and "large" F the there is a linear (or "ohmic") relation between v and F, but the slope dF/dv is much smaller for high F than in the limit of $F \to 0$, see Fig. 14.16.

Finally, let us address the problem of creep motion for flux line systems [14.22,40,41]. Even if the flux lines are in a completely pinned state for $T = 0\,\mathrm{K}$, i.e., $v = 0$ for $F < F_\mathrm{p}$, at non-zero temperature the flux lines (or rather flux bundles of linear size ξ) can jump by thermal excitation over the barriers in the system, and some sliding motion (creep) will occur for arbitrarily small F. This is of great practical importance, since the technologically most interesting property of type-II superconductors is their ability to carry a bulk current $\boldsymbol{J}$ with essentially no dissipation. But the current $\boldsymbol{J}$ will exert a force $\boldsymbol{F}$ on the flux lines which therefore drift along the direction of $\boldsymbol{F}$ by thermally activated motion, leading to dissipation and a decrease of the current $\boldsymbol{J}$ with time.

Creep motion in flux line systems is very similar to the model of creep studied earlier for adsorbate layers (Chap. 11), if we identify a flux bundle with a block and the elastic properties of the flux lattice with the springs connecting the blocks to each other. The probability rate for a flux bundle to jump over a barrier ΔE is $\sim \exp(-\Delta E/k_\mathrm{B}T)$. However, the barrier height ΔE depends on the external force F and, in particular, must vanish for $F = F_\mathrm{p}$. It is usually assumed that for $F \approx F_\mathrm{p}$,

$$\Delta E = \epsilon(1 - F/F_\mathrm{p})^\alpha \ .$$

In the original proposal by *Anderson* [14.41], $\alpha = 1$, which is the same result as we have obtained in the mean-field treatment of the block-spring model in Chap. 11 [see (11.5) and note that $1 - (\sigma/\sigma_\mathrm{a})^2 = (1 + \sigma/\sigma_\mathrm{a})(1 - \sigma/\sigma_\mathrm{a}) \to 2(1 - \sigma/\sigma_\mathrm{a})$ as $\sigma \to \sigma_\mathrm{a}$], but no theory exists for the exponent α, and its calculation may involve a very interesting problem related to the theory of phase transitions and self-organized criticality [14.22]. Using $\alpha = 1$ gives the famous logarithmic time decay of the electric current:

$$J(t) = J_\mathrm{c}\left[1 - \frac{k_\mathrm{B}T}{\epsilon}\ln\left(1 + \frac{t}{\tau_0}\right)\right] \ .$$

The origin of this logarithmic time dependence is the same as that found in Chap. 11 [see (11.13)] for relaxation dynamics in the block–spring model.

Critical Behavior of the Depinning Transition*

Assume that the defect potential U_d is so small that no plastic deformations occur. In this case, at least in the limit of overdamped motion, the depinning transition is continuous, i.e., the sliding velocity v goes continuously to zero as $F \to F_\mathrm{p}$ from above. In this section we briefly describe the nature of the sliding dynamics and the relation between v and F close to the depinning threshold [14.42–45].

Let us decompose the elastic solid into blocks of the linear size determined by the Larkin length. For the present purpose, we can treat each such block as a particle coupled to the other blocks by elastic springs. In a first approximation, the blocks are pinned individually and the elastic interactions between the blocks are of secondary importance. However, close to the depinning transition the elastic interactions between the blocks becomes important. To study this, let us for simplicity consider a 1D-system, and assume that inertia effects can be neglected, as is a good approximation for flux line systems. In this case the "block-version" of (14.3) takes the form

$$\tilde{\eta}\frac{\partial u_i}{\partial t} = k(u_{i+1} + u_{i-1} - 2u_i) + \tilde{F} + \tilde{g}(u_i) \ , \qquad (14.19)$$

where k is an effective force constant, while $\tilde{\eta}$, $\tilde{F}$ and $\tilde{g}$ are renormalized friction, driving force, and pinning potential. If N is the number of particles (or flux lines) in a block then $\tilde{F} = NF$. Let us first consider the mean field treatment of this equation. This approximation is obtained by replacing u_{i+1} and u_{i-1} with the time average which equals vt. Thus (14.19) takes the form (with $u_i = u$)

$$\tilde{\eta}\frac{\partial u}{\partial t} = 2k(vt - u) + \tilde{F} + \tilde{g}(u). \qquad (14.20)$$

This is the same equation as studied in Sect. 10.4. We can interpret $u(t)$ as the coordinate of a particle pulled by a spring (force constant $\tilde{k} = 2k$, spring velocity v) moving in an external force field $\tilde{F} + \tilde{g}(u)$. The motion is overdamped. Thus at low sliding velocity the particle performs a stick-slip motion, staying in a local minimum for a long time before jumping to the next local minima.

Let us now discuss the nature of the solution to (14.19) beyond the mean field approximation. Equation (14.19) is very similar to the model studied in Chap. 11 in the limit where the spring constant $k_1 \to 0$. In this limit the force in the spring k_1 will not change during the (rapid) slip, i.e., this spring force is effectively a constant independent of time and the spatial location along the chain. We know from the study in Chap. 11 that when $k_1/k_2 \to 0$, if the sliding velocity v is "low", when a block rapidly moves forwards during slip, it can itself pull neighboring blocks over their barriers, giving rise to a wide distribution of avalanche sizes (see Fig. 11.11e, f). The same sliding scenario should prevail also in the present case. Thus, when v is very small, strong fluctuations will occur on all length scales, as is typical for continuous phase transitions. These long-wavelength (i.e., small $\boldsymbol{q}$) fluctuations will give rise to (infrared) divergences in perturbation expansions of various quantities. Let us illustrate this with the high-velocity expansion of the sliding friction $\bar{\eta}$.

In the limit of overdamped motion, the "inertia" term $q_x^2 v^2$ in the denominators in the expressions for $\bar{\eta}$ derived in Sect. 8.10.2 can be neglected, and for a D-dimensional solid we get

$$\bar{\eta} = \eta + \frac{(2\pi)^D U_d^2 n_d}{m^2} \sum_s \int d^D q \frac{\eta q_x^2 (\boldsymbol{q} \cdot \boldsymbol{e}_s)^2 \mid f(q) \mid^2}{\omega_s^4(q) + \eta^2 q_x^2 v^2}$$

$$\approx \eta + \frac{(2\pi)^D U_d^2 n_d}{m^2} \sum_{sG} \int_{BZ} d^D q \frac{\eta G_x^2 (\boldsymbol{G} \cdot \boldsymbol{e}_s)^2 \mid f(G) \mid^2}{c_s^4 q^4 + \eta^2 G_x^2 v^2} , \quad (14.21)$$

where the latter integral is over the first Brillouin Zone. For $D < 4$ this integral has an infrared (i.e., small q) divergence as $v \to 0$, while it converges for $D \geq 4$. Performing the $\boldsymbol{q}$-integral in (14.21) gives

$$\bar{\eta} = \eta + A v^{(D-4)/2}.$$

Thus if $D < 4$, close enough to the depinning transition the expansion above will break down. This break down originates from the long-wavelength ($q \to 0$) fluctuations (avalanches) discussed above.

Divergent expansions resulting from infrared ($q \to 0$) singularities, and infinite-range fluctuations, occur at all normal (thermodynamic) continuous phase transition points. These transitions can be analyzed using the renormalization group (RG) procedure. This technique can also be applied to the depinning transition. The basic idea is as follows.

Consider the equation of motion (14.3). We are interested in the long-wavelength properties of this equation which result after eliminating, or "integrating out", the short-wavelength fluctuations as follows: We first eliminate the wave-vector components $\Lambda(1-\epsilon) < \mid \boldsymbol{q} \mid < \Lambda$, where Λ is a short wavelength

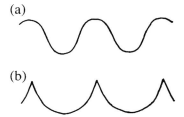

Fig. 14.18. (a) The microscopic pinning potential as a function of the displacement u of the elastic solid. (b) The "fix-point" pinning potential, as obtained after eliminating "short" wavelength fluctuations.

cut-off and ϵ a small number. This will result in a new effective equation of motion with renormalized damping $\eta \to \eta' = \eta + \epsilon f_\eta(\eta, K, g)$, renormalized elasticity $K \to K' = K + \epsilon f_K(\eta, K, g)$ and a renormalized pinning potential $g \to g' = g + \epsilon f_g(\eta, K, g)$. By iterating this process many times one finally obtains an equation of motion which describes the long-wavelength ($q \to 0$) properties of the system. Like any continuous phase transition, the properties of the depinning transition are determined by the very-long-wavelength properties of the system, while the very-short-wavelength properties are unimportant (and almost unaffected). Hence, it is expected that the effective equation of motion obtained after eliminating all short-wavelength fluctuations, directly exhibit the nature of the depinning transition.

The most important results of the RG-study of the depinning transition are the following [14.42–45]. First, the elasticity K remains unchanged, i.e., the short wavelength fluctuations will not modify the long-range elastic properties. However, we know from the arguments given earlier that on the length scale determined by the Larkin length, an elastic instability occurs *independent* of how stiff the elastic solid is. Since the elasticity K is unchanged (and could be arbitrary large), this is possible only if the (fix-point) pinning potential g^*, obtained after many (infinite) iterations, has cusp-like "singularities" at $u = 2\pi n a$ ($n = 1, 2, \ldots$) as indicated in Fig. 14.18. This result is confirmed by the RG-calculation and, in fact, represents the most important result of the RG-study. The RG-theory also shows that for F close to F_p, $v = (F - F_p)^\theta$ where $\theta \approx 0.64$ and 0.81 for $D = 2$ and 3, respectively.

Finally, let us comment on inertia effects. The 2D Frenkel-Kontorova model studied in Sect. 8.9.2 exhibits a discontinuous depinning transition when the damping η is small enough. Similar behavior has been observed for the 1D Frenkel-Kontorova model [14.46]. These results indicate that, in the underdamped case, an elastic solid sliding under the influence of pinning centers will exhibit a discontinuous depinning transition.

14.4.2 Charge Density Waves

Some quasi-one-dimensional metals, such as $NbSe_3$ and TaS_3, make a transition to a new state, called a charge density wave (CDW) state, when the temperature is decreased below a critical value T_c [14.23]. In the new state the ion lattice is distorted and a gap is formed at the Fermi surface, thereby causing a transition to an insulating state. The term "charge density wave" refers to the fact that the ion lattice and the electron charge density forms a new periodic structure, with a wavelength λ that is longer than the original lattice period a. If the charge density wave is incommensurate, that is, if a/λ is an irrational number, then the entire charge density wave structure can slide through the lattice, thereby contributing to the electrical current.

The existence of a charge density wave ground state in quasi one- and two-dimensional systems turns out to be the rule rather than the exception. However, in most systems the charge density wave is so strongly pinned by impurities or by being commensurate with the ion lattice that, in response to an applied electric field E, no sliding of the charge density wave occurs. However, in some materials, such as $NbSe_3$ and TaS_3, beyond a certain threshold electric field, which can be as low as a few mV/cm, the electric current becomes non-linear and begins to increase, asymptotically approaching the same linear dependence on E as expected if no charge density wave were present; this is exactly what is expected for a moving charge density wave.

The sliding dynamics of charge density waves is usually described by an equation similar to (14.3). That is, the charge density wave is treated as an elastic solid [14.47] and everything stated in Sect. 14.4.1 concerning the pinning force and the sliding dynamics of flux line systems is also valid for charge density wave systems. In particular, if the charge density wave "lattice" is stiff compared with the strength of the pinning potential, the elastic coherence length ξ is very long, and the threshold force (per particle) to start sliding very small [see (14.8) and (14.9)]. However, if the pinning potential is strong enough, phase slip processes may occur, which are equivalent to plastic deformations of the charge density wave system. In the latter case the relation between the current I (proportional to the drift velocity) and the electric field E (proportional to the driving force) may exhibit hysteresis and discontinuous dynamical phase transitions, as has indeed been observed in some cases. For example, Fig. 14.19a shows the relation between I and E (or the applied voltage $V \sim E$) for a $NbSe_3$ crystal at $T = 26.5$ K. Note that an "ohmic" electric current (i.e., a current proportional to the applied voltage) occurs below the onset of sliding of the charge density wave; this contribution is due to the normal carriers in the system (note: not all the conduction electrons participate in the charge density wave condensate, e.g., for $T > 0$ K some will always be thermally excited above the gap). Since the "ohmic" current acts in parallel with the intrinsic non-linear response of the crystal, we may represent the crystal by a resistor R connected in parallel with a

14.4 Sliding of Flux-Line Systems and Charge Density Waves   463

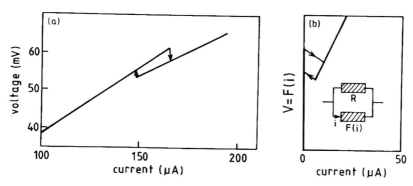

Fig. 14.19. (a) Current–voltage characteristic observed in NbSe$_3$ at $T = 26.5$ K. (b) An equivalent representation consisting of a resistance R parallel with a non-linear element with the current–voltage characteristic $V = F(i)$. From [14.23].

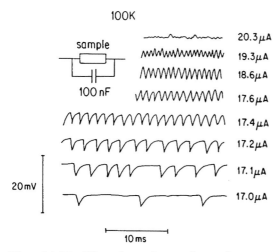

Fig. 14.20. Time-dependent voltage fluctuations at a fixed direct current I. From [14.23].

non-linear element $V = F(i)$ with the current–voltage ($i - V$) characteristic shown in Fig. 14.19b.

When a CDW crystal is exposed to a constant current it sometimes exhibits oscillatory behavior. Figure 14.20 shows the voltage oscillations across a NbSe$_3$ crystal when it is exposed to a constant current I. In the diagram a capacitor is shown parallel to the crystal (see inset in Fig. 14.20). Let us show that these current oscillations are analogous to the stick-slip motion of a block. Assume for simplicity that the resistivity $R = \infty$. In this case, if $i = \dot q$ denotes the current flowing through the nonlinear element, then the charge on the capacitance equals $It - q(t)$ and the voltage $V = (It - q)/C$

which must equal $F(\dot{q})$, i.e.,

$$0 = (It - q)/C - F(\dot{q}) \,.$$

Compare this equation with the equation of motion of a spring–block system

$$M\ddot{x} = k_s(v_s t - x) - F(\dot{x}) \,.$$

Thus, if inertia can be neglected (overdamped motion), and if we identify

$$q \leftrightarrow x, \quad 1/C \leftrightarrow k_s, \quad I \leftrightarrow v_s,$$

we have a complete correspondence between the two systems. From the spring–block analogy we know that with a non-linear relation between i and V of the form shown in Fig. 14.20b one expects oscillatory behavior when the current I (which corresponds to the spring velocity v_s in the mechanical analogy) is below some critical value, which in the experiment shown in Fig. 14.20 is about 20 μA. We also note that the capacitor in Fig. 14.20 does not need to be an external capacitance but could represent the intrinsic capacitance of the CDW crystal (or stray capacitances). The situation is completely analogous to the spring–block model where the spring k_s does not need to be an external spring but could represent the elastic properties of the block itself.

Within the model (14.3) it has been shown that when the interaction with the pinning centers is weak, the transition from stick to slip at $F = F_p$ is continuous, and that at zero temperature it can be described as a dynamic critical phenomenon. The transition can be characterized using v as an order parameter; neglecting creep, v is strictly zero when $F < F_p$ and obeys $v \propto (F - F_p)^\theta$ for F close to but above F_p. Theory predicts [14.32–34] $\theta \approx 0.64$ for 2D-systems and $\theta \approx 0.81$ for 3D-systems. However, the exponent θ deduced from experimental data is larger than 1 (typically $\theta = 1.1 - 1.2$). This suggests that the model (14.3) is not accurate enough to describe all aspects of the sliding dynamics of charge density waves. Perhaps, as has been suggested by Coppersmith [14.48], plastic deformations [which cannot occur in the model (14.3)], *always* occur in real systems and this leads to a destruction of the critical behavior.

In the discussion above about sliding of flux line lattices we pointed out that slow thermally induced relaxation processes occur, leading to a logarithmic decay with time of a super current. Similar slow relaxation processes occur for charge density wave systems. For example, assume that for $t < 0$ an electric field $E > E_p$ acts on a charge density wave, which therefore slides with some velocity v. Assume that at time $t = 0$ the external electric field is abruptly removed. The state of the system at time $t = 0$ is expected to be critical, i.e., some "blocks" (charge density wave domains of linear size ξ) are just below the top of the barrier separating them from the sliding state. These "blocks" will rapidly be thermally excited over the barriers. In this way, just as for the block–spring model studied in Chap. 11, the charge density wave system will relax towards the "ground state" in a way which depends logarithmically on time. This relaxation process can be probed, e.g., by registering the time dependence of the resistivity R of the normal electrons that are excited across the gap. This assumes that the local changes of the charge density wave phase also lead to change in the resistance R. Experiments [14.49] have shown that

$$R(t) - R(0) \propto \ln\left(1 + \frac{t}{\tau_0}\right)$$

This relation holds over many decades in TaS$_3$, with a similarly sluggish time response being found in other materials.

14.5 Frictional Coulomb Drag Between Two Closely Spaced Solids

In Chap. 5 we have shown that the area of real contact between two bodies is usually a very small fraction of the apparent area of contact. Nevertheless, the friction force is determined almost entirely by the area of real (or atomic) contact. Thus, we will show in this section that surfaces separated by more than $\sim 10\,\text{Å}$ give an extremely small contribution to the friction force, even if this non-contacting surface area is many orders of magnitude larger than the area of atomic contact.

We consider two semi-infinite metallic bodies with flat surfaces separated (with vacuum) by a distance d. The upper body moves parallel to the lower body with the velocity v. We show below that the friction force is $F = \gamma A v$, where A is the surface area. Note that the friction force is proportional to the velocity v (for small velocities). The origin of this result is that in the present case no rapid motion occurs at the interface at any stage of the sliding (as would be the case if, e.g., an elastic instability were to occur) and we therefore expect $F \propto v$ as $v \to 0$. In this section we consider only metals (or conducting) bodies as the electronic friction vanishes for insulating bodies (no low-energy electronic excitations are possible for insulators).

Let us first present a qualitative discussion of the origin of the friction force. At first it may seem paradoxical that there should be friction forces between perfectly flat metal surfaces separated by vacuum, since on *average* there is no electric field outside such surfaces, which could couple the surfaces together. However, due to *thermal or quantum fluctuations*, local charge imbalance will occur *temporarily* in the metals giving rise to electric field patches extending from one solid into the other solid, see Fig. 14.21. The fluctuating electric field will induce electric currents in the solids which are damped due to, e.g., normal "ohmic" processes such as scattering of conduction electrons from imperfections. For stationary surfaces (no sliding) this energy transfer will occur in both directions and no net energy transfer from one solid to the

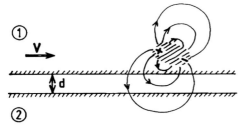

Fig. 14.21. Two semi-infinite metals, *1* and *2*, separated by a vacuum "slab" of thickness d. Solid *1* moves with velocity v relative to solid *2*. The figure illustrates the thermal or quantum fluctuations which give rise to a temporary charge imbalance and an electric field. The electric field penetrates into solid *2* where it creates electron–hole pair excitations.

other will occur (thermal equilibrium). However, during sliding a net energy and momentum transfer will occur from the sliding body to the stationary body which corresponds to a friction force. A detailed study shows that there is a fundamental difference between quantum and thermal fluctuations. Thus, thermal fluctuations will give a contribution to the friction force even at lowest order in the (electric field) coupling between the solids, while quantum fluctuations only contribute in second-(or higher) order of perturbation theory. Furthermore, within the jellium model there is no contribution to the sliding friction from quantum fluctuations so that the sliding friction vanishes at zero temperature in this model. (Note: in the jellium model the ion cores of the solid is replaced by a uniform positive background.) For real (crystalline) solids the contribution from quantum fluctuations is nonzero, but decays extremely rapidly with increasing separation d between the surfaces, $\gamma \sim \exp(-2Gd)$, where $G = 2\pi/a$ is the (smallest) reciprocal lattice vector of the surface unit cell [14.50]. The contributions from thermal fluctuations to the friction force is more long ranged, asymptotically falling of as $1/d^6$. On the other hand, the thermal contribution depends on temperatures as $\sim T^2$ so that at low enough temperature the contribution from quantum fluctuations will dominate for any fixed separation d. We note that the thermal contribution to the friction force requires that *both* solids are metallic, while the contribution from quantum fluctuations only require that at least *one* of the solids is metallic.

Let us define the linear response function $g(q,\omega)$ which is of central importance for a large class of surface processes [14.51]. Assume that a semi-infinite metal occupies the half space $z \leq 0$. A charge distribution in the half space $z > d$ gives rise to an (external) potential which must satisfy Laplace's equation for $z < d$ and which, therefore, can be written as a sum of evanescent plane waves of the form

$$\phi_{\text{ext}} = \phi_0 e^{qz} e^{i q \cdot x - i\omega t},$$

14.5 Frictional Coulomb Drag Between Two Closely Spaced Solids

where $q = (q_x, q_y)$ is a 2D-wavevector. This potential will induce a charge distribution in the solid (occupying $z < 0$) which in turn gives rise to an electric potential which must satisfy the Laplace's equation for $z > 0$, and which therefore can be expanded into evanescent plane waves which decay with increasing $z > 0$. Thus the total potential for $0 < z < d$ can be expanded in functions of the form [14.52]

$$\phi = \phi_0 \left(e^{qz} - g e^{-qz}\right) e^{i q \cdot x - i\omega t},$$

where the reflection factor (or linear response function) $g = g(q, \omega)$. The g-function introduced above plays the same central role for dynamical processes at surfaces as the dielectric function $\epsilon(\omega, k)$ does for bulk processes. During the last 20 years or so a large effort has been devoted to calculating $g(q, \omega)$ for simple metals [14.53]. There are several contributions to Im g which can be distinguished by the source of the momentum involved in the excitation process. For a semi-infinite jellium for small frequencies ($\omega \ll \omega_F$) only the surface can supply momentum and [14.54,55]

$$(\text{Im } g)_{\text{surf}} = 2\xi(q) \frac{\omega}{\omega_p} \frac{q}{k_F}, \tag{14.22}$$

where $\xi(q)$ depends on the electron density parameter r_s but typically $\xi(0) \sim 1$. For real metals, in addition to this *surface* contribution, there will be a *bulk* contribution to Im g derived from "normal" ohmic processes (e.g., scattering of the conduction electrons from imperfections), as characterized by a bulk mean free path l, and given by [14.54,55]

$$(\text{Im } g)_{\text{bulk}} = 4 \frac{\omega_F}{\omega_p} \frac{1}{k_F l} \frac{\omega}{\omega_p}. \tag{14.23}$$

Let g_1 and g_2 be the g-functions of the two solids in Fig. 14.21. Using first-order perturbation theory, and taking into account screening, one can show that thermal fluctuations give the following contribution to the friction parameter [14.56]

$$\gamma = 0.109 \frac{(k_B T)^2}{\hbar} \int_0^\infty dq q^3 \frac{e^{-2qd}}{|1 - g_1(q,0) g_2(q,0) e^{-2qd}|^2}$$
$$\times \lim_{\omega \to 0} \left(\frac{\text{Im } g_1(q,\omega) \text{ Im } g_2(q,\omega)}{\omega^2} \right). \tag{14.24}$$

If Im g_1 and Im g_2 are of the form (14.22) and if we approximate $g_1(q,0) = g_2(q,0) = 1$, which are good approximations for those q involved when the separations d beyond is a few Angstroms, then

$$\gamma = 0.845 \left(\frac{k_B T}{\hbar \omega_p}\right)^2 \frac{\hbar \xi^2(0)}{k_F^2 d^6}. \tag{14.25}$$

As an example, consider two silver bodies, treated as $r_s = 3$ semi-infinite jellium bodies. In this case for $d = 10$ Å equation (14.25) gives $\gamma \sim 10^{-5}$ Ns/m^3, which at the sliding velocity $v = 1$ m/s correspond to a frictional stress $F/A \sim 10^{-5}$ N/m^2. This stress is extremely small compared with the frictional stress $\sim 10^8$ N/m^2 occurring in the areas of atomic contact even for (boundary) lubricated surfaces. Note that the bulk contribution to Im g gives a contribution to the friction force which scales as $1/d^4$ with d. Thus, for large enough d, the bulk contribution dominates over the surface contribution.

In the discussion above we have assumed perfect (single crystal) solids. Most real solids consist of grains which may expose different facets with different work functions. This gives rise to a (static) inhomogeneous electric field distribution in the vicinity of the surfaces. Similarly, adsorbed atoms (e.g., alkali atoms) may give rise to strong local electric fields at the surface. It is clear that when two macroscopic bodies (with static inhomogeneous electric field distributions at their surfaces) are slid relative to each other without direct contact, a finite contribution to the friction force will arise from surface imperfections. However, even in these cases the contribution to the friction force from the non-contacting surface area is completely negligible compared to the contribution from the area of atomic contact.

In spite of its small magnitude, the friction force associated with thermal fluctuations has been observed in several elegant experiments [14.57]. The samples used for these experiments are modulation-doped semiconductor heterostructures grown by molecular beam epitaxy. The experiments involve two thin slabs of electron (or hole) gases separated by a barrier which is high and wide (typically $d \sim 100$ Å) enough to prevent tunneling of electrons, while thin enough to allow strong Coulomb interaction between carriers on opposite sides of the barrier. In the experiments the frictional drag of one electron-gas layer (layer *1*) on the other (layer *2*) was probed by studying how an electric current in one layer tends to induce a current in the other layer. If no current is allowed to flow in layer *2* an electric field E_2 develops whose influence cancels the friction force between the layers. If the current in layer *1* is denoted by $J_1 = n_1 e v$, where v is the drift velocity and n_1 the carrier concentration (per unit area), then the friction force $F = \gamma A v$ will act on layer *2*. This must equal the force $F = A n_2 e E_2$, where E_2 is the electric field in layer *2* induced by layer *1*. Thus $\gamma = n_2 e E_2 / v = n_1 n_2 e^2 E_2 / J_1$. Experiments show that E_2/J_1 is independent of J_1, i.e., the friction force is proportional to the sliding velocity v. Furthermore, γ is found to be nearly proportional to T^2 as expected from (14.24). The dots in Fig. 14.22 show the observed temperature dependence of the friction factor γ. To compare the theory above with the experimental data in Fig. 14.22 it is necessary to specify the g-functions. The electron gas layers can, to a good approximation, be treated as 2D-electronic systems for which [14.58]

$$g = 1 - \frac{1}{1 - \chi v_q}, \tag{14.26}$$

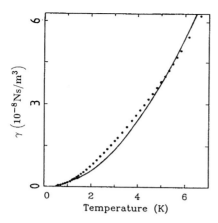

Fig. 14.22. Temperature dependence of the observed friction factor γ for a sample with a 175 Å barrier. The solid line is the theoretical prediction based on intralayer Coulomb scattering (14.24). From [14.57].

where $\chi = \chi(q,\omega)$ is the density–density correlation function for a 2D-electron gas and where $v_q = 2\pi e^2/\epsilon q$ (ϵ is the dielectric function of the surrounding material). Substituting (14.26) into (14.24) and using the well-known expression for χ gives a theoretical result (solid line in Fig. 14.22) in relatively good quantitative agreement with the experimental data.

*Technical Comments**

a) Equations (14.24) is strictly valid only for 3D-electronic systems. For 2D-systems the electron–electron scattering cross section [14.59] and the friction factor γ have logarithmic corrections, e.g., $\gamma \sim T^2[a + b\ln(T/T_{\rm F})]$. However, the logarithmic correction arises from backscattering contributions ($q \approx 2k_{\rm F}$), i.e., from processes involving a large momentum transfer, and because of the factor $\exp(-2qd)$ in the coulomb coupling between the layers, the backscattering contribution is strongly suppressed [$b \sim \exp(-2k_{\rm F}d)$], leading to a nearly pure T^2 dependence of the friction parameter γ.

b) Since the two electron gas slabs are separated by a *solid* layer rather than vacuum, it is also possible for the conduction electrons in the two layers to "communicate" by emission and absorption of phonons. It has been shown [14.57,60] that the exchange of *virtual phonons* gives a small contribution to the coupling of the two conducting layers, which can explain the (small) deviation of the observed temperature dependence from the expected T^2 dependence, see Fig. 14.22. For large separation d between the conducting layers, the phononic contribution will dominate, since it falls off much more slowly with increasing d than the electronic contribution (which is proportional to $1/d^4$ for 2D-systems).

14.6 Muscle Contraction*

In this section we consider muscle contraction. During muscle contraction "thick" filaments slide actively relative to "thin" filaments. This is made possible by the existence of *molecular motors* at the interface between the thick and thin filaments which, in the presence of fuel molecules, displace the two filaments relative to each other, producing directed movement against an external load. Because of their small size, molecular motors are *Brownian machines*, which perform mechanical work by consuming chemical energy. We discuss the topic of muscle contraction here since it has some similarities with sliding friction. Furthermore, it illustrates how nature has outclassed even the most brilliant scientists and engineers, by producing molecular machines with a diameter of just a few 100 Å, to be compared with the smallest man-made machines, which currently have a diameter of $10 - 100$ μm (Fig. 5.6). By understanding the construction and function of molecular machines, it may ultimately be possible for man to produce motors of similarly small dimensions to those developed by nature using the principle of natural selection over a time span of ~ 100 million years. We point out, however, that muscle contraction is not fully understood, and that the microscopic picture and model calculation presented below may need modifications and extensions in order to accurately describe muscle contraction [14.61,62].

The most fundamental problem in the theory of muscle contraction is to relate the macroscopic response of a muscle to its microscopic, molecular construction. The macroscopic response has been studied by *Hill* [14.63] and by *Huxley* and *Simmons* [14.64]. Hill studied the response of a muscle which was continuously stimulated to contract and which had a load F attached to it. He found that the relation between F and the velocity of contraction v can be well described by the equation (see dashed line in Fig. 14.26)

$$F = F_0 \frac{1 - v/v_0}{1 + v/v_1} , \qquad (14.27)$$

where v_0, v_1 and F_0 are constants. Note that F_0 is the maximum load (which occurs when $v = 0$) and v_0 the maximum speed (which occurs when $F = 0$). For the frog sartorius $v_1 \approx 4v_0$. While F_0 and v_0/v_1 are practically temperature independent, v_0 (and v_1) increase by a factor of 2.05 as the temperature is increased from 273 K to 283 K. Such a strong temperature dependence can only be explained by the existence of a potential barrier which must be overcome by a thermally activated process. The standard formula $w = \nu \exp(-\beta \Delta E)$ for the transition rate with $\nu \sim 10^{10} \, \text{s}^{-1}$ gives a barrier height $\Delta E \sim 0.5 \, \text{eV}$. Overcoming this barrier may be the rate limiting process in muscle contraction and thus of crucial importance for a fundamental understanding of the various macroscopic responses of a muscle.

Microscopic Basis of Muscle Contraction

During muscle contraction an array of thick filaments, composed mainly of myosin, slides actively past an array of thin filaments, composed mainly of actin. During the sliding action, the high-energy molecules ATP (adenosine triphosphate) undergo hydrolysis and the energy released is converted into directed movement.

The myosin molecule has a highly asymmetric shape, consisting of a globular head attached to a long tail (actually, a myosin molecule has *two* heads, but we will neglect this complication here). The globular head contains the ATP-binding sites and the actin-binding regions, whereas the long rod-like portion of myosin forms the backbone of the thick filament. The myosin molecule's tail is about ~ 1600 Å long and the head is approximately 165 Å long and 50 Å across.

During muscle contraction the heads of the myosin molecules perform cyclic motions and during each cycle of the motor the thin filament moves a distance $L \approx 100$ Å relative to the thick filament. Several hundred myosin heads are connected to each thick filament (Fig. 14.23). However, the spacing between the heads is incommensurate with respect to the binding sites on the thin filament. This implies that, even in the absence of elasticity, the motors will not simultaneously pass over the barrier ΔE during sliding, as would be the case if the spacing were commensurate with respect to the binding sites on the thin filament.

We will now discuss in more detail the origin of the cyclic motion of the myosin heads during muscle contraction. We first note that the joints connecting the neck to the head and to the rest of the thick filament implies that the head, when not attached to the thin filament, will perform a rapid, large amplitude, Brownian motion. Indeed, it is partly from experimental observation of this Brownian motion that the existence of these joints has been inferred. It is possible to estimate the time-scale involved in the myosin head motion by simply applying the Einstein relation for the diffusivity of a spherical particle (radius R) in a liquid with viscosity μ,

$$\langle x^2 \rangle = \frac{1}{\pi} \frac{k_B T}{R \mu} t \approx \left[10^{10} \text{ Å}^2/\text{s} \right] \times t$$

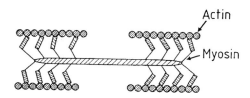

Fig. 14.23. A schematic picture of the smallest contractile unit in the muscle fiber, the so-called sarcomere. During muscle contraction, the "thin" actin filament moves relative to the "thick" myosin filament. The thick and the thin filaments consist of several hundred myosin and actin molecules of which only a few are shown in the figure.

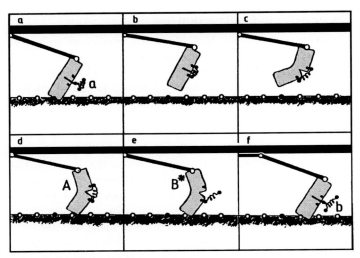

Fig. 14.24a–f. Schematic representation of the contractile cycle. A more accurate picture is presented in [14.65].

if $R \sim 50$ Å and $\mu = 10^{-3}$ Ns/m^2 (the viscosity of water at room temperature). Hence, it takes about $t \sim 10^{-6}$ s for the head to perform a root-mean-square displacement of order 100 Å, which is much shorter than the time involved in the cycle of muscle contraction (of order 10^{-3} s). This is probably important for the speed of muscle contraction because, whenever a binding site becomes available on an actin molecule, the head of the myosin molecule will rapidly find it.

With reference to Fig. 14.24 we now present a schematic picture of what is known today about the microscopic processes involved in muscle contraction (for a more accurate picture, see [14.65]). Let us represent the high energy state of a fuel molecule (ATP) by a "compressed" spring and the low energy state (ADP+P) by a spring which has its natural length. Figure 14.24a shows a head of a myosin molecule bound to an actin binding site on the thin filament. When a fuel molecule binds to an active site on the head, a configurational change occurs in the head which weakens the bond between the head and the thin filament (Fig. 14.24b). Next the fuel molecule makes a transition to its low energy state (elongated spring) while the energy originally stored in the spring is converted into deformation energy of the head (Fig. 14.24c). (In the figure we have assumed that the head has a cleft and that the deformation is localized mainly to the region around the cleft.) The deformed head performs Brownian motion until it finds a new actin binding site to which it will bind (state A), see Fig. 14.24d. Next the fuel molecule partially dissociates (Fig. 14.24e). In this new state B^* the head wants to "straighten out", i.e., because of the elastic deformation of the head it exerts a force on the thin filament which therefore moves relative to the thick filament while the head of the myosin molecule "straightens out" (configuration B). Finally, the fuel molecule detaches from the head (Fig. 14.24f). During a full cycle the thick and thin filaments move relative to each other by a "lattice constant" of the thin filament, i.e., by a distance given by the separation between neighboring actin binding sites. During this process work is usually done against an external force, while a fuel molecule makes a transition from its high energy state (compressed spring) to its low energy state (relaxed spring).

Model

Consider the head of a myosin molecule which is attached to an active site on the thin filament. If the thin filament (displacement coordinate x) moves relative to the thick filament this must be associated with either a displacement of the head along the reaction coordinate q or by an elastic deformation of, e.g., the head of the myosin molecule. The equation of motion for q is taken to be

$$m\ddot{q} + m\eta\dot{q} = k(x - q) - U'(q) + f \ . \tag{14.28}$$

Here m is an effective mass, $-m\eta\dot{q}$ is a friction or damping force due to the viscosity of the intra-cellular fluid and f is the corresponding fluctuating force which, according to the fluctuation–dissipation theorem, must satisfy

$$\langle f(t)f(0)\rangle = 2mk_\mathrm{B}T\eta\delta(t) \ .$$

The potential $U(q)$ is assumed to have the general form shown in Fig. 14.25, with a well A corresponding to the initial state of the myosin–actin bound complex (Fig. 14.24d), separated by a barrier C of height $\Delta E \sim 0.5\,\mathrm{eV}$ from state B. The term $k(x - q)$ in (14.28) arises from an elastic element in the cross-bridge, e.g., a deformation mode of the elongated head of the myosin molecule. This elastic element makes it possible for the head of the myosin molecule to switch state $A \to B$ or $B \to A$ without any relative motion occurring between the thick and thin filament (transition d $\leftrightarrow$ e in Fig. 14.24).

The equation of motion for the thin (rigid) filament, which is assumed to have the mass M, is taken to be

$$M\ddot{x} = -\bar{N}\bar{F} + \sum_i k(q_i - x) \ . \tag{14.29}$$

Here q_i is the reaction coordinate of the head of the ith myosin molecule attached to the thin filament. $\bar{N}$ is the number of attached heads, and $\bar{F}$ is the external load per attached head. It has been found experimentally that inertia effects are negligible in muscle contraction so that (14.29) reduces to

$$\bar{F} = k\sum_i (q_i - x)/\bar{N} \ .$$

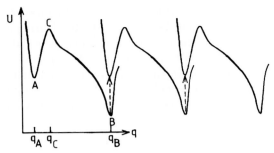

Fig. 14.25. The actin–myosin interaction potential as a function of the reaction coordinate q. We denote the potential energy minima by $q_A = 0$ and $q_B = L$ and the maxima by $q_C = L_1$.

Now, let us focus on one molecular motor and define the effective potential

$$U_{\text{eff}}(q,x) = U(q) + \frac{1}{2}k(q-x)^2 \ .$$

If k is "small", this potential also has two wells A and B separated by a potential barrier C. The activation energies ΔE_A and ΔE_B, to go over the barrier C from $A \to B$ and from $B \to A$ depend on x.

Consider a molecular motor acting on the thin filament, which moves with the velocity v relative to the thick filament. Assume that at time $t = 0$ the head is attached in state A. Let $P_A(t)$ and $P_B(t)$ be the probabilities that the head is in state A and B at time t, respectively. Due to thermal "kicks" from the surroundings, a given head can jump from $A \to B$ with the probability rate $w = \nu \exp(-\beta \Delta E_A)$ where ν is the prefactor and $1/\beta = k_\text{B} T$. But a head can also be "forced" to go over the barrier C by the motion of the thin filament caused by the "pulling" by other attached heads. This results in a *collective behavior* of the system, which is crucial for the understanding of the relation between the force F and the sliding velocity v, as discussed in the next section.

Theory

The theory presented below has some similarities to the discussion of adhesion hysteresis at the end of Chap. 5 [see (5.9,10)]. Thus, we focus on the probabilities P_A and P_B that a molecular motor is in state A and B, respectively, so that $P_A + P_B = 1$. If w_A and w_B denote the jumping rates from $A \to B$ and from $B \to A$, respectively, then we have

$$\frac{dP_A}{dt} = w_B P_B - w_A P_A \ . \tag{14.30}$$

Substituting $P_B = 1 - P_A$ into (14.30) gives

$$\frac{dP_A}{dt} = w_B - [w_A + w_B] P_A.$$

This equation is easy to integrate:

$$P_A(t) = P_A(0) e^{-\int_0^t dt'[w_A(t') + w_B(t')]}$$
$$+ \int_0^t dt' w_B(t') e^{-\int_{t'}^t dt''[w_A(t'') + w_B(t'')]}. \quad (14.31)$$

Let us first consider a sliding (contraction) where the thin filament moves with the constant velocity v relative to the thick filament. We note that this motion is possible only because the sequence of heads of the thick filament is incommensurate relative to the binding sites on the thin filament. We assume that $U(q)$ has the qualitative form shown in Fig. 14.25, and we neglect "back-transfer" of myosin heads from well B to well A, i.e., we assume that $w_B \approx 0$. In that case (14.31) reduces to

$$P_A(t) = P_A(0) e^{-\int_0^t dt' w_A(t')}. \quad (14.32)$$

In order to calculate the relation between the force $\bar{F}$ and the sliding velocity v we must average the force of a single motor, $k(\langle q \rangle - x)$, over the time t_0 for which the head is attached to the linear filament. We assume that a head attaches at time $t = 0$ in state A with the spring k at its natural length, so that the spring force vanishes at $t = 0$. If $q(0) = q_A = 0$ this requires $x(0) = 0$, i.e., $x = vt + x_0$ with $x_0 = 0$. If $L \approx 100$ Å denotes the relative displacement between the thick and the thin filament during one attachment period, then $t_0 = L/v$. We obtain

$$\bar{F} = \frac{1}{t_0} \int_0^{t_0} dt k [\langle q(t) \rangle - x(t)]. \quad (14.33)$$

Now, note that

$$\langle q \rangle \approx q_A P_A + q_B P_B = L P_B = L(1 - P_A), \quad (14.34)$$

where $q_A \approx 0$ and $q_B \approx L$ are the two minima of the effective potential $U_{\text{eff}}(q, x)$. Substituting (14.34) into (14.33) gives

$$\bar{F} = \frac{1}{t_0} \int_0^{t_0} dt k [vt - L + L P_A(t)] = \frac{1}{2} kL \left[1 - \frac{2}{t_0} \int_0^{t_0} dt P_A(t) \right]. \quad (14.35)$$

To proceed further we assume that the potential $U(q)$ has the general form shown in Fig. 14.25. If the spring k is soft, the location of the maxima and the minima of the effective potential $U_{\text{eff}} = U + k(x - q)^2/2$ will be nearly the same as of $U(q)$. Thus if E_A^0 and E_C^0 denote the potential energy at the minima A and the maxima C of $U(x)$, then the corresponding energies for U_{eff} are

$$E_A \approx E_A^0 + kx^2/2$$

and

$$E_C \approx E_C^0 + k(x - L_1)^2/2 .$$

Thus the activation barrier $\Delta E_A = E_C - E_A$ becomes

$$\Delta E_A = \Delta E_A^0 + k(L_1^2 - 2L_1 x)/2 . \qquad (14.36)$$

We assume that

$$w_A = \nu e^{-\beta \Delta E_A} . \qquad (14.37)$$

Using (14.36) with $x = vt$ gives

$$w_A(t) = \nu e^{-\beta\left(\Delta E_A^0 + kL_1^2/2\right)} e^{\beta k L_1 vt} = w_0 e^{\beta k L_1 vt} . \qquad (14.38)$$

Substituting (14.38) into (14.32) and assuming $P_A(0) = 1$ gives

$$P_A(t) = e^{-\int_0^t dt' w_0 \exp(\beta k L_1 vt')} = e^{-(v^*/\lambda v)[\exp(\lambda vt/L) - 1]} , \qquad (14.39)$$

where $\lambda = \beta k L L_1$. Substituting (14.39) into (14.35), introducing $\xi = vt/L$, and using $vt_0 = L$ gives

$$\bar{F} = F_0 \left(1 - 2 \int_0^1 d\xi e^{-(v^*/\lambda v)[\exp(\lambda \xi) - 1]}\right) , \qquad (14.40)$$

where $F_0 = kL/2$ and $v^* = w_0 L$. Note that $\bar{F} \to F_0$ as $v \to 0$, i.e., F_0 is the largest force a cross-bridge can generate.

Analysis of the Experimental Data of Hill

Let N be the total number of molecular motors and $\bar{N}$ the number of attached motors. Thus,

$$\bar{N} = \frac{t_0}{t_0 + t_1} N ,$$

where t_1 is the average time a myosin head stays unattached. We assume that t_1 is independent of v. If $\bar{F}$ denotes the load per attached myosin head and F the load per myosin head, then $NF = \bar{N}\bar{F}$ so that

$$F = \frac{\bar{N}\bar{F}}{N} = \frac{t_0 \bar{F}}{t_0 + t_1} = \frac{L\bar{F}}{L + vt_1} = \frac{\bar{F}}{1 + w_0 t_1 v/v^*} , \qquad (14.41)$$

where $v^* = w_0 L$. Using (14.40,41):

$$\frac{F}{F_0} = \frac{1 - 2 \int_0^1 d\xi e^{-(v^*/\lambda v)[\exp(\lambda \xi) - 1]}}{1 + w_0 t_1 v/v^*} . \qquad (14.42)$$

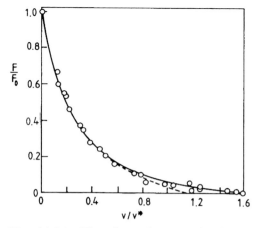

Fig. 14.26. The relation between the sliding velocity v and the force F. The circles are the experimental data (from [14.63]), while the solid line is the theoretical result based on (14.42). The dashed line has been calculated using the empirical formula (14.27).

Note that F/F_0 as a function of v/v^* depends on $\lambda = \beta k L_1 L$ and on $w_0 t_1$. We will show in the next section from independent experimental data that $\lambda \approx 3$ at room temperature. Thus $w_0 t_1$ is the only adjustable parameter and using $w_0 t_1 = 2.5$ gives the solid line in Fig. 14.26, in very good agreement with the experimental data. Note that the maximum sliding velocity, which occurs when the load $F = 0$, is $v_{\max} = av^*$, where $a = 1.6$. Thus for an unloaded muscle, $t_0 = L/v_{\max} = L/(av^*) = 1/aw_0$ so that the fraction of the cycle for which a myosin head is attached to the thin filament is of order $t_0/(t_1 + t_0) = 1/(w_0 t_1 a + 1) \approx 0.2$. This result for the "duty ratio" agrees quite well with independent experimental data [14.66].

Analysis of the Experimental Data of Huxley and Simmons

Huxley and Simmons have studied the response of a muscle to a step change of its length, $x \to x + \Delta x$. During a time period which is short compared with the time scale involved in the cycle of attachment, pulling and detachment, the length of the muscle was changed by a small amount ($\sim 1\%$). It was found that the initial change in tension in the muscle was proportional to the length change of the muscle which Huxley and Simmons interpreted as the stretching of some elastic element. Furthermore, they showed that this elastic element is located somewhere in the cross-bridge, i.e., the thick and the thin filaments themselves have negligible elasticity. After this initial change of tension in the muscle, a fast relaxation process occurs. The relaxation rate depends strongly on the change of length of the muscle as is shown in Fig. 14.27. We can understand this result as follows: Let us assume that x is changed by Δx at time $t = 0$. For $t < 0$ the jumping rates $A \to B$ and $B \to A$ are w_A and

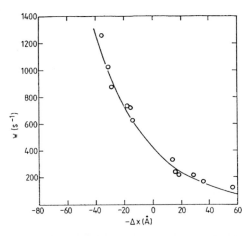

Fig. 14.27. The rate w of the initial (fast) relaxation occurring immediately after a stepwise change in the length of a muscle. The circles are the experimental data of Huxley and Simmons (from [14.64]), while the solid line is the theoretical result calculated using (14.43).

w_B, and the probability of finding a head in state A is $P_A = w_B/w$ where $w = w_A + w_B$. Next for $t > 0$ the jump rates become $w_A^* = w_A \exp(\beta k L_1 \Delta x)$ and $w_B^* = w_B \exp(\beta k L \Delta x)$ and the probability $P_A(t)$ will relax towards the new thermal equilibrium value w_B^*/w^* (where $w^* = w_A^* + w_B^*$) as time proceeds. The way this relaxation occurs is determined by (14.31); the integrals in (14.31) are trivial to perform since the jump rates are time independent for $t > 0$:

$$P_A(t) = \frac{w_B^*}{w^*} + \left(\frac{w_B}{w} - \frac{w_B^*}{w^*}\right) e^{-w^* t} .$$

Thus the relaxation process is characterized by the rate constant $w^* = w_A^* + w_B^*$. We assume that $w_B^* \ll w_A^*$ so that

$$w^* \approx w_A e^{\beta k L_1 \Delta x} . \tag{14.43}$$

The solid line in Fig. 14.27 is calculated using this equation with $\beta k L_1 = 0.028 \text{ Å}^{-1}$, i.e., with $\beta k L_1 L \approx 3$. In a more accurate treatment one must take into account that w_A depends on x, and average $P_A(t)$ over x. This is likely to give only a small numerical change of the decay rate, which we neglect in the present treatment.

Discussion

The simple model studied above is in remarkably good agreement with the experiments of Hill and of Huxley and Simmons. As a further test of the theory, let us discuss some recent measurements [14.67] where an actin filament interacts with a small number of myosin heads fixed to a glass surface in an *in vitro* motility assay (Fig. 14.28). In these measurements it was observed that, at low myosin densities, the actin filament exhibits quantized velocities, depending on the number of myosin molecules which interacted with the filament. This is illustrated in Fig. 14.29 which shows (schematically) the displacement of an actin filament as a function of time, when acted on by one, two, and very many heads. The ratio between the drift velocity of the actin filament when acted on by very many myosin molecules and by a single myosin molecule was found to be about 10. In the model studied above, the drift velocity for zero load ($F = 0$) is $v_{\text{max}} = av^* = 1.6w_0 L$. If, instead, a single head is interacting with the actin filament, then the drift velocity is $v_{\text{min}} = L/(t_1 + 1/w_0)$, where $1/w_0$ is the average time it takes for an attached myosin head to jump over the barrier C from $A \to B$ when $x = 0$ (we assume that the head attaches to the thin filament when $x \approx 0$ corresponding to no elastic deformation). Thus $v_{\text{max}}/v_{\text{min}} = a(1 + w_0 t_1) \approx 6$ in relatively good agreement with the experimental data. Since the experiments have been performed *in vitro* under quite different conditions than *in vivo*, it is clear that one cannot expect perfect agreement between the two numbers. In particular, in order to slow down the diffusion of the actin filament (which was necessary in order for the filament not to diffuse away from the myosin molecules during the time when no myosin heads were attached to it) the actin–myosin system was surrounded by a very high viscosity fluid. But this will also reduce the diffusivity of the myosin head and consequently increase

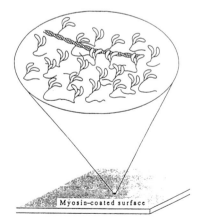

Fig. 14.28. An actin filament, propelled by myosin molecules (motor proteins) over a glass surface covered by myosin molecules.

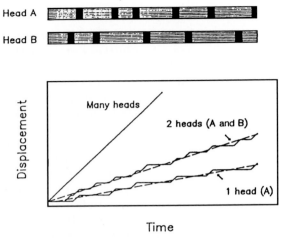

Fig. 14.29. *Top*: Consecutive cycles of molecular motor (myosin) heads. The dark strips indicate the periods when heads A and B are in contact with the actin filament. *Bottom*: The dependence of the displacement of an actin filament on time, for three different cases, where one, two and very many motor heads act on the actin filament. From [14.67].

the time t_1 that a head needs to "find" a binding site on the actin filament, and hence increase the ratio $v_{\max}/v_{\min}$.

14.7 Internal Friction and Plastic Stick-Slip Instabilities in Solids

In this section we discuss the physical origin of internal friction and plastic stick-slip instabilities which have been observed in solids. Figure 14.30 shows the relation between the stress σ and the strain ε for a low-carbon steel when the strain is increased at a constant rate $\dot{\varepsilon}_0$ [14.68]. Up to point A (stress σ_a) the solid deforms (mainly) elastically, after which it deforms plastically. Note that the stress drops sharply (from σ_a to σ_b) when the deformation changes from elastic to plastic. σ_a and σ_b, which are analogous to the static and kinetic friction force, are called the upper and lower yield points. There may be several reasons why $\sigma_a > \sigma_b$. First, we have seen in Chap. 11 that immediately after plastic deformation of a solid, residual stresses occur in the solid which slowly relax via thermal excitation. Another very important reason may be as follows [14.69]. In metals, plastic deformation is caused by the motion of dislocations. The dislocations are always pinned, e.g., by impurity atoms and by the periodic crystal potential, and the yield stress σ_a reflects the force necessary to pull the dislocations over their pinning

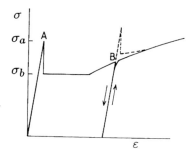

Fig. 14.30. The relation between the stress σ and the strain ε for low-carbon steel.

barriers. Now, consider a fresh dislocation, e.g., created during plastic deformation. For temperatures $T > 0\,\mathrm{K}$, the impurity atoms (in steel mainly carbon and nitrogen) will diffuse, and because of the stress field around dislocations they often experience long-range attractions to the dislocations, and they tend to accumulate along dislocation lines (this lowers the free energy of the system). Thus, the upper yield stress σ_a will have a contribution from the stress necessary to move the dislocations away from the "atmosphere" of impurity atoms. The lower yield stress corresponds to the stress necessary to propagate the dislocations across imperfections such as impurity atoms and grain boundaries, and the periodic crystal potential. The strong rise in the stress σ for "large" ε ("work hardening"), is due to the interaction between the dislocations, e.g., dislocation entanglement. This occurs predominately at large ε because of the high concentration of dislocations which are generated during plastic deformation.

It is important to note that the motion of individual dislocations during plastic deformations is not smooth but rather of stick-slip nature: a dislocation is temporarily trapped by obstacles until the stress becomes high enough to pull the dislocation over the pinning barrier or until it gets thermally exited over the barrier. In general, a short segment of the dislocation will rapidly slip a short distance, before becoming trapped by new pinning centers. Thus, during plastic deformation the motion of the dislocation involves a sequence of elastic instability transitions (Sect. 10.1), and during the rapid slip the external energy input is "dissipated", e.g., by emission of phonons (elastic waves) or excitation of electron–hole pairs. If instead the velocity of the dislocations were nearly constant during plastic deformation, i.e., nearly equal to its time average $\sim 2\dot\varepsilon_0/\rho b$ (where ρ is the number of dislocations per unit area and b the Burger's vector), then the stress σ would be proportional to $\dot\varepsilon_0$, contrary to what is observed experimentally.

From the discussion above it is clear that the yield stress σ_a depends on the time τ^* which has passed since the last plastic deformation of the solid. If τ is the time it takes for a fresh dislocation to obtain its atmosphere of impurity atoms, then if $\tau^* \ll \tau$ negligible relaxation has occurred and $\sigma_\mathrm{a} \approx \sigma_\mathrm{b}$. Thus

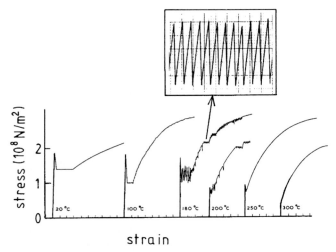

Fig. 14.31. Stress–strain curves for low-carbon steel during uniaxial tension (strain rate $\dot{\varepsilon}_0 = 5 \times 10^{-5}\,\mathrm{s}^{-1}$). Stick-slip plastic deformation is observed for temperatures $100°\mathrm{C} < T < 250°\mathrm{C}$. The inset shows the clock-like regularities of the stick-slip oscillations which were observed in some cases. From [14.70].

if the external force is removed at point B in Fig. 14.30, and kept at zero for a time $\tau^* \ll \tau$, and then returned back to the value which gives the same strain rate $\dot{\varepsilon}_0$ as before stop, one obtains the solid curve in Fig. 14.30. However, if $\tau^* \sim \tau$ (or longer), impurity atoms have time to accumulate along the dislocation lines and $\sigma_\mathrm{a} > \sigma_\mathrm{b}$, as indicated by the dashed line in Fig. 14.30. In general, σ_a increases monotonically with increasing "rest" time t. In the simplest case one may assume that

$$\sigma_\mathrm{a} = \sigma_\mathrm{b} + \Delta\sigma\left(1 - \mathrm{e}^{-t/\tau}\right) \,. \tag{14.44}$$

This formula correctly gives a maximum upper yield stress $\sigma_\mathrm{b} + \Delta\sigma$ and also the $\sigma_\mathrm{a} \to \sigma_\mathrm{b}$ behavior as $t \to 0$, but the actual time dependence of σ_a may be more complicated than indicated by (14.44). For example, (14.44) gives $\sigma_\mathrm{a} - \sigma_\mathrm{b} \propto t$ for $t \ll \tau$ while, if the increase in the yield stress is due to the diffusion of impurity atoms, one expects (for $t \ll \tau$) $\sigma_\mathrm{a} - \sigma_\mathrm{b} \propto t^\alpha$, where $\alpha = 1/2$ or $2/3$ depending on the nature of the interaction potential between the dislocations and the impurity atoms [14.69]. Note also that in general $1/\tau$ has an activated temperature dependence, $1/\tau \propto \exp(-\epsilon/k_\mathrm{B}T)$, because the motion of the impurity atoms is thermally activated.

Many metals, e.g., steel, aluminum and zinc, exhibit multiple yielding (or stick-slip) behavior when they are elongated at low strain rates ($\dot{\varepsilon}_0$) in certain temperature (T) ranges as shown in Fig. 14.31 for low-carbon steel under uniaxial tension (strain rate $\dot{\varepsilon}_0 = 5 \times 10^{-5}\,\mathrm{s}^{-1}$) [14.70]. This phenomenon may be explained by the dynamics of the pinning of the dislocations by impurity

14.7 Internal Friction and Plastic Stick-Slip Instabilities in Solids

atoms as follows. Assume that the plastic deformation suddenly stops (at time $t = 0$). If $\dot{\varepsilon}_0 = v_0/L$ denotes the constant strain rate imposed on the sample by the cross-heads of the tensile machine (we consider only small strain so that the dependence of the sample length L on time can be neglected), then the (elastic) stress in the block will increase with time as $\sigma(t) = \sigma(0) + 2G\dot{\varepsilon}_0 t$ where $2G = E/(1+\nu)$ (we assume that the tensile machine is "stiff", i.e., that its elasticity can be neglected). At the same time the yield stress for plastic deformations will increase with time t because of the increased pinning of the dislocations, resulting, e.g., from the accumulation of impurity atoms along the dislocation lines. Following the arguments presented several times before (see, e.g., Fig. 9.8), if the initial rate of increase of the (upper) yield stress with time t is higher than the rate of increase of the elastic stress (which equals $2G\dot{\varepsilon}_0$), stick-slip plastic deformation will occur. However, if $\dot{\varepsilon}_0$ is high enough this is not the case and steady plastic deformation occurs instead. Since τ depends sensitively on the temperature T, a small change of T can change the deformations from stick-slip to steady. If the temperature is very high, the cloud of pinning centering around the dislocation lines will "evaporate", in which case no stick-slip process occurs.

We now present the simplest possible mathematical description of the stick-slip plastic instability shown in Fig. 14.31. Let $\dot{\varepsilon}_0 = v_0/L$ be the constant strain rate imposed on the sample by the cross-heads of the tensile machine. Assume first that the plastic deformations in the rod are uniform, i.e., that the plastic strain $\varepsilon(x,t)$ is independent of the position $0 < x < L$ along the rod. We then have

$$\sigma = 2G\varepsilon_{\text{el}} = 2G\left(\dot{\varepsilon}_0 t - \varepsilon\right) , \qquad (14.45)$$

where ε_{el} is the elastic strain and $G = E/2(1+\nu)$ the shear modulus. Next, let us assume that

$$\sigma = \sigma_{\text{b}} + \Delta\sigma\left(1 - e^{-\phi/\tau}\right) + A\varepsilon + B\dot{\varepsilon} , \qquad (14.46)$$

where the age variable ϕ satisfies

$$\dot{\phi} = 1 - \alpha\dot{\varepsilon}\phi . \qquad (14.47)$$

In (14.46) the term $A\varepsilon$ describes work hardening while $B\dot{\varepsilon}$ describes friction damping due to, e.g., phonon emission. The coefficient α in (14.47) can be related to the (average) distance $\bar{d}$ that a dislocation moves during the rapid slip process which occurs between two consecutive pinning positions. If $\bar{v}$ denotes the average speed of a dislocation during plastic deformation, then $\dot{\varepsilon} = \rho b\bar{v}/2$ or $\bar{v} = 2\dot{\varepsilon}/\rho b$, where ρ is the number of dislocations per unit area and b the Burger's vector. Now, during steady plastic deformation (14.47) gives $\phi = \dot{\varepsilon}/\alpha$ which is the time that a dislocation remains in a pinned position before (rapidly) slipping to the next pinning position. At "low" strain rate $\dot{\varepsilon}$, the dislocations spend much more time in the pinning positions than in

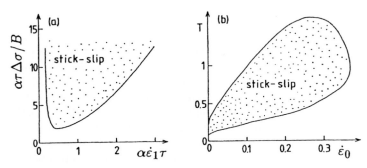

Fig. 14.32. The dotted areas indicate the stick-slip regions in (a) the $(\alpha\dot{\varepsilon}_1\tau, \alpha\tau\Delta\sigma/B)$-plane and (b) the $(\dot{\varepsilon}_0, T)$-plane [from equations (14.47) and (14.48)]. In (b) T is measured in units of ϵ/k_B and $\dot{\varepsilon}_0$ in units of $\Delta\sigma/B$ and we have assumed $A = 0$ and $B = \alpha\tau_0\Delta\sigma$.

rapid slip and we can write $\bar{v} \approx \bar{d}/\phi = \alpha\dot{\varepsilon}\bar{d}$. Comparing this with $\bar{v} = 2\dot{\varepsilon}/\rho b$ gives $\alpha = 2/\rho b\bar{d}$.

Combining (14.45,46), we have

$$B\dot{\varepsilon} = 2G\left(\dot{\varepsilon}_0 t - \varepsilon\right) - \sigma_b - \Delta\sigma\left(1 - e^{-\phi/\tau}\right) - A\varepsilon,$$

or

$$B\dot{\varepsilon} = 2G_1\left(\dot{\varepsilon}_1 t - \varepsilon\right) - \sigma_b - \Delta\sigma\left(1 - e^{-\phi/\tau}\right), \tag{14.48}$$

where $G_1 = G(1 + A/2G)$ and $\dot{\varepsilon}_1 = \dot{\varepsilon}_0/(1 + A/2G)$. Thus the only role of the strain hardening (if it is linear in ε) is to renormalize the shear modulus G and the strain rate $\dot{\varepsilon}_0$. Now, (14.48) is formally identical to the equations of motion studied in Sect. 12.2 if we identify $x \leftrightarrow \varepsilon$, $k_s \leftrightarrow 2G_1$, $v_s \leftrightarrow \dot{\varepsilon}_1$, $M\gamma \leftrightarrow B$, and $D \leftrightarrow 1/\alpha$, and take the $M \to 0$ limit (but with $M\gamma = B$ fixed), where the sliding motion is overdamped. Thus, from (12.34) we obtain the boundary line, which separates the region where stick-slip plastic instabilities occur, from the region where the plastic deformations occur steadily:

$$\dot{\varepsilon}_1^2 B\tau\alpha = \Delta\sigma e^{-1/(\alpha\tau\dot{\varepsilon}_1)}. \tag{14.49}$$

This relation between $\dot{\varepsilon}_1$ and τ is shown in Fig. 14.32a. If we assume $\tau = \tau_0\exp(\epsilon/k_B T)$ then Fig. 14.32a gives the $(\dot{\varepsilon}_0, T)$ dynamical phase diagram shown in Fig. 14.31b. Note that, in accordance with experiments, stick-slip deformations occur only in a finite temperature and strain-rate interval. In practice, the stick-slip region may disappear at much lower temperature than predicted by the theory because of the evaporation of the cloud of impurity atoms centered around the dislocation lines.

14.7 Internal Friction and Plastic Stick-Slip Instabilities in Solids

*Technical Comments**

a) If the tensile machine has a non-negligible elasticity, which may be represented by two external springs (spring constant k), the calculation above is still valid, but with a renormalized shear modulus $G \to G/(1 + 4A_0 G/kL)$ where A_0 is the cross sectional area of the sample.

b) We have assumed that the stick-slip plastic instability occurs uniformly over the length L of the rod, i.e., that the plastic strain $\varepsilon(x,t)$ is independent of x. This assumption is, in general, not correct. For a non-uniform plastic strain one must replace (14.45) with

$$\sigma = 2G \left(\dot{\varepsilon}_0 t - \frac{1}{L} \int_0^L dx \varepsilon(x,t) \right) . \tag{14.50}$$

Thus if, e.g., the plastic deformation were to occur uniformly within a segment ΔL of the rod, and be zero elsewhere, then (14.50) would take the form

$$\sigma = 2G(\dot{\varepsilon}_0 t - f\varepsilon) = 2Gf(\dot{\varepsilon}_0 t/f - \varepsilon) ,$$

where $f = \Delta L/L$. It is clear that in this case the analysis above is again valid but with renormalized shear modulus $G \to Gf$ and strain rate $\dot{\varepsilon}_0 \to \dot{\varepsilon}_0/f$. Experimentally, it is often found that stick-slip plastic deformations propagate as a wave along the rod, but at present there is no rigorous theory which can describe this phenomenon [14.71,72].

c) A condition for stick-slip plastic deformation has been derived in Refs. [14.71,72] by assuming that the relation between the stress σ and the strain rate $\dot{\varepsilon}$, derived for constant strain rate, is valid also when $\dot{\varepsilon}$ varies with time. This is analogous to the analysis of stick-slip presented in Sect. 7.6, where the friction force $F(\dot{x})$ was assumed to depend only on the instantaneous velocity $\dot{x}$. But we know that such a treatment, which neglects memory effects, does not, in general, describe friction dynamics correctly (see Sect. 7.6). However, in the present case where the motion is overdamped, both treatments give the same result for the boundary line (14.49). To see this, note that during stationary plastic deformation ($\dot{\varepsilon}$ = const.) (14.47) gives $\phi = 1/\alpha\dot{\varepsilon}$, and substituting this into (14.46) gives

$$\sigma = \sigma_b + \Delta\sigma \left(1 - e^{-1/\tau\alpha\dot{\varepsilon}} \right) + A\varepsilon + B\dot{\varepsilon} . \tag{14.51}$$

Substituting (14.51) into (14.45) and performing a linear instability analysis (Sect. 7.6) gives (14.49). However, the form of the stick-slip spikes differs in the two treatments, and a complete analysis of the stick-slip dynamics must be based on a formalism which contains memory effects, such as those of (14.45–47).

14.8 Rolling Resistance

Rolling resistance is the resistance to motion which occurs when an object is rolled over a surface. If F is the force necessary to keep an object rolling at some constant velocity v, the rolling friction coefficient μ_R is defined by

$$\mu_R = F/L,$$

where L is the load. We note, however, that while the sliding friction coefficient μ is usually nearly independent of the load L, this is not the case for the rolling resistance coefficient μ_R, which may either increase or decrease with increasing load. In most practical applications $\mu_R \sim 10^{-3} - 10^{-5}$, i.e., much lower than for sliding under boundary lubrication conditions. *Knothe* and *Miedler* [14.73] have shown that if an elastic cylinder (without internal friction) is rolling on a plane of the same material, there is no rolling resistance. If the materials of both bodies are different (but without internal friction), the rolling resistance may be non-zero, but it is usually extremely small. In the case of a polymer roller on a nearly rigid plane, the rolling resistance result mainly from the internal friction of the polymer.

There are at least two very important processes which contribute to the rolling resistance:

Slip at the Surface of Contact

If a perfectly rigid ball were rolling on a perfectly rigid flat substrate, a single contact point occur at any moment in time, and the rolling resistance vanish. However, in practice all solids have a finite elasticity, and the bodies will deform elastically (or plastically, if the local pressure reaches the yield stress of any of the solids), and the actual contact area will be finite. It follows from the theory of elasticity (see also Fig. 7.24) that a tangential stress may develop at the interface between the ball and the substrate. If the local tangential stress becomes high enough local slip will occur between the surface of the ball and the substrate. During this local slip external energy will be converted into heat motion, i.e., the slip will contribute to the rolling resistance. The slip velocities are generally small (usually less that 1% of the rolling velocity), but nevertheless produce in many cases a major part of the total resistance to rolling. The theory of *Carter* [14.74], which considers purely elastic solids, shows that during free rolling (i.e., no applied traction) the tangential stress, and hence the rolling resistance, vanishes if (a) both bodies consist of the same material, or (b) both bodies are incompressible, or (c) one body is rigid and the other incompressible.

The basic theory of the slip contribution to rolling resistance was published by *Carter* in 1926 [14.74,75]. He was interested in the action of a locomotive driving wheel, but considered the simpler problem of two cylinders (radius R, length l) of like materials pressed together and rolling on one

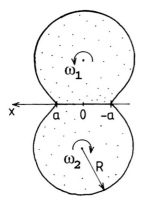

Fig. 14.33. Two cylinders (radius R, length l) pressed together and rolling on one another at different angular velocities ω_1 and ω_2.

another, one being subjected to a torque and the other to an equal counter-torque (Fig. 14.33). Thus, any state of stress or strain in one member, due to tangential tractive forces only, is matched by an equal and opposite state in the other, and the distribution of pressure between the members is unaffected by the traction, since the radial displacements of the surfaces in contact are complementary. Assuming that both bodies can be considered as half spaces and that the theory of elasticity is valid, the normal stress distribution in the contact area $-a < x < a$ is given by the theory of *Hertz* [14.76] $P = (4P_0/\pi)(1 - x^2/a^2)^{1/2}$, where $P_0 = L/2al$ is the average pressure in the contact area. Carter used the Coulomb friction law, but assumed that the static and kinetic friction coefficients were equal. Thus in the contact area, if the tangential stress τ is smaller than $\mu P(x)$ no slip occurs, but if the tangential stress reaches $\mu P(x)$, local slip occurs. The results of the analysis are summarized below.

The solid line in Fig. 14.34a shows $\mu P(x)$, which is equal to the maximum tangential stress possible in the contact area. The dashed line shows the actual tangential stress distribution $\tau(x)$. No slip has occurred for $c < x < a$ where $\tau < \mu P(x)$. In the other part of the contact region, $-a < x < c$, slip occurs and $\tau = \mu P(x)$. Note that when the friction force F increases, so does the size of the slip region, until F reaches $F_1 = \mu L$, the maximum friction force, at which point slip occurs in the whole contact area. In general, the friction force is given by

$$F = F_1 \left[1 - \left(\frac{a-c}{2a} \right)^2 \right], \qquad (14.52)$$

so that $F \to F_1$ as $c \to a$ (complete slip) and $F \to 0$ as $c \to -a$ (no slip). Thus, for a "free" rolling cylinder (no applied torque), there is no rolling resistance due to local slip. When a torque is applied (with constant rolling

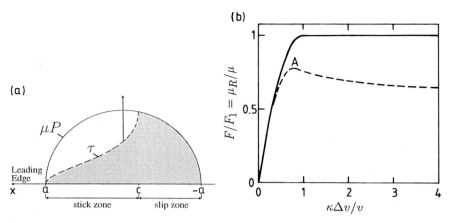

Fig. 14.34. (a) Stress distribution in the contact area $-a < x < a$. The dashed curve shows the dependence of the shear stress τ on x, while the solid curve shows the maximum possible shear stress before slip, as given by the product of the (static) friction coefficient μ and the local pressure $P(x)$. (b) The solid line shows the friction force F (in units of the maximal friction force $F_1 = \mu L$), as a function of the "slip velocity" Δv. The dashed curve is obtained under the assumption that the kinetic friction coefficient decreases monotonically with increasing sliding velocity [14.74,75,77].

velocity) the friction force $F \neq 0$, and the rolling speed of the two cylinders differ. If $R\omega_1$ (ω_1 is the angular velocity) and $R\omega_2$ are the rolling speeds of the two cylinders, then the ratio $\Delta v/v$, where $v = R(\omega_1 + \omega_2)/2$ is the average velocity and $\Delta v = R(\omega_1 - \omega_2)$ the difference in the rotation velocities, increases in proportion to the fraction $(a+c)/2a$ of the contact area where slip occurs,

$$\frac{\Delta v}{v} = \frac{4(1-\nu)\mu P_0(a+c)}{\pi G a}, \tag{14.53}$$

and where G is the shear modulus. Using (14.52,53) one can express the friction force as a function of $\Delta v/v$:

$$F = F_1 \left[1 - (1 - \kappa \Delta v/v)^2\right], \tag{14.54}$$

where $\kappa = \pi G/[8(1-\nu)\mu P_0]$. The solid line in Fig. 14.34b shows F/F_1 as a function of $\kappa \Delta v/v$. Note that $F = F_1$ when $\Delta v/v = 1/\kappa = 8(1-\nu)\mu P_0/\pi G$. If only elastic deformations occur in the contact area, the pressure P_0 must be below the plastic yield stress σ_c of the solids. Since for steel $G/\sigma_c \sim 100$ and since typically $8(1-\nu)\mu/\pi \sim 1$, it follows that at the onset of complete slip $\Delta v/v < 0.01$. In the wheel–rail contact area the pressure P_0 is usually close to the plastic yield stress and in a typical case $\Delta v/v \sim 0.005$ at the onset of complete slip. Thus if, e.g., $v \sim 20\,\text{m/s}$ then $\Delta v \sim 0.1\,\text{m/s}$.

Let us comment on the validity of the theory presented above. First, it is clear that the static friction coefficient is, in general, higher than the kinetic friction coefficient. Furthermore, the steady sliding friction coefficient may be velocity dependent, usually decreasing with increasing sliding velocity. Taking into account these two effects [14.77] results in a relation between F and Δv of the qualitative form indicated by the dashed line in Fig. 14.34b. Another important effect may be the dependence of the static friction force on the time of stationary contact. If a train moves with the velocity v and if the contact area between the wheel and the rail has the diameter $2a$, then the maximum time of stationary contact (for no slip), will be $\tau^* = 2a/v \sim 10^{-4}\,\mathrm{s}$, where we have used $2a \sim 0.01\,\mathrm{m}$ and $v \sim 100\,\mathrm{m/s}$. This is a very short contact time and it is clear that the static friction coefficient will, in general, not have reached its maximum value before the contact is broken. The actual dependence of the static friction force on the time of stationary contact is not known for the wheel–rail system, and is probably not very well defined, as it depends sensitively on the nature of the contamination layer, which varies with the spatial and temporal location. It is well-known, however, that the maximum traction of a locomotive decreases with increasing rolling speed, as is expected on the basis of this picture.

The (dashed) curve in Fig. 14.34b forms the basis for designing anti-skid braking systems. Since the friction force is maximal at point A in Fig. 14.34b, an optimally designed braking system should keep the motion of the wheel close to this point during emergency braking. However, even with a feedback system it is hard to continuously stay close to point A, since the dashed curve is not constant in time as, e.g., the coefficient of friction varies due to changing surface conditions. Most anti-skid systems for automobiles use a pulsed system where the wheel–substrate slip oscillates around the maximum A, continuously adjusting itself to the road conditions. In fact, the most important effect is not the increase in the sliding friction compared with complete slip (locked wheels), but rather the fact that rolling wheels allow steering or directional control which is lost during complete slip.

Internal Friction

During rolling, different regions on the ball and the substrate are first stressed, and then the stress is released as rolling continues. Each time a volume element in either body is stressed elastic energy is stored by it. Most of the elastic energy is later released as the stress is removed, but the pulsating deformations cannot occur entirely adiabatically and some energy will be converted into heat by the internal friction. This transfer of energy from the rotational (and translational) motion to heat will contribute to the rolling resistance.

There are a large number of physical processes which contribute to the (dynamical) internal friction of a solid, usually described by a complex shear modulus $G = G_1 + iG_2$. For example, in a metal, various modes of switching of

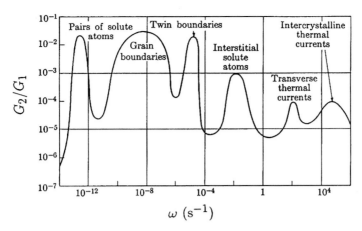

Fig. 14.35. Typical damping spectrum of a crystalline material such as iron at low stress (linear response). The ratio between the imaginary and real part of the complex shear modulus $G = G_1 + iG_2$ as a function of the frequency ω. From [14.79].

segments of (pinned) dislocations can occur. In chain polymers the viscoelastic properties may be due to switching of chain segments between two configurations A and B. In both cases, the stress does not itself force the switch. Rather, its result is to bias the activation energies of the switch in the favorable and unfavorable directions; the jump over the barriers occurs by thermal excitation. For a weak applied external stress which varies very slowly in time, the deformations occur adiabatically and the deformation vanishes as the external load is removed. (Permanent plastic deformations may also occur, but in this section we assume that the external stress is so weak and the frequency of oscillation so high that such processes can be neglected.) However, at finite frequencies this is not the case and external energy is converted into heat in the solid as described by the model calculation at the end of Sect. 5.2. Any specific mechanism of internal friction will, in general, give rise to a characteristic peak in the frequency-dependent dissipative response $G_2(\omega)$ of a solid at a frequency $\omega \sim 1/\tau$, where τ is a relaxation time which characterizes the average time a molecule (or a segment of a molecule) spends in state A before flipping to state B. Most solids have many such peaks (Fig. 14.35) at different resonance frequencies. Note that τ depends strongly (exponentially) on temperature (Sect. 5.2), and the viscoelastic properties (at a fixed frequency) of a solid depend on the (inverse) temperature T in a similar way as they do on the frequency, often exhibiting several peaks as a function of T.

The magnitude of the internal friction varies strongly from one material to another. Thus, for a steel ball rolling on a steel substrate, the contribution from the internal friction is usually rather small. Using the theory of viscoelasticity one can estimate (see Sect. 5.3) $\mu \sim P_0 \mathrm{Im}[1/G(\omega)] = P_0 G_2/(G_1^2 + G_2^2)$, where P_0 is the (average) pressure in the contact area and where $G(\omega)$

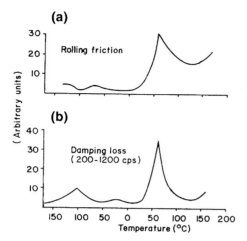

Fig. 14.36. (a) Rolling resistance of a steel ball on a nylon surface as a function of temperature. (b) Low-frequency viscoelastic data (internal friction) for the same polymer as a function of temperature [14.80].

is evaluated at the frequency $\omega \sim v/R$, where v is the rolling velocity and R the radius of the ball. For steel on steel, P_0 cannot be larger than the plastic yield stress $\sigma_c \sim 0.01 G_1$ so that $P_0/G_1 < 0.01$. Furthermore from Fig. 14.35, the ratio G_2/G_1 for steel is typically smaller than 0.001 for the frequencies of interest, we get $\mu < 10^{-5}$. However, for a steel ball rolling on polymer or a wheel rolling on the road, the internal friction may contribute a substantial part of the rolling resistance. Figure 14.36a shows the temperature dependence of the rolling friction when a steel ball is rolling on a nylon surface. Figure 14.36b shows the low-frequency viscoelastic data for the same polymer as a function of temperature. The similarities of the curves should be noted.

As pointed out above, the internal friction of solids depends on the frequency ω of the external oscillating stress, and it will in general exhibit several peaks (resonances) at well defined resonance frequencies. This fact is utilized in many practical applications. For example, in order to minimize the friction losses during steady driving of an automobile, the tire-rubber is designed so that, during rolling at normal speeds, the deformations of the wheels are mainly elastic, while during breaking they behave viscoelastically with a high internal friction. This can be realized since the typical deformation frequencies during rolling are different than from those during sliding; in the latter case the relevant deformations result from surface asperities on the road sliding relative to the tires. This generates pulsating forces acting on the tires, at characteristic frequencies $\omega_S \sim v/r$, where $r \sim 0.1 - 1 \, \text{mm}$ is the typical extent of a surface asperity on a road. With $v \sim 10 - 100 \, \text{m/s}$ this gives $\omega_S \sim 10^4 - 10^6 \, \text{s}^{-1}$. During rolling the characteristic frequency is

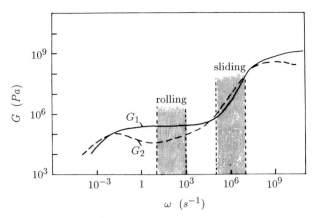

Fig. 14.37. The dependence of the complex shear modulus $G = G_1 + iG_2$ on the frequency ω for a viscoelastic amorphous polymer [14.81,82].

instead the rolling frequency $\omega_R \sim v/R$, where $R \sim 0.1 - 1$ m is the radius of the wheel, which gives $\omega_R \sim 10 - 10^3 \, \text{s}^{-1}$. Figure 14.37 shows (schematically) the complex shear modulus $G(\omega) = G_1(\omega) + iG_2(\omega)$ of a viscoelastic amorphous polymer. The dotted regions indicate the frequency intervals typically involved in rolling and sliding friction. Note that the loss moduli $G_2 > G_1$ in the sliding region while $G_2 < G_1$ in the rolling region. Using $\mu \sim -C \, \text{Im} \, G / |G|$ [see (5.34a)] it is clear that the friction coefficient will be much larger during sliding than during rolling. The emphasis in tire development has been on the improvement of skid properties with the lowest possible trade-off in terms of rolling resistance. In the future the minimization of the rolling resistance without sacrificing skid properties is likely to gain importance [14.78].

14.9 Friction Dynamics for Granular Materials

Sliding friction for solids separated by a layer of granular material, e.g., wear particles produced during sliding, has been studied intensively for many years, mainly because of its relevance for understanding earthquake dynamics. Here slip events occur along faults that are often separated by a granular "gouge". The solid surfaces are squeezed together by a very high pressure, typically of order 100 MPa, and at high temperatures. Under these conditions, strong (time-dependent) plastic deformation is expected to occur, and this results in an increase in the contact area with the time of stationary contact. The sliding dynamics in these cases seems to be similar to that of "clean" surfaces, described (approximately) by the equations of motion presented in Chap. 13.

14.9 Friction Dynamics for Granular Materials

Drastically different friction dynamics has been observed for granular material at low pressures and at room temperature. Under these conditions plastic deformation is less important. At the onset of sliding the solid surfaces separate by a small amount Δh, while a thin layer of particles at the sliding interface fluidize. The dilation Δh is a consequence of the fact that the fluidized region has a slightly lower particle density than the densely packed equilibrium structure. It is found that the dilation Δh decreases as the normal applied stress increases, and may be so small at the high pressures which occur along faults that no fluidization of the gouge occurs; this is probably the reason for the difference in the friction dynamics at high and low pressures.

In this section we discuss friction dynamics for granular materials at room temperature and at low squeezing pressures, $P \sim 10$–100 Pa, i.e., typically a factor of 10^6 lower than along faults. Gollub and coworkers [14.83] have performed friction experiments where a solid block is located on a bed of small particles consisting of smooth spherical glass particles (diameter $\sim 100\,\mu\mathrm{m}$), or rough sand particles with an average diameter of order a few $100\,\mu\mathrm{m}$. The block was made of a transparent material so that the particle distribution at the interface could be studied using an optical microscope. As usual, a spring k_s is connected to the block and the free end is pulled with velocity v_s. Experiments were performed for both "dry" particles at reduced relative humidity (20%), and for wet granular material (the particles are completely surrounded by water). We note that even at 20% relative humidity, a small amount of water is condensed near points of contact forming capillary bridges, which produce substantial strengthening of the granular material (see Sect. 7.5). The capillary bridges are eliminated if the pore spaces are entirely filled with water as in the experiments with submerged particles. We note that the kinetic friction coefficient for the submerged particles, $\mu_\mathrm{k} \approx 0.24$, is significantly smaller than that measured for the same materials under "dry" conditions (i.e., at 20% relative humidity) where $\mu_\mathrm{k} \approx 0.5$. This difference can be explained by the additional energy input necessary in the latter case in order to break the capillary bridges between the particles, and between the particles and the block.

The kinetic phase diagram is of the general form shown in Fig. 3.3. That is, steady sliding is observed if the spring velocity v_s, or the spring constant k_s, is high enough. The kinetic frictional force $F_\mathrm{k} = Mg\mu_\mathrm{k}$ is found to be velocity independent (in the velocity interval studied, $0.1 < v_\mathrm{s} < 1000\,\mu\mathrm{m/s}$), and proportional to the external load or pressure for $7 < P < 100$ Pa.

Figure 14.38 shows (a) the spring elongation $d(t)$ (proportional to the spring force), and (b) the vertical position $h(t)$ of the block, as a function of time t, starting with $d = 0$ at $t = 0$. Note that the spring elongation $d(t)$ initially increases linearly with time, while the block is stationary. The vertical position of the block shown in (b) increases by $\Delta h \approx 5\,\mu\mathrm{m}$ during the transition from a stationary to a sliding block. It is found that h tends to its asymptotic value roughly exponentially with the sliding distance x. Thus,

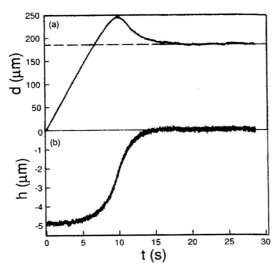

Fig. 14.38. Typical behavior of (a) the spring elongation $d(t)$ and (b) the vertical position $h(t)$ as a function of time. Spring constant $k_s = 189.5$ N/m, spring velocity $v_s = 28.17$ µm/s and mass of block $M = 14.5$ g. From [14.83].

if we define $h(\infty) = 0$,

$$h(x) \approx -\Delta h \, e^{-x/l},$$

or

$$\frac{dh}{dt} \approx -\frac{h}{l}\frac{dx}{dt},$$

where l is the characteristic distance over which the layer dilates, which is found to be of order the mean radius of the particles. The total dilation $\Delta h = h(\infty) - h(0)$ is independent of the driving velocity v_s, but decreases when the applied pressure is increased. Note that the dilation Δh is only of order 5% of the average diameter of a particle. Nevertheless, this small dilation is enough for a few (top) particle layers to fluidize. During sliding, the mobile particles may be organized in horizontal layers in order to facilitate the sliding, as has been observed in the shearing of colloidal systems (see Sect. 7.2). Experiments shows that Δh scales with the particle size.

The initial overshoot of the frictional force (see Fig. 14.38a) is, at least in part, related to the additional energy the system requires to dilute. During the transient, while the layer is dilating significantly, the friction depends roughly linearly on the dilation rate $dh/dt \equiv \dot{h}$. Thus the time evolution of x and h are approximately given by

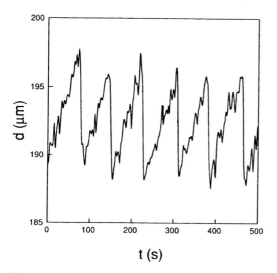

Fig. 14.39. Typical stick-slip motion observed for very small spring velocities. The parameters are $k_s = 189.5$ N/m, $v_s = 0.11\,\mu\text{m/s}$ and $M = 14.5$ g. From [14.83].

$$M\ddot{x} = k_s(v_s t - x) - F_k - \alpha \dot{h} \tag{14.55}$$

$$\dot{h} = -h\dot{x}/l, \tag{14.56}$$

where α is an empirical constant. These equations have been tested by comparing with experimental data for the case where the block is initially at rest, $x(0) = \dot{x}(0) = 0$, while the free end of the spring is moving with a constant velocity v_s. For this case the model is in qualitative agreement with the experimental data, except in the very early stage of the block motion. However, a trivial linear instability analysis of (14.55,56) shows that if F_k is velocity independent, the steady motion is always stable with respect to small perturbations. Thus, if we write $x = x_0 + v_s t + \xi$ and $h = h_0 + \eta$, where $\xi = \xi_0 \exp(\kappa t)$ and $\eta = \eta_0 \exp(\kappa t)$, then to linear order in ξ and η one obtain from (14.56), $\dot{\eta} = -v_s \eta/l$. Thus, either $\eta_0 = 0$ or else, $\kappa = -v_s/l$ so that the perturbation decays with time. In the latter case the steady motion is stable with respect to small perturbations. In the former case it is still possible that the steady motion is unstable if (14.55) has a nontrivial solutions for ξ, even when $\dot{h} \equiv 0$. However, no such solution exist if F_k is velocity independent. For wet granular materials, stick-slip motion (with extremely small amplitude), has only been observed for $v_s < 0.2\,\mu\text{m/s}$, see Fig. 14.39. This indicates that F_k increases with decreasing velocity when $\dot{x} < 0.2\,\mu\text{m/s}$, and that the static friction force is nearly identical to the kinetic friction force. For sliding dynamics with the same granular materials but at a relative humidity of 20%, stick-slip motion has been observed even at relatively high spring velocities v_s, and the static friction is now typically much larger than the kinetic friction.

Thus, in this case F_k is a decreasing function of $\dot{x}$ even at relatively high velocities. In addition, it has been observed that the static frictional force now increases with the time of stationary contact. These effects are easy to understand: At 20% humidity, at thermal equilibrium capillary bridges occur in the contact areas between the particles. However, as discussed in Sect. 7.5, it will take some time for a capillary bridge to form [14.84]. Thus, one expects that the static friction will increase with the time of stationary contact, and the kinetic friction will decrease with increasing $\dot{x}$. As shown in Sect. 7.5, this contribution to the friction force may have exactly the same form $\sim \ln(\theta/\tau)$ as in dry friction dynamics (see Chap. 13), where the contact age variable θ-variable satisfies (13.6).

Optical imaging has shown that localized particle rearrangements already occur before major slip events take place during stick-slip motion. It is known that when a granular media is exposed to external stress, the stress distribution is not homogeneous in space, but is concentrated in "stress chains". Thus, the most likely origin of the local rearrangements is the breaking of stress chains. For "dry" granular layers, these local microslips events lead to a displacement of the block by $\sim 1\%$ of the total slip involved in a major slip event.

15. Outlook

The topic of sliding friction has experienced a major burst of activity since 1987, much of which has developed quite independently and spontaneously. It is likely that the next few years will result in a deeper understanding of the fundamentals of sliding friction and also a general realization that many different physical phenomena, which in the past have been studied independently of each other in different disciplines, such as earthquake dynamics and the sliding of flux line systems, are closely related and perhaps should be studied together. Such a unification may not only lead to a cross fertilization of ideas, but also represents one of the goals of theoretical physics.

Experimental studies of sliding friction with atomically smooth surfaces and with lubrication films of known thickness (to within ± 1 Å) have only been possible during the last few years. Many fundamental questions, some of which have been raised in this book, remain unanswered. For example, how does the process of squeezing out the lubrication "fluid" occur when only a few monolayers remain trapped between the solid surfaces; e.g., is it a thermally activated process? And what is the exact nature of the rapid processes which must occur in the lubrication film during sliding?

New experimental devices, such as the atomic force microscope and quartz-crystal-microbalance probe other aspects of sliding friction with potentially important applications in nano-tribology and fluid dynamics at surfaces. In short, the study of sliding friction, which is one of the oldest problems in physics, is in a phase of rapid and exciting development.

References

Chapter 1

1.1 S. Weinberg: *The Quantum Theory of Fields* (Cambridge Univ. Press, New York 1995)
1.2 K.G. Wilson, J. Kogut: Phys. Rep. **12**, 75 (1975)
1.3 L.D. Landau, E.M. Lifshitz: *Fluid Mechanics* (Pergamon, New York 1975)
1.4 B.N.J. Persson, R. Rydberg: Phys. Rev. **B32**, 3586 (1985)
1.5 A.O. Caldeira, A.J. Legget: Ann. Phys. (N.Y.) **149**, 374 (1983)
1.6 J.B. Sokoloff: Phys. Rev. Lett. **71**, 3450 (1993)
   M.V. Berry, J.M. Robbins: Proc. Roy. Sec. (London) A **442**, 659 (1993)
1.7 J.D. Beckerle, M.P. Casassa, R.R. Cavanagh, E.J. Heilweil, J.C. Stephenson: J. Chem. Phys. **90**, 4619 (1989)
1.8 B.N.J. Persson: Phys. Rev. B **44**, 3277 (1991)
1.9 J. Ford: Phys. Rep. **213**, 271 (1992)
   V.I. Arnold, A. Avez: *Ergodic Problems of Classical Mechanics* (Benjamin, New York 1968)
1.10 E. Fermi, J. Pasta, S. Ulam: In *Collected Papers of Enrico Fermi*, Vol. 2 (Chicago Univ. Press, Chicago 1955) p. 978
1.11 R.P. Feynman, R.B. Leighton, M. Sands: In *The Feynman Lectures on Physics* (Addison-Wesley, Reading, MA 1963) p. 444
1.12 G. Hagen: Monatberichte der Preuss. Akad. d. Wiss. (Berlin) (1952) p. 35
   H.A. Janseen: 2. VDI **39**, 1045 (1895)
   D.E. Wolf: In *Computational Physics, Selected Methods-Simple Exercises-Serious Applications,* ed. by K.H. Hoffmann, M. Schreiber (Springer, Berlin, Heidelberg 1996)
   G. Gudehus: Siloprobleme aus bodenmechanischer Sicht, Beitrag zum Abschlussband des SFB 219 (1997)
1.13 F.P. Bowden, D. Tabor: *Friction and lubrication* (Methuen, London, 1967)
1.14 D.F. More: *Principles and Applications of Tribology* (Pergamon, London 1975)
1.15 E. Rabinowicz: *Friction and Wear of Materials* (Wiley, New York 1965)
1.16 I.L. Singer, H.M. Pollock (eds.): *Fundamentals of Friction: Macroscopic and Microscopic Processes* (Kluwer, Dordrecht 1992)
1.17 B.N.J. Persson, E. Tosatti (eds.): *Physics of Sliding Friction* (Kluwer, Dordrecht 1996)
1.18 B. Bhushan (ed.): *Micro/Nanotribology and its Applications* (Kluwer, Dordrecht 1997)

Chapter 2

2.1   D. Dowson: *History of Tribology* (Longman, New York 1979)

Chapter 3

3.1   J.N. Israelachvili: Surf. Sci. Rpt. **14**, 109 (1992)
3.2   H. Yoshizawa, J. Israelachvili: J. Phys. Chem. **97**, 11300 (1993)
3.3   H. Yoshizawa, Y.-L. Chen, J. Israelachvili: Wear **168**, 161 (1993)
3.4   H. Yoshizawa, Y.-L. Chen, J. Israelachvili: J. Phys. Chem. **97**, 4128 (1993)
3.5   B.N.J. Persson: Phys. Rev. B **51**, 13568 (1995)
3.6   A.D. Berman, W. A. Ducker, J, N. Israelachvili: In *Physics of Sliding Friction*, ed. by B.N.J. Persson, E. Tosatti (Kluwer, Dordrecht 1996)
3.7   R. Erlandsson, G. Hadziioannou, C.M. Mate, G.M. McClelland, S. Chiang: J. Chem. Phys. **89**, 5190 (1988)
      C. Mathew, G.M. McClelland, R. Erlandsson, S. Chiang: Phys. Rev. Lett. **59**, 1942 (1987)
3.8   E. Rabinowicz, D. Tabor: Proc. Roy. Sec. (London) A **208**, 455 (1951)
3.9   J. Krim, A. Widom: Phys. Rev. B **38**, 12184 (1988)
3.10 J. Krim, D.H. Solina, R. Chiarello: Phys. Rev. Lett. **66**, 181 (1991)
3.11 J. Krim: Scientific Am. **275**, 48 (October 1996)
3.12 J. Krim, C. Daly: In *Physics of Sliding Friction*, ed. by B.N.J. Persson, E. Tosatti (Kluwer, Dordrecht 1996)
      C. Daly, J. Krim: Phys. Rev. Lett. **76**, 803 (1996)
3.13 J. Krim, D.H. Solina, R. Chiarello: Phys. Rev. Lett. **66**, 181 (1991)
3.14 J. Krim, E.T. Watts, J. Digel: J. Vac. Sei. Technol. A **8**, 3417 (1990)
3.15 E.T. Watts, J. Krim, A. Widom: Phys. Rev. B **41**, 3466(1990)
3.16 J. Krim, R. Chiarello: J. Vac. Sei. Technol. A **9**, 2566 (1991)
3.17 E. Meyer, R. Lüthi, L. Howald, M. Bammerlin, M. Guggisberg, H.-J. Güntherodt, L. Scandella, J. Gobrecht: In *Physics of Sliding Friction*, ed. by B.N.J. Persson, E. Tosatti (Kluwer, Dordrecht 1996)
3.18 U.D. Schwarz, H. Bluhm, H. Hölscher, W. Allers, R. Wiesendanger: In *Physics of Sliding Friction*, ed. by B.N.J. Persson, E. Tosatti (Kluwer, Dordrecht 1996)
3.19 U.D. Schwarz, R. Wiesendanger: Priv. Commun. (1996)
3.20 M. Rosso, D. Schumacher: Priv. commun. (1996)
3.21 A. Dayo, W. Alnasrallah: J. Krim, Phys. Rev. Lett. **80**, 1960 (1998)

Chapter 4

4.1   A.L. Barabasi, H.E. Stanley: *Fractal Concepts in Surface Growth* (Cambridge Univ. Press, Cambridge 1995)
      T. Vicsek: *Fraktal Growth Phenomena*, 2nd edn. (World Scientific, Singapore 1992)
4.2   R.T. Cundill: Ball Bearing J. **241**, 26 (1993)

1.19 T. Baumberger: In *Physics of Sliding Friction*, ed. by B.N.J. Persson, E. Tosatti (Kluwer, Dordrecht 1996)

4.3   G. Baumann, H.D. Grohmann, K. Knothe: ETR **45**, 792 (1996)
4.4   G. Baumann, H.J. Fecht, S. Liebelt: Wear **191**, 133 (1996)
4.5   K. Knothe, S. Liebelt: Wear **189**, 91 (1995)
4.6   J.A. Greenwood, J.B.P. Williamson: Proc. Roy. Sec. (London) Ser. A **295**, 300 (1966)
4.7   J.A. Greenwood: In *Fundamentals of Friction: Macroscopic and Microscopic Processes,* ed. by I.L Singer, H.M. Pollock (Kluwer, Dordrecht 1992)
4.8   Courtesy of D. Schumacher (1996)
4.9   Courtesy of J. Krim (1996)
4.10 B. Kehrwald: Dissertation, Universität Karlsruhe (1998)

Chapter 5

5.1   L.D. Landau, E.M. Lifshitz: *Theory of Elasticity* (Pergamon, New York 1975)
5.2   F.A. McClintock, A.S. Argon: *Mechanical Behavior of Materials* (Addison-Wesley, Reading, MA 1966)
5.3   J.A. Greenwood: In *Fundamentals of Friction: Macroscopic and Microscopic Processes,* ed. by I.L. Singer, H.M. Pollock (Kluwer, Dordrecht 1992)
5.4   I.L. Singer, H.M. Pollock (eds.): *Fundamentals of Friction: Macroscopic Processes* (Kluwer, Dordrecht 1992) p. 228
5.5   F. Pobell: *Matters and Methods at Low Temperature,* 2nd edn. (Springer, Berlin, Heidelberg 1996)
5.6   J.H. Dieterich, B.D. Kilgore: PAGEOPH **143**, 283 (1994)
5.7   L.D. Landau, E.M. Lifshitz: *Electrodynamits of Continuous Media* (Pergamon, London 1960)
5.8   K.L. Johnson: In *Contact Mechanics* (Cambridge Univ. Press, Cambridge 1985)
5.9   B.V. Derjaguin, V.M. Muller, Y.P. Toporov: J. Coll. Interface Sci. **53**, 314 (1975)
5.10 V.M. Muller, V.S. Yushenko, B.V. Derjaguin: J. Coll. Interface Sci. **77**, 91 (1980)
      B.D. Hughes, L.R. White: J. Mech. Appl. Math. **32**, 445 (1979)
5.11 D. Maugis: J. Coll. Interface Sci. **150**, 243 (1992)
      D.S. Dugdale: J. Mech. Phys. Solids **8**, lOO (1960)
5.12 U. Landman, W.D. Luedtke, B.E. Salisbury, R.L. Whetten: Phys. Rev. Lett. **77**, 1362 (1996)
5.13 U. Landman, W.D. Luedtke, J. Gao: Langmuir **12**, 4514 (1996)
5.14 U. Dürig, A. Stalder: In *Physics of Sliding Friction,* ed. by B.N.J. Persson, E. Tosatti (Kluwer, Dordrecht 1996)
5.15 U. Landman, W.D. Luedtke, E. Ringer: Wear **153**, 3 (1992)
5.16 S.A. Joyce, R.C. Thomas, J.E. Houston, T.A. Michalske, R.M. Crooks: Phys. Rev. Lett. **68**, 2790 (1992)
      R.C. Thomas, J.E. Houston, T.A. Michalske, R.M. Crooks: Science **259**, 1883 (1993)
5.17 J.D. Ferry: *Viscoelastic Properties of Polymers* (Wiley, New York 1980)
      P.J. Flory: *Principles of Polymer Chemistry* (Cornell Univ.Press, Ithaka, NY 1953)
5.18 M. Doi, S.F. Edwards: *The Theory of Ploymer Dynamits* (Oxford Univ. Press, Oxford 1986)
5.19 D.F. More: *The Friction and Lubrication of Elastomers* (Pergamon, London 1972)

5.20 K.A. Grossch: Proc. Roy. Sec. (London) A **274**, 21 (1963)
5.21 A. Schallamach: Wear**6**, 375 (1963)
5.22 U. Dürig, A. Stalder: *In Sliding Friction,* ed. by B. Bhushan (Kluwer, Dordrecht 1997)
5.23 H. Yoshizawa, J. Israelachvili: J. Phys. Chem. **97**, 11300 (1993)
5.24 J. Halaunbrenner, A. Kubisz: ASLE-ASME Lubrication Conf., Chicago (1967) Paper No. 67-Lub-25
5.25 A.D. Roberts: Rubber Chem. Techn. **65**, 673 (1992)
5.26 K. Mori, S. Kaneda, K. Kanae, H. Hirahara, Y. Oishi, A. Iwabuchi: Rubber Chem. Techn. **67**, 798 (1994)
5.27 M. Barquins: Mater. Sci. Eng. **73**, 45 (1985)
5.28 S.G. Corcoran, R.J. Colton, E.T. Lilleodden, W.W. Gerberich: Phys. Rev. B **55**, R16057 (1997)
5.29 B.N.J. Persson: Phys. Rev. Lett. **81**, 3439 (1998); J. Chem. Phys. **110**, 9713 (1999)
5.30 A. Schallamach: Rubber Chem. and Technol. **209**, 69 (1968)

Chapter 6

6.1 F.P. Bowden, D. Tabor: *Friction and Lubrication* (Methuen, London 1967)
6.2 E. Rabinowicz, D. Tabor: Proc. Roy. Sec. (London) A **208**, 455 (1951)
6.3 T.-M. Hong, S.-H. Chen, Y.-S. Chiou, C.-F. Chen: Thin Solid Films **270**, 148 (1995)
6.4 C. Zuiker, A.R. Krauss, D.M. Gruen, X. Pan, J.C. Li, R. Csencsits, A. Erdemir, C. Bindel, G. Fenske: Thin Solid Films **270**, 145 (1995)
6.5 B. Bhushan (ed.): *Tribology and Mechanics of Magnetit Storage Devices* (Springer, New York 1966)
6.6 C.-J. Lu, D.B. Bogy, S.S. Rosenblum, G. J. Tessmer: Thin Solid Films **268**, 83 (1965)
6.7 B. Marchon, N. Heiman, M.R. Khan, A. Lautie, J.W. Ager, D.K. Veirs: J. Appl. Phys. **69**, 5748 (1991)
6.8 Z. Jiang, C.-J. Lu, D.B. Bogy, C.S. Bhatia, T. Miyamoto: Thin Solid Films **258**, 258 (1995)
6.9 C.F. McFadden, A.J. Gellman: J. Chem. Phys.
6.10 C.F. McFadden, A.J. Gellman: Tribology Lett. **1**, 201 (1995)
6.11 C.F. McFadden, A.F. Gellman: Langmuir **11**, 273 (1995)
6.12 A.J. Gellman: J. Vac. Sci. Techn. A **10**, 180 (1992)

Chapter 7

7.1 A.E. Norton: *Lubrcation* (McGraw-Hill, New York 1942)
7.2 V.N. Constantinescu: *Lubrication in Turbulent Regime,* U.S. Atomic Energy Commision/Division of Technical Information, National Bureau of Standards, U.S. Department of Commerce, Springfield, VA (1968)
7.3 M. Doi, S.F. Edwards: *The Theory of Polymer Dynamits* (Oxford, Univ. Press, Oxford 1956)
7.4 A. Berker, S. Chynoweth, U.C. Klomp, Y. Michopoulos: Chem. Sec. Faraday Trans. **88**, 1719 (1992)
7.5 L. Cai, S. Granick: In *Micro/Nanotribology and its Applications,* ed. by B. Bhushan (Kluwer, Dordrecht 1997)

7.6   A.L. Demirel, S. Granick: Phys. Rev. Lett. **77**, 2261 (1966)
      A.L. Demirel, S. Granick: Phys. Rev. Lett. **77**, 2261 (1966)
7.7   P.L. Maffettone, G. Marrucci, M. Mortiert, P. Moldenaers, J. Mewis: J. Chem. Phys. **100**, 7736 (1994)
7.8   L.B. Chen, C.F. Zukoski, B.J. Ackerson, H.J.M. Hanley, G.C. Straty, J. Barker, C.J. Glinka: Phys. Rev. Lett. **69**, 688 (1992)
7.9   A. Bartel: *Oberflächenschaden durch Schmierstoffanomalien* (Koehler-Druck, 1964)
7.10 J. Happel, H. Brenner: *Low Reynolds Number Hydrodynamits* (Prentice-Hall, Engelwood-Cliffs, NJ 1965)
7.11 M.L. Gee, P.M. McGuiggan, J.N. Israelachvili: J. Chem. Phys. **93**, 1895 (1990)
7.12 B.N.J. Persson (ed.): Surface Science Rpts. **15**, 55-58 and 67 (North-Holland, Amsterdam 1992)
7.13 F.P. Bowden, D. Tabor: *Friction and Lubrication* (Methuen, London 1967) pp. 17-18
7.14 J.F. Archard: Proc. Roy. Sec. (London) A **243**, 190 (1957)
7.15 J. N. Israelachvili: *Course on molecular adheasion and tribology* (1992)
7.16 H.-W. Hu, G.A. Carson, S. Granick: Phys. Rev. Lett. **66**, 2758 (1991)
7.17 B.N.J. Persson, E. Tosatti: Phys. Rev. B **50**, 5590 (1994)
7.18 J.G. Dash: *Films on Solid Surfaces* (Academic, New York 1975)
7.19 P.G. de Gennes: Rev. Mod. Phys. **57**, 827 (1985)
7.20 W. Zisman: In *Contact Angle, Wettability and Adhesion,* ed. by F.M. Fowkes, Adv. Chemical Ser., No. 43 (Am. Chem. Sec., Washington, DC 1964) p. 1
7.21 L.D., Landau. E.M. Lifshitz: *Theory of Elasticity* (Pergamon, New York 1975)
7.22 R. Rabinowicz, D. Tabor: Proc. Roy. Sec. (London) A **208**, 455 (1951)
7.23 A.W. Adamson: *Physical Chemistry of Surfaces* (Wiley, New York 1990)
7.24 J. Israelachvili: *Intermolecular and Surface Fortes* (Academic, San Diego 1992)
7.25 J. Gao, W.D. Luedtke, U. Landman: In *Physics of Sliding Friction,* ed. by B.N.J. Persson, E. Tosatti (Kluwer, Dordrecht 1996)
7.26 W.T. Tysoe. In *Physics of Sliding Friction,* ed. by B.N.J. Persson, E. Tossatti (Kluwer, Dordrecht 1996)
7.27 A.N. Chaka, The Lubrizol Corp.: Priv. commun. (1996)
7.28 A.N. Chaka, J. Harris, X.-P. Li: In *Chemical Applications of Density Functional Theory,* ed. by B.B. Laird, R.M. ROSS, T. Ziegler (Am. Chem. Sec., Washington, DC 1996) p. 368
7.29 W.E. Campbell: Trans. ASME **61**, 633 (1939)
7.30 A. Sonntag: Electro-Technol. (New York) **66**, 108 (1960)
7.31 G. Davies (ed.): *Properties and Growth of Diamond* (Inst. Electr. Eng., London 1994) p. 70
7.32 H. Yoshizawa, Y.-L. Chen, J. Israelachvili: Wear **168**, 161 (1993)
7.33 Y.-L. Chen, J .N. Israelachvili: J . Phys. Chem. **96**, 7752 (1992)
7.34 H. Yoshizawa, Y.-L. Chen, J. Israelachvili: J. Phys. Chem. **97**, 4128 (1993)
7.35 J. Crassous, L. Bocquet, S. Ciliberto, C. Laroche: Europhys. Lett. **47**, 562 (1999);
      L. Bocquet, E. Charlaix, S. Cilibrto, J. Crassous: Nature **396**, 735 (1998)
7.36 B.J. Hamrock: *Fundamentals of Fluid Film Lubrication,* McGraw Hill, London (1994)

Chapter 8

8.1   B.N.J. Persson: Phys. Rev. B **44**, 3277 (1991)
8.2   B.N.J. Persson, R. Rydberg: Phys. Rev. B **32**, 3586 (1985)
8.3   D. Forster: *Hydrodynamit Fluctuations, Broken Symmetry, and Correlation Functions* (Benjamin, New York 1975)
8.4   R.W. Zwanzig: J. Chem. Phys. **32**, 1173 (1960)
      S.A. Adelman, J.D. Doll: J. Chem. Phys. **64**, 2375 (1976)
      M. Shugard, J.C. Tully, A. Nitzan: J. Chem. Phys. **66**, 2534 (1977)
      B.N.J. Persson, R. Ryberg: Phys. Rev. B **32**, 3586 (1985)
8.5   A.I. Volokitin, B.N.J. Persson: In *Inelastic Energy Transfer in Interactions with Surfaces and Adsorbates,* ed. by B. Gumhalter, A.C. Levi, F. Flores (World Scientific, Singapore 1993) pp. 217-248
8.6   B.J. Hinch, A. Lack, H.H. Madden, J.P. Toennies, G. Witte: Phys. Rev. B **42**, 1547 (1990)
      F. Hofmann, J.P. Toennies: Chem. Rev. **96**, 1307 (1996)
8.7   D. Schumacher: *Surface Stattering Experiments with Conduction Electrons,* Springer Tracts Mod. Phys., Vol. 128 (Springer, Berlin, Heidelberg 1993) Sects. 4.2, 3
8.8   G. Wedler, H. Reichenberger, H. Wenzel: Z. Naturforsch. **26a**, 1444-1452 (1971)
8.9   B.N.J. Persson, A.I. Volokitin: J. Chem. Phys. **103**, 8679 (1995)
      W.L. Schaich, J. Harris: J. Phys. F **11**, 65 (1981)
      J.B. Sokoloff: Phys. Rev. B **52**, 5318 (1995)
8.10 K. Wandelt, W. Jacob, N. Memmel, V. Dose: Phys. Rev. Lett. **57**, 1643 (1986)
8.11 C. Holzapfel, F. Stubenrauch, D. Schumacher, A. Otto: Thin Solid Films **188**, 7 (1990)
8.12 M. Fuhrmann, Ch. Wöll: Surf. Sci. **368**, 20 (1996)
8.13 Y.J. Chabal: Phys. Rev. Lett. **55**, 845 (1985)
      J.E. Reutt, Y.J. Chabal, S.B. Christman: Phys. Rev. B **38**, 3112 (1988)
8.14 B.N.J. Persson, A.I. Volokitin: Surf. Sci. **310**, 314 (1994)
      K.C. Lin, R.G. Tobin, P. Dumas: Phys. Rev. B **49**, 17273 (1994)
      C.J. Hirschmugl, G.P. Williams, B.N.J. Persson, A.I. Volokitin: Surf. Sci. **317**, L1141 (1994)
8.15 H. Ishida: Phys. Rev. B **49**, 14610 (1994)
8.16 P.J. Rous, T.L. Einstein, E.D. Williams: Surf. Sci. **315**, L995 (1994)
8.17 R. Landauer: Phys. Rev. B **14**, 1474 (1976)
      A.K. Das, R. Peierls: J. Phys. C **6**, 2811 (1973)
      R.S. Sorbello: Phys. Rev. B **31**, 798 (1985)
      W.L. Schaich: Phys. Rev. **13**, 3350 (1976)
8.18 H. Ishida: Phys. Rev. B **51**, 10345 (1995)
      P.J. Rous, T.L. Einstein, E.D. Williams: Surf. Sci. **315**, 995 (1994)
8.19 A.V. Latyshev, A.L. Aseev, A.B. Krasilnikov, S.I. Stein: Surf. Sci. **213**, 157 (1989)
8.20 Y.-N. Yang, E.S. Fu, E.D. Williams: Surf. Sci. **356**, 101 (1996)
8.21 R.L. Schwoebel, E.J. Shipsey: J. Appl. Phys. **37**, 3682 (1966)

8.22 B. Voigtländer, A. Zinner, T. Weber, H.P. Bonzel: Phys. Rev. B **51**, 7583 (1995)
8.23 D. Kandel, E. Kaxiras: Phys. Rev. Lett. **76**, 1114 (1996)
8.24 S. Modesti, V.R. Dhanak, M. Sancrotti, A. Santoni, B.N.J. Persson, E. Tosatti: Phys. Rev. Lett. **73**, 1951 (1994)
8.25 K.S. Ralls, D.C. Ralph, R.A. Buhrman: Phys. Rev. B **40**, 11561 (1989)
    D.M. Eigler, C.P. Lutz, W.E. Rudge: Nature **352**, 600 (1991)
    J.A. Stroscio, D.M. Eigler: Science **254**, 1319 (1991)
8.26 D.M. Eigler, E. K. Schweizer: Nature **344**, 524 (1990)
8.27 D.M. Eigler, P.S. Weiss, E.K. Schweizer, N.D. Lang: Phys. Rev. Lett. **66**, 1189 (1991)
8.28 A. Liebsch: Phys. Rev. B **55**, 13263 (1997)
8.29 I.A. Blech: J . Appl. Phys. **47**, 1203 (1976)
    E. Artz, O. Kraft, U.E. Möckl: Phys. Bl. **52**, 227 (1996)
8.30 H.A. Kramers: Physica **7**, 284 (1940)
8.31 H. Risken: *The Fokker-Planck Equation,* 2nd edn., Springer Ser. Syn., Vol. 18 (Springer, Berlin, Heidelberg 1989)
8.32 C. Engdahl, G. Wahnström: Surf. Sci. **312**, 429 (1994) G. Wahnström, A.B. Lee, J. Strömquist: J. Chem. Phys. **105**, 326 (1996)
8.33 B.N.J. Persson: In *Computations for the Nano Scale,* ed. by P.E. Blöchl, C. Joachim, A.J. Fisher, NATO ASI Ser. E 240, 21 (Kluwer, Dordrecht 1993)
8.34 B.N.J Persson: J. Chem. Phys. **103**, 3849 (1995)
8.35 B.N.J. Persson: Phys. Rev. B **48**, 18140 (1993)
8.36 J.A. Barker, D. Henderson, F.F. Abraham: Physica A **106**, 226 (1981)
8.37 A.I. Larkin, Yu.N. Ovchinnikov: J. Low Temp. Phys. **34**, 409 (1979)
8.38 G. Blatter, M.V. Feigelman, V.B. Geshkenbein, A.I. Larkin, V.M. Vinokur: Rev. Mod. Phys. **66**, 1125 (1994)
8.39 B.N.J. Persson, A. Nitzan: Surf. Sci. **367**, 261 (1996)
8.40 J. Braun, D. Fuhrmann, Ch. Wöll: Unpublished (1996)
8.41 S. Aubry, C. Andry: *Proc. Israel Phys. Sec.,* ed. by C.G. Kuper (Hilger, Bristol 1979) vol. 3, p. 133
8.42 J.B. Sokoloff: Phys. Rev. B **42**, 760 (1990)
8.43 J. Krim, E.T. Watts, J. Digel: J. Vac. Sei. Technol. A **8**, 3417 (1990)
8.44 D. Einzel, P. Panzer, M. Liu: Phys. Rev. Lett. **64**, 2269 (1990)
    P. Panzer, M. Liu, D. Einzel: Int'l. J. Mod. Phys. B **6**, 3251 (1992)
8.45 D.A. Ritchie, J. Saunders, D.F. Brewer: Phys. Rev. Lett. **59**, 465 (1987)
8.46 M. Rodahl, B. Kasemo: Sensors and Actuators A **54**, 448 (1996)
8.47 B.N.J. Persson, E. Tosatti: Phys. Rev. B **50**, 5590 (1994)
8.48 M.L. Gee, P.M. McGuiggan, J.N. Israelachvili: J. Chem. Phys. **93**, 1895 (1990)
8.49 L.D. Landau, E. M. Lifshitz: *Theory of Elasticity* (Pergamon, London 1975)
8.50 W. Cooper, W. Nuttal: J. Agrs. Sci. **7**, 219 (1915)
8.51 P.G. de Gennes: Rev. Mod. Phys. **57**, 827 (1985)
8.52 F. Heslot, A.M. Cazabat, P. Levinson: Phys. Rev. Lett. **62**, 1286 (1989)
8.53 F. Heslot, A.M. Cazabat, P. Levinson, N. Fraysse: Phys. Rev. Lett. **65**, 599 (1990)
8.54 F. Heslot, N. Fraysse, A.M. Cazabat: Nature **338**, 1289 (1989)
8.55 P.G. de Gennes, A.M. Cazabat: C.R. Acad. Sci. (Paris) II **310**, 1601 (1990)
8.56 J. Krim, D.H. Solina, R. Chiarello: Phys. Rev. Lett. **66**, 181 (1991)
8.57 B.N.J. Persson: *Ordered Structures and Phase Transitions in Adsorbed Layers,* Surf. Sci. Rpts. 15, 1 (North-Holland, Amsterdam 1992)
8.58 D. Forster, D.R. Nelson, M. J. Stephen: Phys. Rev. A **16**, 732 (1977)
8.59 S. Ramaswamy, G. Mazenko: Phys. Rev. A **26**, 1735 (1982)

8.60 S. Doniach, E.H. Sondheimer: *Green k Functions for Solid State Physisists* (Benjamin, New York 1974) p. 244
8.61 A. Zangwill: *Physics at Surfaces* (Cambridge Univ. Press, New York 1988)
8.62 S. Ramaswamy, G.F. Mazenko: Phys. Rev. A **26**, 1735 (1982)
L.A. Turski: Phys. Rev. A 28, 2548 (1983)
8.63 W.A. Little: Can. J. Phys. **37**, 334 (1959)
8.64 Z. Feng, J.E. Field: J. Phys. D **25**, A33 (1992)
8.65 E. Granato, M.R. Baldan, S.C. Ying: *Physics of Sliding Friction,* ed. by B.N.J. Persson, E. Tosatti (Kluwer, Dordrecht 1996) p. 103
8.66 B.N.J. Persson, A.I. Volokitin: J. Phys. Condensed Matter **9**, 2869 (1997)
8.67 B.N.J. Persson, E. Tosatti, D. Fuhrmann, G. Witte, Ch. Wöll: Phys. Rev. B **59**, 11777 (1999)
8.68 B. Hall, D.L. Mills, J.E. Black: Phys. Rev. B **32**, 4932 (1985);
L.W. Bruch, F.Y. Hansen: Phys. Rev. B **55**, 1782 (1997);
S.P. Lewis, M.V. Pykhtin, E.J. Mele, A.M. Rappe: J. Chem. Phys. **108**, 1157 (1998)
8.69 A. Dayo, W. Alnasrallah, J. Krim: Phys. Rev. Lett. **80**, 1690 (1998)
8.70 B.N.J. Persson, E. Zaremba: Phys. Rev. B**31**, 1863 (1985);
B.N.J. Persson, E. Zaremba: B **30**, 5669 (1984);
B.N.J. Persson, N. Lang: Phys. Rev. B **26**, 5409 (1982)
8.71 B.N.J. Persson: Phys. Rev. B **44**, 3277 (1991);
J.B. Sokoloff: Phys. Rev. B **52**, 5318 (1995);
S. Tomassone, A. Widom: Phys. Rev. B **56**, 4938 (1997);
B.N.J. Persson, A.I. Volokitin: J. Chem. Phys. **103**, 8679 (1995)
8.72 V.L. Popov: JETP Lett. **69** 558 (1999);
T. Novotny, B. Velicky: Phys. Rev. Lett. **83** 4112 (1999);
J.B. Sokoloff, M.S. Tomassone, A. Widom, Phys: Rev. Lett., in press;
B.N.J. Persson: Submitted to Phys. Rev. Lett.
8.73 J.R. Schrieffer: *Theory of Superconductivity* (Benjamin, New York, 1966)
8.74 R.E. Prange: Phys. Rev. **129** 2495 (1963)
8.75 F. Mugele, M. Salmeron: Submitted to Phys. Rev. Lett.
8.76 D. Kessler, J. Koplik, H. Levine: Adv. Phys. **37**, 255 (1988)
8.77 See, e.g., *Fractals and Disordered Systems* ed. by A. Bunde and S. Havlin (Springer, Berlin, Heidelberg 1991)
8.78 B.N.J. Persson, P. Ballone: Submitted to Phys. Rev. Lett.;
B.N.J. Persson, P. Ballone: Submitted to J. Chem. Phys.

Chapter 9

9.1 H. Yoshizawa, J. N. Israelachvili: J. Phys. Chem. **97**, 11300 (1993)
9.2 M.L. Gee, P.M. McGuiggan, J.N. Israelachvili: J. Chem. Phys. **93**, 1895 (1990)
9.3 J.N. Israelachvii: Surf. Sci. Rpts. **14**, 109 (1992)
9.4 S. Granick: Science **253**, 1374 (1991)
9.5 G. Reiter, A.L. Demirel, S. Granick: Science **263**, 1741 (1994)
G. Reiter, A.L. Demirel, J. Peanasky, L. Cai, S. Granick: J. Chem. Phys. **101**, 2626 (1994)
9.6 A.L. Demirel, S. Granick: In *Physics of Sliding Friction,* ed. by B.N.J. Persson, E. Tosatti (Kluwer, Dordrecht 1996)
9.7 G. Reiter, A.L. Demirel, J. Peanasky, L. Cai, S. Granick: J. Chem. Phys. **101**, 2606 (1994)
9.8 B.N.J. Persson: Phys. Rev. B **50**, 4771 (1994)

9.9   P.A. Thompson, M.O. Robbins: Phys. Rev. A 41, 6830 (1990); Science 250, 792 (1990); Science **253**, 916 (1991)
      P.A. Thompson, M.O. Robbins, G.S. Grest: In *Computations for the Nano-Scale*, ed. by P.E. Blöchl, C. Joachim, A.J. Fisher, NATO ASI Ser. E 240, 127 (Kluwer, Dordrecht 1993)
9.10 B.N.J. Persson: Chem. Phys. Lett. **254**, 114 (1996)
9.11 A.D. Berman, W.A. Ducker, J.N. Israelachvili: In *Physics of Sliding Friction*, ed. by B.N.J. Persson, E. Tosatti (Kluwer, Dordrecht 1996)
9.12 L.D. Landau, E.M. Lifshitz: *Theory of Elasticity* (Pergamon, New York 1966)

Chapter 10

10.1   G.M. McClelland: In *Adheasion and Friction*, ed. by M. Grunze, H.J. Kreuzer, Springer Ser. Surf. Sci., Vol. 17 (Springer, Berlin, Heidelberg 1989)
10.2   K. Shinjo, M. Hirano: Surf. Sci. **283**, 473 (1993)
10.3   Y. Braiman, F. Family, G. Hentschel: Phys. Rev. E **53**, R3005 (1996)
       M.G. Rozman, J. Klafter, M. Urbakh: In *Micro/Nanotribology and its Applications*, ed. by B. Bhushan (Kluwer, Dordrecht 1997)
       F.J. Elmer: In *Physics of Sliding Friction*, ed. by B.N.J. Persson, E. Tossati (Kluwer, Dordrecht 1996)
10.4   H. Matsukawa, H. Fukuyama: Phys. Rev. B **49**, 17286 (1994)
10.5   C. Caroli, P. Nozieres: In *Physics of Sliding Friction*, ed. by B.N.J. Persson, E. Tosatti (Kluwer, Dordrecht 1996)
10.6   B.N.J. Persson, E. Tosatti: In *Physics of Sliding Friction*, ed. by B.N.J. Persson, E. Tosatti (Kluwer, Dordrecht 1996) pp. 179-189
10.7   A.I. Larkin, Yu.N. Ovchinnikov: J. Low Temp. Phys. **34**, 409 (1979)
10.8   P.A. Lee: Nature **291**, 11 (1981)
10.9   L.D. Landau, E.M. Lifshitz: *Theory of Elasticity* (Pergamon, New York 1975)
10.10 Y.J.M. Brechet, B. Boucot, H.J. Jensen, A.-C. Shi: Phys. Rev. B **42**, 2116 (1990)
10.11 P.E. Malin, S.N. Blakeslee, M.G. Alvarez, A.J. Martin: Science **244**, 557 (1989)
10.12 R. Abercrombie, P. Leary: Geophys. Res. Lett. **20**, 1511 (1993)
10.13 S.J. Gibowicz, R.P. Young, S. Talebi, D.J. Rawlence: Bull. Seismolog. Sec. Am. **81**, 1157 (1991)
10.14 C.H. Scholz: *The Mechanics of Earthquakes and Faulting* (Cambridge Univ. Press, Cambridge 1990) p. 250
10.15 R. Lüthi, E. Meyer, H. Haefke, L. Howald, W. Gutmannsbauer, H.-J. Güntherodt: Science **266**, 1979 (1994)
10.16 T.A. Jung, R.R. Schlittler, J.K. Gimzewski, H. Tang, C. Joachim: Science **271**, 181 (1996)
10.17 D.M. Eigler, E. K. Schweizer: Nature **344**, 524 (1990)
10.18 R.E. Walkup, D.M. Newns, Ph. Avouris: In *Atomic and Nanometer Scale Modification of Materials: Fundamentals and Applications*, ed. by Ph. Avouris (Kluwer, Dordrecht 1993) p. 97
10.19 B.N.J. Persson, A.I. Volokitin: J. Phys. Condensed Matter **9**, 2869 (1997)
10.20 M.R. Falvo, R.M. Taylor II, A. Helser, V. Chi, F.P. Brooks Jr, S. Washburn, R. Superfine: Nature **397**, 136 (1999)

Chapter 11

11.1 B.N.J. Persson: Phys. Rev. B **51**, 13568 (1995)
11.2 J.H. Dietrich, B.D. Kilgore: PAGEOPH **143**, 283 (1994)
11.3 J.M. Carlson, J.S. Langer: Phys. Rev. Lett. **62**, 2632 (1989)
11.4 B.N.J. Persson: Phys. Ref. B, **61**, 5949 (2000)
11.5 U. Landman, W.D. Luedtke, N.A. Burnham, R.J. Colton: Science **248**, 454 (1990)
11.6 F.A. McClintock, J.B. Walsh: Proc. 4th U.S. Nat'l Cong. Appl. Mech. **2**, 1015 (1962)
T.E. Davidson: In *Advances in Deformation Processing*, ed. by J.J. Burke, V. Weiss, Sagamore Army Mater. Res. Conf. Proc., Vol. 21 (Plenum, New York 1978) p. 535
11.7 The material presented in this section is based on unpublished notes kindly supplied by Prof. G. Gudehus, University of Karlsruhe, Germany
11.8 D.P. Moon, R.C. Simon, R.J. Favor: *The Elevated-Temperature Properties of Selected Superalloys* (ASTM, Philadelphia 1968) p. 335

Chapter 12

12.1 J.M. Carlson, A.A. Batista: Phys. Rev. E **53**, 4153 (1996)
12.2 B.N.J. Persson: Chem. Phys. Lett. **254**, 114 (1996)
12.3 M. Avrami: J. Chem. Phys. **8**, 212 (1940)
R.H. Doremus: J. Appl. Phys. **36**, 2853 (1965)
H.E. Bennet, J.M. Bennet, E.J. Ashley, R.J. Motyka: Phys. Rev. **165**, 755 (1986)
12.4 B.N.J. Persson: Unpublished (1996)
12.5 B.N.J. Persson: In *Micro/Nanotechnology and its Applications*, ed. by B. Bhushan (Kluwer, Dordrecht 1997)
12.6 A.L. Ruina: J. Geophys. Res. **88**, 10359 (1983)
J.R. Rice, A.L. Ruina: J. Appl. Mech. **50**, 343 (1983)
12.7 F. Heslot, T. Baumberger, B. Perrin, B. Caroli, C. Caroli: Phys. Rev. E **49**, 4973 (1994)
12.8 T. Baumberger, F. Heslot, B. Perrin: Nature **367**, 544 (1994)
12.9 T. Baumberger, C. Caroli, B. Perrin, 0. Ronsin: Phys. Rev. E **51**, 4005 (1995)
12.10 H. Yoshizawa, Y.-L. Chen, J.N. Israelachvili: J. Phys, Chem. **97**, 4128 (1993)

Chapter 13

13.1 F. Heslot, T. Baumberger, B. Perrin, B. Caroli, C. Caroli: Phys. Rev. E **49**, 4973 (1994)
13.2 T. Baumbergen, F. Heslot, B. Perrin: Nature **367**, 544 (1994)
13.3 T. Baumbergen, C. Caroli, B. Perrin, O. Ronsin: Phys. Rev. E **51**, 4005 (1995)
13.4 T. Baumbergen, C. Caroli: Commen. Cond. Mat. Phys. **17**, 306 (1995)
13.5 C.H. Scholz, T. Engeldaard: Int'l J. Rock. Mech. Min. Sci. **13**, 149 (1976)
13.6 J.H. Dietrich, B.D. Kilgore: PAGEOPH **143**, 283 (1994)
13.7 T. Baumbergen, C. Caroli: Commen. Cond. Mat. Phys. **17**, 281 (1995)
13.8 A. Johansen, P. Dimon, C. Ellegaard, J.S. Larsen, H.H. Rugh: Phys. Rev. E **48**, 4779 (1993)

13.9 Y. Braiman, F. Family, G. Hentschel: Phys. Rev. E **53**, R3005 (1996)
M.G. Rozman, J. Klafter, M. Urbakh: In *Micro/Nanotribology and its Applications,* ed. by B. Bhushan (Kluwer, Dordrecht 1997)
F.J. Elmer: In *Physics of Sliding Friction,* ed. by B.N.J. Persson, E. Tosatti (Kluwer, Dordrecht 1996)

Chapter 14

14.1 C.H. Scholz: *The Mechanics of Earthquakes and Faulting* (Cambridge Univ. Press, Cambridge 1990) p. 250
14.2 M.L. Blanpied, D.A. Lockner, J.D. Byerlee: Geophys. Res. Lett. **18**, 609 (1991)
14.3 A.L. Ruina: J. Geophys. Res. **88**, 10359 (1983)
J.R. Rice, A.L. Ruina: J. Appl. Mech. **50**, 343 (1983)
14.4 B. Gutenberg, C.F. Richter: Ann. Geofis. **9**, 1 (1956)
14.5 R. Burridge, L. Knopoff: Bull. Seismol. Sec. Am. **57**, 341 (1967)
14.6 S.R. Brown, C.H. Scholz, J.B. Rundle: Geophys. Res. Lett. **18**, 215 (1991)
14.7 J.M. Carlson, J.S. Langer: Phys. Rev. Lett. **62**, 2632 (1989)
14.8 P. Bak, C. Tang: J. Geophys. Res. **94**, 15635 (1989)
14.9 P. Bak, C. Tang, K. Wiesenfeld: Phys. Rev. Lett. **59**, 381 (1987)
14.10 J.R. Rice: J. Geophys. Res. **98**, 9885 (1993)
14.11 P.E. Malin, S.N. Blakeslee, M.G. Alvarez, A.J. Martin: Science **244**, 557 (1989)
14.12 F.P. Bowden: Proc. Roy. Sec. (London) A **217**, 462 (1953)
14.13 F.P. Bowden, T.P. Hughes: Proc. Roy. Sec. (London) A **172**, 280 (1939)
14.14 P. Barnes, D. Tabor, F.R.S. Walker, J.C.F. Walker: Proc. Roy. Sec. (London) A **324**, 127 (1971)
14.15 F.P. Bowden, D. Tabor: *Friction and Lubrication* (Methuen, London 1967)
14.16 F.P. Bowden: Proc. Roy. Sec. (London) A **217**, 462 (1953)
14.17 S.C. Colbeck: J. Glaciology 34, 78 (1988); also J. Sports Sci. **12**, 285 (1994)
14.18 J.F. Nye: In *Physics of Sliding Friction,* ed. by B.N.J. Persson, E. Tosatti (Kluwer, Dordrecht 1996) p. 293
14.19 M.L. Gee, P.M. McGuiggan, J.N. Israelachili: J. Chem. Phys. **93**, 1895 (1990)
14.20 D. Beaglehole, P. Wilson: J. Chem. Phys. **98**, 8096 (1994)
14.21 D.F. More: *Principles andApplications of Tribology* (Pergamon, Oxford 1975)
14.22 G. Blatter, M.V. Feigelman, V.B. Geshkenbein, A.I. Larkin, V.M. Vinokur: Rev. Mod. Phys. **66**, 1125 (1994)
14.23 G. Grüner: Rev. Mod. Phys. **60**, 1129 (1988)
14.24 A.A. Abrikosov: Phys. Chem. Solids **2**, 199 (1957)
14.25 K.H. Fischer: Superconduct. Rev. **1**, 153 (1995)
14.26 J. Bardeen, M.J. Stephen: Phys. Rev. A **140**, 1197 (1965)
14.27 A.I. Larkin, Yu.N. Ovchinnikov: J. LowTemp. Phys. **34**, 409 (1979)
14.28 P.H. Kes, C.C. Tsuei: Phys. Rev. B **28**, 5126 (1983)
R. Wördenweber, P.H. Kes, C.C. Tsuei: Phys. Rev. B **33**, 3172 (1986)
14.29 E.H. Brandt: Phys. Stat. Solidi B **77**, 551 (1976)
14.30 A.M. Cambell, J.E. Evetts: Adv. Phys. **21**, 199 (1972)
14.31 A. Brass, H.J. Jensen, A.J. Berlinsky: Phys. Rev. B **39**, 102 (1989)
14.32 A.A. Middleton, D.S. Fisher: Phys. Rev. Lett. **66**, 92 (1991)
A.A. Middleton, O. Biham, P.B. Littlewood, P. Sibani: Phys. Rev. Lett. **68**, 1586 (1992)
14.33 O. Narayan, D. S. Fisher: Phys. Rev. B 46, 11520 (1992)

14.34 T. Nattermann, S. Stepanow, L.-H. Tang, H. Leschhorn: J. Phys. II France **2**, 1483 (1992)
14.35 S. Bhattacharya, M.J. Higgins: Phys. Rev. Lett. **70**, 2617 (1993)
14.36 U. Yaron, P.L. Gammel, D.A. Huse, R.N. Kleinman, C.S. Oglesby, E. Bucher, B. Batlogg, D.J. Bishop, K. Mortensen, K.N. Clausen: Nature **276**, 753 (1995)
14.37 N.G. Jensen, A.R. Bishop, D. Dominguez: Phys. Rev. Lett. **76**, 2985 (1996)
14.38 G. D'Anna, P.L. Gammel: H. Safar, G.B. Alers, D.J. Bishop, J. Giapintzakis, D.M. Ginsberg: Phys. Rev. Lett. **75**, 3521 (1995)
14.39 A.E. Koshelev, V.M. Vinokur: Phys. Rev. Lett. **73**, 3580 (1994)
14.40 Y.B. Kirn, C.F. Hempstead, A.R. Strand: Phys. Rev. Lett. **9**, 306 (1962)
14.41 P.W. Anderson: Phys. Rev. Lett. **9**, 309 (162)
14.42 O. Narayan, D.S. Fisher: Phys. Rev. Lett. **68**, 3615 (1992)
14.43 T. Nattermann, S. Stepanow, L.-H. Tang, H. Leschhorn: J. Phys. II France **2**, 1483 (1992)
14.44 O. Narayan, A.A. Middleton: Phys. Rev. B **49**, 244 (1994)
14.45 H. Leschhorn, T. Nattermann, S. Stepanow, L.-H. Tang: Ann. Physik **6**, 1 (1997)
14.46 T. Strunz, F.J. Elmer: In *Physics of Sliding Friction,* ed. by B.N.J. Persson, E. Tosatti (Kluwer, Dordrecht 1996)
14.47 P.A. Lee, T.M. Rice: Phys. Rev. B **19**, 3970 (1979)
14.48 S.N. Coppersmith: Phys. Rev. Lett. **65**, 1044 (1990)
14.49 G. Mihaly, L. Mihaly: Phys. Rev. Lett. **52**, 149 (1984)
14.50 We assume that the asymptotic distance dependence of the friction force caused by virtual fluctuations can be obtained by summing independent atomic probability amplitude contributions
14.51 B.N.J. Persson: Phys. Rev. Lett. **50**, 1089 (1983)
14.52 B.N.J. Persson, S. Andersson: Phys. Rev. B **29**, 4382 (1984)
14.53 P.J. Feibelman: Prog. Surf. Sci. **12**, 287 (1982) A. Liebsch: *Electronic Excitations at Metal Surfaces* (Plenum, New York 1997)
14.54 B.N.J. Persson, E. Zaremba: Phys. Rev. B **30**, 5669 (1984)
14.55 B.N.J. Persson, E. Zaremba: Phys. Rev. B **31**, 1863 (1985)
14.56 B.N.J. Persson, Z. Zhang: Phys. Rev. B **57**, 7327 (1998)
14.57 T.J. Gramila, J.P. Eisenstein, A.H. MacDonald, L.N. Pfeiffer, K.W. West: Physica B **197**, 442 (1994)
14.58 B.N.J. Persson, J.E. Demuth: Phys. Rev. B **30**, 5968 (1984)
14.59 C. Hodges, H. Smith, J.W. Wilkins: Phys. Rev. B **4**, 302 (1971)
14.60 T.J. Gramila, J.P. Eisenstein, A.H. MacDonald, L.N. Pfeiffer, K.W. West: Phys. Rev. B **47**, 12957 (1993)
14.61 B. Alberts, D. Bray, J. Lewis, M. Raff, K. Roberts, J.D. Watson: The Molecular Biology of the Cell (Garland, New York 1994)
14.62 F. Jülicher, J. Prost: Phys. Rev. Lett. **75**, 2618 (1995)
14.63 A.V. Hill: First and Last Experiment in Muscle Contraction (Cambridge Univ. Press, London 1970)
14.64 A.F. Huxley, R.M. Simmons: Nature **233**, 533 (1971)
14.65 I. Rayment, H.M. Holden: TIBS **19**, 129 (1994)
14.66 J.T. Finer, R.M. Simmons, J.A. Spudich: Nature **368**, 113 (1994)
14.67 T.Q.P. Uyeda, H.M. Warrick, S.J. Kron, J.A. Spudlich: Nature **352**, 307 (1991)
14.68 O.H. Wyatt, D. Dew-Hughes: *Metals, Ceramics and Polymers* (Cambridge Univ. Press, Cambridge 1974)
N.P. Suh, A.P.L. Turner: *Elements of the Mechanical Behavior of Solids* (McGraw-Hill, New York 1975)

14.69 J. Friedel: *Dislocations* (Pergamon, Oxford 1964)
14.70 D.J. Dingley, D. McLean: Acta Metallurgica **15**, 885 (1967)
14.71 P. Penning: ActaMetall. **20**, 1169 (1972)
14.72 L.P. Kubin, Y. Estrin: Acta Metall. **33**, 397 (1985)
14.73 K. Knothe, U. Miedler: Konstruktion **47**, 118 (1995)
14.74 F. W. Carter: Proc. Roy. Sec. (London) A **112**, 151 (1926)
14.75 J.J. Kalker: *Three Dimensional Elastic Bodies in Rolling Contact* (Kluwer, Dordrecht 1990)
14.76 H. Hertz: Z. reine und angewandte Mathematik **92**, 156 (1882)
14.77 J.B. Nielsen, A. Theiler: Priv. commun. (1996)
14.78 K.H. Nordsiek, J. Wolpers: Kautschuk Gummi Kunststoffe **43**, 755 (1990)
14.79 B.J. Lazan: In *Structural Dumping,* ed. by J.E. Ruzicka (ASME, New York 1959)
14.80 K.C. Ludema, D. Tabor: Wear **9**, 329 (1966)
14.81 G. Heinrich: Kautschuk Gummi Kunststoffe **45**, 173 (1992)
14.82 A. Gabler, E. Straube, G. Heinrich: Kautschuk Gummi Kunststoffe **46**, 941 (1993)
14.83 S. Nasuno, A. Kudrolli, J.P. Gollub: Phys. Rev. Lett. **79**, 949 (1997);
S. Nasuno, A. Kudrolli, A. Bak, J.P. Gollub: Phys. Rev. E **58**, 2161 (1998);
J.C. Geminard, W. Losert, J.P. Gollub: Phys. Rev. E **59**, 5881 (1999)
14.84 L. Bocquet, E. Charlaix, S. Ciliberto, J.Crassous: Nature **396** 735 (1998)

Subject Index

Adhesion 83
–, hysteresis 154
Adsorbate layers 171
Area of real contact 49, 441
Atomic force microscopy 37

Boundary
–, conditions in hydrodynamics 228
–, lubrication 14, 148, 313
–, line 301
Brittle fracture 48
Brownian motion 172

Capillarity 134
Capillary bridges 141
Chaotic motion 205
Charge density waves 447, 462
Clean (dry) surfaces 93
Collective behavior 474
Colloidal systems 115
Commensurate solid adsorbate structures 256
Contact hysteresis 69, 74
Coulomb 11
–, drag 465
Creep 363
–, in metals 389
–, length 437
Critical
–, behavior 459
–, sliding state 367
–, state 53
–, velocity 324

Defects 226, 278, 354
Depinning transition 459
Diffusion 245
Dry friction dynamics 415
Ductility 389
Dynamical phase transitions 255

Dynamics of spreading 141

Earthquakes 344, 435
Effective temperature 271
Elastic
–, coherence length 341
–, deformations 45, 58
–, flow 457
–, instability 454
–, instability transition 337
–, interactions 335
Elastohydrodynamics 119, 162
Elastomer 83
Electromigration 191
–, at surfaces 192
–, in metallic films 196
Electron wind force 191
Electronic friction 176, 287

Frenkel–Kontorova Model 275
Fluid adsorbate structures 255
Fluid rheology 110
Fluid–Incommensurate solid transition 268
Fluidization 265
Flux-line systems 447
Freezing 396
–, transition 267
Friction
–, force microscopy 26
–, heating 439
Frictional Coulomb drag 465

Glass transition 116
–, temperature 88
Granular materials 492
Growth 297
Gutenberg–Richter law 436

Heat flow 251

Subject Index

Heat transfer 248
High-energy surfaces 137
Human and animal joints 445
Hydrodynamic lubrication 13, 102
–, 2D 243

Ice 439
Incomplete wetting 135
Infrared spectroscopy 188
Instability transitions 335
Interdiffusion 405
Internal friction 480, 489

Junctions 93

Kelvin equation 145
Kinetic friction 443

Laminar flow 108
Layering transition 236, 297
–, nucleation 126
Lennard–Jones Model 255
Leonardo da Vinci 10
Linear
–, instability analysis 397, 406, 423
–, sliding friction 208
–, stability analysis 304
Low-energy surfaces 137
Lubricated
–, surfaces 101
–, friction dynamics 395

Memory effects 418
Microscale junctions 45
Micromotor 56
Moisture-induced aging 143
Muscle contraction 470

Nanoindentation 64
Nanoscale junctions 54
Newtonian 110
Non-equilibrium phase diagram 457
Non-linear
–, analysis 430
–, sliding friction 254
Nucleation 297

One-dimensional creep 386
Overdamped motion 330

Penetration hardness 47
Perpendicular creep 53
Phononic friction 182
Pining centers 213

Plastic
–, deformation 46, 59, 382
–, flow 457
–, stick-slip instabilities 480
Polymers 77, 88
Profilometer 37

QCM-Friction measurements 231
Quantum fluctuations 466
Quartz-crystal microbalance 31

Relaxation 363
Renormalization group 460
Reversibility 4
Reversible elastic deformations 455
Rheological model 75
Rolling 351
–, resistance 486
Rubber wear 89

Scanning tunneling microscopy 37
Self-affine fractal 38
Shear thinning 110
Shear-melting 396
–, transition 265
Sliding
–, friction 198
–, of atoms 348
–, of big molecules 347
–, of islands 345
–, on ice 439
Slip
–, at interface 405
–, length 230
–, zone 488
Slip-size distribution 380
Snow 439
Spreading
–, of wetting liquid drops 239
–, pressure 134
Squeezed film 123
Starting 319
Static and kinetic friction 152
Static friction 442
Stick zone 488
Stick-slip motion 324
Stopping 319
–, wave 356
Stress
–, domains 363
–, fluctuations 318
Stress-aided 112
Superconducting 34
Superconductivitiy 286

Surface
–, contaminants 37
–, energies 138
–, forces apparatus 19
–, resistivity 176
–, roughness 230
–, topography 37

Time dependence of contact area 418
Thermal fluctuations 446
Turbulent flow 108

Undercooled liquids 116
Underdamped motion 329

Viscous fingering 303

Wear 43
Wetting 134

Xe on silver 215

Printing: Mercedes-Druck, Berlin
Binding: Buchbinderei Stürtz AG, Würzburg

Location: http://www.springer.de/phys/

*You are one **click** away from a **world of physics** information!*

Come and visit Springer's
Physics Online Library

Books
- Search the Springer website catalogue
- Subscribe to our free alerting service for new books
- Look through the book series profiles

You want to order?   Email to: orders@springer.de

Journals
- Get abstracts, ToC´s free of charge to everyone
- Use our powerful search engine LINK Search
- Subscribe to our free alerting service LINK *Alert*
- Read full-text articles (available only to subscribers of the paper version of a journal)

You want to subscribe?   Email to: subscriptions@springer.de

Electronic Media
- Get more information on our software and CD-ROMs

You have a question on an electronic product?   Email to: helpdesk-em@springer.de

•••••••••• Bookmark now:

http://www.springer.de/phys/

 Springer

Springer · Customer Service
Haberstr. 7 · D-69126 Heidelberg, Germany
Tel: +49 6221 345 200 · Fax: +49 6221 300186
d&p · 6437a/MNT/SF · Gha.